Complex Inorganic Solids

Structural, Stability, and Magnetic Properties of Alloys

A publication of the Third International Alloy Conference (IAC-3)

Complex Inorganic Solids

Structural, Stability, and Magnetic Properties of Alloys

Edited by

Patrice E. A. Turchi

Lawrence National Laboratory
Livermore, California

Antonios Gonis

Lawrence National Laboratory
Livermore, California

Krishna Rajan

Materials Science and Engineering Department
and Information Technology Program
Rensselaer Polytechnic Institute
Troy, New York

and

Annemarie Meike

Lawrence National Laboratory
Livermore, California

Springer

Proceedings of Third International Alloy Conference (IAC-3), held June 30–July 5, 2002 in Estoril Sol, Portugal

ISBN-10: 0-387-24811-0 (HB)
ISBN-13: 978-0387-24811-0

Printed in the United States of America

9 8 7 6 5 4 3 2 1

springeronline.com

Chairpersons/Organizers of IAC-3

TURCHI, PATRICE E. A.
Lawrence Livermore Nat'l Laboratory
C&MS, L-353
P.O. Box 808
Livermore, CA 94551, USA
Tel.: 925-422-9925
Fax: 925-423-7040
Email: turchi1@llnl.gov

RAJAN, KRISHNA
Rensselaer Polytechnic Institute
Acad. Sci. of the Mater. Sci. & Eng.
110 8th Street
Troy, NY 12180-3590, USA
Tel.: 518-276-6126
Fax: 518-276-8554
Email: rajanjk@rpi.edu

GONIS, ANTONIOS
Lawrence Livermore Nat'l Laboratory
C&MS, L-371
Livermore, CA 94551, USA
Tel.: 925-422-7150
Fax: 925-423-7040
Email: gonis1@llnl.gov

MEIKE, ANNEMARIE
Lawrence Livermore Nat'l Laboratory
L-201
Livermore, CA 94551, USA
Tel.: 925-422-3735
Fax: 925-423-8988
Email: meike1@llnl.gov

List of Participants IAC-3

ABRIKOSOV, IGOR
Uppsala University
Angstromlaboratory
Box-530
Uppsala 75121, Sweden
Tel.: 46-18-471-3568
Fax: 46-18-511-784
Email: Igor.Abrikosov@fysik.uu.se

ARDELL, ALAN
University of California at Los Angeles
Materials Science & Engineering Dept.
Metallurgy Division
6531-G Boelter Hall
Los Angeles, CA 90095-1595, USA
Tel.: 310-825-7011
Fax: (310) 206-7353
Email: aardell@ucla.edu3

BINDER, KURT
Dipartimento Di Fisica
Johannes Gutenberg Universitat Mainz
Institut Fuer Physik
Staudinger Weg 7
Mainz D-55099, Germany
Tel.: 49-6131-3923348
Fax: 49-6131-3925441
Email: kurt.binder@uni-mainz.de

BOETTGER, BERND
ACCESS E.V.
Intzestrasse 5
Aachen D-52072, Germany
Tel.: 49-241-809-8008
Fax: 49-241-385-78
Email: b.boettger@access.rwth-aachen.de

BOETTINGER, WILLIAM
NIST
100 Bureau Dr., MS 8555
Room A153, BD 223
Gaithersburg, MD 20899, USA
Tel.: 301-975-6160
Fax: 301-975-4553
Email: wboettinger@nist.gov

BRUNO, EZIO
Universita' Di Messina
Dipartimento Di Fisica
Contrada Papardo
Salita Sperone
Messina, ME I-98166, Italy
Tel.: 39-090-393-713
Fax: 39-090-676-5042
Email: bruno@dsme01.unime.it

CHEN, YING
RACE
University of Tokyo
Komaba 4-6-1
Meguro-ku
Tokyo 153-8904, Japan
Tel.: 81-3-5453-5896
Fax: 81-3-3467-0648
Email: ying@race.u-tokyo.ac.jp

CHEPULSKII, ROMAN
National Academy of Sciences
Institute for Metal
Institute for Metal Physics
36, Acad. Vernadsky Blvd.
Kiev 142 UA-03680, Ukraine
Tel.: 380-44-444-1221
Fax: 380-44-444-2561
Email: r_chepulskii@yahoo.com

CLOUET, EMMANUEL
CEA Saclay
SRMP- Bat.520
Gif-sur-Yvette F-91191, France
Tel.: 33-1-6908-6663
Fax: 33-1-6908-6867
Email: emmanuel.clouet@cea.fr

DRCHAL, VACLAV
Institute of Physics AS CR
Na Slovance 2
Praha 8 CZ-182 21, Czech Republic
Tel.: 420-2-6605-2926
Fax: 420-2-8689-0527
Email: drchal@fzu.cz6

EIKEN, JANIN
ACCESS E.V.
Intzestrasse 5
Aachen D-52072, Germany
Tel.: 49-241-809-8009
Fax: 49-241-385-78
Email: j.eiken@access.rwth-aachen.de

EVERETT, RICHARD
Office of Naval Research IFO
223 Old Marylebone Road
London NW1 5TH, United Kingdom
Tel.: 44-207-514-4413
Fax: 44-207-723-6359
Email: reverett@onrifo.navy.mil

FAULKNER, JOHN
Florida Atlantic University
Dept. of Physics
825 Walnut Terrace
Boca Raton, FL 33486, USA
Tel.: 561-297-3429
Fax: 561-297-2662
Email: faulkner@fau.edu

FENG, YUAN
National University of Singapore
10 Science Drive 4
Lower Kent Ridge Road
Singapore 117542, Singapore
Tel.: 65-6874-2960
Fax: 65-6777-6126
Email: phyfyp@nus.edu.sg

FRIES, SUZANA
Access e.V., RWTH-Aachen
Intzestrasse 5
Aachen D-52072, Germany
Tel.: 49-241-809-8013
Fax: 49-241-385-78
Email: sufries@mf.mpg.de

GAUNE-ESCARD, MARCELLE
Ecole Polytechnique
IUSTI-UMR 6595 CNRS
5 Rue Enrico Fermi
Marseille F-13453, France
Tel.: 33-4-9110-6887
Fax: 33-4-9111-7439
Email: Marcelle.Gaune-Escard@polytech.univ-mrs.fr

GONIS, ANTONIOS
Lawrence Livermore Nat'l Laboratory
C&MS, L-371
Livermore, CA 94551, USA
Tel.: 925-422-7150
Fax: 925-423-7040
Email: gonis1@llnl.gov

GRIMVALL, GORAN
Royal Institute of Technology
KTH-SCFAB
Theory of Materials
Stockholm SE-106 91, Sweden
Tel.: 46-8-5537-8160
Fax: 46-8-5537-8470
Email: grimvall@theophys.kth.se

GUY, BERNARD
Ecole des Mines
158 Cours Fauriel
Saint-Etienne F-42023, France
Tel.: 33-4-7742-0164
Fax: 33-4-7742-0000
Email: guy@emse.fr

HATA, SATOSHI
Kyushu University
Interdisc. Grad. School of Eng. Sci.
6-1 Kasugakouen
Kasuga, Fukuoka 816-8580, Japan
Tel.: 81-92-583-7536
Fax: 81-92-575-2318
Email: hata@asem.kyushu-u.ac.jp

HAYES, FRED
University of Manchester
Materials Science Centre
Grosvenor Street
Manchester M1 7HS, United Kingdom
Tel.: 44-161-200-3566
Fax: 44-161-200-3586
Email: fred.hayes@umist.ac.uk

HEHENKAMP, THEODOR
Georg-August Univ. Gottingen
Institut fur Metallphysik
Hospitalstrasse 3-7
Gottingen D-37073, Germany
Tel.: 49-551-794-903
Fax: 49-551-395-012

HICKERNELL, BARBARA
Engineering Conferences International
3 Park Avenue
27th Floor
New York, NY 10016-5902, USA
Tel.: 1-212-591-7836
Fax: 1-212-591-7441
Email: engfnd@aol.com

KAUFMAN, LARRY
M.I.T.
140 Clark Road
Brookline, MA 02445-5848, USA
Tel.: 617-731-2622
Fax: 617-566-70967
Email: larrykaufman@rcn.com

KHACHATURYAN, ARMEN
Rutgers University
Dept. of M.S. & E.
607 Taylor Road
Piscataway, NJ 08855-0909, USA
Tel.: 732-445-4711
Fax: 732-445-6780
Email: khach@jove.rutgers.edu

KLEIN, BARRY
University of California -Davis
One Shields Avenue
402 Mrak Hall
Davis, CA 95616-8671, USA
Tel.: 530-752-4091
Fax: 530-752-0602
Email: bmklein@ucdavis.edu

KORZHAVYI, PAVEL
Royal Inst. of Technology (KTH)
Materials Science & Engineering
Brinellvaegen 23
Stockholm SE-100 44, Sweden
Tel.: 46-8-790-8832
Fax: 46-8-207-681
Email: pavel@met.kth.se

KUDRNOVSKY, JOSEF
Institute of Physics AS CR
Na Slovance 2
Prague 8 CZ-182 21, Czech Republic
Tel.: 420-2-6605-2905
Fax: 420-2-821-227
Email: kudrnov@fzu.cz

LE BOUAR, YANN
LEM-CNRS
29 Avenue De La Division Leclerc
BP 72
Chatillon F-92322, France
Tel.: 33-1-4673-4592
Fax: 33-1-4673-4155
Email: lebouar@onera.fr

MASUDA-JINDO, KINICHI
Tokyo Institute of Technology
Material Science Department
Nagatsuta 4259, Midori-ku
Yokohama, Kanagawa 226-8503, Japan
Tel.: 81-45-924-5636
Fax: 81-424-75-0650
Email: wmfjindo@din.or.jp

MAUGIS, PHILIPPE
IRSID - Arcelor Group
Voie Romaine - BP 30320
Maizieres-les-Metz F-57283, France
Tel.: 33-387-704779
Fax: 33-387-704712
Email:
philippe.maugis@irsid.usinor.com

MEIKE, ANNEMARIE
Lawrence Livermore Nat'l Laboratory
L-201
Livermore, CA 94551, USA
Tel.: 925-422-3735
Fax: 925-423-8988
Email: meike1@llnl.gov

MENON, MADHU
University of Kentucky
Physics Department
Lexington, KY 40506, USA
Tel.: 859-257-8737
Fax: 859-323-1029
Email: madhu@ccs.uky.edu4

MILICI, ANTONIO
Graduate Student
University of Messina
Department of Physics
Contrada Papardo Salita Sperone
Messina I-98166, Italy
Tel.: 39-090-393-713
Fax: 39-090-676-5042
Email: amilici@dsme01.unime.it

MOHRI, TETSUO
Hokkaido University
Graduate School of Engineering
Kita-13, Nishi-8, Kita-ku
Sapporo 060-8628, Japan
Tel.: 81-11-706-6348
Fax: 81-11-706-6348
Email: tmohri@eng.hokudai.ac.jp

NISHIZAWA, SHUICHI
Tohoku University
Aramaki-Aza-Aoba
Aoba-Ku 01
Sendai, Miyagi 980-8579, Japan
Fax: 81-22-217-6903

PENTCHEVA, ROSSITZA
University of Munich
Inst. of Crytallography
Theresienstrasse 41
Munich D-80333, Germany
Tel.: 49-89-2180-4352
Fax: 49-89-2180-4334
Email: pentcheva@lrz.uni-muenchen.de

PFEILER, WOLFGANG
University of Vienna
Institut fur Materialphysik
Strudlhofgasse 4
Vienna A-1090, Austria
Tel.: 43-1-4277-51309
Fax: 43-1-42-779-513
Email: pfeiler@ap.univie.ac.at

RAJAN, KRISHNA
Rensselaer Polytechnic Institute
Acad. Sci. of the Mater. Sci. & Eng.
110 8th Street
Troy, NY 12180-3590, USA
Tel.: 518-276-6126
Fax: 518-276-8554
Email: rajank@rpi.edu

RUBAN, ANDREI
Tech. University of Denmark
CAMP, Physics Department
Bldg. 307
Lyngby 2800, Denmark
Tel.: 45-4525-3234
Fax: 45-4593-2399
Email: ruban@fysik.dtu.dk

SAITO, YOSHIYUKI
Waseda University
Dept. of Mater. Sci. & Eng.
3-4-1 Okubo
Shinjuku-ku
Tokyo 169-8555, Japan
Tel.: 81-3-5286-3314
Fax: 81-3-5286-3314
Email: ysaito@mn.waseda.ac.jp

SIBERCHICOT, BRUNO
CEA/DAM-DIF
BP 12
Bruyeres-le-Chatel F-91680, France
Tel.: 39-01-6926-7327
Fax: 39-01-6926-7077
Email: bruno.siberchicot@cea.fr

SOB, MOJMIR
Institute of Physics of Materials
Zizkova 22
Brno CZ-616 62, Czech Republic
Tel.: 420-5-3229-0455
Fax: 420-5-4121-2301
Email: mojmir@ipm.cz

STOLOFF, NORMAN
Rensselaer Polytechnic Institute
Materials Sci. & Engrg. Dept.
Troy, NY 12180-3590, USA
Tel.: 518-276-6436
Fax: 518-276-8554
Email: stolon@rpi.edu

THIBAUDEAU, PASCAL
C.E.A./Le Ripault
BP 16
Monts F-37160, France
Tel.: 33-247-344-656
Fax: 33-247-345-154
Email: pascal.thibaudeau@cea.fr

TURCHI, PATRICE E. A.
Lawrence Livermore Nat'l Laboratory
C&MS, L-353
P.O. Box 808
Livermore, CA 94551, USA
Tel.: 925-422-9925
Fax: 925-423-7040
Email: turchi1@llnl.gov

VAKS, VALENTIN
RRC Kurchatov Institute
Kurchatov Square 1
Moscow 123182, Russia
Tel.: 7-095-1969826
Fax: 7-095-8825804
Email: vaks@mbslab.kiae.ru

WATANABE, TADAO
Tohoku University
Aramaki-Aza-Aoba
Aoba-Ku 01
Sendai Miyagi 980-8579, Japan
Tel.: 81-22-217-6902
Fax: 81-22-217-6903
Email:
watanabe@mdie.mech.tohoku.ac.jp

WONG, JOE
Lawrence Livermore Nat'l. Laboratory
C&MS, L-356
P.O. Box 808
Livermore, CA 94551, USA
Tel.: 925-423-6385
Fax: 925-424-4737
Email: wong10@llnl.gov

WUENSCH, BERNHARDT
M.I.T.
Ceramics Division 13-4037
77 Massachusetts Avenue
Cambridge, MA 02139-4307, USA
Tel.: 617-253-6889
Fax: 617-253-5827
Email: wuensch@mit.edu

ZINGALES, LEON
University of Messina
Department of Physics
Salita Sperone Contrada Papardo
Messina I-98166, Italy
Tel.: 39-090-393-713
Fax: 39-090-676-5042
Email: zingales@dsme01.unime.it

PREFACE

This volume contains the proceedings of the Third International Alloy Conference (IAC-3) that was held in Estoril Sol, Lisbon, Portugal, June 30-July 5, 2002. Like its predecessors, and anticipated for those planned for the future, the conference brought together experimental and theoretical/computational scientists involved in the study of alloys taken to mean (solid) materials composed of more than one chemical species.

The study of alloys involves two main but interrelated areas of science: First, the determination of the electronic states in a material and the effect of these states on properties, and second the thermodynamic behavior of the materials as it arises in the study of phase transformation and phase evolution in time. Properties like formation energy, electronic transport, magnetism, and mechanical behavior are intimately tied into the nature of electronic states. Kinetics of transformation, phase evolution and aging are the subject of thermodynamics. Both kinds of properties, however, are interconnected in that information extracted from one study is often applicable to the other.

The papers presented in the conference span the spectrum of activity in the science of alloys. The theoretical presentations ranged in content from fundamental studies of electronic structure, to first-principles calculations of phase diagrams, to the effects of charge transfer, to the temperature dependence of short-range order parameters. They encompassed the study of mechanical properties, the properties of dislocations, of phase evolution, and computer simulations. Experimental studies were presented based on a variety of state of the art experimental techniques, from TEM to synchrotron diffraction. The phenomena studied varied from the precipitation of nitrides in steel, to the wetting of interfaces between two different crystal structures, to the ordering of vacancies in carbides. And the materials whose properties were measured ranged from Transition metals, to the Lanthanides, to the Actinide series of compounds and alloys.

The third conference in the series confirmed, as if there were any doubt, the richness and complexity of phenomena that is embodied in the physics of alloys. An incredible amount of effort combining experimental and theoretical aspects is continuously expended in the attempt to understand this physics and it in a productive way in engineering applications. This intensive effort points toward the importance of understanding and predicting alloy properties but also to the challenging nature of the task.

There is good reason for hoping that significant progress is in the making, however. Recent developments in the theory of studying correlations in fluctuating systems have emerge with a fresh air of unifying power that encompasses both alloys and the Coulomb interaction in solids. We are looking forward to the next International Alloy Conference to hear more about these developments and to record thcm for the scientific community. Interface between *ab initio* and phenomenological modeling of alloy properties, with predictive tools to help the design of new materials with engineering requirements is another topic that is taking momentum and will certainly be discussed more at the next IAC.

In organizing the IAC-3, we are grateful to our sponsors both for financial assistance and for organizational aid. Thc U. S. Army Research Office Physics Division, the Materials Research Institute at Lawrence Livermore National Laboratory, the U.S. Office of Naval Research Materials Division, and the United Engineering Foundation, our main and official sponsors were forthcoming with generous financial contributions. The UEF also provided the logistics and the administrative components of bringing together a significant number of alloy scientists to a magnificent place in Portugal. As organizers we can only express our deep appreciation to them for these efforts.

At last but certainly not least, the staff at Kluwer/Plenum contributed their expertise along with a great exercise of patience in the production of this volume. Because of their efforts the volumes in the series now begin to take the form of a unified work, a feature that will be augmented, as more volumes become available. We thank them and hope that well continue our fruitful collaboration for a long time into the future.

P. E. A. Turchi
A. Gonis
K. Rajan
A. Meike

CONTENTS

KINETICS, DIFFUSION AND TRANSPORT

MAGNETIC AND ELASTIC PROPERTIES

THEORY

MICROSTRUCTURAL PROPERTIES

PRECIPITATION OF DISORDERED Ni-X SOLID SOLUTION PHASES IN OFF-STOICHIOMETRIC ORDERED Ni_3X ALLOYS

Y. Ma[1], J. Joshi[2] and A.J. Ardell[1]

[1]Department of Materials Science and Engineering
University of California, Los Angeles, CA 90095-1595, USA
[2]Aaj Accumulators Pvt. Ltd., W–54 M.I.D.C.
Shiroli. Tal. Hatkanagale, India 416 122

INTRODUCTION

The coarsening behavior of coherent γ precipitates (Ni-X solid solution, X = Al, Ge or Ga) in a matrix of the off-stoichiometric γ' phase (Ni_3X) is currently under investigation in binary Ni-Al, Ni-Ge and Ni-Ga alloys. Of particular interest are the evolution of the precipitate microstructure (morphology and spatial correlations), the scaling behavior of the particle size distributions and the kinetics of coarsening and its dependence on volume fraction. Since the compositions and temperatures of greatest interest are very close to the (γ + γ')/γ' phase boundary, hereafter called the γ solvus, we need to locate it as accurately as possible in each alloy. The γ solvus is retrograde in binary Ni-Al, Ni-Ge and possibly Ni-Ga alloys, but is not known with precision for any of the 3 alloys. In Ni-Al alloys there is a minimum retrograde solubility of ~22 at. % somewhere between 1000 and 1200 °C, but there is considerable scatter in the published data[1-4], as shown in Fig.1. The uncertainty is unacceptably large for our purposes.

Depending on composition, either isothermal aging or continuous cooling of hypostoichiometric Ni_3X produces precipitates of the γ phase. Cornwell and Purdy[5] stated that the γ precipitates in binary Ni-Al alloys were plate-shaped and coherent. However, Liu et al.[6] reported that the precipitate particles in a quaternary alloy were initially spherical and evolved with aging time to cuboidal shapes, becoming plate-like only at longer aging times. In the reverse system, i.e. γ' precipitates in γ matrix, the γ' precipitates evolve in shape from spheres to cuboids to nearly perfect cubes[7], and can become concave cuboids (cuboids with concave interfaces) at large sizes if the volume fraction is small enough (< 0.04 or so)[8,9]. Plate-shaped γ' precipitates usually form as a result of coalescence[7]. One immediate question of interest is why the morphological evolution of γ' in γ should be different from that of γ in γ'. After all, the interfacial free energy is independent of which is the minority or majority phase, and so are the elastic self- and interaction energies. To be sure, the lattice and elastic-constant mismatches are reversed in sign, but they are identical in magnitude. Since elastic energies are proportional to the square of the lattice and elastic constant mismatches, their signs should not matter.

Other important issues involve comparisons of the kinetics of coarsening of the two phases in each other, and how the coarsening kinetics of γ in γ' depend on volume fraction; the dependence is anomalous for the reverse system[10]. In this paper we present some of the preliminary results of our research. Most of the results to date are on Ni-Al alloys, and the majority of the paper concentrates on these. Nevertheless some information has also been obtained on precipitation in Ni_3Ge and Ni_3Ga alloys, and these results are also briefly mentioned.

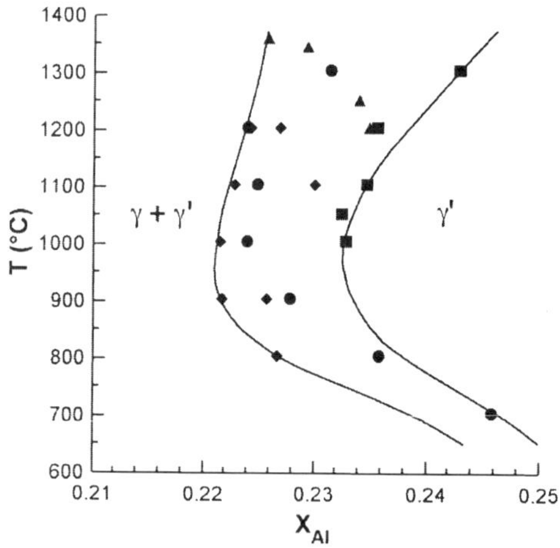

Figure 1. Illustrating the uncertainty in the precise position of the γ solvus (the boundary between the γ + γ' and γ' regions in the Ni-Al phase diagram). The curves are drawn to illustrate the extremes of the range of data reported in the literature. Data of: ■ Janssen[1]; ▲ Verhoeven et al.[2]; ● Jia (cited by Okamoto[3]); ◆ Watanabe et al.[4].

EXPERIMENTAL PROCEDURES

Binary Ni-Al alloys containing 22.0, 22.2, 22.4, 22.6, 22.8 and 23.0 at. % Al were purchased from the Alloy Preparation Facility of the Ames Laboratory, Ames, IA. The alloys were arc-melted and chill cast into water-cooled Cu crucibles and delivered to us in the form of rods approximately 12 mm in diameter. Weight losses during melting and casting of the alloys were at most 0.025%, indicating that the compositions were indeed the compositions of the alloys received. The alloys were annealed in a vacuum of 3×10^{-5} torr at 1200 °C for up to 72 h in order to improve homogeneity. The compositions of the alloys were further verified using energy-dispersive x-ray spectroscopy (EDS) and inductively coupled plasma spectroscopy. We believe that the compositions reported are accurate to within ± 0.03 at. %.

Binary Ni-Ge alloys containing 22.0, 22.5, and 23.0 at. % Ge and binary Ni-Ga alloys containing 22.0, 22.5, and 23.0 at. % Ga were also purchased from the Ames Laboratory, where they were annealed in vacuum at 1000 °C for 100 h prior to shipping. Weight losses during melting and casting of these alloys were small enough to be considered negligible, so the reported compositions are taken as the true compositions.

Slices were made using electric-spark discharge machining, and after polishing were "solution treated" in an inert argon atmosphere at 1100 ± 1 °C for 1.5 h (Ni-Al alloys) or 1000 ± 1 °C for 0.75 h (Ni-Ge and Ni-Ga alloys) and quenched into refrigerated brine. These slices were aged at various temperatures and times in a vertical tube furnace, in which the temperature fluctuated by less than ±0.5 °C. Specimens were cored from the aged slices and electrolytically polished for examination by transmission electron microscopy (TEM) using a JEOL model 100CX TEMSCAN operating at 100 kV.

Particle sizes and shapes were evaluated from dark-field images taken using a {100} superlattice reflection from thin foils in [001] orientation. Other diffracting conditions were tried, but this one produced the best contrast. The sizes and shapes of the γ precipitates were evaluated using image-analysis software. For non-equiaxed γ precipitates, the "radius", r, was defined as $(a + b + c + d)/8$, where a, b, c and d are the sides of the circumscribed polygon.

RESULTS AND DISCUSSION

Dissolution Experiments

Accurate knowledge of the γ solvus is a requirement for determining how the coarsening kinetics of γ in γ' depend on volume fraction. Simply solution-treating the alloys and aging them for various times within the expected 2-phase region of the phase diagram proved to be insufficient to establish the γ solvus. This is exemplified in Fig. 2, which shows the results of two experiments on aging the 22.6 % alloy, one at 650 °C for 240 h and the other at 700 °C for 120 h. Precipitates can be seen clearly in Fig. 2a, but none are present in Fig. 2b. These results would ordinarily lead us to conclude that the solvus temperature for the alloy containing 22.6 % Al is between 650 and 700 °C. However, if the 22.6 % Al alloy is aged first at 650 °C to produce small coherent γ precipitates, and subsequently re-aged at successively higher temperatures, we obtain completely different results. These are exemplified by the micrographs in Fig. 3. Here we observe γ precipitates in specimens re-aged to temperatures as high as 755 °C (Fig. 3c), and it is evident that complete dissolution of the γ precipitates is observed only after re-aging at 765 °C (Fig. 3d). The γ solvus for the alloy containing 22.6 % Al clearly lies between 755 and 765 °C, which is more than 50 °C higher than indicated by the microstructures shown in Fig. 2.

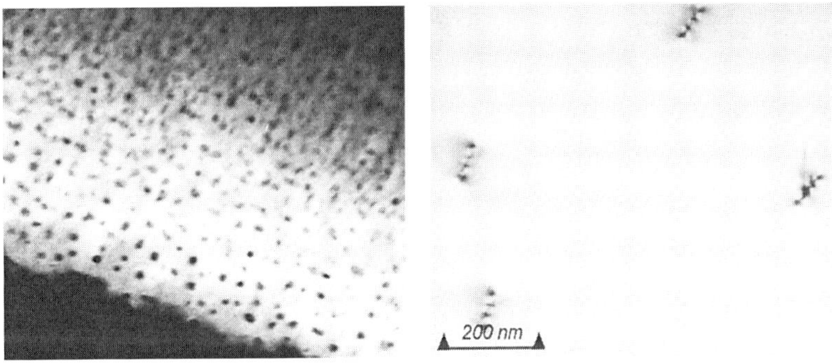

(a) (b)

Figure 2. The microstructures in the alloy containing 22.6 % Al after solution-treating and aging under the following conditions: (a) 650 °C for 240 h (b) 700 °C for 120 h.

The reason that dissolution experiments are needed to accurately determine the γ solvus is simply that nucleation of the γ phase from supersaturated γ' is much slower than nucleation of γ' from supersaturated γ[11]. The principle involved is simple. There is a large energetic barrier to nucleation of the γ phase but no barrier to their dissolution. In addition to dissolution experiments, additional work was done to determine the concentration of Al in the matrix of the alloys containing 22.0 and 22.2 % Al, wherein dendrites of the γ phase were never eliminated by annealing[12]. This work produced the γ solvus shown in Fig. 4.

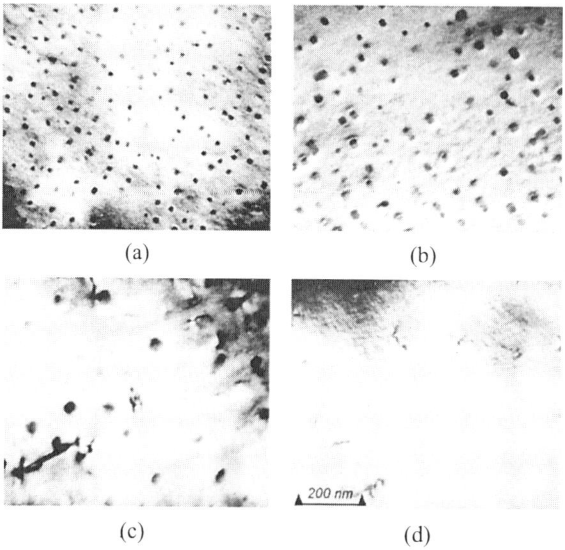

Figure 3. The microstructures in the alloy containing 22.6 % Al after solution-treating and aging under the following conditions: (a) 650 °C for 96 h plus 730 °C for 1.5 h; (b) 650 °C for 144 h plus 740 °C for 27 h; (c) 650 °C for 240 h plus 755 °C for 18 h; (d) 650 °C for 240 h plus 765 °C for 18 h.

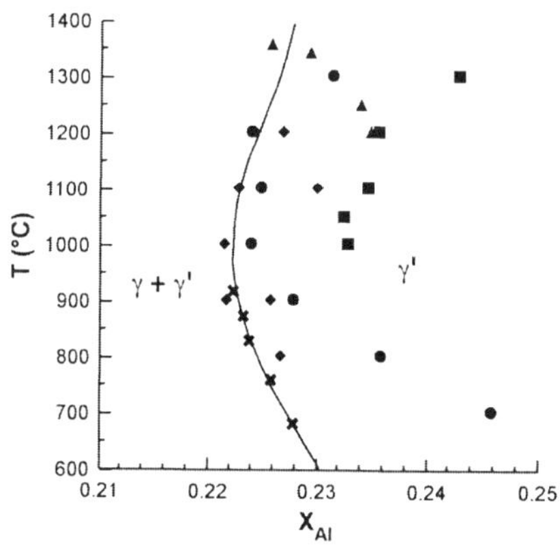

Figure 4. The γ solvus (solid curve) determined from the dissolution experiments of Ma et al.[12]. The symbol ✖ represents the solvus resulting from the dissolution experiments. The other plotting symbols are identical to those in Fig. 1.

Coarsening kinetics

Representative precipitate microstructures during precipitation of the γ phase in the 22.0 at. % Al alloy aged at 700 °C are shown in Fig. 5. The very small number of γ precipitates seen after aging for 8 h is indicative of the difficulty of their nucleation. The morphology of the γ precipitates evolves from spherical to cuboidal even when they are very small (< 50 nm in diameter), as evidenced by the presence of flat interfaces after only 48 h of aging; the flat interfaces are parallel to ⟨001⟩. Many plate-shaped precipitates are evident after 96 h of aging despite their small size. It is not evident at this stage of the research whether the plate shape results from the breaking of equiaxed symmetry, as predicted theoretically[13], or whether adjacent γ precipitates coalesce readily, since there is no issue involving anti-phase ordering relationships, as there is for the coalescence of γ' precipitates[14]. Measurements of the sizes of the precipitates in Fig. 5 and plotting them according to the expected rate law for coarsening[15,16] i.e. $\langle r\rangle^3 \propto kt$, where k is the rate constant for coarsening and t is the aging time, leads to the results shown in Fig. 6. The linearity is quite good, indicating that the γ precipitates evidently grow by diffusion-controlled coarsening.

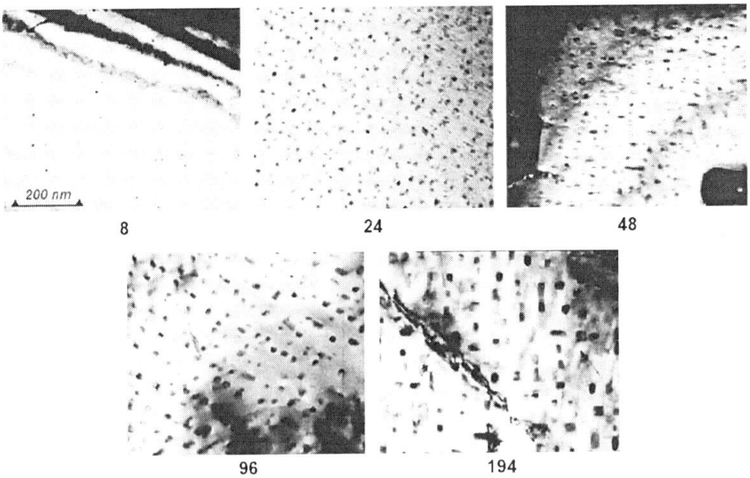

Figure 5. Evolution of precipitates in the 22.0 at % Al alloy at 700 °C. The number under each figure indicates the aging time in h.

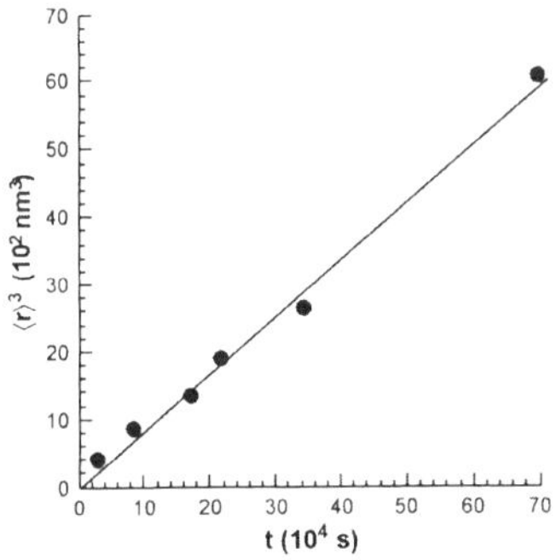

Figure 6. The kinetics of coarsening at 700 °C of the γ precipitates in the γ' matrix of the 22.0 % Al alloy, plotted as the cube of the average radius, $\langle r\rangle$, vs. aging time, t.

Figure 7 illustrates the relative rates of precipitation of γ in γ' and γ' in γ observed in a dendritic region of the alloy containing 22.0 % Al aged for 48 h at 700 °C. It is quite evident that the precipitation of γ in γ' is much slower than the precipitation of γ' in γ. This is most likely due to the faster coarsening kinetics of γ' in γ, but the slower nucleation kinetics of γ in γ' undoubtedly contribute as well. According to Calderon et al.[17], k can be expressed by the equation

$$k = \frac{8 D \sigma V_{m\beta}}{9 G''_{m\alpha}\left(X_{\beta e} - X_{\alpha e}\right)^2}, \tag{1}$$

where D is the coefficient of diffusion of solute in the matrix, $V_{m\beta}$ is the average volume per atomic site in the precipitate (β) phase, $X_{\alpha e}$ is the equilibrium solubility of the solute in the matrix (α) phase, $X_{\beta e}$ is the equilibrium solubility of the solute in the dispersed phase and $G''_{m\alpha}$ is the second derivative of the molar free energy of mixing of the matrix phase with respect to composition, evaluated at $X_{\alpha e}$.

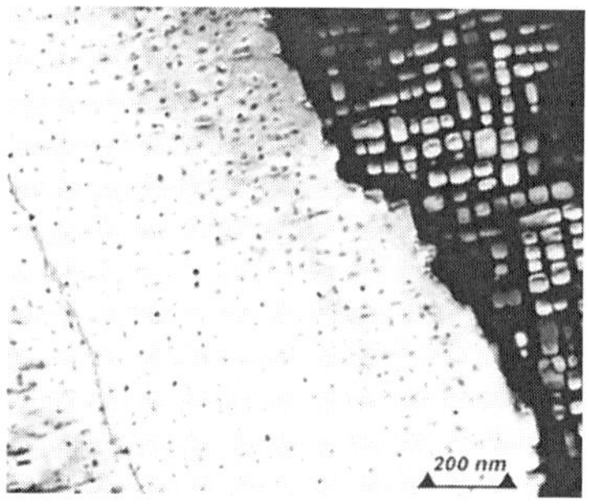

Figure 7. A dark-field TEM micrograph showing the precipitation of γ (dark) in γ' (bright) and γ' (bright) in γ (dark) in a dendritic region of the 22.0 at % Al alloy aged at 700 °C for 48 h.

During coarsening it is necessary not only to transport solute atoms to growing precipitates and away from shrinking ones, but also to transport *solvent* atoms *away* from growing precipitates and *towards* shrinking ones. Assuming that the concentration of vacancies is in equilibrium everywhere during diffusion, it is more appropriate to describe diffusion in terms of the chemical diffusion coefficient, $\tilde{D}$, which can be written as

$$\tilde{D} = S\left\{X D_B^* + (1 - X) D_A^*\right\} X(1 - X)\frac{G''_{m\alpha}}{RT}, \tag{2}$$

where D_A^* and D_B^* are the tracer diffusion coefficients of the solvent (A) and solute (B) a t-oms in the matrix with solute concentration X, R is the gas constant, T is the absolute temperature and S is Manning's vacancy flow factor[18], given by the general equation

$$S = 1 + \left[f_o^{-1} - 1\right]\left\{\frac{X(1 - X)\left[D_A^* - D_B^*\right]^2}{\left[X D_A^* + (1 - X) D_B^*\right]\left[(1 - X) D_A^* + X D_B^*\right]}\right\}, \tag{3}$$

where f_o is the correlation coefficient. Applying equations (2) and (3) separately to the γ and γ' phases, and substituting them into eq. (1) allows us to express the ratio of the rate constants for the coarsening of γ' in γ and *vice versa* as

$$\frac{k_\gamma}{k_{\gamma'}} = \frac{S_\gamma\left[(1 - X_{\gamma e}) D_{Al,\gamma}^* + X_{\gamma e} D_{Ni,\gamma}^*\right] X_{\gamma e}(1 - X_{\gamma e})}{S_{\gamma'}\left[(1 - X_{\gamma' e}) D_{Al,\gamma'}^* + X_{\gamma' e} D_{Ni,\gamma'}^*\right] X_{\gamma' e}(1 - X_{\gamma' e})}, \tag{4}$$

where the subscripts refer to the phase, which acts as the matrix (e.g. k_γ is the rate constant for coarsening of γ' precipitates in the γ phase of equilibrium composition $X_{\gamma e}$). The molar volumes of solute have been taken as equal in eq. (4).

To compare $k_\gamma/k_{\gamma'}$ calculated using eq. (4) with experimental measurement, we analyze data on coarsening at 700 °C. The calculations make use of diffusion coefficients estimated from the sources summarized in Table 1. For the calculation of $k_\gamma/k_{\gamma'}$ we used f_o = 0.781 for diffusion in the γ phase and f_o = 0.689 for diffusion in the γ' phase[19]. The chemical diffusion coefficients are 1.016×10^{-18} m²/s and 8.030×10^{-21} m²/s for the γ and γ' phases. respectively. Experimentally, the slope of the straight line fitted to the data in Fig. 6 yields the value $k_{\gamma'} = 8.49 \times 10^{-3}$ nm³/s. From analysis of data on the coarsening of γ' precipitates[22] at 700 °C we obtain $k_\gamma = 4.23 \times 10^{-2}$ nm³/s. The ratio $k_\gamma/k_{\gamma'}$ from experiment is thus ≈5; the value calculated theoretically is 77. Possible reasons for the discrepancy are not evident, but the assumed value of $D^*_{Ni,\gamma} = D^*_{Al,\gamma}/10$ in Table 1 is an obvious candidate.

Table 1. Tracer diffusion coefficients in the γ and γ' phases.

D (m²/s)	D at 700 °C (m²/s)	Ref.
$D^*_{Al,\gamma} = 7.1\times10^{-4}\, exp\left\{\frac{-276{,}600}{RT}\right\}$	1.01×10^{-18}	20
$D^*_{Ni,\gamma} = \frac{D^*_{Al,\gamma}}{10}$	1.01×10^{-19}	--
$D^*_{Al,\gamma'} = 0.372\, exp\left\{\frac{-375{,}400}{RT}\right\}$	2.63×10^{-21}	21
$D^*_{Ni,\gamma'} = 3.31\times10^{-4}\, exp\left\{\frac{-302{,}200}{RT}\right\}$	1.99×10^{-20}	21

The Ni-Ge and Ni-Ga alloy systems

Precipitates of the Ni-Ge solid solution have been observed in the Ni_3Ge matrix in all 3 Ni-Ge alloys. Figure 8 shows the microstructures in the 22.0 % Ge alloy aged at 700 °C. The morphology changed from cuboidal to plate-like at a fairly short aging time. Observations of precipitates in dendritic regions of the 22.0 % Ge alloy indicate that precipitation of the Ni-Ge solid solution in Ni_3Ge is much slower than in the reverse system, similar to what is observed in Ni-Al alloys. Dissolution experiments are in progress to accurately establish the γ solvus in the Ni-Ge system.

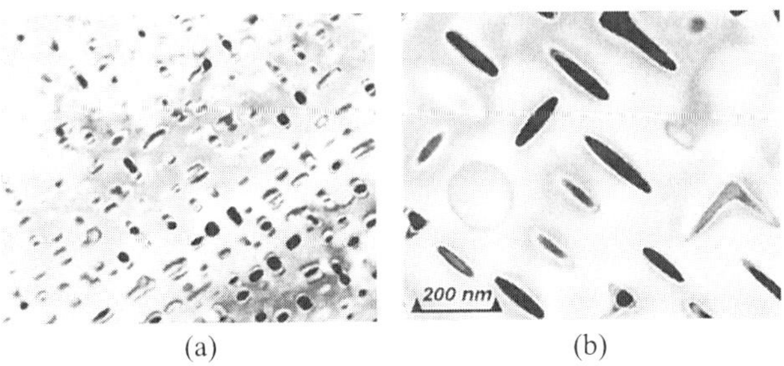

(a) (b)

Figure 8. Precipitates of the Ni-Ge solid solution in Ni_3Ge, observed in an alloy containing 22.0 at. % Ge aged for (a) 8 h and (b) 48 h at 700 °C.

Aging experiments have also been conducted on the Ni-Ga alloys at temperatures in the range 600 to 800 °C. So far we have not observed the precipitation of the solid solution phase, which indicates that the γ solvus in this alloy system does not have the necessary negative slope in the Ni-Ga phase diagram.

SUMMARY

Nucleation of the solid solution γ phase from the ordered γ' phase (Ni_3Al) is quite difficult. Undercooling the 22.6 at.% Al alloy by more than 50 °C to a temperature (700 °C) at which diffusion is reasonably rapid fails to produce observable γ precipitates, even after aging for 120 h. The phase boundary can be best found through the use of dissolution experiments, which take advantage of the absence of a barrier to the dissolution of γ phase.

The γ precipitates are equiaxed when small, then non-equiaxed as they grow and eventually become plate-like shaped. The γ precipitates transform into plates far more readily than γ' precipitates, which is attributed to the absence of anti-phase boundaries in the γ phase. Despite the non-equiaxed shapes, $\langle r \rangle^3$ depends approximately linearly on aging time t. The experiments show that the kinetics of coarsening of γ precipitates is much slower than that in the reverse system. This is consistent with behavior expected from diffusion in the two phases.

The Ni_3Ge phase behaves similarly to Ni_3Al in that it can be aged to produce disordered precipitates of the Ni-Ge solid solution from Ni_3Ge matrix. We do not yet know if this is also true for Ni_3Ga, but preliminary experiments suggest that it is not.

ACKNOWLEDGMENTS

The authors express their appreciation to the National Science Foundation for financial support of this research under Grant # DMR-0209260.

REFERENCES

1. M.M.P. Janssen, *Metall. Trans.* 4;1623 (1973).
2. J.D. Verhoeven, J.H. Lee, F.C. Laabs and L.L. Jones, *J. Phase Equil.* 12;15 (1991).
3. C.C. Jia, Thesis Tohoku (1990), cited by H. Okamoto, *J. Phase Equil.* 14;257 (1993).
4. M. Watanabe, Z. Horita, D.J. Smith, M.R. McCartney, T. Sano and M. Nemoto, *Acta Metall. Mater.* 42;3381 (1994).
5. L.R. Cornwell and G.R. Purdy, *Metall. Trans.* 5;780 (1974).
6. W. Liu, H. Rösner and E. Nembach, *Z. Metallkde* 88;648 (1997).
7. A.J. Ardell and R.B. Nicholson, *Acta Metall.* 14;1295 (1966).
8. A. Maheshwari and A.J. Ardell, *Scripta Metall.* 26;347 (1992).
9. A. Maheshwari and A.J. Ardell, *Phys. Rev. Lett.* 70; 2305 (1993).
10. A. Maheshwari and A.J. Ardell, *Acta Mater.* 40;2661 (1992).
11. Y. Ma, J. Joshi and A.J. Ardell, *XI International Materials Research Congress*, Cancun, Mexico (2002), to be published.
12. Y. Ma, J. Joshi and A.J. Ardell, to be published.
13. M.E. Thompson, C.S. Su and P.W. Voorhees, *Acta Metall.* 42;2107 (1994).
14. A.J. Ardell, *Modelling Simul. Mater. Sci. Eng.* 8;277 (2000).
15. I.M. Lifshitz and V.V. Slezov, *J. Phys. Chem. Solids* 19;35 (1961).
16. C. Wagner, *Z. Elektrochem* 65;581 (1961).
17. H.A. Calderon, P.W. Voorhees, J.L. Murray and G. Kostorz, *Acta Metall.* 42;991 (1994).
18. J.R. Manning, *Acta Metall.* 15;817 (1967).
19. M. Koiwa and S. Ishioka, *Phil. Mag. A* 48;1 (1983).
20. A.J. Ardell, *Interface Sci.* 3;119(1995).
21. K. Fujiwara and Z. Horita, *Acta Mater.* 50;1571 (2002).
22. A.J. Ardell, *Scripta Metall.* 24;343 (1990).

AN ATOMIC SCALE STUDY OF THE PHYSICAL PROPERTIES OF DELTA PLUTONIUM AND Pu:Al ALLOYS

Bruno Siberchicot[1], Gregory Robert[1], Johann Bouchet[1]
and Alain Pasturel[2]

[1]CEA, Département de Physique Théorique et Appliquée
BP 12, 91680 Bruyères-le-Châtel, France.
[2]Laboratoire de Physique Numérique, CNRS,
25 avenue des Martyrs, BP 106, F38042 Grenoble, France.

INTRODUCTION

Plutonium is probably the most intriguing metal in the whole periodic table. For instance, the Pu phase diagram at atmospheric pressure has six stable allotropes, some with very complex open structures (α and β : monoclinic) and others with close-packed structures (γ tetragonal and δ, δ', ε : cubic). No other element displays this complexity of polymorphism. Moreover, the phase transitions are accompanied by large volume changes, with a 25% difference between the cubic (fcc) δ phase and the room temperature monoclinic α phase. In contrast to this phase, which is brittle, the face-centered cubic δ phase is ductile, a property that makes it convenient for engineering applications. Many other physical properties are puzzling like specific heat coefficients or electrical resistivity. The origin of these peculiar properties of metallic plutonium has generally been attributed to the fact that it marks the boundary between the itinerant and the localized *5f* electron systems. From thorium through neptunium, the atomic volume displays the parabolic decrease typical of transition metals as the number of bonding electrons increases. Then the volume of the actinides jumps up sharply from neptunium to americium, ascribed to the localization of the *5f* electrons. From a physical point of view the actinide series could be divided in two sub-series with very different characteristics. Plutonium is exactly located at the center of this discontinuity. The character of the *5f* electrons varies from nearly pure metallic (delocalization) in α-Pu, to varying degrees of localization in the elevated-temperature phases.

The δ phase of pure plutonium is stable from 593 K to 736 K. Small additions of group IIIB metals (Al, Ga, In or Tl) can stabilize this phase at room temperature but the low-temperature limit of the impurity solubility is not well known. The Pu-Ga phase diagram by Chetobarev *et al.*[1] indicates an eutectoid decomposition of the high-temperature δ-phase into the α-Pu and Pu_3Ga. By contrast Ellinger *et al.*[2] report no such decomposition. Similar problems are reported for plutonium-aluminium alloys.

Moreover physics concerning plutonium and its alloys involves predicting its properties under long-term aging in both weapons and storage[3] environnement[4]. The knowledge of all plutonium properties is a major challenge and first-principles studies of pure plutonium, alloys, and finite temperature simulations are needed.

In a first part this paper reports on a study of the electronic structure of pure plutonium for three of its localized phases (δ, δ' and ε) and a finite temperature description of the related phase diagram in a second part. A third part deals with the calculation of solubility limit of aluminium in δ plutonium.

BAND-STRUCTURE CALCULATION RESULTS

The density-functional theory and its local-density approximation (DFT-LDA), which sucessfully describes light actinides underpredicts the δ-phase volume by about 35 %. δ-Pu is not the only material for which LDA fails to reproduce the ground-state properties ; Mott insulators, like *3d* transition metal oxides, and a great variety of materials whose electronic structure contains partially filled valence *d* or *f* shells are not correctly described. The failure of the standard LDA approach has led to attempts to go beyond LDA[5-7] for plutonium. It is well established that this deficiency is linked to the strong on-site repulsion U between electrons in the localized *d* or *f* states. In fact, correlation effects arise when U exceeds or equals the mean conduction bandwith W. Recently the so-called LDA+U approach[8] has been applied to the problem of δ-Pu[9-11]. In these calculations the Hubbard U is treated as an adjustable parameter to fit the calculations to the experimentally observed volume for δ-Pu. Bouchet *et al.*[10] have shown that this method stabilizes the fcc δ-phase versus the bcc one (Bain's paths), reproduces the correct order of elastic constants (C' >0 and negative Cauchy pressure) and improves the relative values of these constants. At the same time, Wang and Sun[12] obtained a correct equilibrium volume by using Generalized Gradient Approximation (GGA) and by considering an antiferromagnetic alignment of spins.

In order to have a general point of view of the effect of different approximations, we performed band-structure calculations for δ, δ' and ε-Pu phases within LDA, GGA and LDA+U in non-magnetic and antiferromagnetic (AF) configurations. The electronic structures and total energies are calculated using the accurate all-electron full-potential linearized augmented plane-wave (FPLAPW) method (WIEN2k[13]) with the von Barth functional and Perdew-Wang 91 for GGA calculations. The relativistic Dirac equation is solved self-consistently for core states and the scalar relativistic one for valence states. Spin-orbit coupling is included, and *6p* local orbitals are used for a better treatment of $6p_{1/2}$ and $6p_{3/2}$ states. For LDA+U calculations the atomic value of 3.13 eV for U has been used.

For δ-Pu, although in the non-magnetic case GGA does not really improve equilibrium properties (volume and bulk modulus), the AF configuration leads to an equilibrium volume close to the experimental one. On the contrary a ferromagnetic order gives a too large volume. Nevertheless, in both cases, ferro and AF, GGA leads to a net magnetic moment of about 2 μ_B per atom in contradiction with experiments. LDA+U also leads to correct equilibrium properties, but moreover, in AF as well as ferromagnetic configuration a perfect cancellation of spin and orbital moment per atom is obtained. This result is in better agreement with scarce experimental data. For a given cristallographic structure, in all cases the more stable state is always obtained for an antiferromagnetic alignment of spins. The main results for AF configurations are displayed in table 1. These results open two main questions : (i) why does GGA give such differences between non magnetic and AF configurations, (ii) do an antiferromagnetic or a more complex magnetic order really exist for the localized phases of plutonium ? According to our results and scarce experimental features, a more probable magnetic configuration for plutonium may be spin-glass like. New calculations in a disordered local moment framework should be a next step toward a better knowledge of plutonium physics.

Table 1. Results of calculations

	Experiment			LDA (T= 0 K)			GGA (T= 0 K)			LDA+U (T= 0 K) U= 3.13 eV		
	850K	750K	650 K									
Phase	BCC	BCT	FCC	BCC	BCT	FCC	BCC	BCT	FCC	BCC	BCT	FCC
V_0 (A^3/atom)	24.29	24.79	24.91	16.54	19.94	20.02	22.61	24.20	24.42	23.95	24	24.05
Bulk modulus GPa	21 (a)	27 (a)	29 (a)	70	70.5	71	40	54	55	52	68	66
Magnetic order	PM	PM	PM	AFM	AFM	AFM	AFM	AFM	AFM	AFM	AFM	AFM
Mag. Mom./ atom (μ_B)	0	0	0	0.38	1.95	1.89	2.03	2.23	2.29	0	0	0
Cohesive Energy (eV)	3.7 (b)	3.8 (b)	3.8 (b)	6.509	6.284	6.269	4.90	4.977	4.981	5.26	5.29	5.3
E_{FCC} -E_{BCT}(eV)				- 15 meV			4 meV			10 meV		
E_{BCT} -E_{BCC}(eV)				- 240 meV			79 meV			59 meV		
Phase order	FCC → BCT → BCC			BCC → BCT → FCC			FCC → BCT → BCC			FCC → BCT → BCC		

a) J.A. Cornet *et al.*, *JNM*, **28**, 303, (1968)
b) M. I. Baskes, *PRB*, 62, 15532, (2000)

PHASE DIAGRAM

Since we are treating high-temperature properties of δ-Pu, δ'-Pu and ε-Pu from calculations at 0K, a reliable thermodynamic model is needed. As Pu-3.6 at. % Ga alloy behaves like a Debye solid at ambiant pressure and low temperature[14], we chose to use the Debye model for plutonium instead of a classical mean-field approach. We first briefly summarize this model.

If we can calculate the Helmholtz free energy as an explicit function of volume and temperature, all other thermodynamic parameters can be derived. The free energy can be written as :

$$F(V,T) = E_c(V,T) + F_{ion}(V,T) + F_{el}(V,T) + F_{mag}(V,T)$$

$E_c(V)$ is the cohesive energy at OK derived from *ab initio* GGA and LDA+U antiferromagnetic calculations. For this analysis, binding curves are fitted to an exponential function mathematically equivalent to a Morse function[15]. $F_{ion}(V,T)$ is the vibrational energy calculated in the framework of Debye model :

$$F_{ion}(V,T) = -k_B T[D(\theta_D / T) - \ln(1 - \exp(-\theta_D / T)] + \tfrac{9}{8} k_B \theta_D$$

with θ_D the Debye temperature related to bulk modulus. $F_{el}(V,T)$ is the thermal electronic contribution :

$$E_{el}(V,T) = \int n(\varepsilon,V) f\varepsilon d\varepsilon - \int^{\varepsilon_F} n(\varepsilon,V)\varepsilon d\varepsilon$$

where *f* is the Fermi distribution. $F_{mag}(V,T)$ is the magnetic free energy[12]

$$F_{mag}(V,T) = -k_B T \ln[M_S(2L - M_S) + 1]$$

where M_S is the total spin magnetic moment and L is the *5f* orbital moment.

From the Helmoltz free energy versus volume curves, equilibrium lines between the three phases (δ-Pu, δ'-Pu and ε-Pu) are calculated and the phase diagram plotted. The corresponding GGA and LDA+U phase diagrams are reported in figures 2 and 3 and could be compared to the experimental pressure-temperature diagram drawn in figure 1.

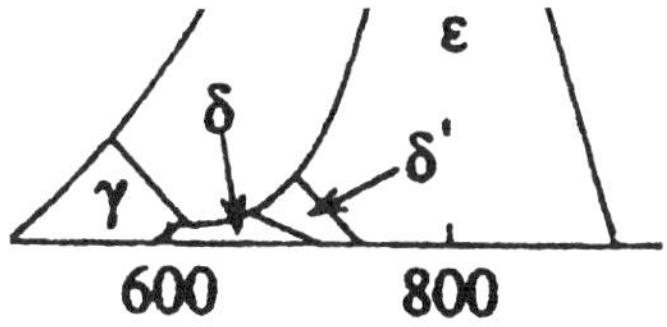

Figure1: Part of the phase diagram of plutonium

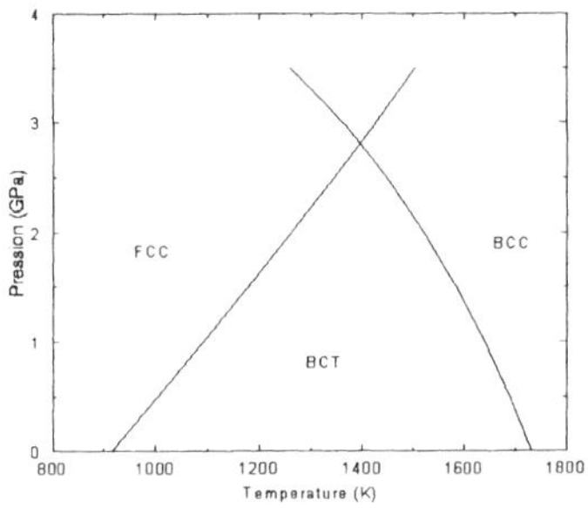

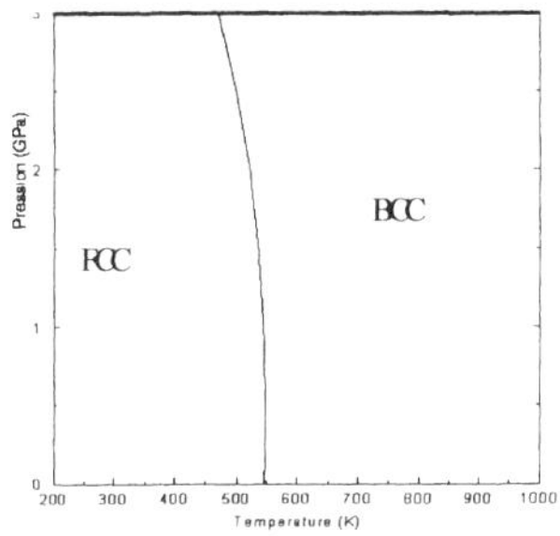

Figure 2. Phase diagram within GGA

Figure 3. Phase diagram within LDA+U (U=3.13 eV)

Although the general shape of the diagram is reproduced within GGA, the absolute values of pressures and temperatures are dramatically too high. The main explanation comes from the large energy differences obtained between the three phases. LDA+U leads to a better phase diagram (only δ-Pu/ε-Pu equilibrium calculated up to now) although the reproduction of the experimental diagram is not perfect.

SOLUBILITY LIMIT OF δ-PLUTONIUM

As discussed in the introduction, the understanding of equilibrium between Ga and Pu_3Ga or Al and Pu_3Al with temperature is one of the main goal in the study of plutonium alloys. Given the crystal structure of an ordered compound, one can calculate its equilibrium properties and the more stable structure between a few possible different crystal forms at 0 K. However, it is necessary to treat solid solutions and to know the evolution of the structural stability as a function of temperature. In this study, the basic tool to determine such properties is based on a generalized three-dimensional Ising model.

The problem consists in the *ab initio* determination of the 3D Ising model parameters[16] and the study of this model at finite temperature.

The total energy of a disordered alloy A_xB_{1-x} can be described in terms of a convergent series of concentration-independent multisite interactions. For a given spin configuration α the total energy is expressed by :

$$E_{tot}^{\alpha}(r) = \sum_{\gamma}^{\gamma \max} J_{\gamma}(r)\xi_{\gamma}^{\alpha}$$

with $J_\gamma(r)$ is the concentation-independent multisite interaction associated with the multisite correlation

$$\xi_{\gamma} = \frac{1}{N_{\gamma}} \sum_{|n_j|} \sigma_{n_1}\sigma_{n_2}...\sigma_{n_\gamma}$$

σ_n equals +1 or –1 when the site is occupated by an atom A or B and N_γ is the number of γ-type clusters. Then, for a finite number of total energies associated with specific ordered structures :

$$J_{\gamma}(r) = \sum_{\alpha} (\xi_{\gamma}^{\alpha})^{-1} E_{tot}^{\alpha}(r), \phi < \gamma < \gamma_{\max} \text{ and } J_{\gamma}(r) = 0, \gamma_{\max} < \gamma < \infty$$

where ϕ is the empty cluster.

The parameters entering the Ising model are derived from first-principles calculations of total energies for a limited set of periodic crystal structures[17]. The following structures were considered : pure Pu and Al (A1 fcc), Pu_3Al and Al_3Pu ($L1_2$ simple cubic), Pu_3Al and Al_3Pu ($D0_{22}$ centered tetragonal), PuAl ($L1_0$ tetragonal and PuAl ($L1_1$ trigonal). We took into account interaction parameters for empty cluster $J_{0,1}$, point $J_{1,1}$, first-neighbor pair $J_{2,1}$, second neighbor pair $J_{2,2}$, three-body cluster $J_{3,1}$, and four-body cluster $J_{4,1}$ terms.

All calculations have been performed within LDA+U with the same value of U as used for pure δ-Pu (3.13 eV) in the framework of the full-potential linear-muffin-tin-orbital (FP-LMTO) method[18,19] with von Barth-Hedin exchange-correlation potential and spin-orbit coupling. We only present preliminary results where the antiferromagnetic configuration was not yet taken into account.

The results of formation energies

$$E_{Pu_xAl_{1-x}}^{f} = E_{Pu_xAl_{1-x}}^{\min} - (xE_{Pu}^{\min} + (1-x)E_{Al}^{\min})$$

are reported in table 2 and plotted in figure 4.

Table 2 . Formation Energies in mRyd/atom

Structure	Composition	LDA	LDA+U
$L1_2$	Pu_3Al	-9.5	-31.2
$D0_{22}$	Pu_3Al	-23	-37.9
$L1_0$	PuAl	1.4	-46.7
$L1_1$	PuAl	-14.5	-24.8
$L1_2$	$PuAl_3$	-9.1	-35.7
$D0_{22}$	$PuAl_3$	1.8	-22.4

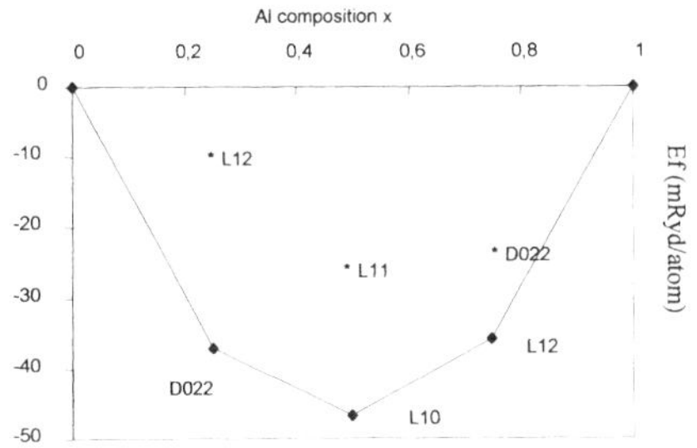

Figure 4. Formation Energies versus Al composition for fcc-based Pu-Al alloy

All the compounds are more stable within LDA+U than within LDA especially for $PuAl_3$ ($D0_{22}$) and PuAl ($L1_0$) which were instable within LDA. For this last compound, LDA stabilizes a tetragonal structure for c/a=1.17 in disagreement with experiment although LDA+U leads to the correct structure. Both LDA and LDA+U suggest $PuAl_3$ ($D0_{22}$) to be the most stable structure instead of $L1_2$. As the energy difference is small, an antiferromagnetic configuration could reverse the stability . This last point will be checked with further calculations.

From these energies, interactions parameters J_γ could be computed and the energy for a given disordered composition x_{Pu} could be written :

$$E_{dis} = J_{0,1} + J_{1,1}\rho + J_{2,1}\rho^2 + J_{2,2}\rho^2 + J_{3,1}\rho^3 + J_{4,1}\rho^4$$

with $\rho = (x_{Pu} - x_{Al})$.

The temperature is then introduced from a simple configuration-entropy term :

$$S = -R(x \ln x + (1-x)\ln(1-x))$$

It is then possible to draw the free energy of the disordered PuAl alloy as a function of composition for different temperatures, and to compare its equilibrium with the $PuAl_3$ ($D0_{22}$) compound. At each temperature a tangent line is drawn between the curve and $PuAl_3$ ($D0_{22}$) point and the solubility limit is calculated. We found a value of 13 at.% very close to the experimental data (13.6 at.%). For high plutonium concentrations, the curve shape indicates that plutonium is not soluble in aluminium. The resulting part of the temperature-composition phase diagram is reported in figure 5.

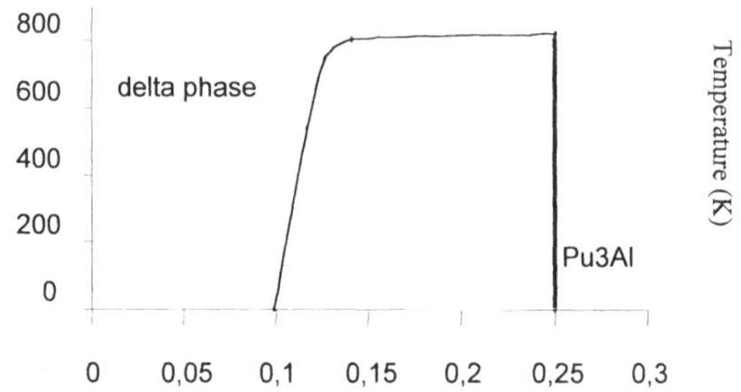

Figure 5. Temperature-composition phase diagram of PuAl alloy

CONCLUSION

The treatment of the strong correlations between *5f* electrons leads to a major improvement in the calculated equilibrium properties of δ, δ' and ε-Pu phases, especially when an antiferromagnetic spin order is taken into account. Beyond these pure phases, LDA+U is also able to describe ordered compounds and their formation energies. Moreover LDA+U could reproduce the main features of the phase diagram of plutonium and the solubility of aluminium in plutonium. This body of results suggests that LDA+U is fairly appropriate to describe energetic properties of correlated plutonium. Concerning phase diagrams, more accurate results require an improved treatment of vibrational properties of plutonium. This will be done by the calculation of phonon spectra in the framework of Projected Augmented Wave+U method (PAW+U, Abinit code). Nevertheless it is clear that the low energy properties governed by band features near the Fermi level are not described by LDA+U. These deficiencies would be removed by *ab initio* Dynamical Mean Field Theory (DMFT[20]) which would be able to describe subtle many-body effects such as the formation of local moments and their quenching via a possible Kondo-like effect.

REFERENCES

1. N. T. Chetobarev, E. S. Smotriskaya, M. A. Andrianov and O. E. Kostyuk, in *Plutonium 1975 and other actinides*, ed. H. Blank and R. Linder (North Holland, Amsterdam, 1976), 37
2. F. H. Ellinger, C. C. Land and V. O. Struebing *J. Nucl. Mater.* 12:226 (1964).
3. S. Drell, R. Jeanloz and B. Peurifoy, *Science* 283:1119 (1999).
4. H. Panosky, *Science* 275:11 (1997).
5. O. Eriksson, J. M. Wills, D. Becker and A. S. Balatsky, *J. Alloys Com.* 198:1 (1999).
6. B. R. Cooper, O. Vogt, Q. G. Sheng and Y. L. Lin, *Philos. Mag. B* 79:683 (1999).
7. P. Söderlind, *Europhys. Lett.* 55:525 (2001).
8. V. I. Anisimov, J. Zaanen and O. K. Andersen, *Phys. Rev. B* 44:943 (1991).
9. P. E. A. Turchi, A. Gonis, N. Kioussis, D. L. Price and B. R. Cooper, in *Electron Correlations and Materials Properties*, ed. A. Gonis, N. Kioussis and M. Ciftan, (Kluwer Academic/Plenum Publishers, New York, 1999), 531
10. J. Bouchet, B. Siberchicot, F. Jollet and A. Pasturel, *J. Phys. Cond. Matter.* 12:1723 (2000).
11. S. Y. Savrasov and G. Kotliar, *Phys. Rev. Lett.* 84:3670 (2000).
12 Y. Wang and Y. F. Sun, *J. Phys. Cond. Matter.* 12(21):L311, (2000).
13. P. Blaha, K. Schwarz, G. K. H. Madsen, D. Kvasnicka and J. Luitz, WIEN2k, An Augmented Plane Wave +Local Orbitals Program for Calculating Crystal Properties (Karlheinz Schwarz, Techn. Universität Wien, Austria), 2001. ISBN 3-9501031-1-2
14. J. E. Lynn, G. H. Kwei, W. J. Trela, V. W. Yuan, R. J. Martinez and F. A. Vigil, *Phys. Rev B* 58:11408 (1998).
15. V. L. Moruzzi and J. F. Janak, *Phys. Rev.* B 37:790 (1988).
16. A. Pasturel, C. Colinet, D. Nguyen Manh, A. T. Paxton and M. van Schilfgaarde, *Phys. Rev.* B 52: 15176 (1995).
17. J. W. D. Connolly and A. R. Williams, *Phys. Rev. B* 27:5169 (1983).
18. S. L Dudarev, G. A. Botton, S. Y. Savrasov, Z. Szotek, W. M. Temmerman and A. P. Sutton *Phys.Status Solidi* 166:429 (1998).
19. S. Y. Savrasov, *Phys. Rev. B* 54:16470 (1996).
20. A. Georges, G. Kotliar, W. Krauth and M. J. Rozenberg, *Rev. Mod. Phys.* 68 :13 (1996).

AB INITIO THERMODYNAMIC AND STRUCTURAL STUDIES OF CATIONIC DISORDERED $MgAl_2O_4$ SPINEL

S.Da Rocha[1,2] and P.Thibaudeau[2]

[1]Laboratoire d'Electrodynamique des Matériaux Avancés
UMR 6157 CNRS CEA
Faculté des Sciences et Techniques
Université François Rabelais
Parc de Grandmont, F-37000 Tours, France
[2]Commissariat à l'Energie Atomique
Le Ripault, BP 16, F-37260 Monts, France

INTRODUCTION

Spinels belong to the ceramic oxide family and have a wide range of applications in geophysics [16], magnetism [26] and irradiated environments [44, 45]. The general formula of a 2-3 spinel is AB_2O_4, where A is a divalent cation (such as Mg^{2+}, Zn^{2+}, Cd^{2+}, etc) and B a trivalent cation (such as Al^{3+}, Ga^{3+}, In^{3+}, etc). Among the spinels family, $MgAl_2O_4$ is often considered as a model of spinel structure where the oxygen atoms form a close-packed pseudo face centered-cubic sublattice. Among all the 96 possible interstices of the anions lattice, 64 are tetrahedral (IV) and 32 are octahedral (VI). In a so-called normal arrangement, one over 8 tetrahedral sites is occupied by divalent cations, whereas trivalent cations occupy half of octahedral sites [30]. Barth and Posnjak [3] have outlined that the normal arrangement cannot reproduce the intensity of X-ray diffraction pattern of some spinels. So, they have suggested an arrangement where equivalent positions are occupied by different atoms according the following general formula $\left[B^{IV}(AB)^{VI}\,O_4\right]$. However, it is now well known that spinels can accommodate cationic disorder [27, 11]. An inversion parameter, usually labelled x, has been introduced to quantify the number of trivalent B ions (here Al^{3+}) located in tetrahedral interstices.

The $MgAl_2O_4$ spinel can then be described by the following formula

$$\left(Mg_{(1-x)}Al_x\right)^{IV}\left[Mg_xAl_{(2-x)}\right]^{VI}O_4,$$

where x ranges from 0 (normal spinel), to 1 (fully inverse spinel). It has been shown that cationic disorder is activated by either pressure [13] or particle irradiations such as neutrons, electrons or Ne^+ ions [29, 31]. However, the usual and probably easiest way to activate disorder by cationic exchange in both natural or synthetic spinel, is to heat the studied specimen [7]. Successive heat pulses can also be applied on samples to synthesize directly disordered spinels [28].

It is now commonly admitted that most of spinels belong to Fd3m (227) space group. However, experimental observations reported by many groups [8, 9, 14, 28, 35] suggest a lower symmetry space group for some of these structures. In particular, Schmocker and Waldner [28] have outlined that in an inverse spinel, domains of reduced symmetry are responsible for observed extra Bragg reflections. This suggests a possible connection between the inversion parameter x and the space group of the structure, whereas Haas has claimed that cation disorder does not induce natural change in structure symmetry [10].

Up to now, most of the reported experimental studies on spinels deal with the determination of x with respect to temperature. Electron spin resonance (ESR) on a natural $MgAl_2O_4$ spinel for temperatures lower than 1278 K [27] and high-resolution ^{27}Al nuclear magnetic resonance (NMR) on quenched natural or synthetic samples [41, 19, 17] have been intensively developed to investigate such behavior. Several other techniques have been reported such as Raman spectroscopy [4], neutron diffraction [23, 25] which allow in situ measurements and X-ray diffraction [42, 2]. Because of various different experimental techniques and samples preparations, discrepancies appear in the reported data.

To complete structural analysis and avoid experimental difficulties, some theoretical investigations have been carried out to study disorder effect on spinel stability. Wei and Zhang [40] have selected 18 members among 2-3 and 4-2 spinels to evaluate the most stable form between normal and inverse configurations. They have also calculated at *ab initio* level, structural parameters such as the lattice constant (a), the average internal oxygen position (u) and band gaps in both configurations. Most of these spinels are found to be more stable in normal structure except for $MgGa_2O_4$, $MgIn_2O_4$ in the 2-3 family and $GeMg_2O_4$ and all Si spinels in the 4-2 group. Moreover, band gaps are found to be smaller in the inverse configuration. To explore cationic disorder continuous variations, Warren et al. [37, 38] have investigated disordered spinel thermodynamic properties with temperature. In their approach, short ranged one-site cluster potentials have been parameterized from a small number of disordered configurations within the Density Functional Theory (DFT) in both Local Density (LDA) and General Gradient Approximation (GGA). These potentials have been applied in Monte Carlo simulations to predict disordering thermodynamics state variables.

Alternatively, the evaluation of volume thermal expansion and specific heat of alumina-magnesium systems with temperature have been evaluated using both molecular dynamics and experimentally derived interatomic potentials [20]. In spite of a well reproduced heat capacity curve, the predicted thermal expansion appears to be significantly lower at high temperature than experimentally observed.

It is important to observe that most of experimental and theoretical investigations are concentrated on spinels cations behavior with temperature. But to our knowledge, it seems that no complete study has been previously carried out to evaluate the variation of both structural and thermodynamical spinel properties with cationic disorder.

In this study, *ab initio* calculations have been performed to evaluate firstly, the influence of cationic disorder on internal energy variation (U) and excess of heat capacity evolution and secondly, the relative density behavior. The cationic disorder parameter versus temperature curve is deduced as a by product, according to an effective thermodynamic model using both a regular solution and a quadratic form of the internal energy [22]. Results are compared to available experimental data.

CALCULATION METHODOLOGY

To approximate the continuous variation of x, a 56 atoms supercell is generated. Among all the cations, N Mg and Al ions are randomly exchanged, with N ranging from 0 to 8, so the inversion parameter x, is defined as $\frac{N}{8}$. Because it is computer time demanding, only 5 different supercells for each cationic arrangement are selected to average cell energy and structural behavior of the disordered crystal.

Simulations are performed using the ABINIT [1] code within Density Functional Theory (DFT) and Local Density Approximation (LDA). The valence electrons are described by pseudopotentials developed on a plane waves basis set. The generation of the cations pseudopotentials follows the scheme proposed by Haman [12], whereas the oxygen follows the Troullier and Martins scheme [36]. The particular choice of these pseudopotential schemes is detailed elsewhere [34]. During calculations, symmetries are turned off to impose P_1 space group.

In a previous study, Warren et al. [37] have selected a finite 2×2×2 Monkhorst-Pack set of irreducible k-points for all their configurations. However, because the variation of geometry induced by disorder has an important influence on the generation of an optimum k-points grid, different configurations must be treated with the same accuracy. Thus, only the first Brillouin zone centre point was considered in this study to integrate the DFT derived properties.

For each disorder rate x and each configuration, cell vectors and atomic positions are relaxed after several self-consistent cycles. The simulation stops when all the forces and resulting stresses are close to a desired precision. As DFT calculations are performed at 0K, an effective thermodynamic model, firstly introduced by Néel [21] and popularized by O'Neill and Navrotsky [22] is used to determine the equilibrium inversion parameter for a given temperature. The main quantity in this model is the Gibbs free enthalpy excess induced by disorder (ΔG). ΔG usually includes a quadratic internal energy (ΔU), a well defined form of configuration entropy (ΔS_c), and a work of external pressure via $P\Delta V$ product are found. As calculations have been carried out without external applied pressure, no extra work induced by pressure is necessary. Then, the Gibbs free enthalpy per molecule can be expressed by equation 1.

$$\Delta G = \Delta U - T\Delta S_c \tag{1}$$

with

$$\Delta U = \alpha x + \beta x^2, \tag{2}$$

$$\Delta S_c = -k_B \left[x \ln x + (1-x) \ln (1-x) + x \ln \frac{x}{2} + (2-x) \ln \left(1 - \frac{x}{2}\right) \right], \qquad (3)$$

where k_B is the Boltzmann constant. The constants α and β are deduced from *ab initio* calculations and the equilibrium condition $\frac{\partial \Delta G}{\partial x} = 0$ is numerically solved to determine the value of the mean equilibrium cationic distribution for a given temperature. Furthermore, using the derived equilibrium curve into the configurational entropy, the heat capacity excess induced by disorder is given as

$$\left(\frac{\partial \Delta H}{\partial T}\right) = T \left(\frac{\partial \Delta S_c}{\partial T}\right)_P = \Delta C_p(T). \qquad (4)$$

RESULTS AND DISCUSSION

Variation of internal Energy

The averaged energy differences $\Delta U(x) = U(x \neq 0) - U(x = 0)$, expressed in eV/molecule, are calculated as shown in Figure 1 and compared with results from Warren et al. [37]. Discrepancies between the two calculations could have their origin in different spannings of the configurational space. Our calculated mean-squared error on ΔU clearly shows that a single calculation per inversion parameter is not probably sufficient to take into account the configurational space variety. As the internal energy differences are small, the convergence of forces on atoms and the resulting stress on the cell have been carefully monitored during calculations.

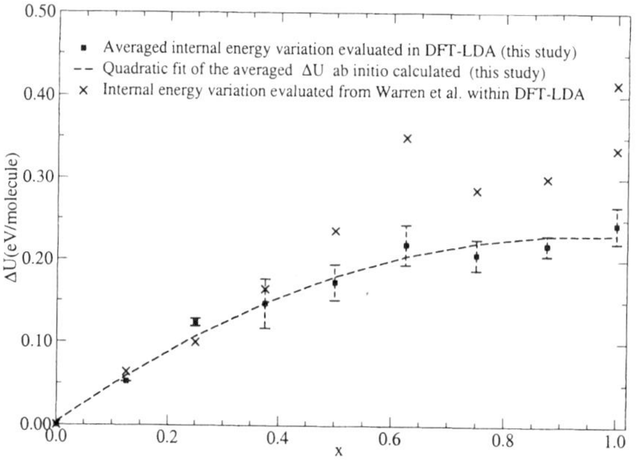

Figure 1: Averaged internal energy variation with cationic disorder expressed in eV/molecule calculated within LDA (□) and compared to LDA Warren et al. results (×) [37].

A quadratic fit of $\Delta U(x)$ provides both averaged values α and β and mean-squared errors $\Delta\alpha$ and $\Delta\beta$. These values are compared with available experimental data as shown in Table 1. In their analysis of cationic distribution, O'Neill and Navrotsky [22] have shown that α and β parameters should be approximately of equal magnitude and opposite in sign. The main suggested reason of this sign difference was induced by the ionic character of the bindings in spinels. As shown by Thibaudeau et al. [34] in the normal spinel, the Born effective charges on atoms are very close to the full ionic charges. So, at least for low disorder rate, there is no obvious reason that the ionic character is deeply altered with in the exchange of ions. As shown in Table 1, this study and some of the reported data [37, 17, 23] fulfill the opposite sign criterion.

Derivation of equilibrium curve and heat capacity

	Reference	α	$\Delta\alpha$	β	$\Delta\beta$	Method
Theory	This study	0.48	0.08	-0.25	0.06	DFT-LDA
	Warren et al. [37]	0.60	-	-0.22	-	DFT-LDA
	Millard et al. [19]	0.26	0.06	0.06	0.10	RMN ^{27}Al
	Maekawa et al. [17]	0.36	0.06	-0.33	0.06	RMN ^{27}Al
Exp.	Peterson et al. [23]	0.32	0.01	-0.10	0.03	neutron diffraction
	Redfern et al. [25]	0.34	0.01	0.05	0.02	neutron diffraction
	Andreozzi et al. [2]	0.24	0.02	0.14	0.05	X-ray diffraction

Table 1: Comparison of experimental and *ab initio* calculated α and β parameters and mean-squared errors .

Variations of the disorder rate with temperature are deduced from the equilibrium condition on the Gibbs free enthalpy. The derived curve is compared to most of available experimental data as shown in Figure 2. It seems that the mean-squared errors induced by $\Delta\alpha$ and $\Delta\beta$ on cationic disorder curve was never reported previously. However Figure 2. shows that significant differences between internal energy parameters are not sufficient to predict accurate variations of x versus temperature. This curve also shows that the uncertainties $\Delta\alpha$ and $\Delta\beta$ can give an insight into conclusions between different experimental and theoretical works. Because the experimental extent of temperatures is limited, an accurate determination of the internal energy variation appears to be more difficult for the quadratic coefficient β than for α.

In spinels, several thermodynamic observables exhibit rapid variations with temperature which suggest possible second-order phase transition. A discontinuity in cell edge and thermal expansion coefficient has been observed by several groups [42, 33], but Kashii et al. [15] using NMR technique have reported, on the basis of kinetic considerations, that the cationic equilibrium disorder reaction is a first-order transition. In this study, the second derivative calculation of the free energy ΔG with respect to the temperature was performed. Once the equilibrium variation of x versus temperature is known, the determination of configurational entropy and its first derivative drop naturally according to Eq.(4). As shown in Figure 3, ΔC_p exhibits a peak which is characteristic of a continuous phase transition. The harmonic part of heat capacity at

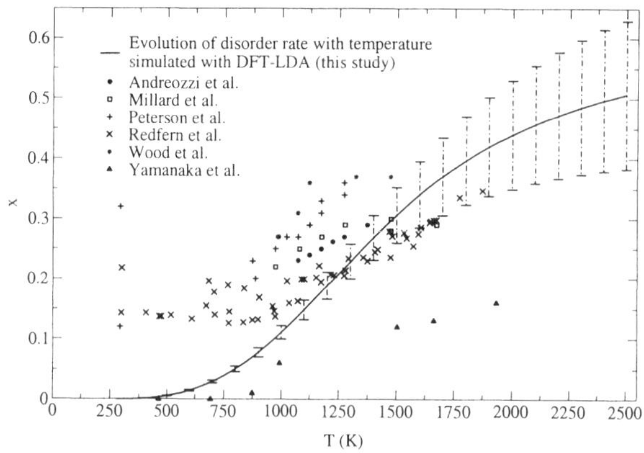

Figure 2: Evolution of disorder rate x obtained by resolution of equation 1 compared to neutron scattering experiments [25]($\times$), [23]($+$), X-ray measurements [42]($\triangle$), [2]($\bullet$), and ^{27}Al NMR experiments [19]($\square$), [41]($*$).

constant pressure is the main contribution to the total capacity in spinels. However, the increase at maximum of heat capacity induced by disorder is about 15 J/mol.K and should be experimentally observed.

This maximum value suggests a definition of a critical temperature for this system. However, an another critical temperature is usually introduced to describe order-disorder transitions [24]. Its determination is strongly connected to the Bragg-Williams model. In this model, the solution of

$$\left(\frac{\partial^2 \Delta G}{\partial Q^2}\right)_{Q=0,T=T_c} = 0, \tag{5}$$

where Q is an order parameter, gives the required temperature T_c. In the case of 2-3 spinels, it is possible to define an order parameter $Q = 1 - \frac{3}{2}x$ connected to the cationic occupation, which varies from 1 (maximum order) to 0 (random disorder). Solving equation 5 with the Gibbs free energy form introduced previously, gives $T_c = -\frac{8\beta}{27k_B}$ where k_B stands for the Boltzmann constant. Using our derived *ab initio* values, $T_c \approx 860K$. This temperature is in fair agreement with the previously reported experimental values of 870 K $\leq T_c \leq$970 K [42], $T_c \approx$ 950 K [39], $T_c \approx$ 930 K [33]. As T_c is positive defined, the sign of β is expected to be negative or zero.

Density with disorder behavior

Discrepancies have been reported on the density behavior with the cationic disorder. Indeed, it is difficult to determine the volume variation induced only by disorder on quenched samples. Wood et al. [41] have carried out ^{27}Al NMR experiments on

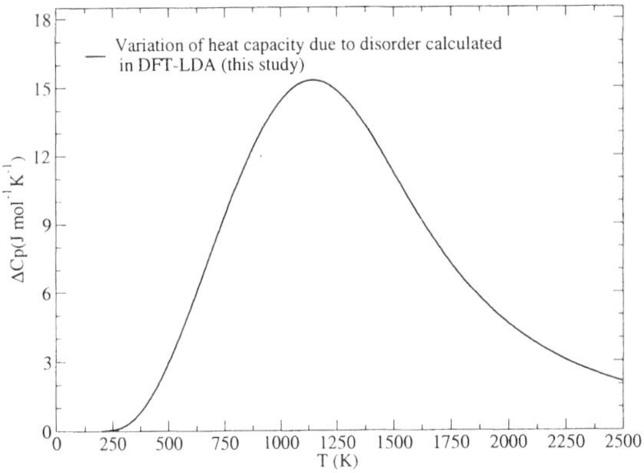

Figure 3: heat capacity induced by disorder curve as a function of temperature

quenched synthetic spinels and found a cell volume increase of 0.03 % for an increase of x from 0.21 to 0.39. An X-ray study performed on a quenched single-crystal has shown a decrease of cell parameters when the disorder increases [42]. Using a description based on ionic radii, O'Neill and Navrotsky [22] have predicted a decrease of volume with disorder for all the 2-3 spinels with cations of similar size to Mg^{2+} and Fe^{3+}.

Ab initio calculations performed in this study validate density increases with cationic exchange. They are compared with recent X-ray measurements on quenched single-crystal [2] on Figure 4. Conclusive agreement is found for the range of temperature experimentally observed.

Infrared vibrational spectra of disordered spinel

Maradudin and coworkers [18] have demonstrated that the macroscopic low-frequency static dielectric permittivity tensor $\epsilon_{ij}(\omega)$ which gives the infrared spectra main contribution, is a sum of both an ionic part and a limit value of a pure electronic contribution. According to Gonze and Lee [6], one has

$$\epsilon_{ij}(\omega) = \epsilon_{ij}^{\infty} + \frac{4\pi}{\Omega_0} \sum_m \frac{\left(\sum_{\kappa j'} Z^{\star}_{\kappa,ii'} U^{\star}_{m,\mathbf{q}=0}(\kappa i')\right)\left(\sum_{\kappa' j'} Z^{\star}_{\kappa',jj'} U_{m,\mathbf{q}=0}(\kappa' j')\right)}{\omega_m^2 - \omega^2} \tag{6}$$

where $1 \leq i,j \leq 3$ and Ω_0 is the cell volume, $Z_{\kappa,ii'}$ are the Born effective tensors, ϵ_{ij}^{∞} is the electronic contribution to the dielectric permittivity tensor. This tensor is given by the knowledge of the mixed derivative of the total electronic energy over the macroscopic electric field. Here ω_m^2 represent the dynamical matrix eigenvalues and

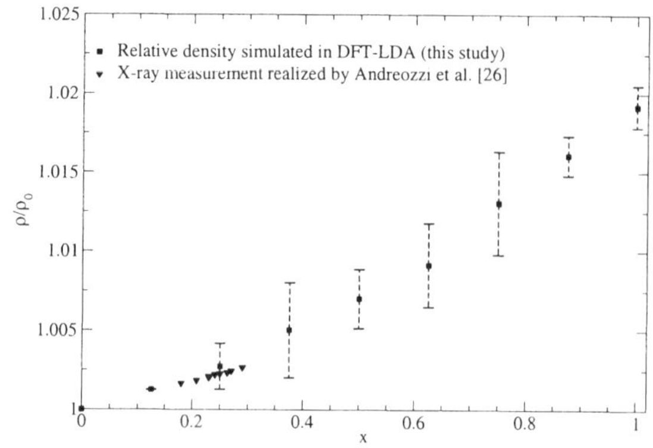

Figure 4: Density evolution with x ab initio calculated (□), compared with X-ray measurements (△) [2]

$U_{m\mathbf{q}}$ the corresponding eigenvectors with the following normalization

$$\sum_{\kappa j} M_k \left[U_{m\mathbf{q}}(\kappa j)\right]^* U_{n\mathbf{q}}(\kappa j) = \delta_{mn}, \tag{7}$$

where M_k are the atomic masses.

Powerful techniques have been developed a few years ago to calculate vibrational properties of extended systems through density functional perturbation theory (DFPT). This technique has been applied successfully to evaluate all vibrational modes at the zone centre for normal cubic $MgAl_2O_4$ spinel [34]. The Born effective charges have also been evaluated within DFPT and compared to a set of new experiments on synthetic spinels [34]. The agreement was found satisfactory, demonstrating that the ionic character was preserved in this spinel even on synthetic and probably small disordered compound.

Many groups have reported extra infrared modes for synthetic $MgAl_2O_4$ spinels [43, 34] and highly non-symmetric peaks in the Raman spectra for quenched samples [5]. The precise origin of these extra modes is unclear [43] but extra Raman mode near 727 cm^{-1} was clearly assigned to the symmetric Al-O stretching vibration of AlO_4 groups created by the redistribution of some aluminum ions from octahedral to tetrahedral sites. However, to explore the possibility that infrared modes originate from cationic disorder, dynamical matrix eigenvectors and eigenvalues have been calculated within DFPT for several inversions parameters and averaged on different configurations. The imaginary parts of the low-frequency dielectric tensor components are reported in Figure 5. Lifetime is required for each mode such as its corresponding damping over its frequency is a constant for a given temperature [32]. As room temperature is high enough to neglect anharmonic residual quantum effect for low temperatures, assum-

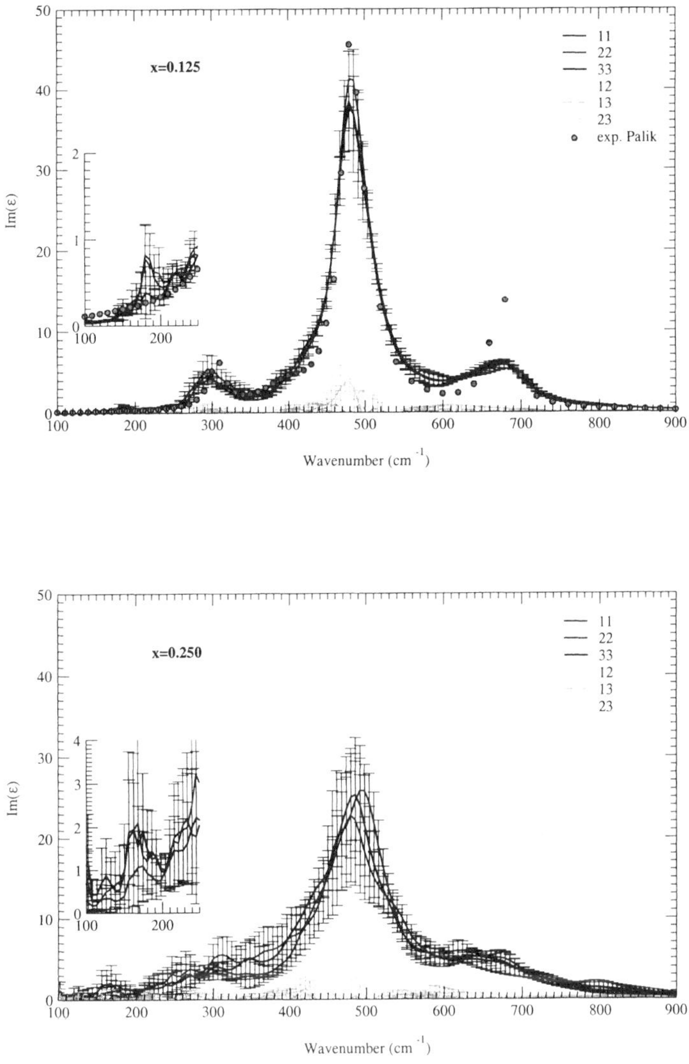

Figure 5: Evolution of infrared spectra for different disorder rate

ing dampings linear behavior with temperature is a natural approximation. Assuming constant interactions of phonons is a crude approximation. However an examination of both LO and TO dampings and frequencies measured for two sets of synthetic spinels provides a linear regression coefficient of 0.8. This value is high enough to consider the proportion between dampings and frequencies constant. The corresponding value of this constant was fitted using previously measured dampings and frequencies [34]. For simplicity, we assume that anharmonicity is sufficiently small for low inversion parameter such as this constant remains unchanged. Using these approximations, the intensity of all the T_{1u} normal spinel infrared modes decreases when the inversion parameter increases as shown in Figure 5. The resulting widths become broader and several modes near 800 cm^{-1} and 180 cm^{-1} become intense. These extra modes have been previously observed in different synthetic spinels and then could have their origin in the defective nature of the spinel lattice.

CONCLUSION

Structural and thermodynamic properties of disordered $MgAl_2O_4$ spinel are calculated by DFT with plane-wave basis set and pseudopotentials. Several configurations are considered to average each property to take account of configurational variety. The evolution of internal energy is calculated as a function of cationic disorder rate, variations of disorder rate with temperature are determined using an effective thermodynamic model. By taking into account the mean-squared errors obtained on α and β parameters, an original approach of the cationic equilibrium curve is introduced. Furthermore, the excess of heat capacity shows a typical behavior which is the sign of a continuous transition and a critical temperature is also evaluated for a perfect random crystal. In addition, it has been shown that the density of the material increases with increasing disorder. Finally, infrared spectra calculation of selected inversion parameters have been carried out within DFPT and have shown extra modes which could originate in this particular defective nature of the spinel lattice.

REFERENCES

[1] The ABINIT code is a common project of the Université Catholique de Louvain, Corning Incorporated, Université de Liège, CEA, Mitsubishi Chemical Corp, and other contributors (url http://www.pcpm.ac.be/abinit).

[2] G. B. Andreozzi, F. Princivalle, H. Skogby, and A. Della Giusta. Cation ordering and structural variations with temperature in $MgAl_2O_4$ spinel: an X-ray single-crystal study. *Am. Mineral.*, 85:1164–1171, 2000.

[3] T. F. W. Barth and E. Posnjak. Spinel structures: with and without variate atom equipoints. *Zeitschr. F. Kristallographie*, 82:325–341, 1932.

[4] H. Cynn, O. L. Anderson, and M. Nicol. Effects of cation disorering in a natural $MgAl_2O_4$ spinel observed by rectangular parallelepiped ultrasonic resonance and Raman measurements. *Pageoph.*, 141:415–444, 1993.

[5] H. Cynn, S. K. Sharma, T. F. Cooney, and M. Nicol. High-temperature Raman investigation of order-disorder behavior in $MgAl_2O_4$ spinel. *Phys. Rev. B*, 45:500–502, 1992.

[6] X. Gonze and C. Lee. Dynamical matrices, born effective charges, dielectric permittivity tensors, and interatomic force constants from density-functional perturbation theory. *Phys. Rev. B*, 55(16):10355–10368, 1997.

[7] E. W. Gorter. Saturation magnetization, crystal chemistry of ferrimagnetic oxides. *Philps Res. Rep.*, 9:403–443, 1954.

[8] N. W. Grimes, P. J. O'Connor, and P. Thompson. Comments on the interpretation of the infrared spectrum from $MgAl_2O_4$ spinel. *J. Phys. C: solid State Phys*, 11:L505–L507, 1978.

[9] N. W. Grimes, P. Thompson, and H. F. Kay. New symmetry and structure for spinel. *Proc. R. Soc. London*, A 386:333–345, 1983.

[10] C. Haas. Phase transitions in crystal with the spinel structure. *J. Phys. Chem. Solids*, 26:1225–1232, 1965.

[11] S. Hafner and F. Laves. *zeitschrift f. Kristalloraphie*, 115:321–330, 1961.

[12] D. R. Hamann. *Phys. Rev. B*, 40, 1989.

[13] R. M. Hazen and A. Navrotsky. Effects of pressure on order-disorder reactions. *Am. Mineral.*, 81:1021–1035, 1996.

[14] L. Hwang, A. H. Heuer, and T. E. Mitchell. On the space group of $MgAl_2O_4$ spinel. *Phil. Mag.*, 28:241–243, 1973.

[15] N. Kashii, H. Maekawa, and Y. Hinatsu. Dynamics of cation mixing of $MgAl_2O_4$ and $ZnAl_2O_4$ spinel. *J. Am. Ceram. Soc.*, 82:1844–1848, 1999.

[16] S. Lucchesi and A. Della Giusta. Crystal chemistry of highly disordered Mg-Al natural spinel. *Mineralogy and petrology*, 59:91–99, 1997.

[17] H. Maekawa, S. Kato, K. Kawamura, and T. Yokokawa. Cation mixing in natural $MgAl_2O_4$ spinel: a high temperature ^{27}Al NMR study. *Amer. Miner.*, 82:1125–1132, 1997.

[18] A. A. Maradudin, E. W. Montroll, G. H. Weiss, and I. P. Ipatova. Solid state physics : Advances in research and applications. volume 4. Academic, 1971.

[19] R. L. Millard, R. C. Peterson, and B. K. Hunter. Temperature dependence of cation disorder in $MgAl_2O_4$ spinel using ^{27}Al and ^{17}O magic-angle spinning NMR. *Am. Mineral.*, 77:42–52, 1992.

[20] S. Morooka, S. Zhang, T. Nishikawa, and H. Awaji. Potential model parameters for molecular dynamics simulation of alumina-magnesia systems. *J. Cer. Soc. of Jap.*, 107:1225–1228, 1999.

[21] L. Néel. *Compt. Rend.*, 230:190, 1950.

[22] H. St. C. O'Neill and A. Navrotsky. Simple spinels: crystalographic parameters, cation radii, lattice energies, and cation distribution. *Am. Mineral.*, 68:181–194, 1983.

[23] R. C. Peterson, G. A. Lager, and R. L. Hitterman. A time of-flight neutron powder diffraction study of $MgAl_2O_4$ at temperatures up to 1273 K. *Am. Miner.*, 76:1455–1458, 1991.

[24] S. A. T. Redfern. Order-disorder phase transitions. In S. A. T. Redfern and M. A. Carpenter, editors, *Transformation Processes in Minerals*, volume 39, pages 105–133. Mineralogical society of America, 2000.

[25] S. A. T. Redfern, R. J. Harrison, H. St.C. O'Neill, and D. R. R. Wood. Thermodynamics and kinetics of cation ordering in $MgAl_2O_4$ spinel up to 1600 c from in situ neutron diffraction. *Am. Miner.*, 84:299–310, 1999.

[26] W. Schiessl, W. Poltzel, H. Karzel, M. Steiner, and G. M. Kalvius. Magnetic properties of the $ZnFe_2o_4$ spinel. *Phys. Rev. B*, 53:9143–9152, 1996.

[27] U. Schmocker, H. R. Boesh, and F. Waldner. A direct determination of cation disorder in $MgAl_2O_4$ spinel by esr. *Phys. Letters*, 40A:237–238, 1972.

[28] U. Schmocker and F. Waldner. The inversion parameter with respect to the space group of $MgAl_2O_4$ spinels. *J. Phys.C: Solid state Phys.*, 9:L235-L237, 1976.

[29] K. E. Sickafus, A. C. Larson, N. Yu, M. Nastasi, G.W. Hollenberg, F.A. Garner, and R.C. Bradt. Cation disorder in high dose, neutron-irradiated spinel. *J. Nucl. Mater.*, 219:128–134, 1995.

[30] K. E. Sickafus and J. M. Wills. Stucture of spinel. *J. Am. Ceram. Soc.*, 82:3279–3292, 1999.

[31] T. Soeda, S. Matsumura, C. Kinoshita, and N. J. Zaluzec. Cation disordering in magnesium aluminte spinel crystals induced by electron or ion irradiation. *J. Nucl. Mater.*, 283-287:952–956, 2000.

[32] G. P. Srivastava. *The Physics of phonons.* Adam Hilger, 1990.

[33] I. Suzuki and M. Kumazawa. Anomalous thermal expansion in spinel $MgAl_2O_4$. *Phys. Chem. Miner.*, 5:279–284, 1980.

[34] P. Thibaudeau and F. Gervais. Ab initio investigation of phonon modes in the $MgAl_2O_4$ spinel. *J. Phys : Condens. Matter.*, 14:3543–3552, 2002.

[35] P. Thompson and N. Grimes. Multiple diffraction in spinel and the space group ambiguity. *J. Appl. Cryst.*, 10:369–371, 1977.

[36] N. Troullier and J. L. Martins. *Phys. Rev. B*, 43, 1991.

[37] M. C. Warren, M. T. Dove, and S. A. T. Redfern. Ab initio simulations of cation ordering in oxides : application to

spinel. *J. Phys.:Condens Matter*, 12:L43–L48, 2000.

[38] M. C. Warren, M. T. Dove, and S. A. T. Redfern. Disordering of $MgAl_2O_4$ spinel from first principles. *Min. Mag.*, 64:311–317, 2000.

[39] R. A. Weeks and E. Sonder. Electrical conductivity of pure and fe-doped magnesium-aluminium

spinel. *J. Amer. Ceramic Soc.*, 63(1-2):92–95, 1979.

[40] S. H. Wei and S. B. Zhang. First-principles study of cation distribution in eighteen closed-shell ab_2o_4 spinel oxides. *Phys. Rev. B*, 63:045112–1 045112–8, 2001.

[41] B. J. Wood, R. J. Kirkpatrick, and B. Montez. Order-disorder phenomena in $MgAl_2O_4$ spinel. *Am. Mineral.*, 71:999–1006, 1986.

[42] T. Yamanaka and Y. Takeuchi. Order-disorder transition in $MgAl_2O_4$ spinel at high temperatures up to 1700 C. *Kristallographie*, 165:65–78, 1983.

[43] J. M. Yeomans, A. Chopelas, and A.M. Hofmeister. Statistical mechanics of phase transitions vibrational spectroscopy of aluminate spinels at 1 atm and of $MgAl_2O_4$ to over 200 kbar. *Phys. Chem. Minerals*, 18:279–293, 1992.

[44] N. Yu, K. E. Sickafus, and M. Nastasi. First observation of amorphization in single-crystal $MgAl_2O_4$ spinel. *Phil. Mag. Letters*, 70:235–240, 1994.

[45] S. J. Zinkle. Effect of irradiation spectrum on the microstructural evolution in ceramic insulators. *J. Nucl. Mater.*, 219:113–127, 1995.

COMPUTER SIMULATION OF MOLTEN AND GLASSY SILICA AND ITS MIXTURES WITH SODIUM OXIDE AND ALUMINIUM OXIDE

Kurt Binder[1)], Jürgen Horbach[1)], Walter Kob[2)], and Anke Winkler[1)]
[1)] *Institut für Physik, Johannes Gutenberg–Universität Mainz, Staudinger Weg 7, D–55099 Mainz, Germany;*
[2)] *Laboratoire des Verres, Université Montpellier II, Place E. Bataillon, Case 069, 34095 Montpellier, France*

1. INTRODUCTION

Molten silica and its mixtures with various other oxides are of central interest in geosciences, silicates that have formed from such melts in the earth crust are very relevant materials. Such melts are also very important for the glass and ceramics industry, and although both of these materials are in their crystalline and amorphous forms in use for many centuries, the understanding of their structure–property relationship on an atomistic level still poses challenging scientific problems. In recent years, important progress has been made possible by atomistic molecular dynamics simulations, and a selection of problems by this method will be presented below.

One characteristic feature of pure SiO_2 is that it can crystallize in many polymorphs (α–quartz, β–quartz, tridymite, β–cristobalite are the low pressure phases, coesite and stishovite follow at higher pressures, etc. [1]). Nevertheless the precise nature of some of the structures [2] and the character of the transitions between them [3, 4] has remained under debate until recently [5, 6]. Of course, the phase diagrams become even much more complicated when mixtures of SiO_2 with other oxides (Na_2O, Al_2O_3, etc.) are considered [7, 8].

There is also great interest in a detailed atomistic understanding of the structure of glassy SiO_2 and its mixtures with other oxides, and of the fluids from which these amorphous materials are formed by appropriate cooling schedules [9, 10].

Although some crude feature of the basic "continuous random network" model of glassy SiO_2 have been known for a long time [11], namely each Si–atom sits in the center of a (slightly irregular) tetrahedron, while the oxygen atoms are at the corners of the tetrahedron, each oxygen being shared by two neighboring tetrahedra, the medium range order sustained by such a structure still is a subject of study [12, 13, 14, 15, 16]. Considering mixtures with other oxides, one either expects that the network is locally broken (such as in the case of the network modifier Na_2O, which causes dangling Si–O bonds to exist in the network) or the additional ions are bound into the network, locally disturbing its coordination (as happens in the case of Al_2O_3). However, the extent to which these ions (Na^+, Al^{+++}, etc.) are randomly distributed throughout the network, or whether "chemical clustering" occurs, has been a longstanding issue, difficult to clarify by experiments (e.g. [17, 18, 19]).

2. MODELS AND SIMULATION DETAILS

In a molecular dynamics simulation [20], the statistical mechanics of condensed matter is reduced to averages along trajectories through the phase space of the chosen model system, generating these trajectories simply from Newton's equations of motion, using simple effective potentials for the interactions between the atoms. Thus, quantum mechanical effects are neglected from the outset, and since in the considered systems the character of the bonding varies from ionic to covalent, it is not a priori clear that a sufficiently realistic description will be possible. In fact, early attempts to simulate molten SiO_2 were clearly hampered not only by too limited computer resources but also by too inaccurate potentials [21]. Significant progress was possible due to the so-called "BKS potential" [22] for SiO_2, later generalized by Kramer *et al.* [23] to zeolites containing Na and Al. This potential is based on a parameterisation of quantum–chemical calculations of small silica(-te) clusters, and the parameters were optimized to correctly reproduce some properties of bulk crystalline materials. Although some limitations of this potential to reproduce very small scale structures are well documented [14, 16, 24, 25], it describes the properties of both the crystalline materials [6, 26, 27] and the glasses and melts [12, 13, 15, 28, 29, 30, 31, 32, 33, 34, 35, 36, 37, 38, 39, 40] surprisingly well.
This BKS–potential uses pseudo–Coulomb interactions between the ions and a Buckingham potential to describe the short range part,

$$\phi_{\alpha\beta}(r) = \frac{q_\alpha q_\beta}{r} e^2 + A_{\alpha\beta} \exp(-B_{\alpha\beta} r) - \frac{C_{\alpha\beta}}{r^6}, \tag{1}$$

where $\alpha, \beta \in$ [Si, O, Na, Al], r being the distance between an ion of species α and an ion of species β. The values of the parameters $A_{\alpha\beta}, B_{\alpha\beta}, C_{\alpha\beta}$ are given in [22, 23]. The effective charges proposed for Si and O are [22] $q_{\mathrm{Si}} = 2.4$, $q_{\mathrm{O}} = 1.2$, whereas for sodium and aluminium effective charges $q_{\mathrm{Na}} = 1.0$, $q_{\mathrm{Al}} = 1.9$ were proposed [23]. While for pure SiO_2 charge neutrality is fulfilled, since $q_{\mathrm{Si}} = -2q_{\mathrm{O}}$, this is not the case for the mixtures studied in the present work, $(Na_2O)x(SiO_2)$, x=2,3,5 (abbreviated as NSx in the following), as well as for $(Al_2O_3)2(SiO_2)$. Therefore the above potential was slightly modified, introducing distant dependent charges $q_\alpha(r)$ for Na and Al as follows,

$$q_{\rm Na}(r) = \begin{cases} \tilde{q}_{\rm Na}\{1+\ln[C_{\rm Na}(r_{\rm Na}-r)^2+1]\}\ , r < r_{\rm Na} \\ \tilde{q}_{\rm Na}, r \geq r_{\rm Na}, \end{cases} \tag{2}$$

where the choice $\tilde{q}_{\rm Na} = 0.6$ ensures charge neutrality, while with $r_{\rm Na} = 4.9$ Å, $C_{\rm Na} = 0.926$ Å^{-2} the charge $q_{\rm Na}(r)$ smoothly increases from $\tilde{q}_{\rm Na} = 0.6$ for $r = r_{\rm Na}$ to the above $q_{\rm Na} = 1.0$ which is reached for $r = 1.7$ Å. Similarly, for Al a somewhat different functional form of $q_{\rm Al}(r)$ was found advantageous, namely

$$q_{Al}(r) = \begin{cases} \tilde{q}_{\rm Al}\{1+\ln[C_{\rm Al}\frac{(r_{\rm Al}-r)^2}{1+(r_{\rm Al}-r)^2}+1]\}\ \exp[-\frac{d_{Al}}{(r-r_{\rm Al})^2}]\ , r < r_{\rm Al} \\ \tilde{q}_{\rm Al}, r \geq r_{\rm Al} \quad . \end{cases} \tag{3}$$

Here $\tilde{q}_{\rm Al} = 1.8$ ensures charge neutrality, while $r_{\rm Al} = 6$ Å, $C_{\rm Al} = 0.0653$ Å^{-2}, $d_{\rm Al} = 2$ Å^2 ensure that $q_{\rm Al} = 1.9$ for $r = 1.25$ Å, where the original potential for the Al–O interaction [23] has an inflection point. It is clear that the choices Eqs. (2), (3) are ad hoc–modifications that lack any quantum–chemistry justification, but they have the merit that for small distances the forces resulting from the original potential [23] are almost perfectly reproduced, and at the same time overall charge neutrality, a necessary condition for the stability of the system, is restored.
In the simulations, a total number of atoms $N = 8064, 8016, 8064, 8016$ were used for NS2, NS3, NS5 and SiO_2, respectively, while for $(Al_2O_3)2(SiO_2)$ 1408 atoms were used. All simulations were done at constant density, which was chosen as $\rho = 2.37$ g/cm^3 for NSx and SiO_2, while for $(Al_2O_3)2(SiO_2)$ a density of $\rho = 2.60$ g/cm^3 (the experimental value at $T = 300$ K [41]) was chosen. The resulting linear dimensions of the cubic simulation box was about 48 Å for NSx and 26.347 Å for $(Al_2O_3)2(SiO_2)$. We applied periodic boundary conditions as usual, and treated the long range Coulomb interactions with the Ewald summation technique. The velocity Verlet algorithm with a time step $\delta t = 1.6$ fs was used. While $(Al_2O_3)2(SiO_2)$ could only be equilibrated for $T \geq 2300$ K, NSx could be equilibrated for $T \geq 2100$ K, and pure molten SiO_2 for $T \geq 2750$ K. In the case of NS2, also $T = 1900$ K could be equilibrated. In order to study glasses at $T = 300$ K, we used the respective fluid around the lowest temperature that still could be equilibrated (note that typically 4.5 million time steps were necessary, corresponding to a real time of 7.5 ns) and then cooled the system rapidly down (with a cooling rate of 1.16×10^{12} K/s). The temperature of the system was always controlled by coupling it to a stochastic heat bath, i.e. by substituting periodically the velocities of the particles with the ones from a Maxwell–Boltzmann distribution with the correct temperature. After the system was equilibrated at the

target temperature, the run was continued in the microcanonical ensemble (i.e. the heat bath was switched off, so energy and momentum was conserved). In order to improve the statistics, 2 independent runs were performed for SiO_2 and all the NSx systems, and 5 independent runs for the (smaller!) $(Al_2O_3)2(SiO_2)$. All these simulations were carried out on CRAY T3E/900 supercomputers, using MPI to parallelize the simulation code (for applying a parallelization of the force calculations), using 32–64 processors in parallel. The MD simulations of crystalline SiO_2, however, were carried out in the constant stress ensemble using [6] the Parrinello–Rahman method [42]. The standard system size used was $N = 1080$ atoms (the typical linear dimensions of the orthorhombic simulation box then is $A = 25$ Å, $B = 26$ Å, $C = 22.1$ Å, with A, B, C parallel to the crystallographic axes a, b, c respectively). Initial conditions were chosen according to the ideal structures of α–quartz and β–quartz, respectively. Temperatures were chosen in the range from zero to $T = 1700$ K. Unlike the case of fluid and amorphous SiO_2, the short range (non–Coulombic) part of the BKS potential was not cut off at $r_{\mathrm{cutoff}} = 5.5$ Å but at $r_{\mathrm{cutoff}} = 9.5$ Å. In order to speed up to the calculations, interactions and forces were tabulated on a grid with a resolution of 5×10^{-4} Å. Near the α–β phase transition, also particle numbers $N = 2160$ and $N = 4320$ were used (with $A, B, C = 30.0$ Å, 34.7 Å, 27.6 Å and 40.0 Å, 43.3 Å, and 33.2 Å, respectively), to check for finite size effects. Note that classical molecular dynamics methods are very unsatisfactory for crystals below the Debye temperature Θ_{D}, of course: the specific heat of a crystal treated by classical statistical mechanics would approach for $T \to 0$ a nonzero constant (Dulong–Petit law) rather than vanish according to the Debye law; similarly, also the temperature derivatives of the lattice parameters and the elastic constants are all non–vanishing. Therefore for $T \leq 1000$ K also quantum–mechanical simulations (with the path integral molecular dynamics technique [43], PIMD) were also performed [27] (note that Θ_{D} for α–SiO_2 when extracted from the specific heat depends on temperature, $\Theta_{\mathrm{D}}(T = 300\ \mathrm{K}) \approx 10^3$ K [44]).

3. CRYSTALLINE QUARTZ AND β–CRISTOBALITE

The analysis of the simulations of crystalline SiO_2 has focused on a study of the lattice parameters and their temperatures dependence, on the elastic constants (which can be gotten from the Parrinello–Rahman fluctuation relations [45]), and quantities characterizing the local aspects of the structure (probability distributions to find an atom a distance r away from its equilibrium position, partial radial pair distributions, average bond angles and their distributions, etc.). Of particular interest is also the global order parameter ϕ of the phase transition between α–quartz and β–quartz, which measures the rotation of the (distorted) tetrahedra about the [100] axis,

$$\phi = \frac{1}{N^*} \sum_{i^*=1}^{N^*} \varphi_{i^*}, \tag{4}$$

where the sum over i^* is confined to a sublattice of the β–quartz structure taken by Si–atoms, such that one can envisage the formation of α–quartz from β–quartz via a rotation of (rigid) units about the [100] axis by an angle φ_{i^*}. Since in reality the SiO_4^{4-} tetrahedra are not rigid, we have used for φ_{i^*} the (averaged) deviation of the four Si–O bonds in the y–z plane from the value in the ideal β–quartz structure [6].

While the lattice parameters a, c exhibit a temperature dependence that resembles the experiment [46] qualitatively, there is no perfect agreement: at $T = 150$ K, experiment yields $a, c \approx 4.905$ Å, 5.401 Å while classical MD yields 4.956 Å, 5.461 Å and PIMD yields 4.965 Å, 5.470 Å; at 300 K, the corresponding numbers are 4.915 Å, 5.406 Å; 4.996 Å, 5.495 Å; 4.974 Å, 5.476 Å; and at 1000 K, 5.000 Å, 5.459 Å; 5.031 Å, 5.527 Å; 5.031 Å, 5.531 Å, respectively. Thus, there is a discrepancy between simulation and experiment of about 0.06 Å for the a–axis and and 0.07 Å for the c–axis which shows that the BKS potential is not perfect, although these deviations are smaller than for most other potentials for SiO_2 discussed in the literature [6, 27]. The same conclusion emerges for β–cristobalite [27]. What is most disturbing is the fact that the ratio c/a when analyzed as a function of temperature in the simulation is perfectly smooth at the transition from α– to β–quartz, implying that the transition is second order, while the experiment reveals a small jump ($\Delta(c/a) \approx 0.0222$). Also the location of this transition comes out at $T_{tr} = 740 \pm 5$ K, while the experiment yields $T_{tr} = 845$ K [27]. Carrying out a finite–size scaling analysis [47] we were able to locate the transition temperature of our model precisely [6], and rule out the possibility that the smooth behavior of c/a near T_{tr} is an artefact of finite size effects.

Apart from these small discrepancies in the character and location of the transition, there is a very good overall agreement in the temperature dependence of elastic constants (see Figs. 3a,b of Ref. [6]), and the local structure in β–quartz for $T > T_{tr}$ could be clarified. The widely accepted view that β–quartz consists of a superposition of microdomains of α–quartz (with different domain orientations) could be refuted by an analysis of partial radial distribution functions: e.g., $g_{SiO}(r)$ has a double peak for $T < T_{tr}$ but a single peak for $T > T_{tr}$ near $r = 6.3$ Å [6]. However, it is important to realize that the system is very anharmonic, and the *ideal* β–quartz positions do not correspond to the positions around which the oxygen atoms actually fluctuate most of the time: both in β–quartz and in β–cristobalite the Si–O distance b_{av} in the "average structure" is somewhat smaller than the distance b_{max} where the probability distribution $p(b)$ has its peak. The value $b_{max} \approx 1.6$ Å from the simulations [6, 27] deduced for β–cristobalite agrees well with recent

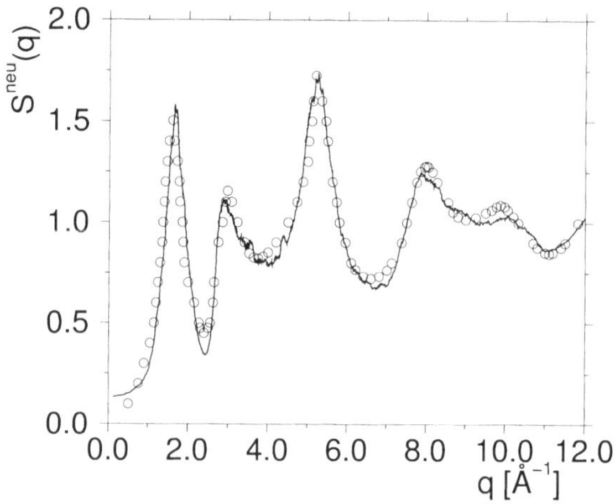

Figure 1. Static neutron structure factor of SiO_2 at room temperature ($T = 300$ K) plotted versus wave–vector q. The full curve is the molecular dynamics simulation from [13], using the experimental neutron scattering lengths for Si and O atoms, while the symbols are the neutron scattering data of [49]. From Horbach and Kob [13].

conclusions from experiments of Dove *et al.* [2] ($b_{av} \approx 1.55$ Å in the ideal β–cristobalite structure). These conclusions are corroborated by a study of the bond angle θ_{SiOSi}: rather than the ideal value $\theta_{SiOSi} = 159.2°$, it was found that $\theta_{SiOSi} \approx 155°$ at T_{tr} and decreases down to about 153° at $T = 1700$ K. This result is again in good agreement with experiments [48].

4. MOLTEN AND AMORPHOUS QUARTZ

Since due to the dramatic increase of the structural relaxation time of molten SiO_2 from the ps–scale at very high temperatures to 10^3 sec at the glass transition temperature $T_g = 1450$ K (defined from the empirical convention [9, 10] that the viscosity $\eta(T = T_g) = 10^{13}$ Poise) the temperatures where SiO_2 can be equilibrated by MD are rather high, $T \geq 2750$ K, as mentioned above. Thus *a priori* it is not clear to what extent results valid for temperatures of experimental interest (such as glass at $T = 300$ K, where the structure of the fluid at the experimental T_g was frozen in) can be obtained. Therefore it is very satisfying that very good agreement is obtained both with respect to the static structure (Fig. 1) [13, 49] and with respect to self–diffusion constants [13, 50, 51] and viscosity [13, 52] (Figs. 2, 3). Also the temper-

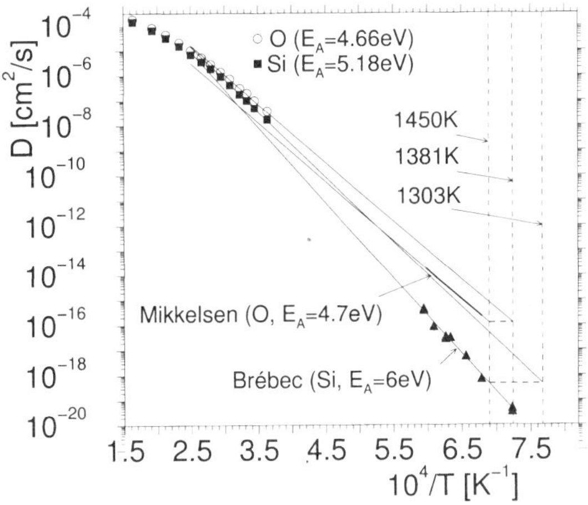

Figure 2. Plot of the self–diffusion constant D of silicon atoms (Si) and oxygen atoms (O) in molten SiO_2 as a function of inverse temperature. The symbols in the upper left part are the results from molecular dynamics simulations and the data in the lower right part stems from experiments [50, 51]. The thin straight lines show simple Arrhenius behavior ($D \propto \exp(-E_A/(k_B T))$) with various choices of the activation energy E_A, as indicated in the figure. The vertical broken lines indicate the experimental glass transition temperature, $T_g = 1450$ K, as well as values for T_g that one obtains if one extrapolates the data from the simulations to low temperatures and then estimates T_g from the experimental value of the O diffusion constant ($D_O(T = T_g^{sim}) = 10^{-16}$ cm²/sec $\Rightarrow T_g^{sim} = 1381$ K) or the Si diffusion constant, respectively ($D_{Si}(T = T_g^{sim}) = 5.10^{-19}$cm² /sec $\Rightarrow T_g^{sim} = 1303$ K). From Horbach and Kob [13].

ature dependence of the specific heat [31] is in very good agreement with the corresponding experimental data, as well as the results for the longitudinal and transverse sound velocities (Fig. 4) [35, 53] and the thermal conductivity, which was found to be nearly independent of temperature, $\lambda \approx$ 2.4 W/(Km) in the range 3000 K $\leq T \leq$ 4700 K [36], while corresponding experiments imply $2 \leq \lambda \leq 3$ W/(Km) if $T \geq 1000$ K [41]. Of course, the simulations of molten and glassy SiO_2 were not only done with the intention to reproduce just as many experimental data as possible, but also in order to go beyond experiment, e.g. by a careful test of the mode coupling theory of the glass transition (which needs to be done at such high temperatures [13, 33, 54] that experiments can no longer be performed, since the critical temperature is $T_c \approx 3330$ K), or by elucidating the behavior of the so–called "Bose peak" [35], and the frequency–dependent specific heat in the high–frequency domain [36], the atomistic structure at the free surface

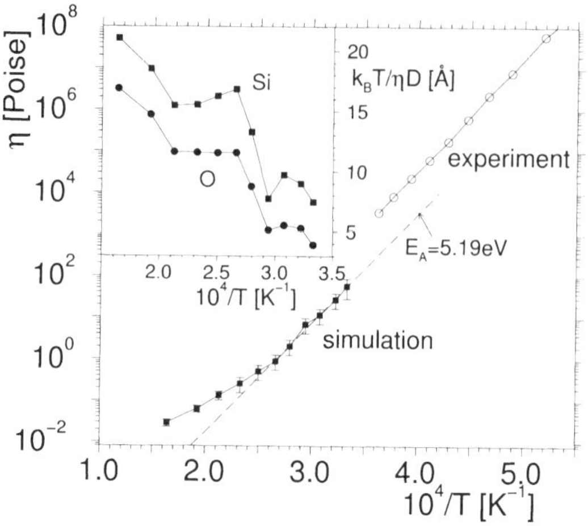

Figure 3. Molecular dynamics results for the viscosity of the BKS model for SiO_2 plotted vs. inverse temperature. The dashed straight line indicates an Arrhenius fit with an activation energy $E_A = 5.19$ eV. Experimental data [52] are compatible with this value but would suggest a slightly different pre–exponential factor. Note that for SiO_2 the analysis of other correlation functions at very high temperature suggests a critical temperature of mode coupling theory $T_c = 3330$ K and $\eta(T = T_c) \approx 8$ Poise. The insert shows the failure of Stokes–Einstein relations. From Horbach and Kob [13].

of SiO_2 (against vacuum [15, 16]), etc. However, these topics are out of the focus of the present paper.

5. STRUCTURE AND DYNAMICS OF SODIUM SILICATE MELTS

As has been emphasized above for the NSx systems, we have used an ad hoc–modification of the potential due to Kramer *et al.* [23] to ensure charge neutrality. Therefore it is appropriate again to test the accuracy of the description by comparing the neutron structure factor to corresponding experimental data [55] (Fig. 5). We see that the overall agreement between simulation and experiment is good. For $q > 2.3$ Å^{-1}, which corresponds to length scales of next–nearest Si–O and Na–O neighbors, the largest discrepancy is at the peak located at $q = 2.8$ Å^{-1}. Rather well reproduced is the "first sharp diffraction peak" at $q = 1.7$ Å^{-1}, which is a prominent feature in pure SiO_2 as well (Fig. 1), and arises from the tetrahedral network structure (the length scale which corresponds to it, i.e. $(2\pi)/1.7$ Å$^{-1} = 3.7$ Å, is approximately the spatial extent of two connected SiO_4 tetrahedra). As expected, this structure

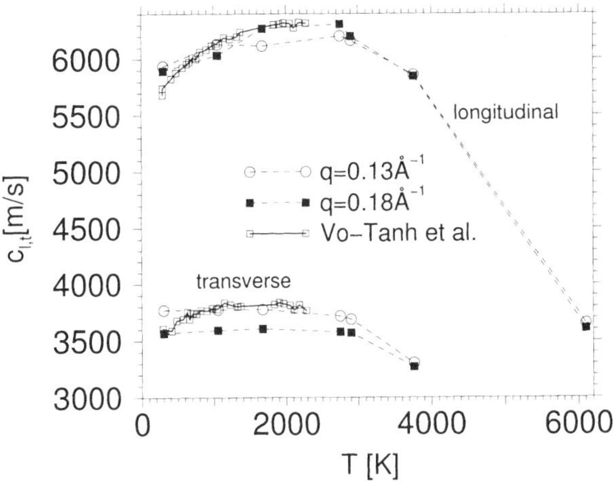

Figure 4. The temperature dependence of the longitudinal (C_ℓ) and transverse (C_t) sound velocities, which were determined from the maxima of the corresponding longitudinal and transverse current correlation functions at $q = 0.13$ Å^{-1} (open circles) and $q = 0.18$ Å^{-1} (filled squares). Also included are the experimental data of Vo–Tanh *et al.* [53] which were multiplied with the factor $(2.2/2.37)^{1/2}$ since the simulation was done at a density of $\rho_{\rm sim} = 2.37$ g/cm^3, while the experiment was done for $\rho_{\rm exp} = 2.2$ g/cm^3. From Horbach *et al.* [35].

is partly present in NSx also. The simulation also shows a weak "prepeak" at $q = 0.95$ Å^{-1}, related to a super–structure which is formed by the Na and Si atoms (the length $(2\pi)/0.95$ Å$^{-1} = 6.6$ Å is twice the mean distance of the nearest Na–Na or Na–Si neighbors, as an analysis of the corresponding partial radial distribution functions shows). While this feature was not seen in the experiment of Misawa *et al.* [55], it is now clear that the discrepancy between the simulation and the experiment was due to lack of resolution in the experiment: a more recent experiment [18] gave a very clear evidence for this prepeak, at the wave number predicted by the simulation. Of course, the simulation can again go beyond experiment by recording all 6 partial structure factors

$$S_{\alpha\beta}(q) = \frac{f_{\alpha\beta}}{N}\left\langle \sum_{i=1}^{N_\alpha}\sum_{j=1}^{N_\beta} \exp(i\vec{q}\cdot\vec{r}_{ij})\right\rangle_T, \quad \alpha,\beta \in \mathrm{Si,Na,O} \qquad (5)$$

where $f_{\alpha\beta} = 1$ for $\alpha = \beta$, $f_{\alpha\beta} = 1/2$ for $\alpha \neq \beta$, and the sum over i extends only over the atoms of species α, the sum over j only over atoms of species β,

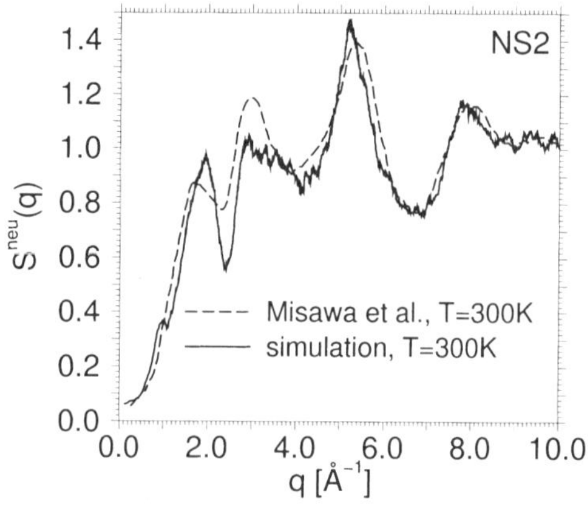

Figure 5. Static structure factor of sodium disilicate $S^{neu}(q)$ at room temperature ($T = 300$ K) plotted versus wave vector q. The full curve is the molecular dynamics simulation of Ref. [32], where the experimental neutron scattering lengths for Si, O and Na atoms were used, so there is no adjustable parameter whatsoever. The broken curve represents the corresponding experimental data of Misawa *et al.* [55]. From Ref. [32].

$\vec{r}_{ij}$ being the distance between atoms i and j. Then the structure factor shown in Fig. 5 (and similarly in Fig. 1) results as

$$S^{neu}(q) = \left(\sum_{\alpha} \frac{N_\alpha}{N} b_\alpha^2\right)^{-1} \sum_{\alpha\beta} b_\alpha b_\beta S_{\alpha\beta}(q). \qquad (6)$$

As an example, Fig. 6 shows two of the $S_{\alpha\beta}(q)$ (more details are given elsewhere [40, 56]). Note that the first sharp diffraction peak (at $q = 1.7$ Å^{-1}) in NSx is mostly due to Si–O and O–O correlations, and thus in the Si–Si correlations it shows up only as a mild shoulder, unlike SiO_2 where there is a clear peak. This already reflects that the rigid network of SiO_4^{4-} tetrahedra present in pure SiO_2 to some extent is broken up in NSx, due to the presence of the network modifier. But the "prepeak" at $q = 0.95$ Å^{-1} now can be very clearly recognized.

The local structure of the modified network can be characterized further by computing partial pair distribution functions $g_{\alpha\beta}(r)$. We now can count all β–atoms as nearest neighbors of an α–atom which are closer in distance from a selected α–atom than the location r_{min} of the first minimum of $g_{\alpha\beta}(r)$. In this way, one can characterize the corresponding coordination number z for

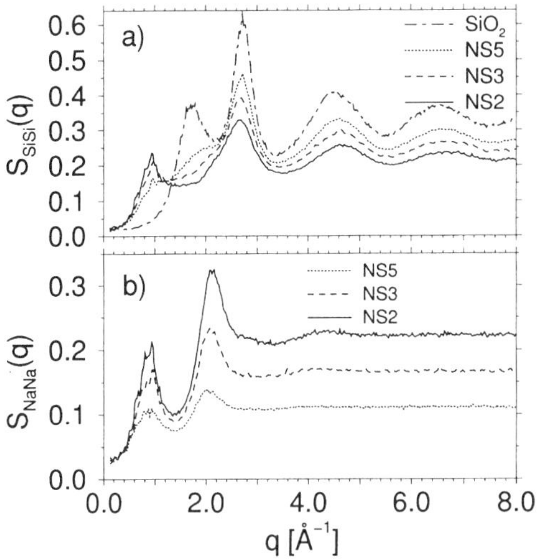

Figure 6. Partial structure factors a) $S_{SiSi}(q)$ and b) $S_{NaNa}(q)$ of NSx at $T = 300$ K plotted vs. q. In part a), the result for pure SiO_2 is included, in order to show that there the prepeak is completely absent. From Winkler *et al.* [56].

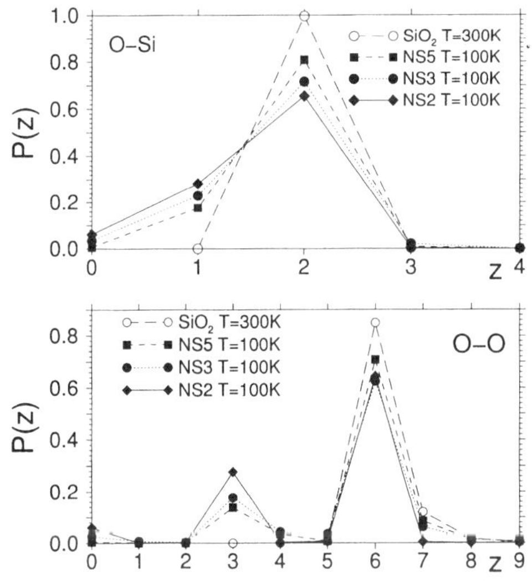

Figure 7. Distribution $P(z)$ of the coordination number z for a) Si neighbors of an oxygen atom and for b) O neighbors of an oxygen atom. Data for SiO_2 are taken at $T = 300$ K, data for NSx are taken at $T = 100$ K. From Winkler *et al.* [56].

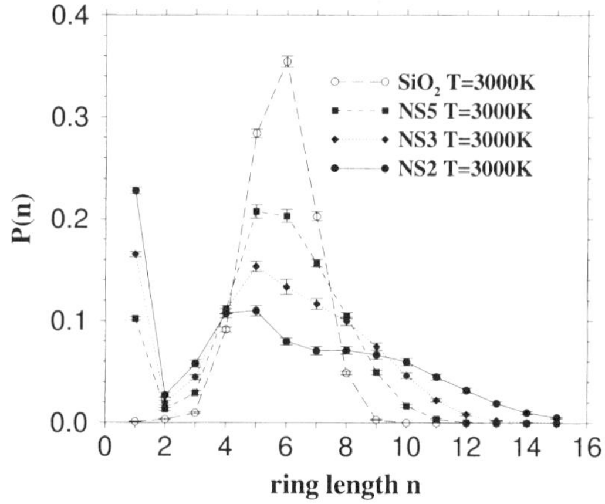

Figure 8. Probability distribution $P(n)$ of the ring length for SiO_2 and various sodium silicate melts at $T = 3000$ K. From Winkler *et al.* [56].

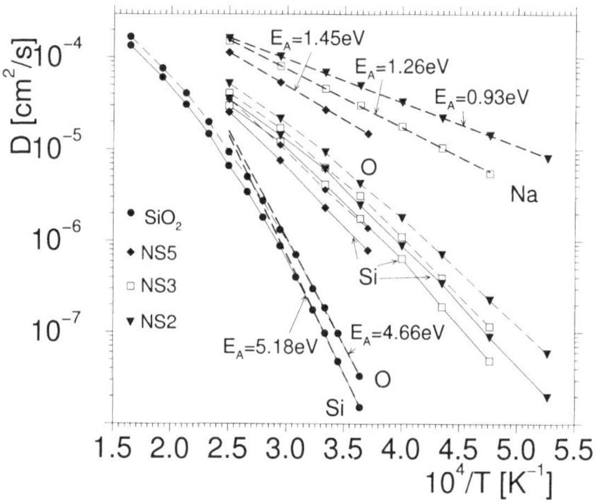

Figure 9. Arrhenius plot of the self diffusion coefficients of Si, O and Na in SiO_2 and sodium silicates plotted vs. inverse temperature. Straight lines indicate the Arrhenius relations, $D \propto \exp\left(-E_A^\alpha/(k_B T)\right)$, and the various activation energies are quoted in the figure. From Winkler *et al.* [56].

each atom (or the corresponding distribution $P(z)$, when all atoms of type α in a considered configuration are studied.) Fig. 7 shows that for α =O, β =Si in pure SiO_2 at $T = 300$ K every O has two Si neighbors, as expected for any perfect network where every SiO_4^{4-} tetrahedron shares a corner with another tetrahedron. In contrast, for NSx $P(z = 2) < 1$, and there is a non–negligible fraction of $(z = 1)$ and $(z = 0)$. While the latter O atoms do not contribute to the network at all, they are still bound to Na, the case $z = 1$ in part can be attributed to “dangling bonds" in the Si–O–network of the glass. This interpretation is corroborated by $P(z)$ when one considers O–O–neighborhoods: while in pure SiO_2 there is a single peak at $z = 6$ (each oxygen which is at a corner shared by two tetrahedra has 2×3 other corners of the same tetrahedra in its neighborhood), we find a second peak at $z = 3$ for NSx, due to “dangling bonds”. These dangling bonds are also responsible for the presence of large rings in the networks (the ring length n of closed loops in the network is defined by starting from an O–atom and seeking the shortest path leading back to the considered atom via O–Si and Si–O bonds, using each bond no more than once). Fig. 8 shows that for NSx $P(n)$ gains weight at n=1, i.e. oxygen atoms which are not part of rings at all, and $P(n)$ gains more and more weight for large n the larger the Na content of the structure. While experimental information on $P(z)$ in principle can be gotten from methods such as EXAFS, information on medium order as it is contained in the ring statistics can be inferred from experiments at best rather indirectly.
Following the mean squared displacements of the various types of atoms with time over large enough times, such that the atoms have travelled many inter-atomic distances and the Einstein relation can be applied, $\langle [\vec{r}_{i,\alpha}(t) - \vec{r}_{i,\alpha}(0)]^2 \rangle = 6D_\alpha t$, for $t \to \infty$, the self–diffusion constants D_α of the various types of ions can be obtained (Fig. 9). Again the fact that the network of SiO_4 tetrahedra is broken up more and more the more Na ions are added is evident, since the apparent activation energies for Si and O diffusion do depend on Na concentration. Particularly interesting is the result, that the Na diffusion is orders of magnitude faster at low temperatures: it turns out that this fast diffusion occurs on time–scales where the remaining SiO_2 network is still frozen. This fast motions of the Na^+ ions move in channels embedded in the SiO^2 matrix [37, 39]; the characteristic distance between the channels is the structural feature reflected in the prepeak at $q \approx 0.95$ Å^{-1} in the structure factor (Figs. 5,6). The lifetime of these channels is given by the structural relaxation time of the (highly viscous) SiO_2 matrix [39]. This interpretation was demonstrated by dividing the simulation box into small cells (of linear dimension of about 1 Å), and studying the probability that a cell that does not contain a Na atom at time $t = 0$ is not visited by a Na atom until time t. The geometrical structure of these sodium–free regions is a kind of percolating "Swiss cheese" structure surrounding the (also percolating) Na–rich channels. The simulations [39] allowed a detailed analysis of this "Swiss cheese" struc-

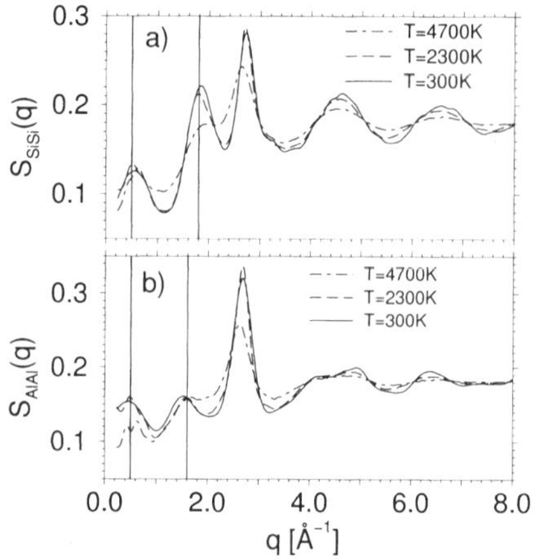

Figure 10. Partial structure factors $S_{\rm SiSi}(q)$, (a), and $S_{\rm AlAl}(q)$, (b), of $(Al_2O_3)2(SiO_2)$ at the three temperatures $T = 4700$ K, $T = 2300$ K and $T = 300$ K plotted vs. q. Vertical lines highlight peaks at $q = 0.5$ Å^{-1}, 1.8 Å^{-1} (a) and 1.6 Å^{-1} (b). From Winkler *et al.* [56].

ture and its relaxation, and hence showed how these very different mobilities seen in Fig. 9 can arise via a micro–segregation, without chemical clustering of compact Na–rich regions.

6. ALUMINIUMDISILICATE

Also for $(Al_2O_3)2(SiO_2)$ we have checked by a comparison with experimental data (in this case only the total X–ray structure factor is available [57]) that the structure predicted from the simulation is reasonably well in agreement with the experiment. Again, we obtain a much more detailed information than is accessible experimentally by a study of the partial pair distribution functions and structure factors (Fig. 10) [56]. At large q, the structure factor $S_{\rm SiSi}(q)$ resembles that of pure SiO_2 and of NSx, cf. Figs. 1,5; in particular, the peak at $q = 2.8$ Å^{-1} reflects the Si–O nearest neighbor distance. However, the "first sharp diffraction peak" which occurs at $q = 1.8$ Å^{-1} as for pure SiO_2 and NSx in $S_{\rm SiSi}(q)$, now shows up at the slightly shifted position in $S_{\rm AlAl}(q)$, namely $q = 1.6$ Å^{-1}. From the analysis of the distribution of coordination numbers $P(z)$, Fig. 11, we can attribute this feature to the presence of AlO_4 tetrahedra in the structure. A very interesting effect is also the pronounced prepeak at $q \approx 0.5$ Å^{-1} (actually, due to the finite size of the simulation box

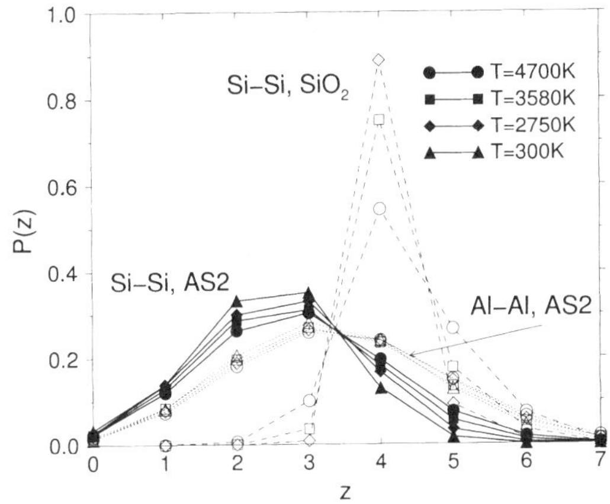

Figure 11. Distribution function $P(z)$ of the coordination number z for nearest neighbor Si–Si pairs and nearest neighbor Al–Al pairs. The corresponding distribution for pure SiO_2 is shown for comparison. From Winkler *et al.* [56].

this is close to our resolution limit for small q!). This peak shows up in all partial structure factors apart from $S_{OO}(q)$: while the oxygens are still rather randomly distributed in the structure, there clearly is a micro–segregation of Al–rich and Si–rich regions, and the characteristic length scale is larger than in the NSx systems. In view of the fact that in the phase diagram one expects a liquid–liquid phase separation in supercooled SiO_2–Al_2O_3 mixtures below $T_c \approx 1920$ K, we tentatively interpret this peak at $q \approx 0.5$ Å^{-1} as a precursor to this phase separation that would occur in metastable equilibrium at $T < T_c$. Of course, MD time scales of nanoseconds are probably not sufficient to fully equilibrate the chemical clusters forming at $T \gtrsim T_c$, and alternative more powerful simulation methods are clearly required.

When one studies $P(z)$ for the various ions, one finds that the introduction of Al in the SiO_2 network does not lead to a breaking–up of the network as in NSx, but rather the topological characteristics of the network change: many oxygens now have either 3 Al or 3 Si neighbors, unlike pure SiO_2 where each O always had 2 Si neighbors ("bridging oxygen"). The "tricluster" oxygens with $z = 3$ for Si(Al) neighborhoods of oxygens have the further effect that $z = 2$ and $z = 3$ occurs frequently in the Si–Si and Al–Al neighborhoods: this happens because Si and Al can substitute each other to some extent in the role as network former. When one does not distinguish between Si and Al as the neighboring atom at the appropriate distance (both pair distributions $g_{SiSi}(r)$

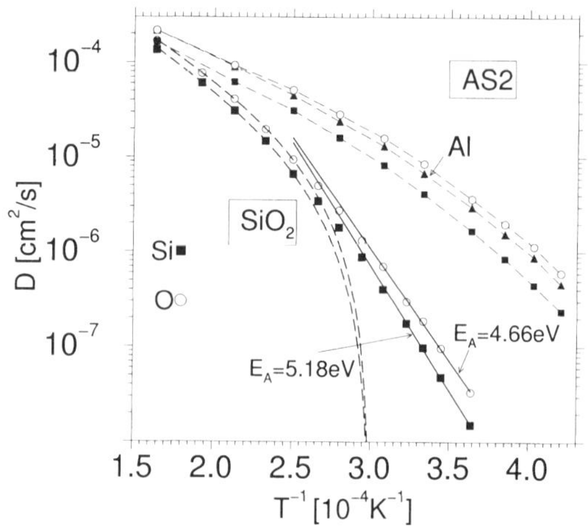

Figure 12. Arrhenius plot of the self–diffusion coefficients of Si, O and Al in SiO_2 and $(Al_2O_3)2(SiO_2)$ melts. In the case of SiO_2, curves shown represent fits according to mode coupling theory ($D \propto (1 - T/T_c)^\gamma$, $T_c = 3330$ K, $\gamma \approx 2$; broken curves) and according to the Arrhenius relation, respectively. In the case of aluminium–disilicate, curves are guides to the eye only. From Winkler *et al.* [56].

and $g_{\mathrm{AlAl}}(r)$ have their nearest neighbor peak at about 3.1 Å [56]), the peak of $P(z)$ is not as broad as in Fig. 11 and occurs at $z \approx 5$–6 rather than at $z = 3$. This different role of Al as an atom that can be incorporated into the SiO_2 network to some extent shows also up in a diffusion behavior that is very different from the one in the NSx system: while the diffusion constants of Si and O are again much larger than in SiO_2 (Fig. 12), since the Al–containing network is much less rigid than pure SiO_2, the diffusion constant of Al is very similar to that of O: it is the lifetime of the covalent bonds between Si and O that controls also the mobility of the Al ions, and although part of the Al is probably segregated in clusters (leading to the prepeak in Fig. 10) these clusters presumably are more compact and hence do not give rise to percolating channels that would allow enhanced Al diffusion.

7. CONCLUDING REMARKS

By the examples presented here, evidence was given that computer simulations (based on classical and quantum Molecular Dynamics and Monte Carlo methods) can give a very useful and detailed information on the structure, dynamics, and thermodynamic properties of solid and liquid silica and its mixtures with various oxides. These simulations can complement experiment in various ways, e.g. data can be obtained at high temperatures and pressures which are not accessible experimentally, and quantities can be estimated which are not available in the experiment (partial structure factors, ring statistics in SiO_2, etc.). Of course, these advantages in principle are by no means restricted to the systems studied here, but apply also to broad classes of other systems, e.g. metallic alloys.

But one important caveat needs to be made: one must keep in mind the limitations of the method, due to the use of effective potentials (with pair and perhaps triplet interactions, etc.) and due to the small size of the simulation box and the restricted time range that is accessible (typically less than 100 ns in the case of MD). In fact, long wavelength phenomena such as spinodal decomposition in mixtures [58] so far can only be simulated with phenomenological coarse–grained meso–scale models, and it remains a challenge to quantitatively bridge the gap between such mesoscopic studies of slow long wavelength phenomena, and the atomistic simulations as described here. Similarly, one needs to bridge the gap between the atomistic simulations and the more accurate "ab initio" methods [14, 16, 24, 25, 59] as well. While the Car–Parrinello method [59] is in principle very superior since it does not need a classical effective potential as an input, the limitation in size (about 100 atoms) and time scales (10 ps) that are accessible would not allow to address many of the questions treated in the present work. But clearly it is of great value to use this approach to validate classical potentials and, if possible, improve them.

With respect to mixtures of SiO_2 with other oxides, there is still a lack of experimental data (in particular on dynamic properties) with which our results can be directly compared (with the notable exception of Ref. [18].) It is hoped that the present work will hence stimulate also further experimental efforts.

Acknowledgments: We are grateful to M. H. Müser, P. Nielaba, and Chr. Rickwardt for a fruitful collaboration on the simulation of crystalline SiO_2, and to K. Vollmayr, A. Roder, and P. Scheidler for a fruitful collaboration on the earlier stages of the simulation of molten SiO_2. This work was partially supported by the Bundesministerium für Bildung und Forschung (BMBF grant No 03N6015), SCHOTT Glas, the Deutsche Forschungsgemeinschaft (DFG) under grants SFB262/D1 and HO 2231/2-1. Finally, generous grants of computing time on CRAY T3E from the NIC Jülich and the HLRS Stuttgart are gratefully acknowledged.

References

1. Klein and Hurlbut, *Manual of Mineralogy* (John Wiley & Sons, New York, 1983); P. J. Heaney, C. T. Prewitt, and G. V. Gibbs (eds.) *Silica . Physical behavior, geochemistry and materials applications* (Mineralogical Society of America, Washington D.C., 1994).
2. F. Liu, S. H. Garofalini, R. D. King-Smith, and D. Vanderbilt, Phys. Rev. Lett. **70**, 2750 (1993); M. T. Dove, D. A. Kreen, A. C. Hannon, and I. P. Swainson, Phys. Chemn. Miner. **24**, 311 (1997).
3. R. A. Cowley and A. D. Bruce, Adv. Phys. **29**, 1 (1980).
4. G. Dolino, Phase Trans. **21**, 59 (1990).
5. M. T. Dove, M. Gambhir, and V. Heine, Phys. Chem. Miner. **26**, 344 (1999).
6. M. H. Müser and K. Binder, Phys. Chem. Miner. **28**, 746 (2001).
7. H. Rawson, *Properties and Applications of Glass* (Elsevier, Amsterdam, 1980).
8. J. F. Mac Dowell and G. H. Beall, J. Am. Ceram. Soc. **52**, 17 (1969).
9. R. Zallen, *The Physics of Amorphous Solids* (John Wiley & Sons, New York, 1983).
10. H. Bach and D. Krause (eds.), *Analysis of the Composition and Structure of Glass and Glass Ceramics* (Springer, Berlin, 1999).
11. W. H. Zachariasen, J. Am. Chem. Soc. **54**, 3841 (1932); R. J. Bell and P. Dean, Phil. Mag. **25**, 1381 (1972).
12. K. Vollmayr, W. Kob and K. Binder, Phys. Rev. B **54**, 15808 (1996).
13. J. Horbach and W. Kob, Phys. Rev. B **60**, 3169 (1999).
14. M. Benoit, S. Ispas, P. Jund, and R. Jullien, Eur. Phys. J. B **13**, 631 (2000).
15. A. Roder, W. Kob, and K. Binder, J. Chem. Phys. **114**, 7602 (2001).
16. C. Mischler, W. Kob, and K. Binder, Comp. Phys. Comm. (2002, in press).
17. G. E. Brown, R. Farges, and G. Calas, Rev. Mineral. **32**, 317 (1995).
18. A. Meyer, H. Schober, and D. B. Dingwell, Europhys. Lett. (2002, in press).
19. B. T. Poe, P. F. McMillan, B. Coté, D. Massiot, and J.-P. Coutures, J. Phys. Chem. **96**, 8220 (1994).
20. M. P. Allen and D. J. Tildesley, *Computer Simulation of Liquids* (Clarendon Press, Oxford, 1987).
21. C. A. Angell, J. H. R. Clarke, and L. K. Woodcock, Adv. Chem. Phys. **48**, 397 (1981).
22. B. H. W. van Beest, G. J. Kramer, and R. A. van Santen, Phys. Rev. Lett. **64**, 1955 (1990).
23. G. J. Kramer, A. J. M. de Man, and R. A. van Santen, J. Am. Chem. Soc. **64**, 6435 (1991).
24. S. Ispas, M. Benoit, P. Jund, and R. Jullien, J. Noncryst. Solids (2002, in press).
25. M. Benoit and W. Kob, Europhys. Lett. (2002, in press).
26. J. S. Tse and D. D. Klug, J. Chem. Phys. **95**, 9176 (1991); Phys. Rev. Lett. **67**, 3559 (1991); J. S. Tse, D. D. Klug, and Y. Te Page, Phys. Rev. B**46**, 5993 (1992).
27. M. H. Müser, J. Chem. Phys. **114**, 6364 (2001); C. Rickwardt, M. H. Müser and K. Binder, Phys. Rev. B **63**, 045204 (2001).
28. J. Horbach, W. Kob, K. Binder, and C. A. Angell, Phys. Rev. E **54**, R5897 (1996)
29. J. Horbach, W. Kob, and K. Binder, Phil. Mag. B **77**, 297 (1998).
30. J. Horbach, W. Kob, and K. Binder, J. Non–Cryst. Solids **235–237**, 320 (1998).
31. J. Horbach, W. Kob, and K. Binder, J. Phys. Chem. B **103**, 4104 (1999).
32. J. Horbach and W. Kob, Phil. Mag. B **79**, 1981 (1999).
33. J. Horbach and W. Kob, Phys. Rev. E **64**, 041503 (2001).
34. J. Horbach, W. Kob, and K. Binder, Chem. Geol. **174**, 87 (2001).
35. J. Horbach, W. Kob, and K. Binder, Eur. Phys. J. B **19**, 531 (2001).
36. P. Scheidler, W. Kob, A. Latz, J. Horbach, and K. Binder, Phys. Rev. B **67**, 104204 (2001).
37. P. Jund, W. Kob, and R. Jullien, Phys. Rev. B **64**, 13403 (2001).

38. S. Ispas, M. Benoit, P. Jund, and R. Jullien, Phys. Rev. B **64**, 214206 (2001).
39. J. Horbach, W. Kob, and K. Binder, Phys. Rev. Lett. **88**, 125502 (2002).
40. A. Winkler, *Dissertation* (Johannes-Gutenberg-Universität Mainz, 2002).
41. O. V. Mazurin, M. V. Stultsina, and T. P. Shvaiko-Shvaikaskaya, *Handbook of Glass Data, Part. A: Silica Glass and Binary Silicate Glasses* (Elservier, Amsterdam, 1983).
42. M. Parrinello and A. Rahman, Phys. Rev. Lett. **45**, 1196 (1980).
43. M. E. Tuckerman, B. J. Berne, G. J. Martyna, and M. L. Klein, J. Chem. Phys. **99**, 2796 (1993).
44. M. E. Striefler and G. R. Barsch, Phys. Rev. B **12**, 4553 (1975).
45. M. Parrinello and A. Rahman, J. Chem. Phys. **76**, 2662 (1982).
46. M. A. Carpenter, E. K. H. Salje, A. Graeme-Barber, B. Wrucki, M. I. Dove and K. S. Knight, Am. Mineral **83**, 2 (1998).
47. K. Binder, Phys. Rev. Lett. **47**, 693 (1981); K. Vollmayr, J. D. Reger, M. Scheucher, and K. Binder, Z. Physik B **91**, 113 (1993).
48. K. Kihara, Eur. J. Miner. **2**, 63 (1990).
49. D. L. Price and J. M. Carpenter, J. Noncryst. Solids **92**, 153 (1987).
50. J. C. Mikkelsen, Appl. Phys. Lett. **45**, 1187 (1984).
51. G. Brebec, R. Seguin, C. Sella, J. Bevenot, and J. C. Martin, Acta Metall. **28**, 327 (1970).
52. G. Urbain, Geochim. Cosmochim. Acta **46**, 1061 (1982).
53. A. Polian, D. Vo–Tanh, and P. Richet, Europhys. Lett. **57**, 375 (2002).
54. F. Sciortino and W. Kob, Phys. Rev. Lett. **86**, 648 (2001).
55. M. Misawa, D. L. Price, and K. Suzuki, J. Non–Cryst. Sol. **37**, 85 (1980).
56. A. Winkler, J. Horbach, W. Kob, and K. Binder, publication in preparation.
57. H. Morikawa, S.-I. Miwa, M. Miyake, F. Maruno, and T. Sata, J. Amer. Ceramic Soc. **65**, 78 (1982).
58. K. Binder and P. Fratzl, in *Phase Transformations in Materials*, G. Kostorz (ed.), p. 409 (Wiley-VCH, Berlin, 2001).
59. R. Car and M. Parrinello, Phys. Rev. Lett. **55**, 2471 (1985).

A COMPUTER MODEL OF CARBONITRIDE PRECIPITATION IN STEEL

Philippe Maugis and Mohamed Gouné
IRSID - Arcelor Group
Voie Romaine, BP 30320, 57283 Maizières-lès-Metz, France
philippe.maugis@irsid.usinor.com

1 INTRODUCTION

In order to monitor the mechanical properties in relation to the microstructure, the knowledge of the precipitation state at the end of a thermo-mechanical treatment is of prime importance. In this purpose, Arcelor develops models that allow for the prediction of the influence of the process parameters on the state of precipitation. The model Multipreci, developed at IRSID is one of them[1]. It predicts the precipitation kinetics of mono- and di-atomic particles in ferrite and austenite as a function of the time-temperature history. It is based on the classical theories for diffusive phase transformation and treats simultaneously the nucleation, growth and ripening phenomena. The state of precipitation that is predicted includes the particle size distribution, their number and volume fraction. From these values, the effect of the precipitates on the mechanical properties can be calculated.

The recurrence of nitrogen in solid solution in steels implies that the precipitation kinetics of the carbonitrides that form from the metallic elements in solid solution be understood and modelled. One of the main difficulties of this task is the accounting of tri-atomic particles having a variable C/N ratio. This paper proposes a simple and fast computer model that treats the precipitation of the vanadium carbonitride V(C,N). This case is chosen here to exemplify the principles a new version of the Multipreci model which is designed to be valid for M(C,N)-type precipitates, M=V, Nb or Ti.

2 THE MULTIPRECI MODEL

2.1 V(C,N) versus VC

It is an experimental fact that nitrogen present in austenite solid solution in addition to carbon during the precipitation of vanadium results in the formation of vanadium carbonitrides. The ratio of C/N in the precipitates depends on the alloy composition and the thermal cycle. A comprehensive model of vanadium precipitation kinetics must take this fact into account, and therefore solve the following points:

- Composition of the nuclei
- Nucleation rate
- Growth rate
- Time evolution of the particle composition
- Gibbs-Thomson effect and ripening

In chapter 2 and 3, we provide analytical equations and a procedure to solve them to answer the above-mentioned questions.

2.2 Physical assumptions

The assumptions of the model are classical for this kind of problem:

- The precipitates are spherical particles.
- The thermodynamics of V(C,N) is an ideal solid solution of VC and VN.
- Nucleation occurs homogeneously according to the classical theory.
- Growth is limited by the diffusion of vanadium in the volume of the matrix.
- Ripening is driven by the Gibbs-Thomson effect.

2.3 The class model

The kinetic equations of nucleation and growth are used to compute the time evolution of the size histogram of the particles. The size histogram is represented by the number per unit volume and the particle composition of each class of size. This approach has been chosen to render the size-composition correlation that is commonly observed during controlled cooling experiments: the larger particles have higher nitrogen content than the smaller. We will see in the following that this approach has also the advantage of treating in a simple way non trivial ripening effects of varying composition particles.

Each class of size is characterised by its radius R and its composition y. At each time step of the calculation, nucleation is rendered by the creation of a new class of radius R' and composition y_g. During the same time-step, growth modifies the radius of each existing class of particles. The growth rate takes into account the Gibbs-Thomson effect. As a result, Oswald ripening occurs by dissolution of the smallest particle to the advantage of the largest ones (see Figure 1).

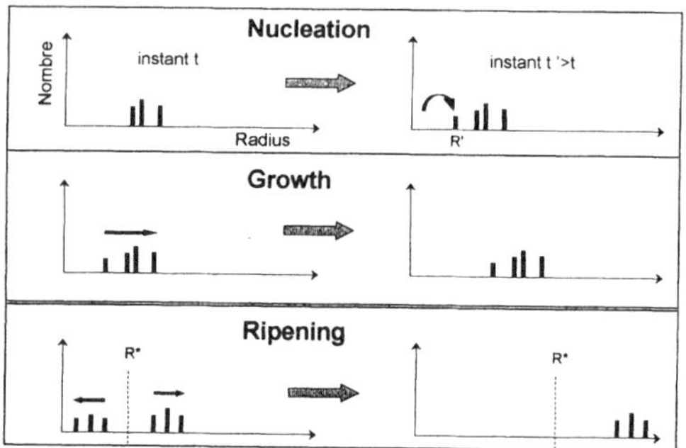

Figure 1. Schematics of the treatment of the classes of size, rendering nucleation, growth and ripening.

3 EQUATIONS OF THE MODEL

3.1 The model Fe-V-C-N steel

The generic case of vanadium precipitation in a model Fe-V-C-N steel is used here to exemplify the approach used in the Multipreci model. The steel has the following composition (Table 1):

Table 1. Composition of the steel, in weight ppm and atomic fraction.

	wt ppm	at. fract
V	2150	$2.36\ 10^{-3}$
C	1900	$8.87\ 10^{-3}$
N	150	$6.0\ 10^{-4}$

The steel is composed of a metallic alloying element, V, and two interstitial elements C and N, the balance being a majority of Fe atoms. The ratio of the interstitial atoms is C/N $\approx$ 15. The ratio of substitutional over interstitial atoms is V/(C+N) $\approx$ 0.25. This ratio being less than 1, the maximum quantity of precipitated vanadium is not limited by the amount of interstitial elements in the steel, but by the total amount of vanadium.

3.2 Thermodynamics

The equilibrium at a given temperature in austenite is the state towards which the system evolves from its initial, out of equilibrium state. It is useful to compute this state as a reference for the precipitation kinetics.

The equilibrium between the austenite matrix and the carbonitride VC_yN_{1-y} is described by the mass action law:

$$RT\left[\ln a_V^e + y\ln a_C^e + (1-y)\ln a_N^e\right] = \Delta G^{\circ}_{VC_yN_{1-y}} . \tag{1}$$

In this equation, a_i^e is the activity of the element i (i=V, C, N) at equilibrium. In the case of infinite dilution, the activity is proportional to the atomic fraction at equilibrium X_i^e (Henry state of reference). This assessment is true for V and N but remains a simplification for C. $\Delta G^{\circ}_{NbC_yN_{1-y}}$ is the Gibbs energy of formation of the carbonitride and is a function of its composition y. VC and VN are both of the fcc NaCl type crystal structure of very similar molar volume (see Table ?). We thus consider the carbonitride as an ideal mix of VC and VN, and following Hillert-Staffanson[2], we can write

$$\Delta G^{\circ}_{VC_yN_{1-y}} = y\Delta G^{\circ}_{VC} + (1-y)\Delta G^{\circ}_{VN} + RT\left[y\ln y + (1-y)\ln(1-y)\right] \tag{2}$$

where ΔG°_{VC} and ΔG°_{VN} are the Gibbs energy of formation of respectively VC and VN. Theses two thermodynamic functions are related to the solubility products for the individual VC or VN in equilibrium with the matrix by the relations

$$\begin{cases} \ln(X_V^e.X_C^e) = \Delta G_{VC}^\circ / RT \\ \ln(X_V^e.X_N^e) = \Delta G_{VN}^\circ / RT \end{cases} \quad (3)$$

By substituting Equation (2) into (1), it is easily shown that the equilibrium equations are[3,4]:

$$\begin{cases} yK_{VC} = X_V^e.X_C^e \\ (1-y)K_{VN} = X_V^e.X_N^e \end{cases} \quad (4)$$

where K_{VC} and K_{VN} are the solubility products of respectively VC and VN. This set of equations will be used later in this chapter for the derivation of the nucleation and growth rates. Relations (4) in association with the conservation equations of the alloying elements alloy for the calculation of the equilibrium concentration of V, C and N in solid solution, as well as the composition y of the carbonitride. The values of the solubility products can be found in the literature, and we have chosen the data selected by Gladman[5] (Table 2 and Figure 3):

Table 2. Logarithm of the solubility products of VC and VN in austenite (T in K, compositions in wt%).

	Solubility product
Log(%V.%C)	$6.72 - 9500/T$
Log(%V.%N)	$3.02 - 7840/T$

From Figure 2, it can be seen that the solubility product of VN is more than a hundred times smaller than that of VC. In other words, vanadium nitride is much more thermodynamically stable than vanadium carbide in austenite. This will have important consequences on the kinetics of nucleation and growth, as will be demonstrated later.

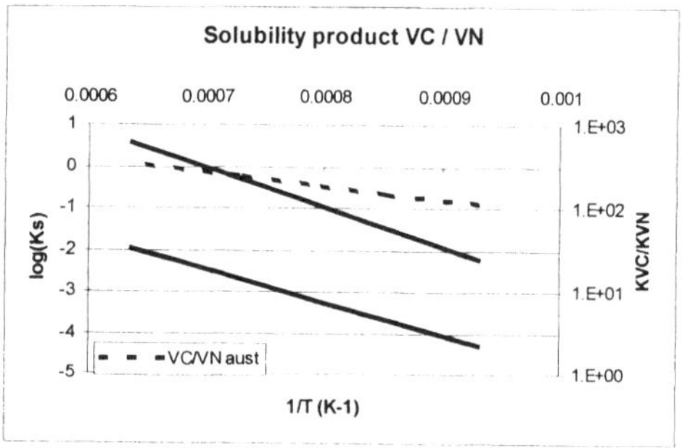

Figure 2. Arrhenius plot of the solubility product of VC and VN in austenite from Table 2 (T in K, compositions in wt%). The ratio K_{VC}/K_{VN} is plotted in dashed line.

The results of the calculation of the equilibrium state as a function of temperature are summarised in Figure 3. The temperature for complete dissolution is 1150°C. Below this temperature, when the temperature decreases:

- Nitrogen precipitation increases.
- The proportion of carbon in the precipitates increases.
- Under 900°C, the full precipitation of nitrogen is achieved.

We note that the total precipitation of vanadium cannot be achieved in austenite ($T > 800$°C).

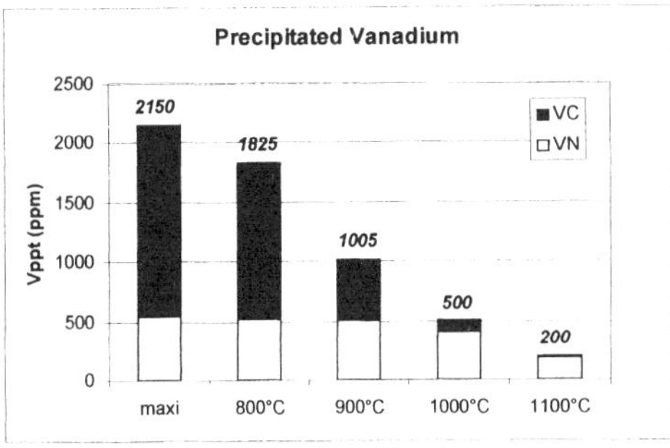

Figure 3. Equilibrium state of precipitation as a function of temperature calculated for the steel in Table 1, with data of Table 2.

3.3 Nucleation

Neglecting the mechanical energy term, the Gibbs energy for the formation of a spherical embryo of vanadium carbonitride from the elements in solid solution is classically expressed as the sum of a volume and an interface term :

$$\Delta G = \Delta g \frac{4}{3}\pi R^3 + \gamma 4\pi R^2 \tag{5}$$

In this equation, Δg is the driving energy for nucleation per unit volume, R is the radius of the embryo. γ is its interface energy with the matrix and is supposed to be isotropic. The expression of the driving energy Δg is a clue to the derivation of the size and composition of the critical nucleus. To find its expression we proceed as follows.

3.3.1 Driving energy for nucleation

In the simple case of a binary compound of formula A_yB_{1-y} , the driving energy for nucleation writes

$$V_m \, \Delta g = \mu_{A_yB_{1-y}}(y) - (y\mu_A + (1-y)\mu_B) \tag{6}$$

In the case of the VC_yN_{1-y} compound, Equation (6) can be generalised. The chemical potentials μ_i are developed as a function of the molar fractions in solid solution X_i^{ss} according to the regular matrix solid solution assumption, and the equilibrium fractions X_i^e are introduced. This leads to the expression[6-7]:

$$\Delta g = -\frac{RT}{V_{VCN}}\left[\ln\left(\frac{X_V^{ss}}{X_V^e}\right) + y\ln\left(\frac{X_C^{ss}}{X_C^e}\right) + (1-y)\ln\left(\frac{X_N^{ss}}{X_N^e}\right)\right] \tag{7}$$

where V_{VCN} is the molar volume of the vanadium carbonitride. Taking into account that the molar volume of VC and VN are almost identical, V_{VCN} is considered here to be independent of the composition y (see Table 3).

3.3.2 *Composition of the critical nucleus*

Using Equations (4), the unknown equilibrium compositions in the solid solution in Equation (7) are replaced by the solubility products K_{VC} and K_{VN} :

$$\Delta g(y) = -\frac{RT}{V_{VCN}}\ln\left[\frac{\left(X_V^{ss}\right)\left(X_C^{ss}\right)^y\left(X_N^{ss}\right)^{1-y}}{\left(yK_{VC}\right)^y\left[(1-y)K_{VN}\right]^{1-y}}\right] \tag{8}$$

Δg is here a function of y. We state that the nuclei that appear in the matrix are those that produce the maximum variation of Gibbs energy during their formation. The composition of the critical nucleus is then the value of y that minimises the driving energy (8) for nucleation. The expression is found to be

$$y_g = \frac{1}{1 + \dfrac{X_N^{ss}}{X_C^{ss}}\dfrac{K_{VC}}{K_{VN}}} \tag{9}$$

This equation shows that the composition y_g of the critical nucleus does not depend on the vanadium concentration in solid solution, and depends exclusively on

- the temperature, via the ratio K_{VC}/K_{VN} ,
- the composition in C and N of the solid solution, via the ratio X_C^{ss}/X_N^{ss} .

Notice that formula (9) relates in a linear way the ratio C/N in the critical nucleus to the ratio C/N in solid solution:

$$\left(X_C/X_N\right)_{nucleus} = \left(K_{VN}/K_{VC}\right)\left(X_C/X_N\right)_{matrix} \tag{10}$$

Figure 4 show the composition of the nucleus $VC_{yg}N_{1-yg}$ as a function of the ratio X_C^{ss}/X_N^{ss} , for three temperatures. It appears that for a given X_C^{ss}/X_N^{ss} ratio, a decrease in temperature leads to an enrichment in carbon of the carbonitride. On the other hand, if carbon and nitrogen in solid solution are of equal order of magnitude, the nucleus will be very rich in nitrogen, similar to a pure vanadium nitride. In particular, for the ratio C/N $\cong$ 15 characteristic of our reference steel, the composition of the carbonitride is $VC_{0.1}N_{0.9}$. The explanation for this is as follows: as we have seen previously, the nitride VN is much more stable thermodynamically than the carbide VC. Then it is the nucleation of VN that reduces most the total Gibbs energy. This is rendered by the ratio $K_{VC}/K_{VN} >> 1$ present in Equation (9). The counterbalance of this effect of relative stability is possible if the carbon concentration in solid solution is increased relatively to nitrogen concentration. Hence at 800°C, nuclei of equimolar composition will appear if the carbon concentration in solid solution is approximately 110 times that of nitrogen (see Figure 4).

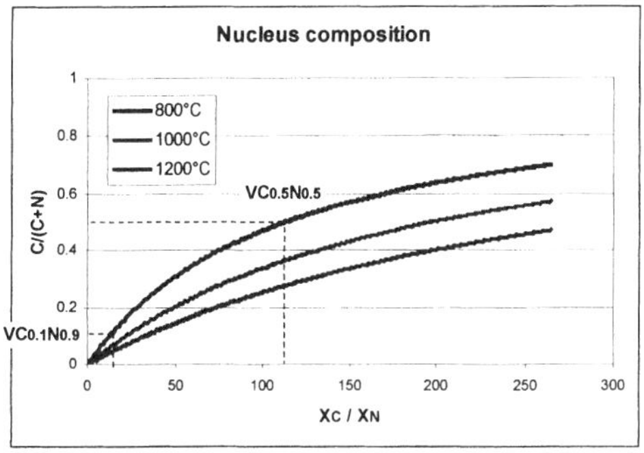

Figure 4. Composition y_g of the nucleus $VC_{y_g}N_{1-y_g}$ as a function of the ratio X_C^{ss}/X_N^{ss}, for three temperatures.

3.3.3 *Radius of the critical nucleus*

Given the composition of the critical nucleus, its radius has to be calculated. The Gibbs energy for the formation of an embryo, ΔG (Equation 1), reaches a maximum value ΔG * for a particular value of *R*, the critical radius $R^* = -2\gamma/\Delta g(y_g)$. *R** corresponds to a nucleus that is in equilibrium with the matrix, taking into account the radius-dependent Gibbs-Thomson effect. For a precipitate to be able to grow effectively, it has to have a radius *R'* slightly higher than *R** such that $\Delta G(R') = \Delta G^* - k_B T$, where k_B is the Boltzmann constant. *R'* is the solution of the following equation:

$$R'^3 + \frac{3\gamma}{\Delta g} R'^2 = \left[\frac{4}{3} \frac{\gamma^3}{\Delta g^3} - \frac{3k_B T}{4\pi \Delta g} \right]. \tag{11}$$

The approximate solution

$$R' = R^* + \frac{1}{2}\sqrt{\frac{k_B T}{\pi\gamma}} \tag{12}$$

proves to be precise enough in practice.

In the Multipreci model, we treat the nucleation stage at each time step as follows: a new class is introduced in the histogram of size of radius *R'* and composition y_g according to Equations (9) and (12). The number of precipitates *N* of size *R'* is given by the classical non stationary nucleation rate equation:

$$\frac{dN}{dt} = N_0 \exp\left(-\frac{\Delta G^*}{kT}\right)\Gamma Z\left[1-\exp\left(\frac{-t}{\tau}\right)\right] \tag{13}$$

where N_0 is the number of substitutional sites per volume in austenite, Γ the absorption frequency of a vanadium atom by the critical nucleus, Z is the Zeldovitch constant and τ is the incubation time. Note that the incorporation of the Zeldovitch constant in this equation is consistent with the choice of R' as the radius of the precipitates introduced by nucleation.

3.4 Growth

For each individual class of particle of radius R, the growth rate has to be established as a function of the composition of the solid solution. Carbon and nitrogen, being interstitial elements, are very fast diffusing species compared to vanadium. As a result the growth of a precipitate is limited by the diffusion of vanadium form the matrix towards the precipitate. A gradient in vanadium concentration builds around the precipitate while the concentration profiles of carbon and nitrogen are almost flat (see Figure 5).

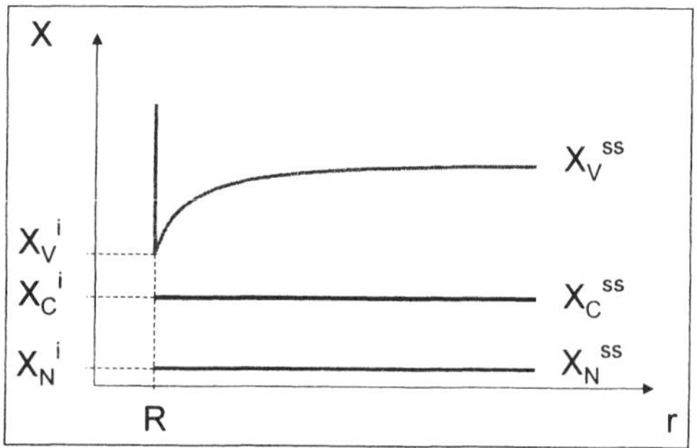

Figure 5. Schematics of the concentration profiles around a growing precipitate. The C and N concentration profiles remain almost flat.

To a very good approximation, the quasi-stationary approximation for the diffusion profiles can be applied. From the mass balance at the precipitate / matrix interface the classical Zener equation for the grow rate is derived. For vanadium, the equation is

$$\frac{dR}{dt} = \frac{D_V}{R} \frac{X_V^{ss} - X_V^i}{\frac{V_{Fe}}{V_{VCN}} - X_V^i}. \tag{14}$$

During an isothermal precipitation treatment, the composition of vanadium in solid solution X_V^{ss} decreases, and consequently the growth rate decreases. In the meantime, the composition of vanadium at the interface X_V^i evolves. Note that Equation (14) is also valid for dissolution, a feature that is used for the modelling of the Oswald ripening phenomenon (see § 3.5).

Growth occurs by accretion of successive shells of carbonitrides. Each shell has its specific composition y_c. The calculation of the shell composition is done together with the calculation of the matrix composition in the vicinity of the interface by the simultaneous resolution of the local equilibrium and a flux-compatibility condition:

- The local equilibrium of the shell with the surrounding matrix writes, according to Equation (4),

$$\begin{cases} yK_{VC} = X_V^i . X_C^i \\ (1-y)K_{VN} = X_V^i . X_N^i \end{cases} \tag{15}$$

- The composition of the precipitating shell has to be compatible with the flux of precipitating species. This can be written in a simplified way as $J_C = yJ_V$ and $J_N = (1\text{-}y)J_V$. According to the quasi-stationary assumption, this leads to

$$\begin{cases} D_C\left(X_C^i - X_C^{ss}\right) = yD_V\left(X_V^i - X_V^{ss}\right) \\ D_N\left(X_N^i - X_N^{ss}\right) = (1-y)D_V\left(X_V^i - X_V^{ss}\right) \end{cases} \tag{16}$$

At each time interval, the resolution of the set of equations (15-16) gives the composition y_c of the shell and the concentrations X_V^i, X_N^i and X_C^i at the interface. From the condition $D_C \gg D_V$ and $D_N > D_V$, a very good approximation of the shell composition can be achieved, that is:

$$y_c = \frac{1}{1 + \dfrac{X_N^{ss}}{X_C^{ss}} \dfrac{K_{VC}}{K_{VN}}} \tag{17}$$

It appears that, at each time interval, the composition of the shell is the same as that of the critical nucleus ($y_c = y_g$), and is only driven by the composition in carbon and nitrogen of the solid solution. This can be understood by the following reasoning. Notice first that the composition of the shell is independent on the diffusion coefficients of carbon and nitrogen (Equation 17). This non trivial result is due to the fact that C and N, being interstitial elements, are very fast diffusing species compared to V. As a result, the characteristic time for C and N diffusion towards the precipitate is much smaller than for V. In other words, C and N have enough time to equilibrate with one another around the particle during the precipitation of vanadium. They do so such that the Gibbs energy of the interfacial region be minimum. This minimisation criterion is exactly the one that led to the expression of the critical nucleus, implying that $y_c = y_g$.

3.5 Ostwald ripening

Ostwald ripening is the process by which the smallest precipitates dissolve to the profit of the bigger ones. This phenomenon is particularly important when the system reaches the equilibrium precipitate fraction, but it is to be noted that ripening occurs at every stage of the precipitation process, even when the matrix is still supersaturated, as will be demonstrated by the numerical results later. The ripening will automatically be dealt with by the class treatment as soon as the growth rate is written in a proper way as a function of the precipitate size. We proceed as follow.

The incorporation of the interface energy in the total energy of formation of a precipitation of radius R introduces an additional curvature-dependent term to the chemical potential of the precipitate, equal to $2\gamma V_{VCN}/R$. We introduce this term in the right-hand side of equation (1) and perform the calculation of the equations of local equilibrium of the shell with the surrounding matrix as in paragraph 3.2. This leads to a similar set of equilibrium equations as Equations (4), where the solubility products of VC and VN have to be replaced by the radius dependent functions $K_{VC}(R)$ and $K_{VN}(R)$:

$$\begin{cases} K_{VC}(R) = K_{VC} \exp(\dfrac{2\gamma V_{VCN}}{RR_g T}) \\ K_{VN}(R) = K_{VN} \exp(\dfrac{2\gamma V_{VCN}}{RR_g T}) \end{cases} \tag{18}$$

In Equations (18), the usual solubility products are just multiplied by a factor of $\exp(2\gamma V_{VCN}/R\,R_g T)$ where R_g is the gas constant. To treat the ripening phenomenon, the radius-dependant solubility products are simply substituted to K_{VN} and K_{VC} in equations (15) and (16). The resolution of this modified system of equation allows for the incorporation of the Gibbs-Thomson effect on the concentrations X_V^i, X_N^i and X_C^i at the interface. the interface compositions are affected in such a way that for supercritical precipitates (R>R*) the condition $X_V^{ss} - X_V^i > 0$ applies. According to Equation (14), those precipitates grow. On the contrary, the undercritical precipitates dissolve since for them $X_V^{ss} - X_V^i$ is negative. This treatment renders in a natural way the ripening phenomenon without any additional hypothesis.

In addition, it can be easily shown that the composition of the shell at a given time does not depend on the size of the precipitate, and is still given by Equation (17).

4 RESULTS OF THE MODEL

4.1 Calculation procedure

The model is programmed in Fortran language. In addition to the physical constants, an initial state of precipitation can be introduced. The model computes the histogram of size at each time along any thermal cycle, including stages of reheating and cooling. The calculation proceeds iteratively: at each time step, nucleation introduces a new class of size in the histogram and the new radius of every existing class is computed. The mass balance in the system gives the new composition of the solid solution, before the next calculation loop (see Figure 6). The calculations presented in this chapter run in less than one minute on a PC computer.

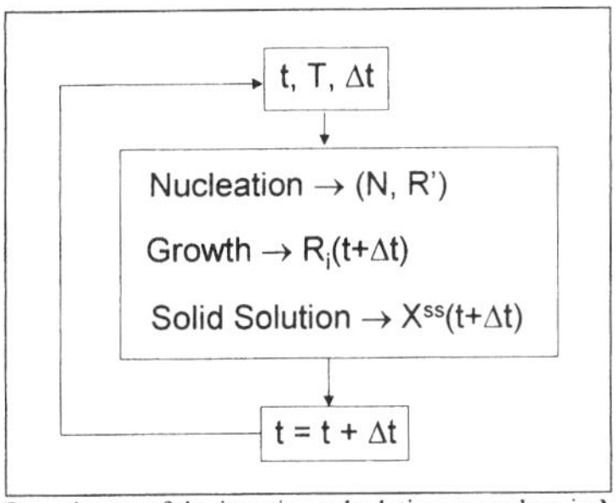

Figure 6. Organigram of the iterative calculation procedure in Multipreci.

4.2 Entry data

The Multipreci model is applied here to isothermal treatments of the model alloy presented above. The entry data of the model are summarised in Tableau 3:

Table 3. Values of the entry data of the calculations.

	Symbol	Value
Vanadium nominal atomic fraction	X_V	$2.36\ 10^{-3}$
Carbon nominal atomic fraction	X_C	$8.87\ 10^{-3}$
Nitrogen nominal atomic fraction	X_N	$6.0\ 10^{-4}$
Carbon diffusion coefficient [cm²/s]	D_C	$0.1\ \exp(-137500/R_gT)$
Nitrogen diffusion coefficient [cm²/s]	D_N	$0.91\ \exp(-168600/R_gT)$
Vanadium diffusion coefficient [cm²/s]	D_V	$0.25\ \exp(-264200/R_gT)$
Interface energy	γ	$0.5\ J.m^{-2}$
Molar volume of VCN	V_{VCN}	$10.65\ cm^3.mol^{-1}$
Molar volume of austenite	V_{Fe}	$7.11\ cm^3.mol^{-1}$
Solubility product of VN	$\text{Log}(K_{VN})$	$3.02 - 7840/T$
Solubility product of VC	$\text{Log}(K_{VC})$	$6.72 - 9500/T$

4.3 Isotherms of 800°C and 900°C

The isotherm at 800°C illustrate the various step of precipitate formation through the stages of nucleation, growth and ripening. Those stages are in fact non fully differentiated as discussed in detail in the following sections. The isotherm as 900°C is used to illustrate the non trivial ripening stage that occurs will composition-varying particles.

4.3.1 Isotherm of 800°C

As can be seen in Figure 7a, the nucleation rate of the particles increases rapidly during the first 5s of transitory nucleation stage. The precipitates that form are nitrogen reach $VC_{0.1}N_{0.9}$ (Figure 7g). It can be shown that the presence of nitrogen increases considerably the nucleation rate of the precipitates, since the driving energy for the nucleation of $VC_{0.1}N_{0.9}$ is much higher than that of VC (Cf. paragraph 3.2). During the nucleation stage, nitrogen and vanadium are consumed from the matrix, which decreases the driving energy and trends to slow down the nucleation rate. The initial critical radius of nucleation is about 0.3nm (Figure 7c). After ~20s, the composition of the solid solution reaches a value where the nucleation rate is practically zero : nucleation stops.

From this point, the number of particles begins to decrease (Figure 7e). A transitory ripening stage occurs, during which the critical radius has not yet reached the average radius (Figure 7c). After ~35s, almost all nitrogen initially in solid solution as been consumed from the matrix and growth now proceeds by precipitation of carbon and vanadium in the form of VC shells (Figure 7f). As a result, the average carbon composition of the precipitates grows slowly.

A time t = 500s, the critical radius has reached the mean radius of the particles and ripening develops in a quasi-steady state manner, leading to an endless growth of the mean radius of the particles and a steady decrease of their total number per unit volume (Figures 7c, d). The composition of the precipitates reaches its equilibrium value of $y = 70\%$ after 10^6s of annealing. In the meantime, the volume fraction of the precipitates has increased towards its equilibrium value of 0.3%.

Figure 7b shows the time evolution of the size histogram. Nucleation builds up the histogram initially around small radii. The whole distribution shifts to the right by growth. After a while, ripening is visible in the decrease of the total number of particles and the shift of the mean radius. Notice that the shape of the histogram is none of the usually measured log-normal distribution, nor the theoretical LSW distribution. This fact is extensively discussed in ([8]) and will be published in the future.

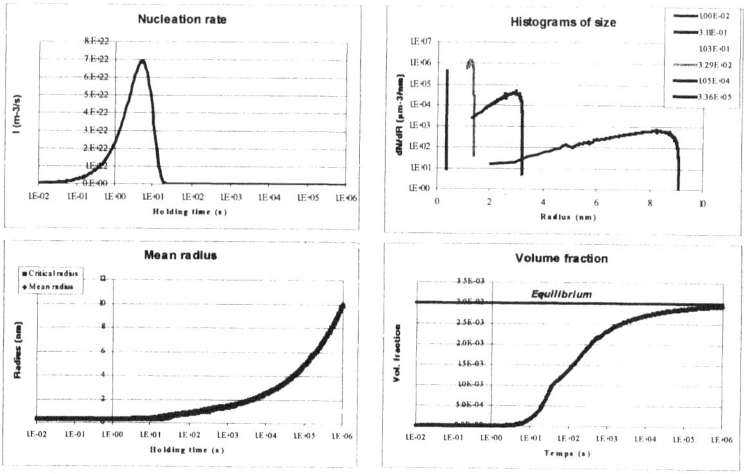

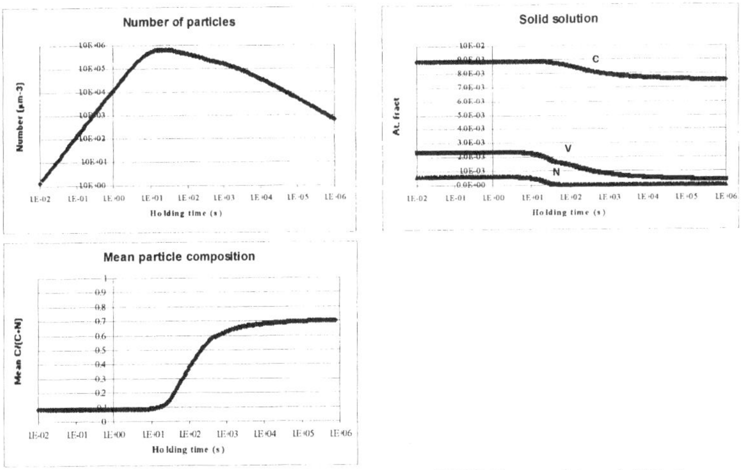

Figure 7. Results of the calculations at a temperature of 800°C. The entry data are in Table 3.

Figure 8 is a schematic representation of the evolution of the successive shell accretions that form the precipitates. Starting from a nitrogen rich core, the successive shells are richer and richer in carbon. The overall composition of the particle is that of a carbonitride V(C,N). During the growth of the precipitates, their composition is likely to homogenise somewhat by internal interdiffusion of carbon and nitrogen. This phenomenon is not taken into account in the model, but it is supposed to have a effect on the kinetics of growth and the final state of precipitation.

Figure 8. Schematic representation of the accretion sequence that forms a V(C,N) precipitate.

4.3.2 *Isotherm of 900°C*

At 900°C, precipitation occurs in three steps: the first step, corresponding to time $t < 200$s, is the nucleation and growth of nitrogen-rich particles (see Figure 9). A stasis period follows ($200s < t < 500s$), during which the size histogram evolves at constant total number and mean radius of the particles. After 500s, a non usual ripening phenomenon is at work: indeed, the total number of particles decreases while the mean radius increases, as a results of the growth of the big particles to the expense of the smallest. However, this occurs while in the meantime the mean composition of the particles evolves from y = 6% to 50%, and while the volume fraction increases from 0.1% towards its equilibrium value of 0.165%. This leads to a deviation form the usual time-scaling laws of $R^3 \propto t$ and $N^{-1} \propto t$.

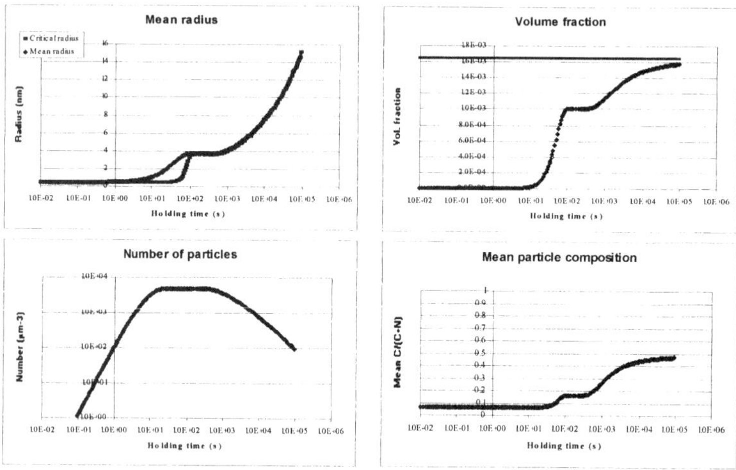

Figure 9. Results of the calculations at a temperature of 900°C. The entry data are in Table 3.

4.4 TTT diagram

The Multipreci program allows for rapid calculation of Time Temperature Transformation (TTT) diagrams for different sets of alloy composition. For the alloy under study in this paper, the TTT diagram is represented in Figure 10. The percentages in the legend refer to the maximum quantity of interstitials that can possibly precipitate according to the alloy concentration (i.e. 150ppm N and 505ppm C here). The diagram exhibits the usual C-shape curves, the maximum of precipitation rate being around 850°C. It is clear that the precipitation of the interstitial elements occurs in two successive stages: a fast precipitation of nitrogen under the form of nitride particles, followed by a longer precipitation of carbon that transforms the initial nitrides VN into carbonitrides V(C,N).

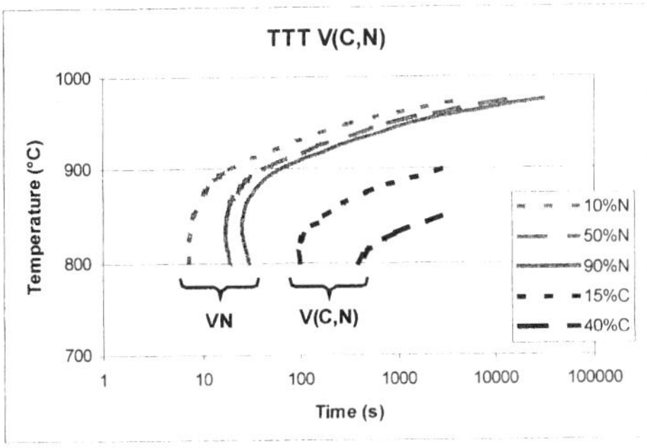

Figure 10. TTT diagram of the isothermal kinetics of precipitation of N and C with V.

4.5 CCT diagram

Diagrams of the Transformation kinetics during Continuous Cooling (CCT) are of high practical interest. A CCT diagram for the alloy under study is shown on Figure 11. The cooling rates range from -0.01 to -10°C/s in the temperature interval from 950 to 800°C. As expected, the beginning of transformation (10%N precipitated) occurs at a lower temperature when the cooling rate is higher. Under such conditions of continuous cooling, carbon precipitation is not expected to occurs in the temperature range 950-800°C, unless the cooling rate is very slow (<-0.01°C/s).

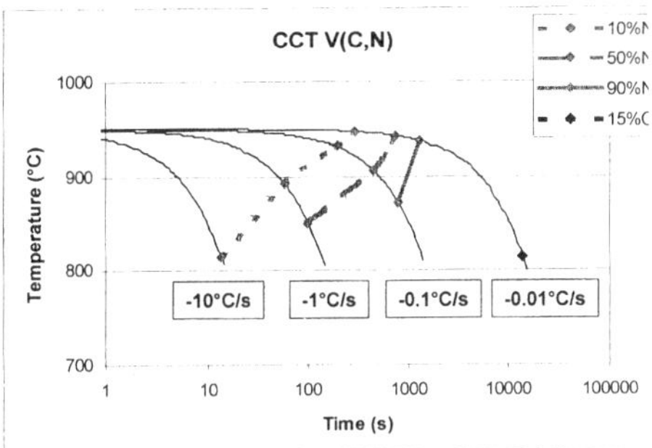

Figure 11. CCT diagram of the kinetics of precipitation of C and N with V during continuous cooling.

5 CONCLUSION

The model presented here proposes a methodology that solves the difficulty of dealing with tri-atomic and composition-varying precipitates. This approach allows for the simple account of nucleation, growth and ripening of M(C,N)-type carbonitrides. The application of the model to a Fe-V-C-N alloy illustrates the prominent role of nitrogen in the precipitation sequence and kinetics. This approach will be extended in the future to the case of (M,M')(C,N) mix carbonitrides (M,M' = V,Nb,Ti).

6 REFERENCES

1. D. Gendt et al, in Mathematical Modeling in Metals Processing and Manufacturing, Met. Soc., P. Martin ed. (2000).
2. M. Hillert, L. I. Staffansson, *Acta Chem Scand*, **24**, p.3618 (1970).
3. P. R. Rios, Method for the determination of mole fraction and composition of a multicomponent fcc carbonitride, *Materials Science and Engineering*, p.87 (1991).
4. P. Rios, Expression for solubility product of niobium carbonitride in austenite, *Materials Science and Technology*, p.324 (1988).
5. T. Gladman, The Physical Metallurgy of Microalloyed Steels, The University Press, Cambridge (1997).
6. W. J. Liu, J. J. Jonas, Nucleation kinetics of Ti carbonitride in microalloyed austenite, *Metallurgical Transactions* A, p.89 (1989).
7. A. Samoilov, B. Buchmagr, H. Cerzak, A thermodynamic model for composition and chemical driving force for nucleation of complex carbonitrides in microalloyed steel, *Steel Research*, p.298 (1994).
8. D. Gendt, Cinétiques de précipitation du carbure de niobium dans la ferrite, phD thesis, Orsay (2001).

ORDERING, PHASE STABILITY AND PHASE DIAGRAMS

CURRENT AND FUTURE APPLICATIONS OF CALPHAD TECHNOLOGY

Larry Kaufman
140 Clark Road, Brookline MA 02445-5848,USA. E-mail: larrykaufman@rcn.com

1. INTRODUCTION

During the past year the author has had the opportunity of participating in two international conferences held in the U.S. devoted to exploring recent examples of methods for predicting multiphase equilibria in diverse materials. The first of these, held at the 130th TMS meeting in New Orleans was organized by Zi-Kui Liu entitled "*Computational Thermodynamics and Materials Design*", Kaufman[1]. The second, organized by Patrice Turchi was "*CALPHAD and Alloy Thermodynamics*" held in Seattle at the 131st TMS meeting, Kaufman[2].Most of the papers presented at these meetings have been published. In May of 2002 CALPHAD XXXI held near Stockholm provided additional examples of rapid progress in this field. Finally, the second issue of the volume 26 (2002) of the CALPHAD Journal contains descriptions of the variety of software available for applying the modern methods of "Computational Thermodynamics" (CT) to practical problems encountered in dealing with commercial materials. The author has chosen several examples of recent work on Nb alloys, Kaufman[3], giant magneto-resistance, Bamberger et al[4], metallic glasses [5,6,7], lithium batteries, Thackeray et al.[8], Kaufman[9], zirconia ceramics, Jacobson[10], Arroyave et al.[11], and multi-component rhenium alloys Erymenko et al.[12], Velikanova et al.[13] for discussion in order to note current problems that require attention and to foster future progress in this field. At the above-mentioned conferences Kaufman[1,2] reviewed progress in this field since 1966 and compared the limited capability for making useful predictions of phase equilibria in commercial processes and multi-component alloys which were used in 1966 in contrast with the current practice where CT methods are used with success in such systems. This was attributed to the development of **CALPHAD THERMODYNAMICS (CT)** that aims to describe all possible phases in a system over wide ranges of temperatures, pressures and compositions to an extent that is uncommon in classical thermodynamics. The latter feature presents many opportunities for future applications as well as some problems that require attention. The inclusion of this feature grew naturally in **CT** from the realization that commercial processing always endeavors to increase the rate of production to become more profitable. By contrast, thermodynamic measurements are performed under equilibrium conditions. Thus since **CT** applies the results of scientific measurements and observations, which are made under conditions where equilibrium prevails, to commercial practice where non-equilibrium or quasi-equilibrium persists, the **CT** framework must be broader in scope than that used in classical thermodynamics. The success achieved by pursuing this track is due to the explicit definition of the stability of unstable and metastable phases and the functional descriptions of the compositional, temperature and pressure dependences of the Gibbs energies and entropies of such phases. This feature has permitted many workers worldwide to understand and apply this framework to many important problems. These investigators have been able to communicate their results to others who are active in this field. The ready availability of fast, powerful PC's and efficient software programs for performing such calculations contributed to this progress. Another important outcome of this effort has been the interface that has been established between the CALPHAD descriptions of stable, metastable and unstable phases and derived by *ab initio* methods. While good agreement has been attained between these methods in many instances some differences still exist. The discussion below will review the current applications of **CT** to several systems as well as CALPHAD and *ab initio* lattice stabilities.

2. RECENT APPLICAIONS OF CALPHAD TECHNOLOGY

As noted earlier, Kaufman[2], the principle driving force for the recent growth of "Computational Thermodynamics" (**CT**) has been the availability of powerful PC computers, databases and a language that is generally accepted within the community so that independent workers, worldwide, can correspond and contribute to progress in this field. Figures 1 and 2 are examples taken from a Niobium database currently under development, Kaufman[3], that can be used for a wide variety of practical and scientific investigations. The thermodynamic calculations were applied to multi-component alloys to calculate equilibrium and Scheil simulation of the freezing of alloys in order to illustrate how the method can be used in practice to anticipate segregation during the casting of niobium alloys. In addition, it was shown how the formation of a ternary miscibility gap in the Nb-Cr-Mo system could be predicted merely by using the data available for binary edge components. The database presented covers the Nb-Al-Cr-Ti system. However it is quite possible to expand the description to many other elements since many more Nb-based binary databases exist in the literature and in commercial databases.

Interest is growing in extension of solid solubility in alloys exhibiting miscibility gaps. Bamberger et al.[4] Systems such as Co-Cu, Cu-Fe, and Co-Cu-Fe are of special interest, because one or more of the elements is ferromagnetic and are known to exhibit "Giant Magneto Resistance" corresponding to a large drop in the electrical resistivity when a magnetic field is applied. Previous work has shown that supercooling of Cu-Fe, Co-Cu or Co-Cu-Fe alloys can result in metastable separation of the melt into two liquids. Because there is limited experimental data for the ternary system calculations were first carried out using the binary description as is illustrated in Figure 3. Under these conditions an isolated ternary gap is computed at 1450 °C. By contrast, experimental studies of this system using levitation melting techniques showed that a single-phase liquid is formed at 1450 °C in this system. Accordingly, small ternary interaction parameters were added to the liquid in order to suppress the miscibility gap as shown in Figure 3. Subsequent measurement of the melting points over a wide range of ternary compositions were made and found to compare favorably with the calculations as shown in Figure 3, when a similar correction was applied to the fcc phase.

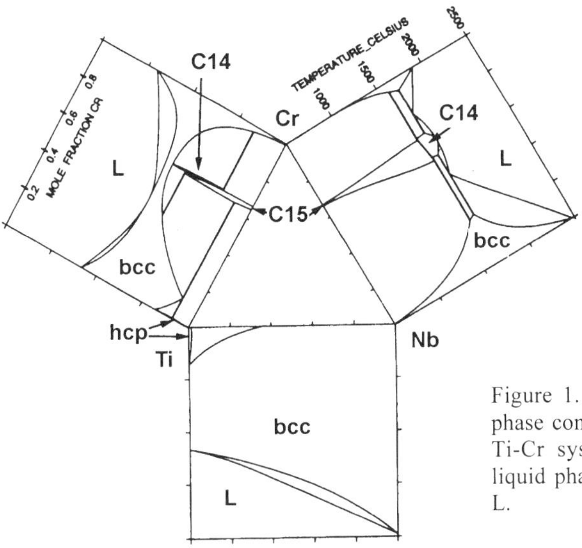

Figure 1. Calculated binary phase components of the Nb-Ti-Cr system (Ref. 3). The liquid phase is designated by L.

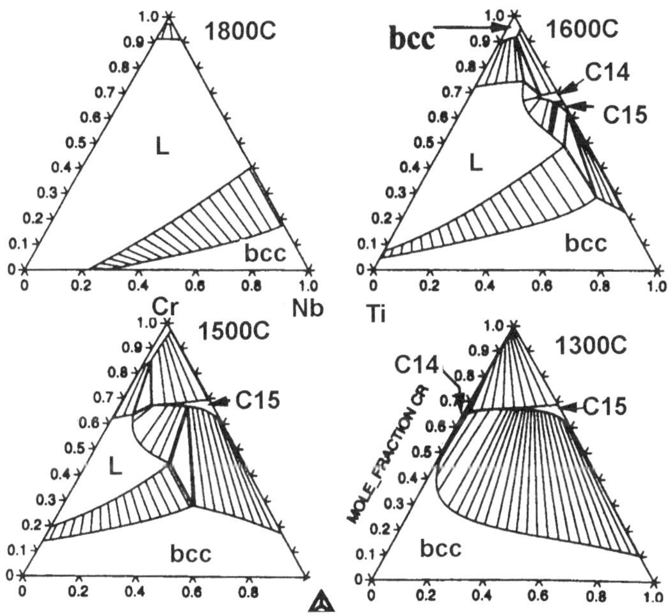

Figure 2. Calculated isothermal sections in the Nb-Ti-Cr (Ref. 3). Single-phase regions are labeled. Two-phase fields are identified by tie lines, and three-phase regions are referred to by triangles.

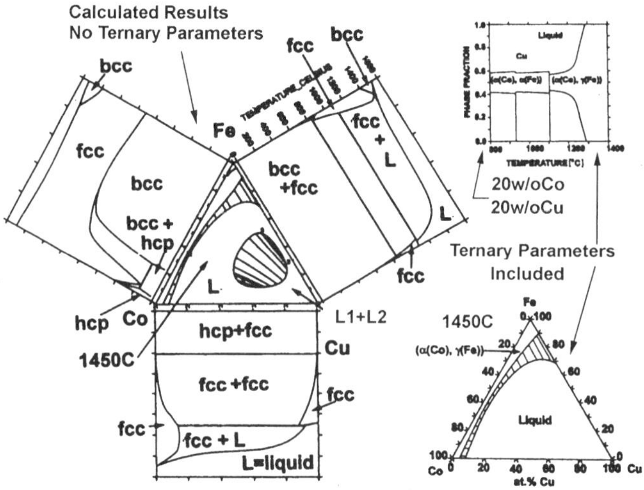

Alloy No	Composition in Wt.%			Melting Temperature(°C)	
	Cu	Co	Fe	Measured	Calculated
1	96	2	2	1000	1090
2	90	5	5	1143	1210
3	84	8	8	1230	1245
4	80	10	10	1240	1250
5	80	5	15	1300	1310
6	75	10	15	1280	1270
7	72	5	23	1315	1350
8	65	25	10	1290	1280
9	64	18	18	1280	1280
10	60	30	10	1280	1295
11	60	20	20	1300	1290
12	55	30	15	1330	1300
13	55	10	35	1365	1380
14	49	16	36	1340	1380
15	45	37	18	1325	1330
16	38	20	42	1300	1390

Figure 3. Calculated and experimental results for the Co-Cu-Fe system (Ref. 4). Small corrections were made to the liquid (L) and fcc phases. The calculated and measured melting points agree well as shown in the table.

The discovery of metallic glasses in 1988 by He et al.[5], based on Al-transition metal-rare earth metal metallic glasses with high specific strengths (ratio of fracture strength to density) has generated substantial scientific and technical interest. Accordingly a database, AlGl5r4 comprising Al-Fe-Gd-Ni-Y has been constructed and applied in elucidating the phase relations in Al-Gd-Fe and Al-Gd-Ni alloys experimentally, Hackenberg et al.[6], Ferro et al.[6], and Gao et al.[7] These results were compared with observations made on melt spun and devitrified ribbons in continuous-heating DSC scans and by isothermal annealing. X-Ray diffraction and TEM were used to verify the wholly amorphous state of the as-spun ribbons and to track the devitrification process. Figure 4 shows the calculated binary phase diagram components of the Al-Gd-Ni system computed from the AlGl5r4 database. This database was then used to calculate the Al-Gd-Fe system, Hackenberg et al.[6], and Al-Gd-Ni, Ferro et al.[6], Gao et al.[7], ternary systems. The latter shows three ternary compounds based on the compilation of Ferro et al.[6] Figure 5 shows ternary sections for the Al-Gd-Ni ternary system calculated from the AlGl5r4 database. The Al-Ni-Gd ternary has been identified as a good glass forming system extending over a wide range of composition, Gao et al.[7], ideal for experimental studies of primary crystallization. This study of the Al-rich corner was performed by conducting an extensive study of seventeen alloys at 250 °C which were compared with calculations of the driving force for nucleation of crystalline phases originating from the undercooled liquid with good results. The latter temperature was chosen because it is very close to the primary crystallization peaks for all the alloys, and still sufficiently removed in temperature from the second crystallization peaks. These studies illustrate how coupling **CT** and experimental studies of a complex problem can enhance the level of scientific understanding and increase the efficacy of both aspects of the investigation. The ternary sections in Figure 5 were calculated from the edge binary components in Figure 4 by first fixing the stability of the M_2R and $M_{17}R_2$ phases in accordance with the compilation due to Ferro et.al[6]. Once these phases were defined the remaining ternary phases that appear in Figure 5 (i.e., Al_4MR, $Al_7M_7R_2$, and $Al_{16}M_3R$) were constrained by the results presented in, Ferro et.al[6]. Calculated melting points for these phases were consistent with the observed results[7]. Moreover, the description of the edge binary compounds Al_7R_2 and M_2R that appear in the other ternary components of the Al-Fe-Gd-Ni-Y database Hackenberg et al.[6] are all characterized by thermodynamic parameters that are similar in magnitude.

The development of the AlGlr4 database was initiated about 12 months BEFORE the detailed experimental work was started with good descriptions of the Al-Fe and Al-Ni binaries. A number of iterations (i.e., AlGlr1, AlGlr2, and AlGlr3) were made before most of the actual experimental data had been evaluated. Thus the experimental study had the benefit of the detailed descriptions afforded by the AlGlr4 database as a guide. The general agreement between the predictions and the experiments was quite good. However it is likely that an even better database, i.e. AlGlr5 will result once all of the experimental results, Gao et al.[7], are fully analyzed.

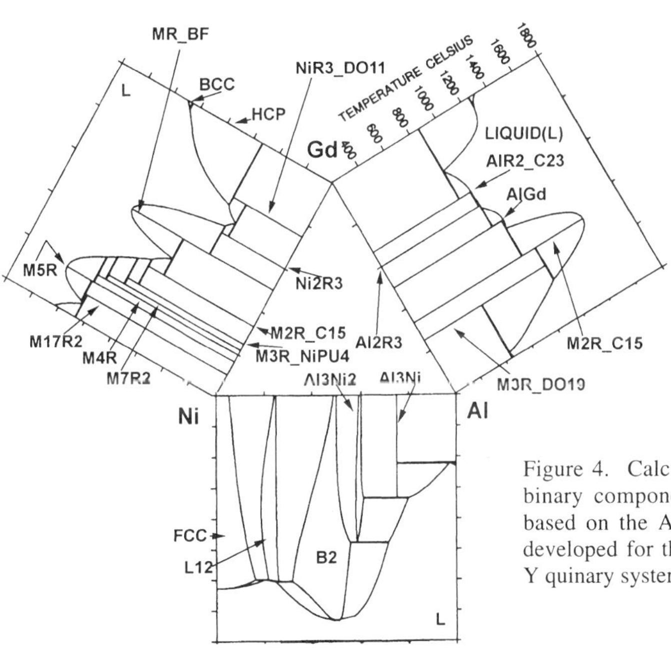

Figure 4. Calculated Al-Gd-Ni binary components (Refs. 5-7) based on the AlGl5r4 database developed for the Al-Fe-Gd-Ni-Y quinary system.

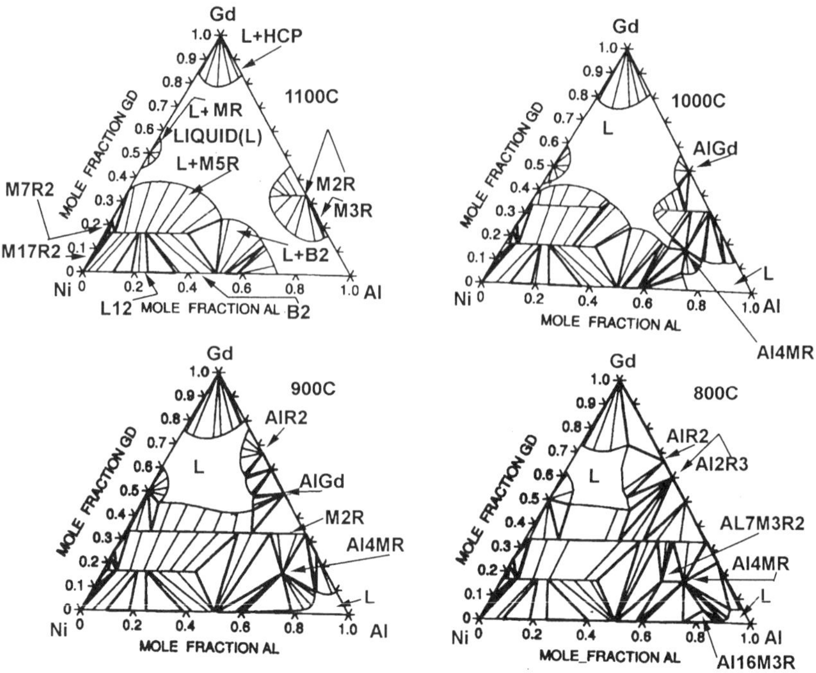

Figure 5. Calculated Isothermal Sections in the Al-Gd-Ni System based on the AlGl5r4 database (Refs. 5-7).

Lithium-ion batteries, preferred for their high energy and power, also present several challenges. Thackeray et al.[8] reviewed recent development and enumerated some of the problems that state-of-the-art batteries can pose in large-scale applications. Extensive scientific and practical studies are being conducted world wide to find alternative anode materials to replace the current materials that are a carbon anode and a $LiCoO_2$ cathode. Alloys or intermetallic compounds are attractive alternatives to graphite because they can be tailored to operate between 0 and 1 volt above the potential of metallic lithium. Many compounds in binary lithium systems (with Al, Si, Sn, Sb, and Pb) have already been investigated and rejected due to severe crystallographic changes during charging and discharging. Thackeray et al.[8], have suggested that intermetallic structures that could accommodate Li with a minimal change could mitigate this problem and have suggested a number of interesting examples including Cu_6Sn_5 and InSb. Investigation of the performance of such anodes is a complex problem requiring the solution of many processing and performance evaluation steps. Consequently a **CT** analysis of the potential multi-component candidate systems could be quite valuable as a screening tool and/or companion to experimental synthesis of such materials. To illustrate this procedure a database for the Li-In-Sb system was constructed and employed, Kaufman[9], to compute the binaries and room temperature isotherm in Figure 6. Two ternary compounds, Li_6InSb_3 and Li_3InSb_2, Thackeray et al.[8], were included. The results are shown in Figures 6-8. Figure 6 shows the isopleth trace from Li to InSb while Figure 8 identifies the phases that are present along the trace starting at Li and heading toward InSb. The voltage difference between Li and any point on the trace is calculated by applying Faraday's Law to the chemical potential of Li that is computed at that point. Figure 8 shows a series of steps starting near Li, where the difference in voltage is zero, as the trace in Figure 6 crosses the three-phase Li+$InLi_7$+ Li_3Sb field. In this phase field the chemical potential of Li is constant. Each time a phase boundary is crossed and the identity of the coexisting phases change the chemical potential of Li changes. This results in the steps shown in Figure 8 as the Li concentration drops along the isopleth. The calculations suggest that this behavior persists until the Li concentration crosses into the ternary field where the ternary intermetallic phases form. From that point the voltage difference remains constant near 0.9 volts. Figure 7 shows a Phase Fraction vs. Temperature curve for an alloy with 0.32 mole fractions of Sb and 0.55 Li (balance In). This corresponds to a composition between the ternary compounds. Figure 7 suggests that at low temperatures a mixture of the compounds is stable but that above 510 °C Li_3InSb_2 decomposes to form Liquid, Li_2Sb, and Li_6InSb_3. Once the temperature exceeds 520 °C the latter compound forms Liquid and Li_3Sb. These **CT** calculations show how this approach can be applied to the synthesis of anode materials for Li-ion batteries. Yet another area within this field is application of this method to analysis and design of cathode compositions. According to Thackeray et al.[8], the preferred cathode material is based on $LiCoO_2$. A fully lithiated graphite electrode, LiC_6, provides a high specific capacity and a very high volumetric capacity based on the low density of LiC_6. The $Li_xC_6/Li_{1-x}CoO_2$ cells operate at 4.2-3.5 V.A **CT** description of the "$LiCoO_2$" phase in the Li-C-.5O system should be performed in order to explore methods for increasing the stability of this phase and reducing the propensity for catastrophic battery failures. For example, Mn oxides are much more stable than their Co and Ni counterparts and should be investigated as a safety enhancement.

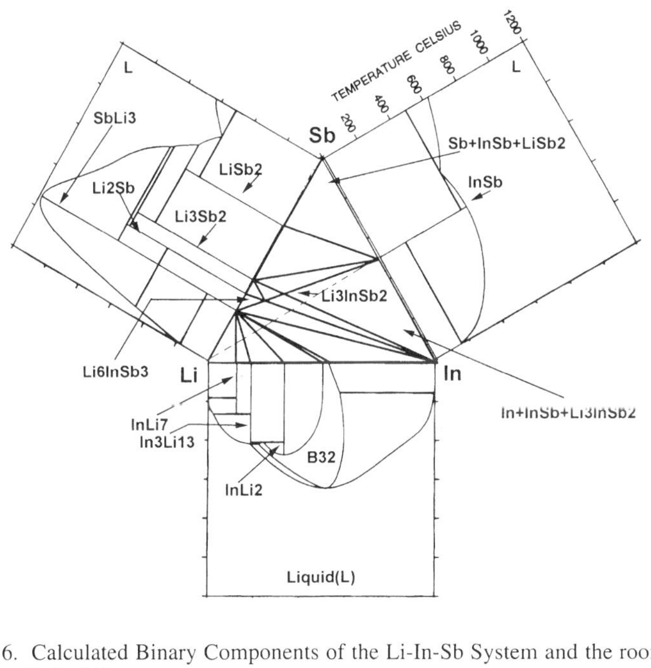

Figure 6. Calculated Binary Components of the Li-In-Sb System and the room-temperature isotherm (Refs. 8,9).

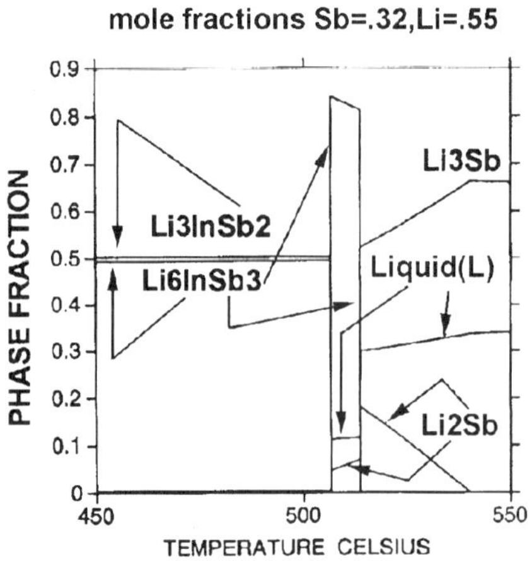

Figure 7. Calculated phase fraction versus temperature curves for Li-In-Sb that shows the decomposition of the ternary compounds shown just above 500 °C (Ref. 9).

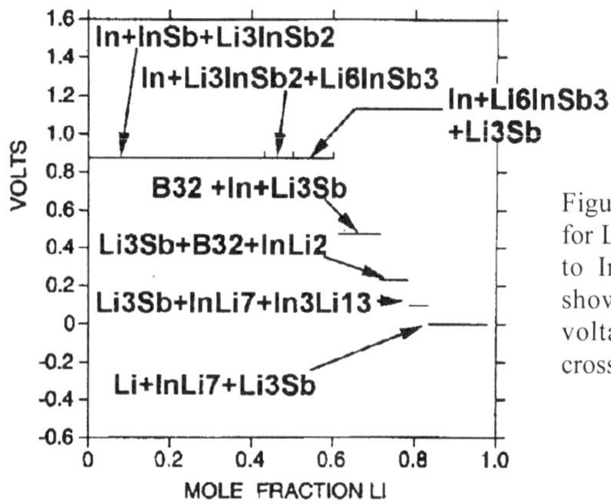

Figure 8. Calculated voltage drop for Lithium along the trace from Li to InSb at the room temperature shown in Fig. 6. Each of the voltage steps corresponds to a crossing of a phase boundary.

The description of a three-sublatice oxide phase that could be applied to treat a Li $(Co,Mn,Ni)O_2$ phase could result in the kind of behavior previously illustrated for the Hf-N and W-Hf-N sytems, Kaufman[1], which is shown in Figure 9. In the binary phase diagrams displayed as a function of pressure between 0.01 and 80 atmosphere the melting point increases by 1000 °C. This extremely large increase is the result of the fact that the stability of the HfN phase (with respect to the gas) is strongly dependent on composition. Thus, small changes in the composition of the nitride can have profound effects on its melting point as a function of pressure. This behavior carries over into the ternary in Figure 9 with substantial changes in the configuration of the condensed and gaseous phases. The analysis of the ZrO_2-.5Y_2O_3-.5Yb_2O_3 system[1] displayed in Figure 10 has been extended to include 5Nd_2O_3 and .5Sc_2O_3, Jacobson et al.[10] Figure 11 shows isothermal sections and phase fraction versus temperature curves for compositions of commercial interest in the ZrO_2-.5Y_2O_3-.5Sc_2O_3 system. The essential feature of this oxide database is the treatment of the ZrO2-.5RE_2O_3 cubic solid solution as one where a miscibility gap separates the ZrO2-rich and the .5Re_2O_3-rich phases. This in turn is considered to be .5(Zr_2O_4-Y_2O_3) where there is a continuous replacement of oxygen atoms by vacancies as the yttrium concentration increases.

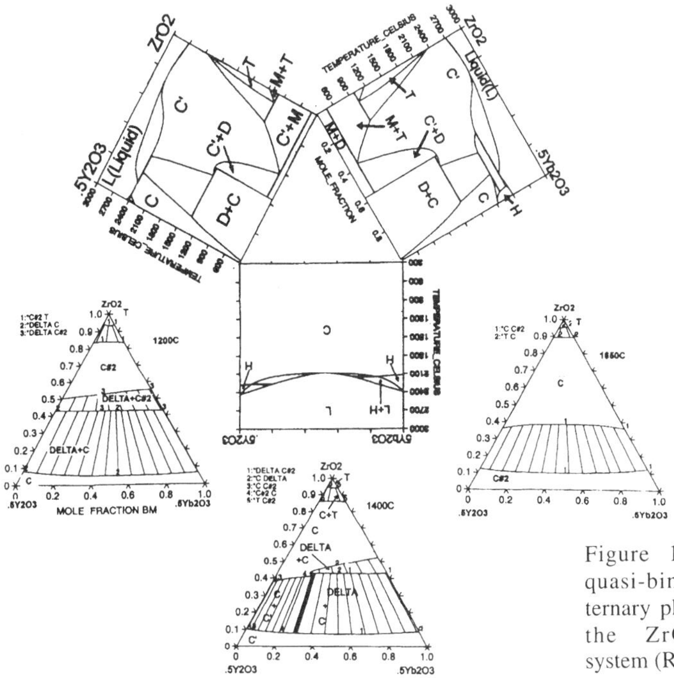

Figure 10. Calculated quasi-binary and quasi-ternary phase diagrams for the ZrO_2-$.5Y_2O_3$-$5Yb_2O_3$ system (Refs. 1,9).

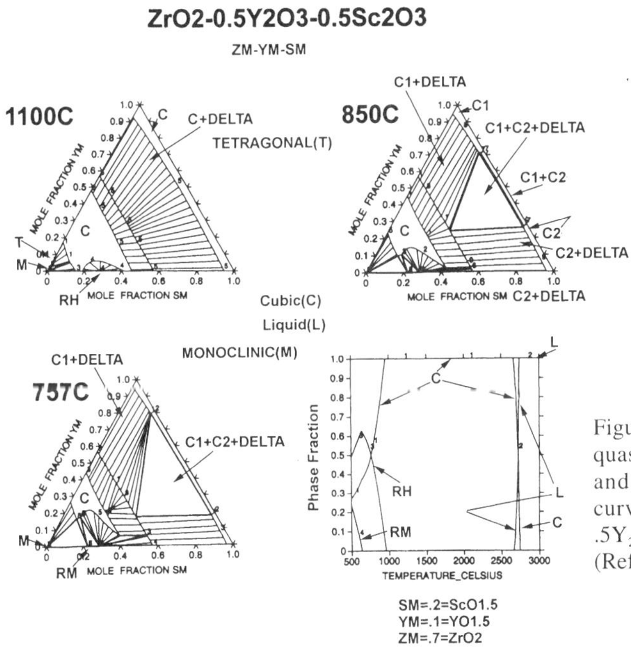

Figure 11. Calculated quasi-ternary sections and phase fraction curves for the ZrO_2-$.5Y_2O_3$-$.5Sc_2O_3$ system (Ref. 10).

This definition constrains the two cubic phases that are treated as end members of a miscibility gap. The other feature of this model that is required to perform the calculations is the estimation of the "lattice stability parameters" for all of the crystal structures of the end members in direct analogy to the values for the lattice stabilities established for the metallic structures. Recently a detailed analysis of the Zr-O system was performed, Arroyave et al.[11], for the Zr-O system as shown in Figures 12 and 13 as a direct analogue to the Hf-N case shown in Figure 9. It is anticipated that this detailed formulation of the Zr-O system[11] can be applied to other metal-oxygen systems and used to reconsider the ZrO_2-.5RE_2O_3 systems as Zr-Re-O affording a greater level of accuracy and utility.

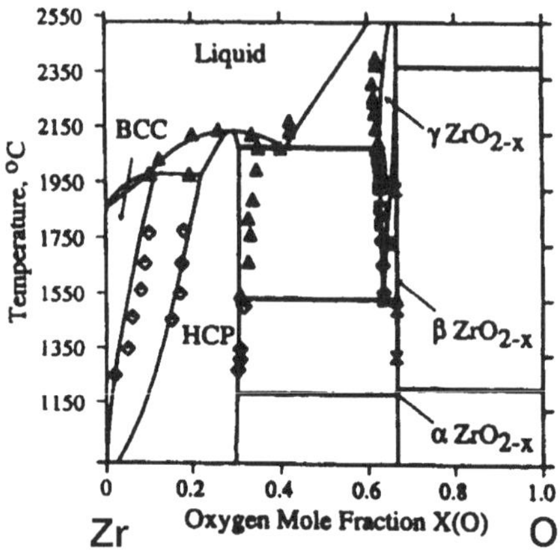

Figure 12. Experimental and calculated Zr-O phase diagram (Ref. 11).

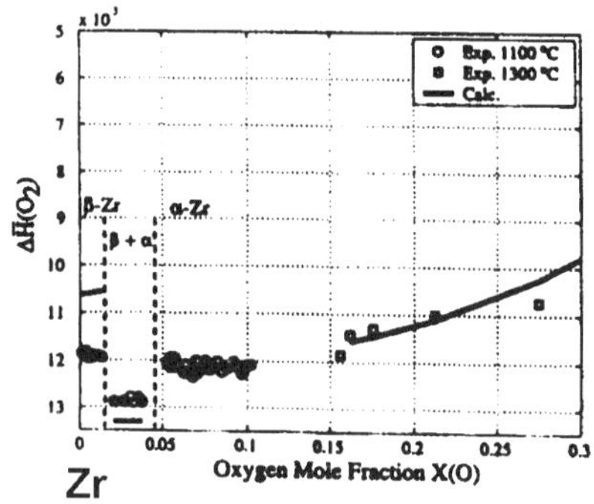

Figure 13. Experimental data and calculated partial molar enthalpies of mixing for oxygen in the bcc, hcp, and bcc+hcp phase fields for the Zr-O System (Ref. 11).

The final example for discussion is shown in Figure 14 that repeats the earlier results of calculations for the W-Re-C system at 2200 °C and 2300 °C as compared with experimental results at 2000 °C, Kaufman[1]. The important feature of these calculations was the unusual finding that the close packed hexagonal phase runs continuously from the well known W_2C phase to the stable close packed hexagonal Re. The W_2C phase is described by a two-sublattice model for the hcp lattice of W atoms with an interstitial lattice containing C atoms and vacancies. Extensive experimental studies have shown that as Re is added, the Re atoms substitute for the W atoms in a continuous fashion until pure Re is reached. The c/a ratio as well as the c and a parameters observed by X-rays is a continuous function of composition. Recent experimental studies of the Mo-Re-C system, Erymenko et al.[12], Velikanova et al.[13], shows the same behavior as the hcp Mo_2C phase merges continuously into Re as shown in the lower right panel of Figure 14. What is more surprising is that a fcc MoC_{1-x} exhibits nearly the same behavior as it extends nearly 75% of the way across the diagram toward Re.

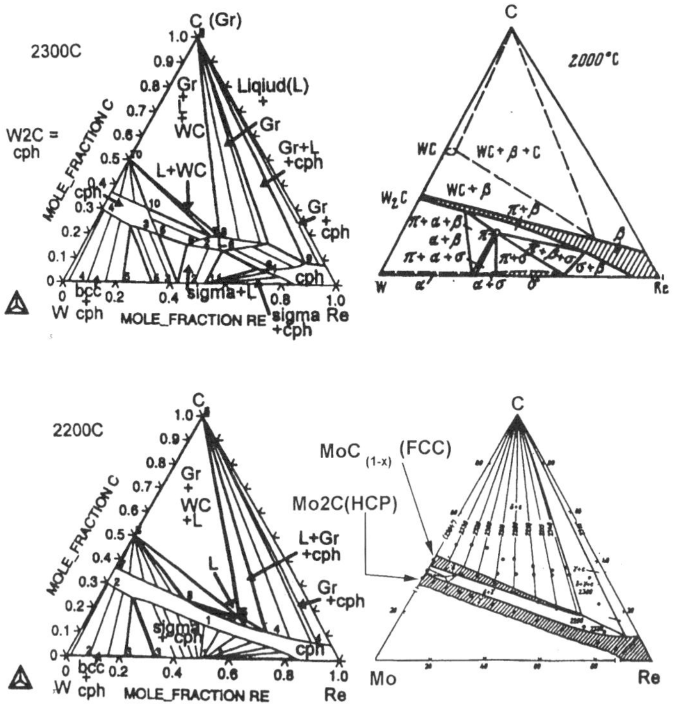

Figure 14. Calculated and experimental phase diagrams for the C-Re-W (Ref. 1) and C-Re-Mo (Refs. 12,13) systems.

This can only mean that the difference in stability between hcp and fcc is small. Much smaller than is suggested by *ab initio* calculations. At the outset of this discussion attention was drawn to the fact that the **CT** framework required the definition of the stability of unstable and metastable phases in addition to stable phases. Clearly the definition of the stability of the former phases is difficult if not impossible if conventional methods are used exclusively. It is for this reason that the interface between **CT** and *ab initio* calculations is so important. In a recent review of this subject, Grimvall[14] noted that substantial differences exist between the **CT** and *ab initio* results for a number of cases (the *ab initio* results are 3-6 times larger) and that in many cases the higher energy phase is mechanically unstable. Hence there is no Gibbs energy, $G = H - TS$, for the unstable phase. Grimvall[14] surveys some of the instabilities as predicted in *ab initio* calculations, and an example of how the Gibbs energy varies when an instability is approached through changing alloy composition. Surprisingly, Grimvall[14] argues that the *ab initio* calculations and CALPHAD approaches can be reconciled, with no essential changes in the CALPHAD method. Recently, Chen and Sundman[15], modeled the thermodynamic properties of bcc, fcc, liquid, and amorphous phases using all the data available.

The results for the Gibbs energy difference between fcc and bcc iron between 0 and 1800 K agree closely with those of Kaufman et al.[16,17]. The energy difference at absolute zero is 5800J/mol versus 5440 in the earlier work. By comparison there are three recent *ab initio* calculations for Fe performed using the FLAPW method. The results for the energy differences at absolute zero are 9754 J/mol by Herper et al.[18], 19560 J/mol by Cho et al.[19], and 18250 J/mol by Singh et al.[20]. The most recent calculations suggest that differences between the earlier studies[19,20] and Herper et al.[18] is the treatment of the fcc phase within the "generalized gradient approximation (GGA) applied to the full-potential linearized augmented plane wave (FLAPW) method of density functional theory". A close reading of Herper et al.[18] shows how the stability differences can be substantially changed depending on the variety of methods employed to perform the calculation. Indeed the fourfold differences between Cho et al.[19], Singh et al.[20] and the Kaufman et al.[16,17] thermo-chemical values derived in the 1960's are reduced to a twofold difference by Herper et al.[18] It is of further interest to note that the "Two-Gamma States" detailed in the early thermo-chemical analyses Kaufman et al.[16,17], describing metastable fcc iron over a wide range of temperature (0-1200 K, 0-130 kbars) has been included in the recent calculations, i.e., as a low-moment, low volume fcc (AFM-1) Herper et al.[18], anti-ferromagnetic fcc iron and a high moment, high volume (FM/HS)[18] fcc iron. Moreover the volumes of these structures are remarkably close. Thus at absolute zero the volume in cm^3/gat or cm^3 per mol of atoms for bcc Fe is 7.061, while the low and high spin (volume) fcc forms are 6.695 and 7.216, respectively, Kaufman et al.[16,17]. The corresponding volumes calculated by Herper et al.[18] are 6.892 for the bcc phase and 6.441 and 7.244 for the low and high spin/volume forms of the fcc phase. The experimental result for bcc iron at absolute zero and one atmosphere is 7.046 according to Herper et al.[18] Nevertheless the *ab initio* calculations are still about twice the CALPHAD results. It is essential that the **CT**/*ab initio* interface be clarified. From the CALPHAD side this has already been initiated via an awareness and a willingness to incorporate *ab initio* results into CALPHAD databases where possible. From the *ab initio* side it is necessary to make comparisons of the calculated bcc/fcc (and bcc/hcp) equilibria in those systems where the greatest discrepancies exist. The Cr-Ni case is detailed in Figure 10 of Kaufman[2]. The latter discloses that a six-fold discrepancy (-3kJ/mol versus -17kJ/mol) exists between the calculated TB-CPA iso-strucural *ab initio* enthalpy of the fcc phase at $Cr_{.67}Ni_{.33}$ and the value estimated by requiring that the lattice stability difference beteen bcc and fcc Cr is 37 kJ/mol at absolute zero. The latter value (37kJ/mol) is the *ab initio* value. The CALPHAD value is 7 kJ/mol! It is clear that the TB-CPA calculations performed for the lattice stabilities and iso-structural energies nearly 20 years ago for the fcc (and bcc) phase in this system can certainly be improved upon. They should be! The point is that the proponents of *ab initio* calculations of these lattice stabilities and heats of formation of phases who insist that the entropy of metastable/unstable cannot be defined and that the Gibbs energy is not defined should examine their current calculations of lattice stabilities and iso-structural enthalpies at absolute zero to see if they are consistent with the binary phase diagrams. This should be done for the binary systems that are formed when elements in the (Nb/Mo) group are alloyed with elements in the (Ru/Rh/Pd) group.

REFERENCES

1. L. Kaufman, "Computational Thermodynamics and Materials Design", CALPHAD **25**, 141 (2001)

2. L. Kaufman, "Hume-Rothery and CALPHAD Thermodynamics", <u>CALPHAD and Alloy</u>

Thermodynamics, ed. by P. E .A. Turchi, A. Gonis, and R. D. Shull (TMS, Warrendale PA, 2002) 3-19.
3. L. Kaufman, "Calculated Multicomponent Phase Diagrams for Niobium Alloys", Proceedings of the International Conference "Niobium 2001", ed. by D. Howden, to be published.
4. M. Bamberger, A. Munitz, L. Kaufman, and R. Abbaschian, submitted for publication to CALPHAD (2002)
5. Y. He, S. J.Poon, and G. Shiflet, Science **241**, 1640-42 (1988).
6. R. Hackenberg, M. C. Gao, L. Kaufman, and G. J. Shiflet, Acta Materialia **50**, 2245-58 (2002), and R. Ferro, G. Zanicchi, and R. Marazza, in "Al-Gd-Ni", Ternary Alloys, ed. by G. Petzow and G. Effenberg , 5, 677-81 (1993).
7. M. C. Gao and G .J. Shiflet, CALPHAD and Alloy Thermodynamics, ed. by P. E. A.Turchi, A. Gonis and R. D. Shull (TMS, Warrendale PA, 2002) 215-23.
8. M. M. Thackery,J. T. Vaughey, and L. M. L. Frannson, "Recent Developments in Anode Materials for Lithium Batteries", J. Metals, March ,20 (2002).
9. L. Kaufman, "Research Conducted under U.S.ATP Cooperative Agreement 70ANBOH3023" (2001) Polystore Corp. Livermore, CA.
10. N. S. Jacobson, E. H. Copeland, and L. Kaufman, "Thermodynamic Database for the NdO1.5-YO1.5-YbO1.5-ScO1.5-ZrO1.5 System", NASA/TM-2001-21073, September 2001.
11. R. Arroyave, L. Kaufman, and T. Eagar, CALPHAD **26**, 95 (2002).
12. N. Erymenko, T. Ya. Velikanova, and A. M. Kharkova, Kiev. IPM An USSR, 3-28 (1981).
13. T. Ya. Velikanova, A. A. Bondar, A. Grystiv, and O. L. Dovbenko, J. of Alloys and Compounds **320**, 341-52 (2001).
14. G. Grimvall, "CALPHAD and *Ab-Initio* Approaches to Lattice Stabilities" CALPHAD and Alloy Thermodynamics, ed. by P. E. A. Turchi, A. Gonis ,and R. D. Shull (TMS, Warrendale, PA, 2002) 81.
15. Q.Chen and B. Sundman, J. Phase Equilibria **22**, 63 (2001).
16. L. Kaufman, E. V. Clougherty, and R. J. Weiss, Acta Met. **82**, 281 (1963).
17. L. Kaufman and H.Bernstein, Computer Calculation of Phase Diagrams (Academic Press New York, N.Y., 1970) 18.
18. H. C. Herper, E. Hoffman, and P. Entel, Phys. Rev. B **60**, 3839 (1999).
19. J.-H. Cho and M. Scheffler, Phys. Rev. B **53**, 10685 (1996).
20. D. J. Singh, W. E. Pickett, and H. Krakauer, Phys. Rev. B **43**, 11628 (1991).

PHASE STABILITY AND ORDERING IN (Ga,Mn)As ALLOYS

Václav Drchal[1,2], Josef Kudrnovský[1,2], František Máca[1], Jan Mašek[1],

Ilja Turek[2,3,4], and Peter Weinberger[2]

[1]Institute of Physics, Academy of Sciences of the Czech Republic
Na Slovance 2, CZ-182 21 Praha 8, Czech Republic
[2]Center for Computational Materials Science, Technical University
of Vienna, Getreidemarkt 9/134, A-1060 Vienna, Austria
[3]Institute of Physics of Materials, Academy of Sciences of the Czech
Republic, Žižkova 22, CZ-616 62 Brno, Czech Republic
[4]Department of Electronic Structures, Charles University
Ke Karlovu 5, CZ-121 16 Praha 2, Czech Republic

INTRODUCTION

Diluted magnetic III-V semiconductors (DMS) such as $Ga_{1-x}Mn_xAs$ alloys represent a new class of promising materials with potential applications in spin electronics[1]. The samples are usually prepared by molecular beam epitaxy (MBE) on GaAs substrates at temperatures ranging from 200 °C to 300 °C. Higher temperatures or a Mn content higher than 7 at.% can lead to precipitation of MnAs. According to the measurements[2,3] of Hall resistivities in strong magnetic fields the conductivity is of p-type. Extended X-ray absorption fine structure (EXAFS) studies have shown that Mn is substitutionally incorporated into the Ga sublattice[4] and acts as acceptor. These materials are highly compensated, i.e., the experimentally observed number of holes in the valence band is considerably smaller than the concentration of Mn impurities. This indicates the presence of other lattice defects acting as donors.

The most probable candidate for such a compensation are As antisites, but also other possibilities (e.g., Mn interstitials[5,6]) are conceivable. The highest Curie temperature reported[2,7] for $Ga_{1-x}Mn_xAs$ alloys is 110 K, while for $Ga_{1-x}Mn_xN$ alloys it is 348 K[8].

It is often assumed that the defects are randomly distributed, however, experiments[9] suggest that impurities can diffuse rapidly at 250 °C, which is a typical growth and annealing temperature for $Ga_{1-x}Mn_xAs$ samples prepared by MBA. In such a case, the spatial distribution of impurities is not random any more, but it becomes (at least partially) correlated[10]. These correlations can substantially modify the transport and magnetic properties of the DMSs.

Here we investigate the spatial distribution of impurities in the DMSs from first principles. Based on *ab initio* electronic structure calculations, we determine the total energies of disordered alloys as a function of their chemical composition. In a next step, by using the generalized perturbation method (GPM), we evaluate effective interatomic interactions between impurities by taking into account also the electrostatic interactions. Finally, by employing the methods of statistical physics, namely, by using the linearized version of the concentration wave method we determine the ordering temperature $T_{\rm ord}$ and the k-vector of the transition from the disordered to the ordered state. For temperatures above the $T_{\rm ord}$ we finally calculate the Warren-Cowley short-range order parameters which yield information about the spatial correlations of impurities. As an illustration, we present the results for $Ga_{1-x}Mn_xAs$ alloys.

METHODOLOGY

Electronic structure

The electronic structure is determined using the *ab initio* all-electron scalar-relativistic tight-binding linear muffin-tin orbital (TB-LMTO) method in the atomic-sphere approximation (ASA)[11]. The underlying lattice, zincblende structure, refers to an fcc Bravais lattice with a basis which contains a cation site (at $a(0,0,0)$), an anion site (at $a(\frac{1}{4},\frac{1}{4},\frac{1}{4})$), and two interstitial sites occupied by empty spheres (at $a(\frac{1}{2},\frac{1}{2},\frac{1}{2})$ and $a(\frac{3}{4},\frac{3}{4},\frac{3}{4})$) which in turn are necessary for a correct description of open lattices[12,13].

The anion sublattice is occupied only by As atoms, while the cation sublattice is occupied by Ga and Mn atoms, and also by As antisite defects. We consider only substitutional disorder on the cation sublattice which in turn is described within the coherent potential approximation (CPA)[13,11]. We thus neglect local environment effects and lattice relaxations.

In addition to substitutional randomness, the DMSs are characterized also by some degree of magnetic disorder[14,15] which we treat in terms of the disordered local moment (DLM) method[11,16] that can be naturally incorporated in the CPA: the Mn atoms have collinear, but random spin-up ($Mn^{\uparrow}$) and spin-down ($Mn^{\downarrow}$) orientations. It is worthwhile to mention that for total energy calculations it is not necessary to consider all possible $2S+1$ projections of the atomic spin: it is sufficient to carry out the calculations for only two extremal values, namely $\pm S$[16]. The corresponding concentrations $x^{\uparrow}$ and $x^{\downarrow}$ fulfill the condition $x = x^{\uparrow} + x^{\downarrow}$. The degree of magnetic disorder is then characterized by the parameter $r = x^{\downarrow}/x$, $0 \leq r \leq \frac{1}{2}$. For example, a GaAs mixed crystal with Mn- and As-atoms on the Ga-sublattice, with respective concentrations x, y, is treated as a multicomponent alloy $(Ga_{1-x-y}Mn^{\uparrow}_{(1-r)x}Mn^{\downarrow}_{rx}As_y)As$. In the saturated ferromagnetic (FM) state, $r = 0$, i.e., all magnetic moments are pointing in the direction of a global magnetization. The paramagnetic (PM) state, $r = \frac{1}{2}$, is characterized by complete disorder of spin-directions. Besides the FM and PM states[17] a partial ferromagnetic (pFM) state, $0 < r < \frac{1}{2}$, is possible in which spin orientations are partially disordered.

Ising Hamiltonian

A particular configuration of a homogeneous disordered multicomponent alloy can be characterized by a set of occupation indices $\{\eta_{\mathbf{R}}^{Q}\}$, where $\eta_{\mathbf{R}}^{Q}=1$ if the site $\mathbf{R}$ is occupied by an atom of type Q, and $\eta_{\mathbf{R}}^{Q}=0$ otherwise. Configurational averaging of occupation indices $\langle\eta_{\mathbf{R}}^{Q}\rangle = c^{Q}$ yields the concentrations c^{Q}. The energy of such configuration can be expressed in the form of an effective Hamiltonian of Ising type,

$$H = E_0 + \sum_{\mathbf{R}Q} D_{\mathbf{R}}^{Q}\,\eta_{\mathbf{R}}^{Q} + \frac{1}{2}\sum_{\mathbf{RR}'}\sum_{QQ'} V_{\mathbf{RR}'}^{QQ'}\,\eta_{\mathbf{R}}^{Q}\,\eta_{\mathbf{R}'}^{Q'} + \dots\,, \tag{1}$$

whose parameters are the configurationally independent part of the alloy internal energy E_0, the on-site energies $D_{\mathbf{R}}^{Q}$, the interatomic pair interactions $V_{\mathbf{RR}'}^{QQ'}$, and generally, interatomic interactions of higher order. In homogeneous alloys, the first two terms in (1) do not depend on the alloy configuration and will therefore be omitted in the following. For matters of simplicity we limit ourselves to pair interactions only, although in principle triplets, quadruplets, etc., can be included.

The pair interactions $V_{\mathbf{RR}'}^{QQ'}$ in DMSs consist of two contributions,

$$V_{\mathbf{RR}'}^{QQ'} = v_{\mathbf{RR}'}^{QQ'} + \phi_{\mathbf{RR}'}^{QQ'}\,, \tag{2}$$

where the $v_{\mathbf{RR}'}^{QQ'}$ result from a mapping of the band part of the total energy onto the Ising Hamiltonian (1) and the $\phi_{\mathbf{RR}'}^{QQ'}$ refer to the electrostatic interaction energy of a given pair of atoms Q, Q' located at sites $\mathbf{R}, \mathbf{R}'$

$$\phi_{\mathbf{RR}'}^{QQ'} = \frac{e^2 q_{\mathrm{eff}}^{Q} q_{\mathrm{eff}}^{Q'}}{|\mathbf{R}-\mathbf{R}'|}\,, \tag{3}$$

where $q_{\mathrm{eff}}^{Q} = q^{Q} - \langle q\rangle$ is the effective net charge of atomic species Q defined as a difference of the net charge q^{Q} of atomic species Q and the averaged charge $\langle q\rangle$. The band term contribution is calculated using the Generalized Perturbation Method[11,18]

$$v_{\mathbf{RR}'}^{QQ'} = \frac{1}{\pi}\mathrm{Im}\int_{E_{\mathrm{min}}}^{E_{\mathrm{F}}} \mathrm{d}E\,\mathrm{tr}\left[t_{\mathbf{R}}^{Q}(z)\bar{g}_{\mathbf{RR}'}(z)t_{\mathbf{R}'}^{Q'}(z)\bar{g}_{\mathbf{R}'\mathbf{R}}(z)\right], \tag{4}$$

where $z = E + i0$, E_{F} is the CPA Fermi energy, E_{min} is a suitably chosen energy below the valence energy spectrum, $\bar{g}_{\mathbf{RR}'}(z)$ corresponds to the block of the averaged auxiliary Green function between sites $\mathbf{R}$ and $\mathbf{R}'$, and $t_{\mathbf{R}}^{Q}(z)$ is the t-matrix for atomic species Q.

It is advantageous to eliminate one of the atomic species, say Ga, (further denoted with a superscript 0) using the transformation $\eta_{\mathbf{R}}^{0} = 1 - \sum_{Q}' \eta_{\mathbf{R}}^{Q}$ which converts Eq. (1) into the form

$$H = E_0' + \frac{1}{2}\sum_{\mathbf{RR}'}\sum_{QQ'}{}' \tilde{V}_{\mathbf{RR}'}^{QQ'}\,\eta_{\mathbf{R}}^{Q}\,\eta_{\mathbf{R}'}^{Q'}\,, \tag{5}$$

where the primed sum denotes summation over $Q \neq 0$, $Q' \neq 0$, and

$$\tilde{V}_{\mathbf{RR}'}^{QQ'} = V_{\mathbf{RR}'}^{QQ'} + V_{\mathbf{RR}'}^{00} - V_{\mathbf{RR}'}^{Q0} - V_{\mathbf{RR}'}^{0Q'}\,. \tag{6}$$

Method of linearized concentration waves

The ordering temperature T_{ord} and the type of the ordered structure that appears below T_{ord} can be studied in terms of the concentration-wave method[19,18] Here we employ its linearized version[20] extended to a quaternary alloy since we consider possible ordering of four atomic species (Ga, $\text{Mn}^{\uparrow}$, $\text{Mn}^{\downarrow}$, As) on the fcc cation sublattice. The free energy in a mean field approximation (i.e., assuming a Bragg-Williams form of the entropy) is expressed in terms of local concentrations $c_{\mathbf{R}}^{Q}$

$$F = F_0 + \frac{1}{2} \sum_{\mathbf{RR}'} \sum_{QQ'} V_{\mathbf{RR}'}^{QQ'} c_{\mathbf{R}}^{Q} c_{\mathbf{R}'}^{Q'} + k_{\text{B}} T \sum_{\mathbf{R}} \sum_{Q} c_{\mathbf{R}}^{Q} \ln(c_{\mathbf{R}}^{Q}) , \tag{7}$$

where k_{B} is the Boltzmann constant, T is temperature, and F_0 is the free energy in the absence of concentration waves ($c_{\mathbf{R}}^{Q} = c^{Q}$ for all $\mathbf{R}$). Starting from the disordered state, the free energy can be expanded up to quadratic terms in concentration fluctuations $\delta c_{\mathbf{R}}^{Q} = c_{\mathbf{R}}^{Q} - c^{Q}$,

$$F = F_0 + \frac{1}{2} \sum_{\mathbf{RR}'} \sum_{QQ'} \Big[V_{\mathbf{RR}'}^{QQ'} + \frac{k_{\text{B}} T}{c^{Q}} \delta_{\mathbf{RR}'} \delta_{QQ'} \Big] \, \delta c_{\mathbf{R}}^{Q} \, \delta c_{\mathbf{R}'}^{Q'} \tag{8}$$

with the terms linear in $\delta c_{\mathbf{R}}^{Q}$ vanishing because $\sum_{\mathbf{R}} \delta c_{\mathbf{R}}^{Q} = 0$ for all Q and $\sum_{\mathbf{R}'Q'} V_{\mathbf{RR}'}^{QQ'} c^{Q'}$ is a constant for all $\mathbf{R}$ and Q. The equation (8) can be rewritten in terms of a lattice Fourier transform as

$$F = F_0 + \frac{1}{2} \sum_{\mathbf{k}}^{\text{BZ}} \sum_{QQ'} \Big[V^{QQ'}(\mathbf{k}) + \frac{k_{\text{B}} T}{c^{Q}} \delta_{QQ'} \Big] [\delta c^{Q}(\mathbf{k})]^{*} \, \delta c^{Q'}(\mathbf{k}) , \tag{9}$$

or, in a matrix notation, as

$$\Delta F = F - F_0 = \frac{1}{2} \sum_{\mathbf{k}}^{\text{BZ}} Y^{\dagger}(\mathbf{k}) \big[V(\mathbf{k}) + k_{\text{B}} T C^{-1} \big] Y(\mathbf{k}) |\epsilon(\mathbf{k})|^{2} , \tag{10}$$

where $[V(\mathbf{k})]_{QQ'} = V^{QQ'}(\mathbf{k})$, and the matrix C is defined as $[C]_{QQ'} = c^{Q} \delta_{QQ'}$. In Eq. (10) the concentration fluctuations $\delta c^{Q}(\mathbf{k})$ are expressed in terms of a vector Y and the order parameter $\epsilon(\mathbf{k})$ as $\delta c^{Q}(\mathbf{k}) = Y^{Q}(\mathbf{k}) \epsilon(\mathbf{k})$. At sufficiently high temperatures ΔF is positive definite, because the hermitian matrix $V(\mathbf{k}) + k_{\text{B}} T C^{-1}$ has only positive eigenvalues and thus the high temperature state is completely disordered ($\epsilon(\mathbf{k}) = 0$ for all $\mathbf{k}$). With decreasing temperature it can become indefinite at T_{ord} because of a vanishing eigenvalue for a critical vector $\mathbf{k}_0$ which then determines the period of the concentration wave. The components $Y^{Q}(\mathbf{k})$ of such a critical eigenvector determine the amplitude of the concentration wave for each alloy component Q. We remark that the minimization of ΔF, and thus the eigenvalue problem, are, for each $\mathbf{k}$, subject to a subsidiary condition $\sum_{Q} Y^{Q}(\mathbf{k}) = 0$.

Short range order parameters

The Warren-Cowley short-range order parameters

$$\alpha_{\mathbf{RR}'}^{QQ'} = - \frac{\langle \eta_{\mathbf{R}}^{Q} \eta_{\mathbf{R}'}^{Q'} \rangle - \langle \eta_{\mathbf{R}}^{Q} \rangle \langle \eta_{\mathbf{R}'}^{Q'} \rangle}{\langle \eta_{\mathbf{R}}^{Q} \rangle \langle \eta_{\mathbf{R}'}^{Q'} \rangle} = 1 - \frac{\langle \eta_{\mathbf{R}}^{Q} \eta_{\mathbf{R}'}^{Q'} \rangle}{c^{Q} c^{Q'}} \tag{11}$$

can be used to visualize the ordering tendencies of impurities in the DMSs. The matrix of the Warren-Cowley parameters can be approximately calculated by means of the Krivoglaz-Clapp-Moss (KCM) formula in the reciprocal space

$$\alpha(\mathbf{k}) = -D[M + \beta\widetilde{V}(\mathbf{k})]^{-1}D^T\,, \tag{12}$$

where $\beta = (k_\mathrm{B}T)^{-1}$, the matrix M is defined as

$$[M]_{QQ'} = \frac{1}{c^0} + \delta_{Q,Q'}\frac{1}{c^Q}\,, \tag{13}$$

and the matrix D is introduced to ensure a correct normalization of α for $\mathbf{R} = \mathbf{R}'$ which follows from the definition (11). Note that in contrast to Taggart[21], we use a symmetric normalization. An inverse lattice Fourier transform yields the Warren-Cowley parameters in the real space

$$\alpha^{QQ'}_{\mathbf{RR}'} = \frac{1}{\Omega_\mathrm{BZ}}\int_{(\mathrm{BZ})} \mathrm{d}^3\mathbf{k}\,[\alpha(\mathbf{k})]_{QQ'}\,\exp[-i\mathbf{k}(\mathbf{R}-\mathbf{R}')]\,. \tag{14}$$

RESULTS AND DISCUSSION

Computational procedure

We neglect changes of the lattice constant due to the varying composition and assume the lattice constant $a = 5.652$ Å. We use equal radii of the Wigner-Seitz spheres for all atomic species and for the empty spheres ($R_\mathrm{WS} = 2.63$ bohr), a set-up, which leads to a considerable charge transfer among the alloy constituents (Tab. 1), and the Vosko-Wilk-Nusair parameterization of the exchange and correlation energy[22].

Table 1. The valence, averaged number of valence electrons N^Q_el, net charges q^Q, and effective charges of atoms q^Q_eff in the $(\mathrm{Ga}_{0.9325}\mathrm{Mn}^{\uparrow}_{0.040625}\mathrm{Mn}^{\downarrow}_{0.009375}\mathrm{As}_{0.0175})\mathrm{As}$ alloy.

site	atomic species	valence	N^Q_el	q^Q	q^Q_eff
cation	Ga	3	2.34628	−0.65372	0.00280
	Mn↑	7	6.34350	−0.65650	0.00002
	Mn↓	7	6.35124	−0.64876	0.00776
	As	5	4.18989	−0.81011	−0.15358
anion	As	5	4.22693	−0.77307	0.00000
interstitial 1	empty	0	0.77633	0.77633	0.00000
interstitial 2	empty	0	0.65310	0.65310	0.00000

The energy integration is performed along a contour in the complex energy plane, and the k-space integration over the irreducible wedge of the Brillouin zone using usually 280 points. We have verified that 1638 points lead to very similar results, the difference in the E_tot being less than 2 μRy.

The computation of the lattice Fourier transform of the $v^{QQ'}_{\mathbf{RR}'}$ is trivial, but that of the $\phi^{QQ'}_{\mathbf{RR}'}$ has to be done by employing the Ewald summation technique[23] which leads to a formula

$$\sum_{\mathbf{R}\neq\mathbf{0}}\frac{\exp(i\mathbf{kR})}{|\mathbf{R}|} = \frac{4\pi}{\Omega_0}\sum_{\mathbf{K}}\frac{\exp[-\frac{(\mathbf{k}+\mathbf{K})^2}{4\sigma^2}]}{|\mathbf{k}+\mathbf{K}|^2} + \sum_{\mathbf{R}\neq\mathbf{0}}\exp(i\mathbf{kR})\frac{\mathrm{erfc}(\sigma|\mathbf{R}|)}{|\mathbf{R}|} - \frac{2}{\pi}\,, \tag{15}$$

where Ω_0 is a volume of an elementary cell and the optimal value for σ is $\sigma = \sqrt{\pi}/\Omega_0^{1/3}$.

Alloy formation energy

We have calculated the total energy for alloys with conductivity of p-type, $0 \leq y \leq x/2$, $0 \leq x \leq 0.08$, within steps of $\Delta x = 0.01$, $\Delta y = 0.0025$. In order to find the ground state, we varied the ratio r in steps of $\Delta r = 0.025$ for each composition (x, y) and evaluated the total energy. The local moments $m(\mathrm{Mn}^{\uparrow})$ and $m(\mathrm{Mn}^{\downarrow})$ have opposite signs and are nearly of the same value (about 4 μ_B), i.e., depend only weakly on the composition. As a result, the magnetization in the pFM state, $m \approx x^{\uparrow} m(\mathrm{Mn}^{\uparrow}) + x^{\downarrow} m(\mathrm{Mn}^{\downarrow})$, is strongly reduced[15] as compared to the FM state. The magnetization is zero in the PM region due to a complete orientational disorder of local moments.

The calculated total energies (per elementary cell) enable us to investigate the stability of $(\mathrm{Ga}_{1-x-y}\mathrm{Mn}_x\mathrm{As}_y)\mathrm{As}$ alloys with respect to segregation into alloys with extremal chemical composition. Let us first consider segregation into an alloy without As-antisites $(\mathrm{Ga}_{1-x}\mathrm{Mn}_x)\mathrm{As}$ and an alloy with the highest possible concentration of As-antisites which is still not overcompensated, $(\mathrm{Ga}_{1-3x/2}\mathrm{Mn}_x\mathrm{As}_{x/2})\mathrm{As}$,

$$\begin{aligned}\Delta E_2(x, y) &= E[(\mathrm{Ga}_{1-x-y}\mathrm{Mn}_x\mathrm{As}_y)\mathrm{As}] - \frac{x-2y}{x} E[(\mathrm{Ga}_{1-x}\mathrm{Mn}_x)\mathrm{As}] \\ &- \frac{2y}{x} E[(\mathrm{Ga}_{1-3x/2}\mathrm{Mn}_x\mathrm{As}_{x/2})\mathrm{As}] \,. \qquad (16)\end{aligned}$$

The results are given in Fig. 1 and clearly show the stabilizing effect of the As antisites.

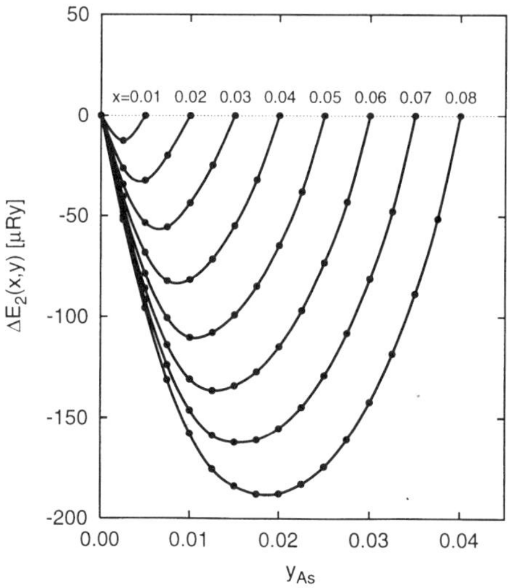

Figure 1. The energy difference $\Delta E_2(x, y)$ for $(\mathrm{Ga}_{1-x-y}\mathrm{Mn}_x\mathrm{As}_y)\mathrm{As}$ alloys, see Eq. (16).

Let us next consider the segregation into three different compounds, namely, GaAs, and two alloys, $(\mathrm{Ga}_{1-x_0}\mathrm{Mn}_{x_0})\mathrm{As}$ and $(\mathrm{Ga}_{1-3x_0/2}\mathrm{Mn}_{x_0}\ \mathrm{As}_{x_0/2})\mathrm{As}$, where $x_0 = 0.08$, i.e., the highest concentration for which we have data available. Then

$$\Delta E_3(x, y) = E[(\mathrm{Ga}_{1-x-y}\mathrm{Mn}_x\mathrm{As}_y)\mathrm{As}] - \frac{x_0 - x}{x_0} E[\mathrm{GaAs}]$$

$$- \frac{x-2y}{x_0} E[(Ga_{1-x_0}Mn_{x_0})As]$$
$$- \frac{2y}{x_0} E[(Ga_{1-3x_0/2}Mn_{x_0}As_{x_0/2})As] \,. \tag{17}$$

The results are shown in Fig. 2. Besides the energy lowering by the As-antisites, the regions of instability are visible, particularly close to the lines $y = 0$ and $y = x/2$.

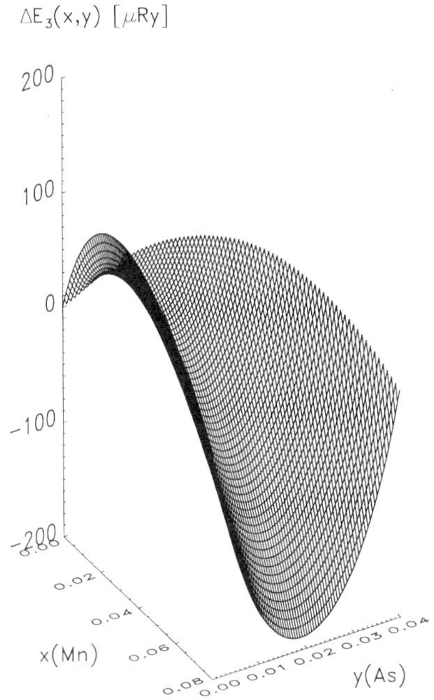

Figure 2. The energy difference $\Delta E_3(x, y)$ for $(Ga_{1-x-y}Mn_xAs_y)As$ alloys, see Eq. (17).

Impurity formation energy

The formation energy E^{A} of a substitutional impurity A_X, i.e., the energy needed to replace a host atom X by an atom A, is the reaction energy for the substitution

reference solid + A $\longrightarrow$ solid with one additional impurity + X.

The formation energies of impurities and their compositional dependence can be easily obtained[24] from the concentration dependence of the total energy $E(x_A, ...)$ of the alloy

$$E^{A}(x_A, ...) = \frac{\partial E(x_A, ...)}{\partial x_A} + E_{at}\{X\} - E_{at}\{A\} \,. \tag{18}$$

The last two terms are the total energies of free atoms X and A. The additional constant $E_{at}\{X\} - E_{at}\{A\}$ is not important for the concentration–dependent effects we have in mind.

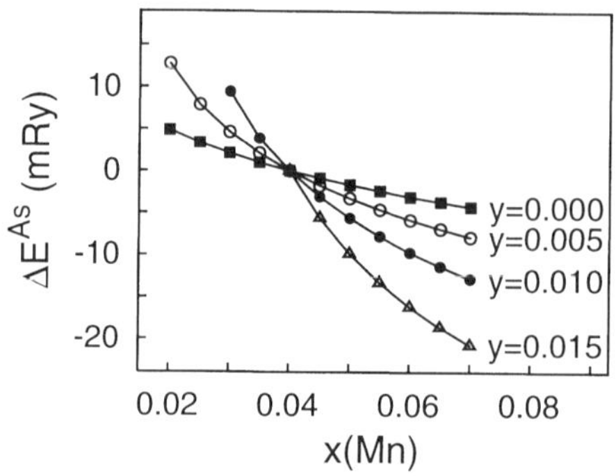

Figure 3. Dependence of the As antisite formation energy on the concentration x of Mn acceptors and on the concentration y of the antisite defects. The formation energy for $x = 0.04$ is used as a reference.

Fig. 3 shows the formation energy of the As antisite defect as a function of $x(\mathrm{Mn})$. The curves correspond to various concentrations of the antisite defects. The formation energy for $x = 0.04$ is used as a reference. The formation energy E^{As} is in all cases decreasing function of x. It is reduced approximately by 0.01 Ry, if the Mn concentration increases by a few atomic percent. This means that the number of the antisite defects can be considerably enhanced in the presence of Mn. This effect may contribute to the self-compensation behavior of (Ga,Mn)As alloys.

Similarly, also the formation energy of the substitutional Mn (Fig. 4) decreases with an increasing concentration y of As antisites. The changes of the formation energy $E^{\mathrm{Mn}}(y)$ are again of order of 0.01 Ry. This means that the presence of As antisites (and probably also of other donors[24]) is important for an improved solubility of Mn in III-V materials.

The decreasing character of both $E^{\mathrm{Mn}}(y)$ and $E^{\mathrm{As}}(x)$ indicates a tendency to correlation between these two impurities. The correlation is symmetrical because the slope of both $E^{\mathrm{Mn}}(y)$ and $E^{\mathrm{As}}(x)$ equals to the second (mixed) derivative of the total energy with respect to x and y. This quantity is negative in the present case of preferential co-doping and it would be positive, if the two impurities tend to segregate.

Ordering temperature

We made detailed calculations for an partially ferromagnetic alloy of a typical composition $(\mathrm{Ga}_{0.9325}\mathrm{Mn}^{\uparrow}_{0.040625}\mathrm{Mn}^{\downarrow}_{0.009375}\mathrm{As}_{0.0175})\mathrm{As}$. We have found $T_{\mathrm{ord}} = 797.94$ K. The ordering vector is $\mathbf{k}_0 = 0.1455661(1,1,1)$, which corresponds to a formation of domains of cubic form with sides approximately equal to 22 lattice constants a of the original zincblende structure, i.e., approximately 124 Å. The centers of domains are incident with lattice points of a simple cubic lattice with a lattice constant of 124 Å. The eigenvector $Y(\mathbf{k}_0)$ gives the following amplitudes: $Y(\mathrm{Ga}) = -0.787$, $Y(\mathrm{Mn}^{\uparrow}) = 0.604$, $Y(\mathrm{Mn}^{\downarrow}) = 0.102$, and $Y(\mathrm{As}) = 0.081$. Consequently, domains of two types are formed: in the first type the concentration of impurities Mn and As is increased, while in the

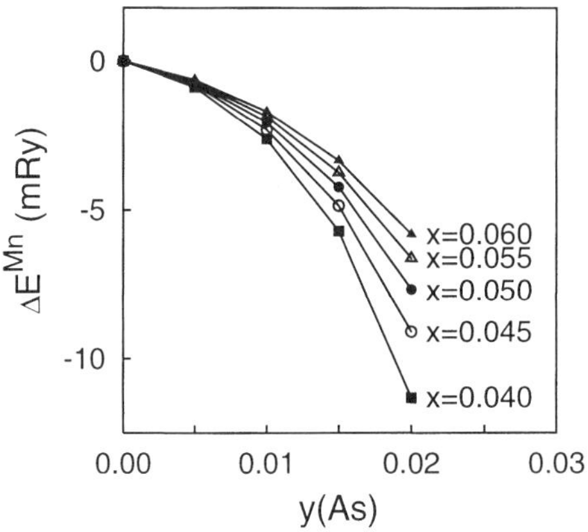

Figure 4. Changes ΔE^{Mn} of formation energy of substitutional Mn in $(\mathrm{Ga}_{1-x-y}\mathrm{Mn}_x\mathrm{As}_y)\mathrm{As}$ alloys due to the As antisite defects.

second type the impurity concentrations are diminished.

It is interesting to estimate separately the role of the electrostatic interactions $\phi^{QQ'}_{\mathbf{RR}'}$ and the interatomic pair interactions $v^{QQ'}_{\mathbf{RR}'}$, see Eq. (2). If only electrostatic interactions were considered, we find $T_{\mathrm{ord}} = 150.34$ K and an ordering vector $\mathbf{k}_0 = \frac{\pi}{a}(2, 0.892, 0)$, while the amplitudes amount to $Y(\mathrm{Ga}) = 0.697$, $Y(\mathrm{Mn}^{\uparrow}) = 0.0003$, $Y(\mathrm{Mn}^{\downarrow}) = 0.019$, and $Y(\mathrm{As}) = -0.717$. This low value of the ordering temperature is a consequence of small net charges. The impurities order on an fcc lattice according to the sign of their charges. On the other hand, if only pair interactions $v^{QQ'}_{\mathbf{RR}'}$ are considered, we find $T_{\mathrm{ord}} = 820.00$ K and $\mathbf{k}_0 = \frac{\pi}{a}(0, 0, 0)$, which corresponds to segregation. The amplitudes $Y(\mathrm{Ga}) = -0.790$, $Y(\mathrm{Mn}^{\uparrow}) = 0.599$, $Y(\mathrm{Mn}^{\downarrow}) = 0.100$, and $Y(\mathrm{As}) = 0.091$ then show that the preferred mode of segregation is into a pure GaAs and the rest with a high impurity concentration. Quite obviously, the Coulomb interactions, even though they are weak, play an important role because they can change qualitatively the ordering behavior of the system.

Warren-Cowley parameters

The Warren-Cowley parameters provide more detailed information about the mutual correlation of impurities; they also can be directly used in calculations of transport and magnetic properties.

The results for the Warren-Cowley parameters, summarized in Tab. 2, show that $\mathrm{Mn}^{\uparrow}$ atoms have always a tendency to attract each other, while Mn atoms of opposite orientation of spin repel each other. In the first coordination sphere all impurities repel each other with the exception of $\mathrm{Mn}^{\uparrow}$-$\mathrm{Mn}^{\uparrow}$ and $\mathrm{Mn}^{\downarrow}$-As pairs. The attraction of $\mathrm{Mn}^{\downarrow}$-$\mathrm{Mn}^{\downarrow}$ and As-As pairs in the second and third coordination spheres is remarkable.

Separate estimates of the role played by the electrostatic and band-term interactions as above show that if only electrostatic interactions are included, due to the weakness of net charges, the α's are close to zero. The band-term interactions, on the other hand, yield very similar results to those when both types of interactions are included. This result is to be expected because weak long-range forces should not change the short range order considerably.

Table 2. Warren-Cowley short-range order parameters for impurities in the $(Ga_{0.9325}Mn^{\uparrow}_{0.040625}Mn^{\downarrow}_{0.009375}As_{0.0175})As$ alloy at $T = 1400$ K.

neighbor	Mn↑-Mn↑	Mn↓-Mn↓	As-As	Mn↑-Mn↓	Mn↑-As	Mn↓-As
$(0\frac{1}{2}\frac{1}{2})$	−0.480	0.603	0.835	0.958	0.457	−0.709
(001)	−0.148	−0.204	−0.230	0.069	0.041	−0.019
$(\frac{1}{2}\frac{1}{2}1)$	−0.166	−0.111	−0.115	0.084	−0.038	0.020
(011)	−0.122	0.086	0.229	0.090	0.133	−0.190
$(0\frac{1}{2}\frac{3}{2})$	−0.030	−0.019	0.066	0.026	0.013	−0.012
(111)	−0.074	−0.018	0.081	0.010	−0.016	−0.010

CONCLUSIONS

We have studied the phase stability and the possible ordering in ferromagnetic semiconducting alloys (Ga,Mn)As on an *ab initio* level. The main conclusions of our study can be summarized as follows:
(i) The alloys are thermodynamically unstable with respect to segregation into related compounds or alloys with extremal chemical composition.
(ii) The As-antisites have a stabilizing effect and make the incorporation of Mn atoms energetically favorable.
(iii) At a critical temperature of about 800 K domains of two types, namely, with an enhanced and with a lowered concentration of impurities, may appear. The expected form of these domains is a cube with side length of approximately 120-130 Å, arranged to form a simple cubic lattice.
(iv) At short distances, the Warren-Cowley parameters indicate mutual attraction of $Mn^{\uparrow}$-$Mn^{\uparrow}$ and $Mn^{\downarrow}$-As pairs, and repulsion of all other pairs of impurities.

Acknowledgements

Financial support for this work was provided by the Grant Agency of the Academy of Sciences of the Czech Republic (Project A1010203), the Grant Agency of the Czech Republic (Project 202/00/0122), the Ministry of Education, Youth, and Sports of the Czech Republic (COST P5.30 and MSM113200002), the Center for Computational Materials Science in Vienna (GZ 45.504), and the RT Network *Computational Magnetoelectronics* of the European Commission (Contract HPRN-CT-2000-0143).

REFERENCES

1. H. Ohno, J. Mag. Mag. Mat. **200**, 110 (1999).
2. F. Matsukura, H. Ohno, A. Shen, and Y. Sugawara, Phys. Rev. B **57**, R2037 (1998).
3. H. Ohno, F. Matsukura, T. Omiya, and N. Akiba, J. Appl. Phys. **85**, 4277 (1999).
4. R. Shioda, K. Ando, T. Hayashi, and M. Tanaka, Phys. Rev. B **58**, 1100 (1998).

5. K.M. Yu, W. Walukiewicz, T. Wojtowicz, I. Kuryliszyn, X. Liu, Y. Sasaki, and J.K. Furdyna, Phys. Rev. B **65**, 201303 (2002).
6. F. Máca and J. Mašek, Phys. Rev. B **65**, 235209 (2002).
7. S.J. Potashnik, K.C. Ku, R. Mahendiran, S.H. Chun, R.F. Wang, N. Samarth, and P. Schiffer, arXiv:cond-mat/0204250.
8. M.L. Reed, M.K. Ritums, H.H. Stadelmaier, M.J. Reed, C.A. Parker, S.M. Bedair, and N.A. El-Masry, Mat. Lett. **51**, 500 (2001).
9. S.J. Potashnik, K.C. Ku, S.H. Chun, J.J. Berry, N. Samarth, and P. Schiffer, Appl. Phys. Lett. **79**, 1495 (2001).
10. C. Timm, F. Schäfer, and F. von Oppen, arXiv:cond-mat/0201411.
11. I. Turek, V. Drchal, J. Kudrnovský, M. Šob, and P. Weinberger, *Electronic Structure of Disordered Alloys, Surfaces and Interfaces*, (Kluwer, Boston, 1997).
12. D. Glötzel, B. Segal, and O.K. Andersen, Solid State Commun. **36**, 403 (1980).
13. J. Kudrnovský, V. Drchal, M. Šob, N.E. Christensen, and O.K. Andersen, Phys. Rev. B **40**, 10029 (1989).
14. J. Schliemann, J. König, and A.H. MacDonald, Phys. Rev. B **64**, 165201 (2001).
15. P.A. Korzhavyi, I.A. Abrikosov, E.A. Smirnova, L. Bergqvist, P. Mohn, R. Mathieu, P. Svedlindh, J. Sadowski, E.I. Isaev, Yu.Kh. Vekilov, and O. Eriksson, Phys. Rev. Lett. **88**, 187202 (2002).
16. B.L. Gyorffy, A.J. Pindor, J. Staunton, G.M. Stocks, and H. Winter, J. Phys. F: Met. Phys. **15**, 1337 (1985); H. Akai and P.H. Dederichs, Phys. Rev. B **47**, 8739 (1993).
17. H. Akai, Phys. Rev. Lett. **81**, 3002 (1998); T.C. Schulthess and W.H. Butler, J. Appl. Phys. **89**, 7021 (2001).
18. F. Ducastelle, *Order and Phase Stability in Alloys* (North-Holland, Amsterdam, 1991).
19. A.G. Khachaturyan, *Theory of Structural Transformations in Solids* (Wiley, New York, 1983).
20. S.K. Bose, V. Drchal, J. Kudrnovský, O. Jepsen, and O.K. Andersen, Phys. Rev. B **55**, 8184 (1997).
21. G.B. Taggart, Phys. Rev. B **19**, 3230 (1979).
22. S.H. Vosko, L. Wilk, and M. Nusair, Can. J. Phys. **58**, 1200 (1980).
23. M. Born and K. Huang, *Dynamical Theory of Crystal Lattices* (Clarendon, Oxford, 1954).
24. J. Mašek, I. Turek, V. Drchal, J. Kudrnovský, and F. Máca, to appear in Acta Phys. Polon. A (2002).

VACANCY ORDERING AND NON-STOICHIOMETRY IN $TiC_{1-x}\square_x$ and $TiN_{1-x}\square_x$

Gus L. W. Hart,[1] Barry M. Klein,[2] and Shanadeen Begay[3]

1. INTRODUCTION

Nearly two centuries ago, Dalton discovered that most compound materials occur in precise integer stoichiometry.[1] As pointed out by Lewis,[2] the existence of this integer ratio of the constituent atoms in a compound (or even in elemental solids) results from the fact that atoms bonded in a solid tend to donate, accept, or share an integer number of electrons with their neighbors. Although deviations from precise stoichiometry (due to intrinsic vacancies or interstitials) are known to exist, such deviations are small and are *entropically* stabilized and thus will disappear as $T \to 0$. However, there exist several classes of binary compounds for which the existence of vacancies actually *reduces* the free energy, even at low temperatures, and for which the deviations can be astonishly large, 50% or more. Categorizing these materials by their properties and behavior, vacancy ordering and complexes, homogeneity range etc., one can mention three classes of these "intrinsically" non-stoichiometric materials: (i) transition metal oxides (such as VO or TiO), (ii) transition metal nitrides and carbides (such as HfC, NbN, etc.), and (iii) early transition metal chalcogenides[3] (such as ScS or ZrSe). All three groups have been studied extensively but the first two groups have garnered more interest because of the broad range of their technological applications.[4]

In this paper we focus on two materials in the second group, namely TiC and TiN. Both of these materials, along with the others in their class, have been studied extensively (see, e.g., Refs. [4–8] and references therein); they are of particular interest for their industrial applications as well as for their basic science appeal owing to there unusual properties (e.g., hardness and high melting temperatures). The materials in this class are characterized by strong mechanical properties and high melting temperatures. Titanium carbide, for exam-

[3]Gus Hart, Department of Physics and Astronomy, Northern Arizona University, Flagstaff AZ 86011-6010, gus.hart@nau.edu

[3]Barry M. Klein, Department of Physics, University of California, Davis CA 95616, bmklein@ucdavis.edu

[3]Shanadeen Begay, Northern Arizona University, Flagstaff AZ 86011-6010, scb8@dana.ucc.nau.edu

ple, is a major component in cemented carbides and is widely used for surface coatings.

We have studied how the electronic structure of the TiC and TiN systems stabilizes such a large range of non-stoichiometry ($0 \le x \lesssim 50\%$), leading to the predicted low temperature, vacancy-ordered structures. Although the low temperature structures are not likely to be realized experimentally due to the sluggish diffusion, understanding the ordering tendencies gives insight into the chemical mechanisms responsible for the non-stoichiometry and how the mechanical properties are affected by the existence of vacancies.

Since the vacancy configuration, that is the crystal structures, for these materials are not known experimentally and the vacancy concentration-dependent behavior is difficult to guess based on past experience, we cannot resort to the prevalent but unreliable approach of assuming a limited set of "intuitive" trial structures and then performing first-principles calculations to determine the lowest energy structure in the trial set. The limitation of this approach is particularly acute in vacancy-ordering compounds where the ground state structures tend to have relatively large unit cells and low crystallographic symmetry. Many times in the literature, conclusions have been draw from first-principles calculations of high-symmetry, small-unit-cell structures. Though convenient from the point of view of computational cost, these structures often turn out to have energies that are far above that of the true ground states and thus do not represent the true "chemistry" responsible for the formation and ordering of vacancies in the compound. This approach of selective sampling is driven by the high computational cost of these calculations and the prohibitive cost of exploring the all possible configurations.

Performing calculations for a material when the thermodynamic states are actually not known and for which large-unit-cell structures are common presents a formidable problem for the conventional first-principles approach. The difficulty arises both from the size of unit cells that must be treated and the literally more-than-astronomical number of configurations that must be considered. For this reason, we have used a cluster expansion (CE) method[9–14] in which first principles, total-energy calculations for a relatively small database of ordered superstructures is mapped onto a 3D Ising model. Once this CE Hamiltonian has been successfully constructed, efficient and accurate ground state search techniques (simulated annealing, direct enumeration, genetic algorithms, parallel tempering) can be applied, without guessing, to identify the minimum-energy configurations and Monte Carlo simulations can be used to study finite-temperature effects. Once the low energy structures are determined, we analyze their electronic structure using the LDA/GGA pseudopotential approach. Thus we have a significant advantage over the conventional approach of rounding up a few suspect structures because we analyze those structures that are actually energetically stable, without limiting ourselves to small unit cells or artificially high symmetry structures.

2. THE CLUSTER EXPANSION

2.1. General Formalism

The cluster expansion (CE) expresses the excess energy of any lattice configuration σ (a particular occupation of the N lattice sites by A or B atoms) as

$$\Delta H_{\mathrm{CE}}(\sigma) = J_0 + \sum_{f}^{N_f} D_f \overline{\Pi}_f(\sigma) J_f \tag{1}$$

where J_f is the effective atom-atom interaction for cluster type f (pair, triangle, tetrahedron, etc.), N_f is the number of clusters of type f in configuration σ, D_f is the number of clusters of type f per lattice site, and $\overline{\Pi}_f$ are the averaged spin products[15] for configuration σ. In the case of TiC/$N_{1-x}\square_x$, the up and down "spins" represent the carbon/nitrogen atom and its vacancy. Because it is present on every anion site and does not constitute a configurational degree of freedom, the titanium atom is not explicitly included in the expansion (but it is included, of course, in the LDA calculations of the total energy.)

Sanchez, Ducastelle, and Gratias showed[9] that a single set of interactions can exactly reproduce the directly calculated total energies of all possible configurations σ. Of course, this is only true for the untruncated expansion ($N_f = 2N$) so the validity of the truncated expansion naturally depends on the vanishing nature of higher order terms. The J's are determined by fitting the expansion $\Delta H_{\text{CE}}(\sigma)$ to the excess total energies $\Delta H_{\text{LDA}}(\sigma)$ of a set of N_σ "input structures" calculated by first-principles methods. We use the pseudopotential planewave method,[16] as implemented in the VASP[17, 18] code. In each case, we relax both cell-external and cell-internal degrees of freedom to obtain $\Delta H_{\text{LDA}}(\sigma)$. The interactions were chosen by first eliminating from the fit several of the input structures and choosing the interactions that result in an accurate fit to the structures retained as well as accurate predictions for the eliminated structures. The process is repeated using different sets of eliminated structures to ensure a set of interactions that work well *generally*.[4] The process of determining a good fit is discussed in detail in Ref. [13].

2.2. Ground State Searches

The most common way to find the thermodynamically stable ground states of alloy systems modeled via cluster expansion approaches is Monte Carlo-based searches (i.e., simulated annealing).[20, 21] Although simulated annealing has been quite successful for simple intermetallic and semiconductor alloys,[22–26] such an approach does not guarantee that the states found are *global* minima of the energy functional (they may not be true ground states).

In our studies, we have found that, although the simulated annealing (SA) approach to ground state searches is useful in many cases, there are cases (non-stoichiometric systems in particular) where a complementary search technique is helpful, or in some cases, essential. The SA approach suffers from three problems. (1) Critical slowing down: When the simulation suffers from this problem, the necessary computer time to reach an answer becomes prohibitive. This problem seems to be particularly acute for the cases (such as the refractory carbide and nitride compounds) where the ordered states have relatively large unit cells. (2) Failure to converge: In some cases, the low energy states at a given concentration (including the ground state) are so closely spaced (in energy) that the SA approach fails to converge. (3) False ground states: Occasionally we have found ground states with SA that appear to be robust, when in fact, a direct enumeration of ground states (restricted to a limited cell size, of course) will turn up a ground state *lower* in energy than that found by the SA approach.

The most robust solution to these obstacles is to perform the ground state searches by direct, exhaustive enumeration, that is, a direct brute-force determination of all possible configurations. Formally, however, such an approach is impossible (for very large unit cells) because there are a prohibitive number of states to enumerate. Direct enumeration is, however, very useful when applied to relative small unit cell sizes (about 16 to 20 atoms)

[4]The fit is optimized by requiring the "maximum smoothness criteria (Eqs. 24 26 of Ref. [19]). The parameters t and λ are simultaneously optimized to yield both a good fit for fitted structures and accurate predictions for "eliminated" structures.

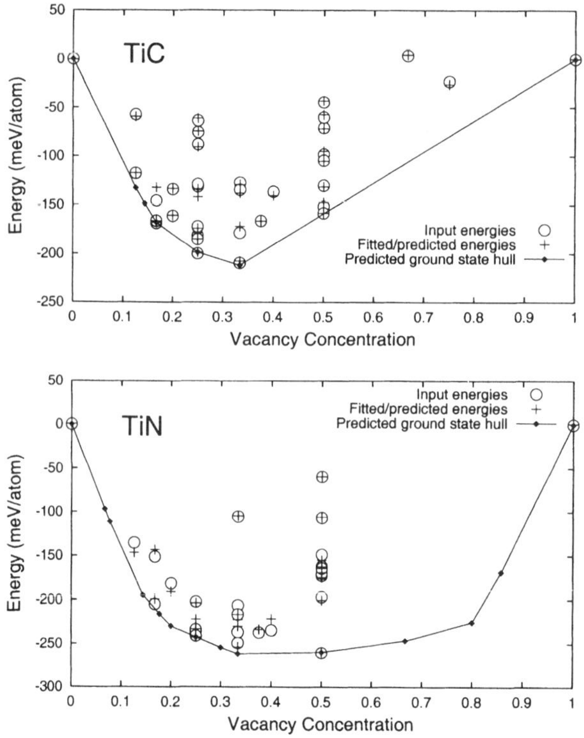

Figure 1: Cluster expansion fits for TiC (top) and TiN (bottom). CE-predicted values are very close to the LDA values, indicating an accurate fit. The solid lines show the ground state hull and the solid points show the ground states. The ground states are discussed in Sec. 3.

and we have applied this method in this work. We have enumerated all possible configurations for unit cell sizes $\leq$ 20 atoms/cell using the method of Ferreira et al.[27] There are more than 3 million unique configurations for the fcc lattice in this size range. Results of ground state searches using this direct enumeration method are shown in the following section.

2.3. Cluster Expansion Fits for $TiC_{1-x}\square_x$ and $TiN_{1-x}\square_x$

Figure 1 demonstrates the accuracy of the fits we developed for the case of $TiC_{1-x}\square_x$ and $TiN_{1-x}\square_x$ using the procedure discussed in Sec. 2.1. Ultimately, we included about 30 first-principles energies in the fitting. The CE fitting parameters were checked not only for *fitting* accuracy (how well the energies of the input structures are reproduced) but also for their accuracy in *prediction* (how accurately the energies of structures *not* included in the

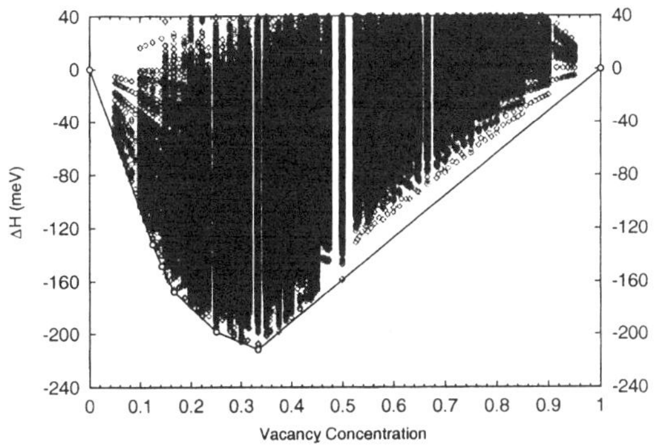

Figure 2: Ground state search results (via direct enumeration) for TiC. The five stable configurations occur at concentrations $x \leq 1/3$. However, the lowest configuration at $x = 1/2$ is very close to the ground state hull and may actually be stabilized by effects not considered in the model (such as vibrational entropy).

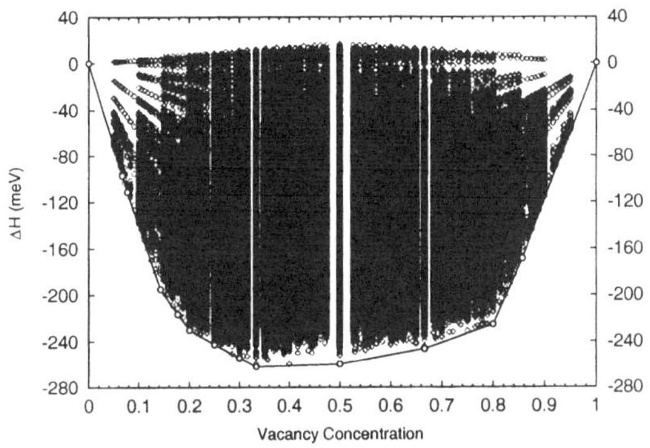

Figure 3: Ground state search results (via direct enumeration) for TiN. In this case the ground states cover a broader vacancy concentration range than for TiC. Also, the formation enthalpies are about 25% lower than for the case of TiC.

fit are predicted by the CE). When a fit is accurate both for fitting and for predictions, we refer to that fit as *robust*. As mentioned above, an in-depth discussion of how to determine the optimum interaction energies (i.e., the fitting parameters) is given in Ref. [13].

The results displayed in Fig. 2 show that the input energies (calculated via first principles) are reproduced very accurately by the CE; in most cases, the error is only a few percent. These robust fits were obtained with ~17 pair and 7 three-body and four-body

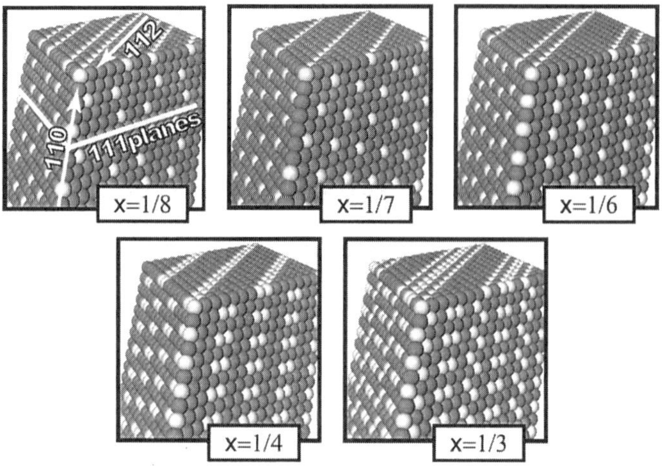

Figure 4: Ground state structures for TiC. Only the occupation of the carbon sites is shown (black occupied, white unoccupied); the titanium atoms are not shown. Note that the pictures are not visualizations of a cubic supercell but rather stacked (111) planes of the FCC lattice. Note the common motif of (112) rows of vacancies in each (111) plane and that planes are generally stacked so that rows of alternating vacancies are formed in the (110) direction.

interactions. Typical fitting and prediction errors (rms) for TiC and TiN are $\sim$4 meV/atom, whereas the average ΔH is approximately -150 and -200 meV/atom for TiC and TiN, respectively. In the simulations discussed below, we used a final fit that included *all* of the input structures.

3. RESULTS AND DISCUSSION

Figure 2 shows a ground state search (via direct enumeration) of the ordered vacancy states for the case of TiC. Our search finds five ground states (stable vacancy-ordered configurations). The most stable structure occurs at $x = 1/3$. However, the lowest energy configuration at $x = 1/2$ is so near the ground state hull that it may actually be realized if effects not included in our model (such as vibrational entropy) are properly taken into account. Consequently, the homogeneity range for $TiC_{1-x}\Box_x$ may actually be 0–50% rather than 0–33%.

Figure 3 shows the same information as Fig. 2 except for TiN. The most stable structure again occurs at a vacancy concentration of $x = 1/3$ but the ground state at $x = 1/2$ is almost as low in energy. In all, twelve ground states are predicted by the CE. The predicted homogeneity range of $TiN_{1-x}\Box_x$ is much larger in this case. This is in qualitative agreement with experimental observations which show a homogeneity range of $0 \leq x \lesssim$ 52% for TiC and $0 \leq x \lesssim 62\%$ for TiN.[3] In fact, for the class of transition-metal carbide

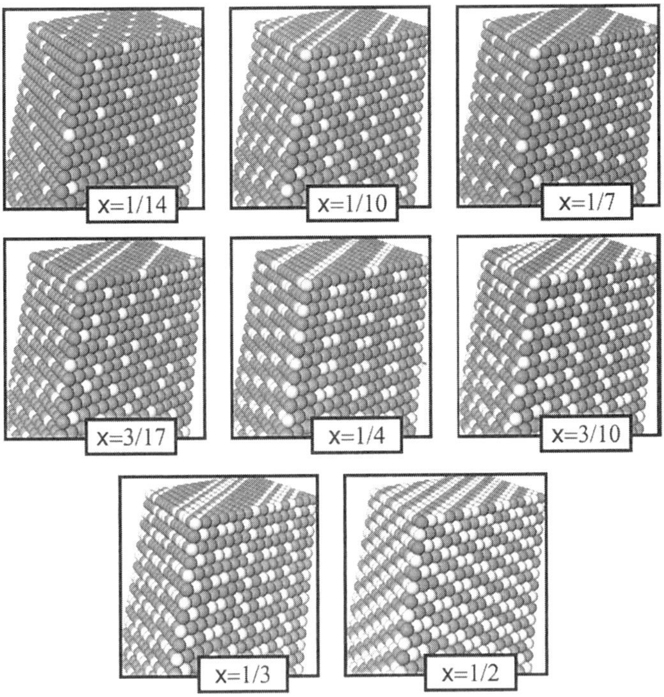

Figure 5: Ground state structures for TiN. See the caption for Fig. 4 for labeling of states and a description of the common motifs observed. TiN yields more ground state structures than TiC, due in part to the larger homogeneity range.

and nitride binary compounds that exhibit intrinsic non-stoichiometry, it is always the case that the nitride has a larger homogeneity range than the corresponding carbide.

The most striking result of the ground state searches is the common motifs that are observed. The motifs are the same for both TiC and TiN and exist across a large concentration range in each system. In Figures 4 and 5, the ground states are visualized as a stacking of (111) planes of the FCC lattice. Within the (111) planes, vacancies are arranged in (112) rows (i.e., adjacent vacancies are on third nearest neighbor sites). Also, the rows are often stacked so that, in the (110) direction, vacancies occur on every other site (fourth nearest neighbor distance). The motif of (112) vacancy rows persists even for low vacancy concentrations, i.e., configurations which do not maximize the inter-vacancy distance are the rule. This implies that vacancy-vacancy interactions at some distance beyond first nearest neighbors become attractive.

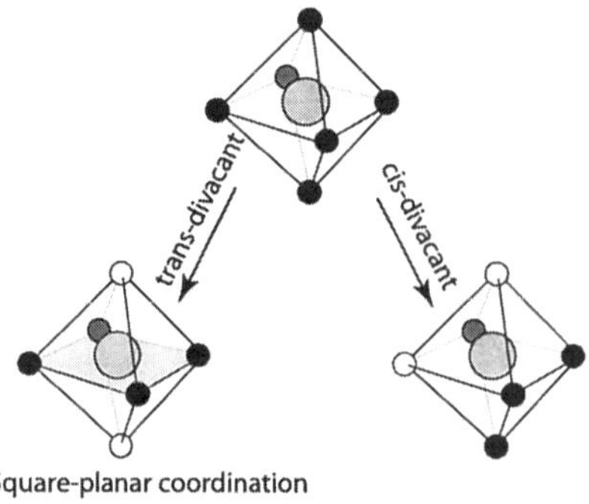

Figure 6: Titanium octahedra in TiC and TiN. In the NaCl structure, titanium atoms (large gray spheres) are surrounded by six carbon (or nitrogen) atoms (black spheres). If two vacancies (white spheres) are present, only two arrangements are possible: vacancies on opposite corners (trans-divacant) or vacancies as nearest neighbors (cis-divacant). If the octahedron is trans-divacant, then the titanium atom in the center will be square-planar coordinated by the remaining non-metal atoms.

The structure at $x = 1/3$ is identical for both TiC and TiN. This structure is known as "Sc_2S_3."[28] (ScS is also a non-stoichiometric binary system, albeit with vacancies on the metal, rather than non-metal, site.) The Sc_2S_3 structure has a large unit cell (12 Sc sites and 12 S/□ sites) and occurs very commonly in the early transition-metal chalogenide binary compounds (which are also intrinsically non-stoichiometric).[3] This is intriguing: despite the chemical differences between ScS and TiC/N and the fact that vacancies occur on the opposite sublattice (metal vs. non-metal), TiC and TiN display a set of vacancy-ordered configurations very similar to ScS and other transition-metal chalogenide binary compounds. Indeed, at several concentrations, the configurations are identical.

We also note that in most cases, the size of the unit cells of the ground states are relatively large (12–16 C/N sites/cell); larger than the unit cells of intermetallic compounds (typically 4–8 sites/cell). And the unit cells have relatively low symmetry.

Owing to the NaCl structure, each atom in these binary compounds is six-fold coordinated with its nearest neighbors (which are of the opposite atomic type). One can imagine a octahedron around the titanium atoms where carbon (or nitrogen) atoms sit on the six corners of the octahedron, as shown in the top of Fig. 6. If the vacancy concentration is $x \leq 1/3$ then the average number of vacancies per octahedron will be ≤ 2. Similarly, for $x \leq 1/6$ the number of vacancies per octahedra will be ≤ 1. If we examine the predicted ground state structures from the point of view of this picture of the octahedra around the metal atoms, we uncover two noteworthy facts: (i) generally, these structures minimize the number of vacancies per octahedra for a given concentration, e.g., at $x = 1/3$ each and every octahedron has exactly two vacancies, and (ii) when an octahedron has two vacancies present, the vacancies are *never* present on opposite corners of the octahedra, rather the two vacancies are always nearest neighbors.

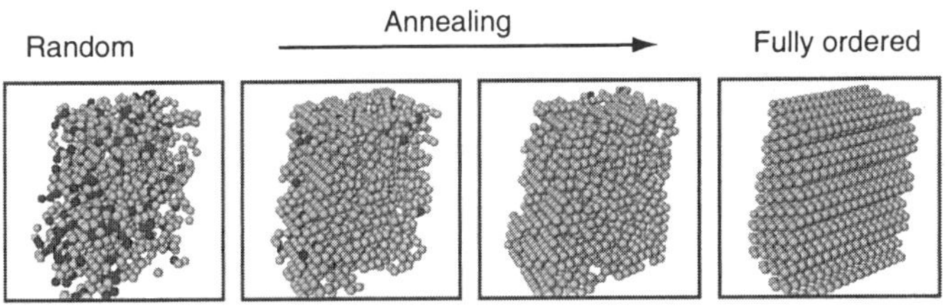

Figure 7: Simulated annealing simulation of TiC at $x = 1/4$. Only the titanium atoms in *divacant* octahedra are shown; carbon atoms and other titanium atoms are not shown. Titanium atoms in "cis-divacant" octahedra are shown in gray and those in "trans-divacant" octahedra are shown in black. Note that *trans*-divacant Ti octahedra disappear well before significant long-range order is observed.

The first observation implies that vacancies are repulsive for first and second nearest neighbor distances but the second observation shows that a *first*-nearest-neighbor coordination of the vacancies is far more energetically favorable than a *second*-nearest-neighbor coordination. For the case of ScS, Burdett pointed out[29] that when scandium vacancies are on opposite corners of the octahedron ("trans-divacant" arrangement) the sulfur atom becomes square-planar coordinated (see Fig. 6), a situation never observed in solid state chemistry. Without a single exception, all of the vacancy-ordered structures presented here (as well as those of ScS in Ref. [30]) have this so-called "cis-divacant" arrangement for all double vacancy octahedra.

Using the CE Hamiltonian in Monte Carlo simulations, we found that the trans-divacant octahedra naturally occurring in a random arrangement of vacancies are quickly replaced by cis-divacant octahedra. In fact, most of the trans-divacant octahedra are eliminated well before the structure acquires significant long range order, as shown in Fig. 7.

In a previous study of non-stoichiometry in ScS, [30] an analysis of the charge densities and partial densities of states for a series of low-energy structures with increasing vacancy concentration revealed very clearly the mechanism of intrinsic non-stoichiometry in that system. Despite the very obvious similarities of the ground states in these two systems compared to those of ScS, the details of the mechanism for non-stoichiometry appears to be different. An analysis of the partial densities of states(not shown here) for non-stoichiometric TiC and TiN did not reveal a mechanism for vacancy formation similar to that found for ScS—a somewhat puzzling result given the fact that the ground states in ScS are so closely related to those in TiC/N. At this point, we have not been able to determine the electronic mechanism that drives the formation of vacancies in TiC and TiN. This topic is a focus of our current research and will be discussed in a future publication.

4. SUMMARY

Using a database of first-principles calculated formation enthalpies, we constructed cluster expansions[9–14] for the vacancy systems TiC and TiN. The ground state hulls predicted by the cluster expansion are consistent with the experimentally known homogeneity ranges[3] for these two compounds. Using a direct enumeration method,[27] we determined the ($T = 0$) ground state structures and found that they are very similar in these two systems. A common feature of the ground states is that vacancies prefer to arrange themselves in

(112) rows, almost independent of the vacancy concentration. In all of the ground states, vacancies completely avoid any arrangement that would lead to square planar coordination of the titanium atoms, even if this results in nearest-neighbor vacancies. Surprisingly, the ground states we discovered and their common structural motifs are very similar (often identical) to those discovered in the ScS system.[30] However, an analysis of the charge density and densities-of-states for TiC and TiN did not indicate a mechanism for vacancy formation similar to that found for ScS despite the very obvious structural similarities of the stable configurations in all three systems. The precise electronic mechanism driving the formation of vacancies is still unclear and is the subject of ongoing research.

5. REFERENCES CITED

[1] John Dalton. *A New System of Chemical Philosophy*. R. Bickerstaff, 1808.

[2] G. N. Lewis. *Valence*. The Chemical Catalog Company, 1923.

[3] Jeremy K. Burdett and John F. Mitchell. Nonstoichiometry in early transition metal compounds with the rock salt structure. *Prog. Solid St. Chem.*, 23:131–170, 1995.

[4] L. E. Toth. *Transition Metal Carbides and Nitrides*. Academic Press, New York, 1971.

[5] V. A. Gubanov. *Electronic Structure of Refractory Carbides and Nitrides*. Cambridge University Press, 1994.

[6] K. Schwarz. *CRC Crit. Rev. Solid State Mater. Sci.*, 13:3, 1987.

[7] Hakan W. Hugosson, Pavel Korzhavyi, Ulf Jansson, Borje Johansson, and Olle Eriksson. Phase stabilities and structural relaxations in substoichiometric TiC_{1-x}. *Phys. Rev. B*, 63:165116, 2001.

[8] Gus L. W. Hart and Barry M. Klein. Phonon instabilities in MoC and MoN. *Phys. Rev. B*, 61:3151–3154, 2000.

[9] J. M. Sanchez, F. Ducastelle, and D. Gratias. Generalized cluster description of multicomponent systems. *Physica A*, 128:334–350, 1984.

[10] D. de Fontaine. Cluster approach to order-disorder transformations in alloys. *Solid State Physics*, 47:33–176, 1994.

[11] D. de Fontaine. In H. Ehrenreich, F. Seitz, and D. Turnbull, editors, *Solid State Physics*, pages 73–, New York, 1979. Academic.

[12] Alex Zunger. First-principles statistical mechanics of semiconductor alloys and intermetallic compounds. In P. E. A. Turchi and A. Gonis, editors, *Statics and Dynamics of Alloy Phase Transitions*, NATO ASI Series, Ser. B, pages 361–419, New York, 1994. Plenum Press.

[13] Alex Zunger, Ligen Wang, Gus L. W. Hart, and Mahdi Sanati. Obtaining Ising-like expansions for binary alloys from first principles. *Modelling and Simulation in Materials Science and Engineering*, 10:685–706, 2002.

[14] Axel van de Walle and G. Ceder. Automating first-principles phase diagram calculations. *J. Phase Eq.*, 23:348, 2002.

[15] L. G. Ferreira, S.-H. Wei, and A. Zunger. First-principles calculation of alloy phase diagrams: the renormalized-interaction approach. *Phys. Rev. B*, 40:3197–3231, 1989.

[16] J. Ihm, Alex Zunger, and M. Cohen. *J. Phys. C*, 12:4409, 1979.

[17] G. Kresse and J. Furthmüller. *Phys. Rev. B*, 54:11169, 1996.

[18] G. Kresse and J. Furthmüller. *Comput. Mat. Sci.*, 6:15, 1996.

[19] D. B. Laks, L. G. Ferreira, S. Froyen, and A. Zunger. Efficient cluster expansion for substitional systems. *Phys. Rev. B*, 46:12587–12605, 1992.

[20] K. Binder. *Monte Carlo Methods in Statistical Physics: An Introduction*. Springer-Verlag, 1992.

[21] K. Binder. *Applications of the Monte Carlo Method in Statistical Physics*. Springer-Verlag, 1987.

[22] Z. W. Lu, D. B. Laks, S.-H. Wei, and A. Zunger. First-principles simulated-annealing study of phase transitions and short-range order in transition-metal and semiconductor alloys. *Phys. Rev. B*, 50:6642–6661, 1994.

[23] Chris Wolverton, Vidvuds Ozoliņš, and Alex Zunger. First-principles theory of short-range order in size-mismatched metal alloys: Cu-Au, Cu-Ag, and Ni-Au. *Phys. Rev. B*, 57:4332–4348, 1998.

[24] Chris Wolverton and Alex Zunger. Ni-Au: A testing ground for theories of phase stability. *Computational Materials Science*, 8:107–121, 1997.

[25] Chris Wolverton and Alex Zunger. Ising-like description of structurally relaxed ordered and disordered alloys. *Phys. Rev. Lett.*, 75:3162–3165, 1995.

[26] Chris Wolverton, G. Ceder, D. de Fontaine, and H. Dreysse. Ab-initio determination of structural stability in fcc-based transition metal alloys. *Phys. Rev. B*, 48:726–747, 1993.

[27] L. G. Ferreira, S.-H. Wei, and Alex Zunger. Stability, electronic structure, and phase diagrams of novel inter-semiconductor compounds. *The International Journal of Supercomputer Applications*, 5:34–56, 1991.

[28] J. P. Dismukes and J. G. White. The preparation, properties, and crystal structures of some scandium sulfides in the range Sc_2S_3–ScS. *Inorg. Chem.*, 2:1220, 1964.

[29] Jeremy K. Burdett and John F. Mitchell. Electronic origin of nonstoichiometry in early-transition-metal chalcogenides. *Chem. Mat.*, 5:1465–1473, 1993.

[30] Gus L. W. Hart and Alex Zunger. Origins of non-stoichiometry and vacancy ordering in $Sc_{1-x}\Box_xS$. *Phys. Rev. Lett.*, 87:275508, 2001.

VACANCY-MEDIATED PHASE TRANSFORMATIONS: HOMOGENEOUS OR HETEROGENEOUS?

Wolfgang Püschl,[1] William A. Soffa,[2] and Wolfgang Pfeiler[1]

[1]Institut für Materialphysik, University of Vienna, Strudlhofgasse 4 A-1090 Vienna, Austria
[2]Department of Materials Science and Engineering, University of Pittsburgh, 842 Benedum Hall, Pittsburgh, PA 15261, U.S.A.

INTRODUCTION

Phase transitions have traditionally been classified according to Ehrenfest [1] by the lowest order of the derivative of free energy, with respect to an intrinsic parameter like temperature or pressure, that becomes discontinuous at the transition. Higher-order phase transformations where no jump in composition or specific volume takes place can therefore be expected to occur in a more smooth, continuous manner than first-order transitions. In Fig. 1 [2] the schematic equilibrium phase diagrams are shown for systems where on quenching from a higher temperature in case a) a higher order transition b) a first order transition takes place. In the first case qualitatively the same transition will be made in every part of the sample volume irrespective of the depth of the quench and independent of the exact stoichiometry, provided we are in the equilibrium region of the new phase. In the second case where we quench into a two-phase region - which entails a first-order transition - according to classical ideas one of two possible phase-separation scenarios can be encountered: Either even very small composition fluctuations result in a reduction of free energy and grow therefore spontaneously or the fluctuations must overcome an energy barrier by creating a part of the new phase with a critical minimum size. Typical for the two cases are respectively long-wavelength, small-amplitude fluctuations ("homophase fluctuations") or short-range, high-amplitude, particle like fluctuations ("heterophase fluctuations")[1] as shown in Fig. 2 [2]. Which of the two qualitatively different mechanisms will be operative is thought to depend on the properties of the Gibbs free energy per volume g(c) as a function of composition: If the second derivative at the given composition is negative, then phase separation should happen spontaneously and is called spinodal decomposition, the line in the phase diagram where $\partial^2 g/\partial c^2 = 0$ being termed a spinodal.

[1] We remind that in the context of the present study, any phase transformation via nucleation is called „heterogeneous" and must not be confused with the more specific term „heterogeneous nucleation" which means nucleation at a pre-exisiting heterogeneity of the matrix phase like a grain boundary, an inclusion or a defect.

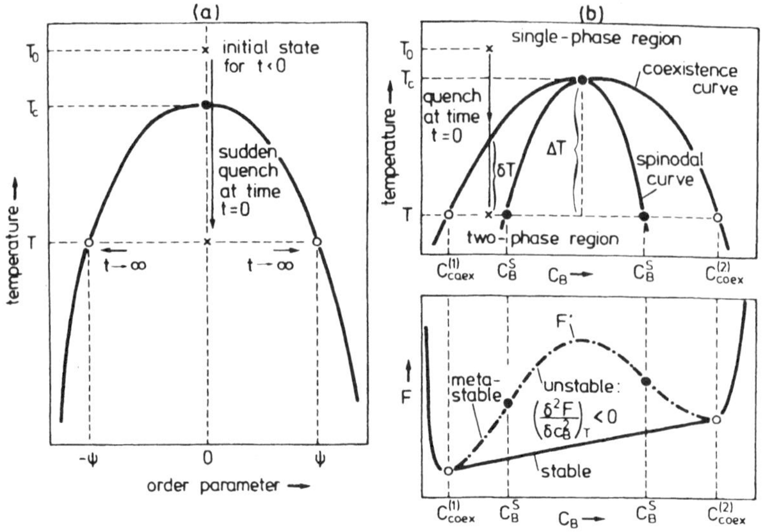

Figure 1. Schematic phase diagrams for the cases of a) a higher order phase transition b) a first-order phase transition [2].

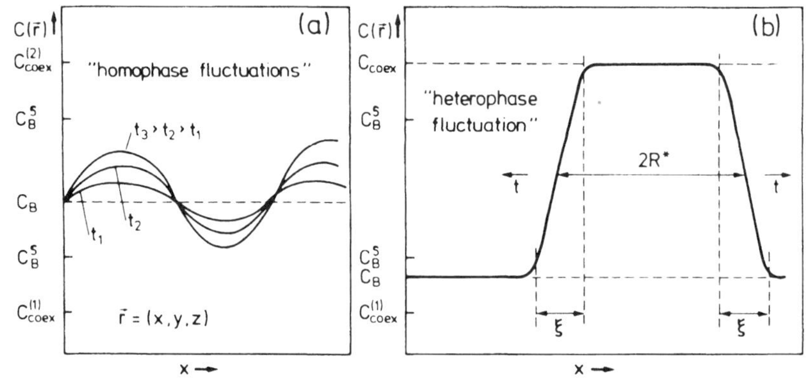

Figure 2. First order phase transition: a) homophase fluctuation, b) heterophase fluctuation [2].

Otherwise an energy barrier has to be overcome by nucleation. A recent review paper [3] discusses several reasons why this clear distinction cannot be made in most real systems. First of all, in order to define a smooth local composition an average has to be taken over a finite volume (coarse graining). The shape of g(c) can be shown to depend sensitively on the ratio of coarse-graining length to characteristic length of concentration change [4] (Fig. 3). The same flattening-out of the free energy curve is observed if in a calculation by the cluster variation method (CVM) cluster size is increased ever more until it finally reaches the whole crystal volume [5]. It follows that a spinodal curve cannot be defined in an unambiguous way, with the asymptotic and unrealistic exception of a system with infinitely weak and at the same time infinitely far-reaching interactions for which the mean-field approximation becomes exact. On the contrary, in practice we require for the concept of a local free energy to be meaningful that the effective interaction distance between atoms should not be large as compared to the coarse-graining length, which in turn is required to be small compared to typical lengths characterizing the fluctuation. Furthermore, it must be kept in mind that using g(c) means transferring an equilibrium concept valid for macroscopic systems to very small parts of a system not in equilibrium. So it is hardly surprising that in real systems the distinction between the two scenarios of phase separation cannot be made as clearly as in the ideal picture and that the boundary is uncertain and in addition shifted owing to kinetic effects and elastic interaction of coherent precipitates. In fact there is a rather gradual than a sharp transition from one mode of phase separation to the other.

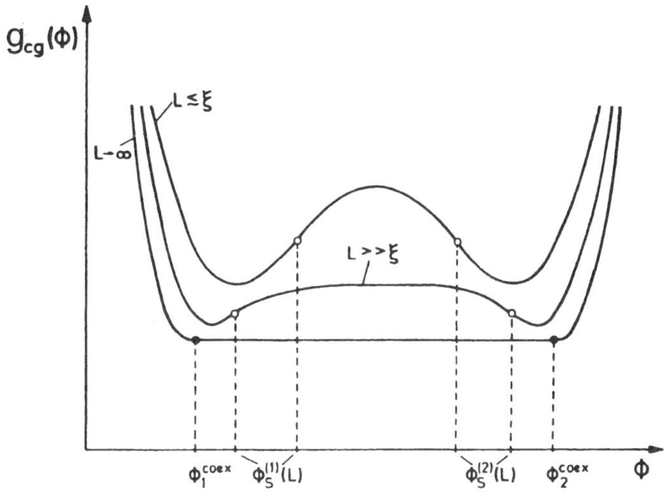

Figure 3. Coarse-grained free energy as a function of composition ($\Phi=(c-c_{crit})/c_{crit}$) for various coarse-graining lengths L [4].

The classical spinodal decomposition theory by Cahn and Hilliard [6-9, see also 10,11] takes the free energy to be a functional of local composition with gradient-dependent terms. This is in itself a problematic concept, as we have seen. The neglect of higher order gradient terms and the linearization required to enable analytic solutions severely restrict the scope of the classical spinodal decomposition theory. Its prediction of a stationary wavelength of composition fluctuations, for instance, is not confirmed in scattering experiments [3]. Any realistic description [12] requires nonlinear and higher-order terms and therefore numerical solution. Some special systems, mostly liquid or composed of polymers, can be specified, though, where pure spinodal decomposition can be observed and its behaviour rendered by theoretical calculations [3,13]. In all continuum theories of spinodal decomposition diffusion fluxes are set proportional to the gradient of chemical potential difference. This implies that a diffusion mechanism be at work which guarantees a sufficiently smooth flux of material down this gradient. (That this condition is not always met and the consequences resulting thereof will be the central message of this paper). For all these reasons more direct Monte Carlo methods working at the atomistic level and taking account of elastic interaction and individual atom jump probabilities have been favoured recently [14,15].

An early alternative to the Cahn-Hilliard concept was the model of Hillert [16,17] who ascribed composition values to parallel lattice planes, keeping the discrete nature of the lattice in the direction normal to them. Configurational energies are computed from the mean-field atomic environment in the so-called zeroth approximation. Dependent on atomic pair interaction potentials, a behaviour of spinodal decomposition via unstable long-wavelength fluctuations, a nucleation process or an ordering process is obtained. In the latter case the wavelength of the composition fluctuation is equal to twice the lattice spacing. From the experimental point of view, this amounts to homogeneous long-range ordering. Though the Hillert method is attractive due to its ability of describing continuous ordering and spinodal decomposition within one concept, it is obviously limited for practical purposes by its one-dimensional formulation and simplified atomic interaction. Cook et al. [18] later published a discrete version of the continuous Cahn-Hilliard model, defining a local composition as an ensemble average of the occupation of a particular lattice site and extending the possibilities of the Hillert model to 3D. The restrictive assumptions of defining the free energy via a local composition and the principal limitation of the method to small deviations from the initially uniform composition remain, however.

A crucial but difficult question is how to identify the experimental signature of a spinodal or homogeneous higher-order phase transition as set against a heterogeneous one involving the nucleation of particles. There is at the moment no technique capable of continuously observing a complete sequence of decomposition. TEM and light microscopical observations reveal characteristic structures far into the transformation like the well-known 'tweed' contrast often erroneously taken to be indicative of spinodal decomposition [13,19,20]. A dispersion of 'particles' on the other hand can, but need not be the result of a foregoing nucleation process. Simulation calculations reveal that in late stages of spinodal decomposition both a particle-like distribution and a network-like interconnected structure may appear, depending on overall composition [3,13]. It seems that a coarsening mechanism is always at work in an advanced stage of the phase separation, in the spinodal as well as in the nucleation scenario. Whereas real-space inspection thus shows the state of the system more or less *post factum*, various scattering methods (SAXS, SANS, light scattering for optically transparent systems) respond to critical dimensions in the early phases of the phase-separation [13]. Electrical resisitivity, in being extremely sensitive to small changes of order parameter, enables a fine resolution in time of the transition kinetics [21,22] from which spatial ranges can be inferred, as will beome evident from our experimental results. However, it is questionable anyhow that there are two clearly distinguishable regimes of decomposition, starting from either metastable or unstable regions with sharp boundary in between. In fact, at the atomic scale only one type of mechanism is involved, that is diffusion by atom jumps to neighbouring vacancies.

Whereas a first-order phase transition may or may not start continuously from a uniform composition we should expect a truly homogeneous transformation when in a transition from a one-phase region to another one-phase region composition and crystal structure are conserved and some order parameter (long-range order, magnetization) changes continuously. Such a higher-order transition is usually expected to proceed simultaneously and homogeneously within the whole sample volume, with the possible disturbance of a domain structure of the ordered phase resulting from symmetry breaking.

ORDER-ORDER TRANSITIONS BY VACANCY-MEDIATED DIFFUSION

In intermetallic compounds which show long range order up to the melting point no domain structure appears. A transition between two states of nearly perfect, but slightly different, long range order corresponding to equilibrium at different temperatures should be an ideal model of a homogeneous kinetic process. In a recent paper [23] long-range order kinetics of a B2-ordered Fe-48at%Al alloy was studied by residual resistometry (REST), a well-established method that has been shown to be extremely sensitive to changes in the degree of order. It has successfully been applied to get information on atom jump processes in ordered alloys [21,22,24]. Changes of LRO parameter in the range of 10^{-3} that escape detection by any diffraction method can easily be resolved by REST.

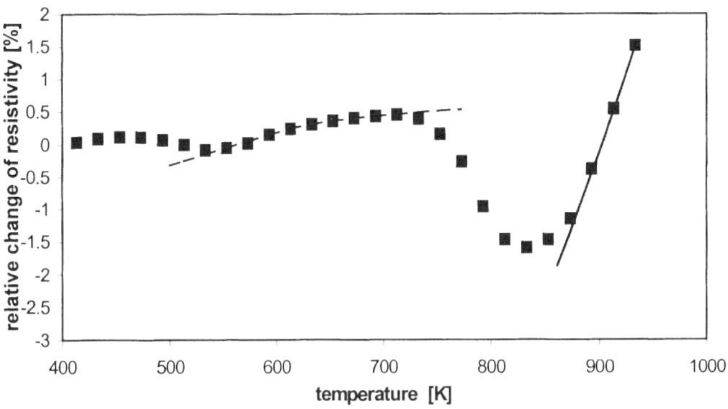

Figure 4. Normalized isochronal resistivity curve for Fe-44.8at%Al (run 4) with local and global equilibrium curves.

The specimen was subjected to four isochronal annealing sequences after different thermal pre-treatment. In all cases the initial state was one of reduced long-range order. Fig. 4 shows the typical qualitative behaviour for one of these isochronal curves (after 3h homogenisation at 1223 K and quenching from 773K): Just below 500K vacancies become mobile and the order increases as resisitivity decreases. Above 710K a strong increase of order is observed as the resistivity falls towards the equilibrium curve (drawn-out in Fig. 4). At still higher temperatures kinetics is fast enough for thermodynamical equilibrium to be established within the isochronal time interval, and the equilibrium degree of order decreases with rising temperature, as thermodynamics requires. It is in an intermediate temperature interval of 530-710K that truly remarkable behaviour occurs: Resistivity increases slightly (long-range order is diminished). Keeping to small isochronal time intervals, this part of the curve can be run backwards in a reversible way. For long enough waiting times (e.g. isothermal long-time annealing), however, resistivity moves down towards the overall equilibrium curve even in this special temperature range.

The key to understanding this counter-intuitive behaviour can be found in vacancy-mediated atom kinetics: In FeAl changes of the atomic configuration take place by a vacancy mechanism. We know from literature that in FeAl the formation energy of vacancies is rather low ~0.7eV [25,26,27], the migration energy however is quite high ~1.5eV [28,29]. This means on the one hand that vacancies, once created, have difficulty reaching sinks and therefore tend to remain in the crystal for a long time, on the other hand they can move only in restricted regions during short time intervals.

Changes in order parameter can take place only in regions visited by vacancies. As these regions for low temperatures and short times are small and disconnected, long range order is at first enhanced and brought to the equilibrium value only in these small parts of the specimen volume. When the temperature is then increased, the equilibrium state of order is adjusted to its new, smaller value within these restricted regions of the sample. Therefore the overall resistivity increases in a certain temperature interval. It is quite plausible to expect that an increase of local order is observed when the global degree of LRO throughout the sample is lower than that corresponding to the actual annealing temperature. It is however not self-evident that an increase of temperature will lead to lower values of local order although the global degree of LRO is still below equilibrium. A vacancy as it travels around in a confined region of space in the crystal may create or destroy antisites, investing or gaining energy, respectively. Let us define as 1 a state where the vacancy sits adjacent to an atom on its regular sublattice, as 2 a state where they have exchanged positions, so that the vacancy now sits next to an antisite atom. The transition rate for this exchange will be given by an appropriate Boltzmann factor

$$\Gamma_+ = \Gamma_0 \exp(-\frac{E^* - E_1}{k_B T}) \tag{1}$$

with E^* denoting an energy barrier between states 1 and 2 . The backward process, returning the antisite atom to its regular sublattice, is assumed to follow an analogous transition probability

$$\Gamma_- = \Gamma_0 \exp(-\frac{E^* - E_2}{k_B T}). \tag{2}$$

Let f_1 and f_2 be the average populations of regular and antisite atoms in the region of crystal under consideration. In equilibrium detailed balance requires

$$\Gamma_+ f_1 = \Gamma_- f_2 \tag{3}$$

so that the relative density of antisites now becomes

$$\frac{f_2}{f_1} = \exp(-\frac{E_2 - E_1}{k_B T}) \tag{4}$$

which is an expression corresponding to what we can derive from equilibrium thermodynamics. The regular position is associated with lower energy than the antisite, so $E_2 > E_1$. For very low temperatures, the fraction f_2/f_1 therefore becomes effectively zero (perfect order). It increases rapidly with temperature, approaching 1 asymptotically for very high temperatures. Kinetically speaking, this means: As the temperature rises the vacancy becomes ever less selective where to jump, the Boltzmann factor becomes less sensitive to the height of the barrier, the transition rates Γ_+ and Γ_- approaching each other. In conclusion we expect for the present case that at low enough temperatures a 'quasi-equilibrium curve' of local order results (dashed line in fig. 4), which with increasing annealing time shifts more and more into the (true) equilibrium line of LRO (full line in fig. 4).

It must be kept in mind that this very simple model loses validity as we depart from perfect order, the real barriers depending on the order parameter. But it helps to visualize how a change of order is brought about, even by a single vacancy making jumps in a restricted volume.

A strong decrease is observed only when the random walk regions of the vacancies begin to impinge on one another and thus reach a percolation limit. As we have seen, this can be achieved by increasing the mobility of vacancies (higher temperature) or by increasing the waiting time. To illustrate this concept, in fig.5 a 2D numerical simulation of vacancy motion is shown concomitant to the isochronal curve. Regions of the crystal that have been visited by a vacancy appear in black. As they become interconnected, long-range order increases in the whole crystal volume, tending towards the overall equilibrium curve.

We believe this example to be representative of a quite general principle with important implications for a large class of configurational changes in solids: Often the kinetics of a phase transformation is described as continuous or "homophase" by means of a generalized diffusion theory, such as the Cahn-Hilliard theory for spinodal decomposition. The tacit assumption is always that *a sufficiently large number of mobile diffusion-mediating defects is available in any small volume element considered* in order to allow a continuum description. Whereas this requirement may always be well fulfilled in

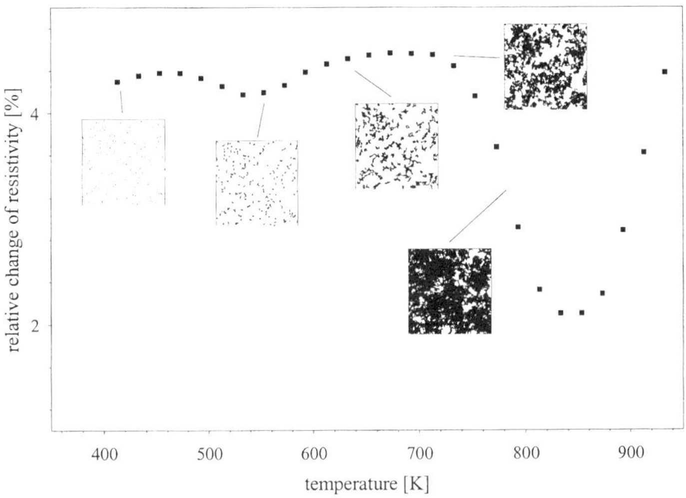

Figure 5. Regions visited by vacancies (shown in black) adjacent to corresponding points on the isochronal resistivity curve.

interstitial alloys with a great number of very mobile defects, in the case of a vacancy mechanism it crucially depends on the number and mobility of the vacancies and the time interval considered. If the number of vacancies is small and/or they move sluggishly and/or we do not observe for a long enough time, any changes in the configuration will take place only in localized regions visited by the vacancies. The resulting picture is therefore a heterogeneous one resembling nucleation of particles rather than the spontaneous growth of long spatial concentration waves. The kinetic model in this case should center around the moving vacancies.

A SIMPLE MODEL

Just to make an order-of-magnitude estimate of the conditions under which a heterogeneous nature is introduced in a phase transformation by the limited mobility of vacancies, we propose a very simple model.

The heterogeneous regime persists at least as long as the average random walk regions of the vacancies do not overlap. Assuming an isotropic random walk, they begin to impinge upon one another at time t_{imp} when

$$\bar{l} \cong 2\sqrt{6D_v t_{imp}} \tag{5}$$

$\bar{l}$ being the average distance of vacancies and D_v their diffusivity. Let the vacancies be equally distributed in space. Their average distance is then connected to the lattice spacing a by the vacancy concentration c_v

$$\bar{l} = \frac{a}{c_v^{1/3}} \tag{6}$$

For the diffusivity of the vacancies we assume the usual Arrhenius law

$$D_v = (Z/6)\, a^2 \nu \exp(-\frac{H_m}{kT}) \tag{7}$$

with coordination number Z, a jump frequency ν and vacancy migration enthalpy H_m. Solving for the impingement time t_{imp} we get

$$t_{imp} = \frac{\exp(\frac{H_m}{kT})}{4\, Z \nu\, c_v^{2/3}} \tag{8}$$

Fig. 6 is a graphical representation of this formula as a contour plot of lg t_{imp} versus H_m/kT and lg c_v. In this approximation, the jump frequency ν has been set so that Z ν amounted to 10^{14} , Z=8 being used for a body centered cubic crystal structure. As can be easily verified from eqn. (8) equal impingement times are repesented by straight lines in Fig. 6, covering five orders of magnitude. However, this plot may serve only for crude estimates since not all atoms necessarily are visited in a sphere of radius $\bar{l}$ around a vacancy and that in addition in a realistic model crystal the vacancies do not perform a true random walk, let alone an isotropic one. Instead the jump probabilities are influenced strongly by the local atomic configuration. Quenching for instance from a completely disordered state into a region of long-range order the vacancies will be preferentially involved in jumps which increase order, and will be drawn to disordered regions, avoiding regions where order has already been established. To get a more realistic view on vacancy motion a Monte Carlo study will be undertaken.

It is nonetheless instructive to apply this graph to our FeAl data. Assuming H_m= 1.5 eV and a vacancy concentration of $c_v=10^{-5}$ not varying with temperature (which are reasonable assumptions for FeAl [27,30]) the positions indicated by the black circles in the diagram (Fig. 6) are reached. At 500K the migration regions of the vacancies would take several hours to touch and are therefore well-separated, giving a heterogeneous character to the ordering process. At 700K a continuous transformation may be safely assumed for the isochronal waiting time, the impingement time being well below one second. These

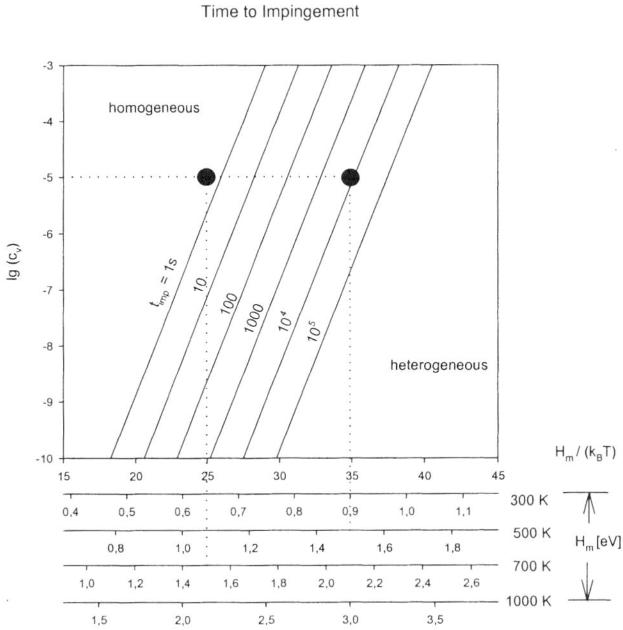

Figure 6. Time needed for the impingement of random walk regions of the vacancies as a function of enthalpy of migration and logarithm of vacancy density.

temperatures can be identified in Fig. 4 with the local ordering regime and the steep decrease of resisitivity associated with global ordering.

CONCLUSIONS AND SUGGESTED FUTURE APPROACHES

As has been demonstrated very clearly by the experimental results on order-order transformations in B2-ordered FeAl, the sample volume the vacancies visit during their walk through the lattice plays an essential role in diffusion controlled, vacancy mediated phase transformations. The simple model given above explains very well our findings that in a certain temperature range and for usual observation times changes in the LRO-parameter occur in a restricted sample volume only, resulting in a heterogeneous, metastable state with a lower average degree of order than corresponding to thermodynamic equilibrium.

A similar reason may explain the findings of Miyazaki et al. concerning the A2→B2 phase transition in FeAl [31]. With their interesting composition gradient method they find a two phase field giving evidence of a first-order phase transformation where a higher-order transformation is generally accepted. It could well be that the B2-ordered particles found in this case result from the restricted vacancy motion and correspond to already transformed regions, leaving the rest of the material still in the untransformed state of A2.

A mechanism of local ordering has in fact been described by previous authors. Allen and Cahn [32] have made an extensive analysis of phase transition/ordering phenomena in the FeAl system. In a region of the phase diagram where continuous ordering was expected distinct ordered particles were observed experimentally. These were surmised to have been created by single vacancies traveling around at the rim of the ordered regions, in keeping with the results of an earlier computer simulation by Beeler [33]. A very careful distinction of the thermodynamical preconditions and the underlying causes of local ordering has however to be made. More recent and more refined Monte-Carlo simulations by Athènes et al [34] deal with ordering in a B2 model alloy in the two-phase region below a tricritical point[2]. As is to be expected in such a system, different morphologies of precipitation were reported according to alloy composition. A part of the temporal sequence of decomposition, however, can be expected to proceed in a homogeneous way: In the area between the two conditional spinodals continuous ordering precedes spinodal decomposition. When in this case certain interaction energies were varied in the simulation, a preferred exchange of vacancies with either A or B atoms led to either localized or homogeneous ordering, a behaviour that could very well explain the observations by Cahn and Allen [32]. It has likewise been demonstrated in Monte-Carlo simulations that changing the vacancy interaction energies can alter the mechanism of precipitate coarsening [35,36]: If the energies are chosen so that the vacancies move preferentially within the matrix phase then Lifhitz-Slyozov-Wagner [37,38] type coarsening prevails. If they prefer to stay within the particles these move as a whole and coagulation predominates. Certainly the possible influence of bias effects of this kind has to be considered when applying criteria of vacancy concentration, mobility, and observation time as we propose. A detailed study of vacancy motion in anisotropic, single and two phase settings with different degrees of atomic order is therefore mandatory.

In conclusion, it is quite difficult, both from a theoretical and an experimental viewpoint, to establish the conditions for a homogeneous *first-order* phase transformation. A homogeneous character is, however, natural to higher-order transforma-tions. It is all the more remarkable that in our example of an order-order transition, which should meet the requirements for homogeneity in an exemplary way, a heterogeneous structure is introduced by restricted vacancy motion. The possibility of such behaviour should therefore always be kept in mind when dealing with vacancy-mediated diffusive phase transformations.

Acknowledgements

The financial support by the Austrian 'Fonds zur Förderung der wissenschaftlichen Forschung' and the 'Hochschuljubiläumsstiftung der Gemeinde Wien' is gratefully acknowledged.

REFERENCES

1. P. Ehrenfest, *Communications Leiden* 1933; Suppl. 756. See also discussion in Epstein, *Thermodynamics*, Wiley, New York 1937, pp. 128-133
2. K. Binder in: *Stochastic nonlinear systems in physics, chemistry and biology*, L. Arnold and R. Lefevre, editors. Springer, Berlin 1981, p. 62
3. K. Binder and P. Fratzl in: *Phase Transformations in Material,* G. Kostorz, editor. Wiley-VCH, Weinheim 2001, Chap. 6.
4. K. Binder, *Rep. Progr. Phys.* **50**, 783 (1987).
5. R. Kikuchi, *J. Chem. Phys.* **47**, 1664 (1967).
6. J.W. Cahn and J.E. Hilliard, *J. Chem. Phys.* **28**, 259 (1958).

---

[2] This is a more complex scenario than our order-order transformation which was done on an FeAl alloy with a composition of 44.8at% Al and at a temperature which is definitely not in a two-phase region of the equilibrium phase diagram.

7. J.W. Cahn and J.E. Hilliard, *J. Phys. Chem.* **31**, 3 and 788 (1959).
8. J.W. Cahn, *TMS AIME* **242**, 166 (1968).
9. J.E. Hilliard in *Phase Transformations,* H.I. Aaronson, editor. ASM, Metals Park, Ohio 1970, p. 497.
10. W.A. Soffa and D.E. Laughlin in *Solid-Solid Phase Transformations,* H.I. Aaronson, editor. Met. Soc. AIME, Warrendale 1982, p. 159.
11. W.A. Soffa and D.E. Laughlin, *Acta Metall.* **37**, 3019 (1989).
12. J.S. Langer, M. Baron, and H.D. Miller, *Phys. Rev. A* **11**, 1417 (1975).
13. R. Wagner, R. Kampmann, and P.W. in *Phase Transformations in Materials,* G. Kostorz, editor. Wiley-VCH, Weinheim 2001, Chap. 5.
14. P. Fratzl and O. Penrose, *Acta Metall. Mater.* **43**, 2921 (1995).
15. C.A. Laberge, P. Fratzl, and L. Lebowitz, *Phys. Rev. Letters* **75**, 4448 (1995).
16. M. Hillert, *D.Sc. Thesis,* MIT 1956.
17. M. Hillert, *Acta Metall.* **9**, 525 (1961).
18. H. E. Cook, D. de Fontaine, and J.E.. Hilliard, *Acta Metall.* **17**, 765 (1969).
19. L.-Q. Chen, Y. Wang, and A.G. Khatchaturyan, *Phil. Mag. Letters* **65**, 15 (1992).
20. A.G. Khatchaturyan, *Theory of Structural Transformations in Solids.* John Wiley & Sons, New York 1983.
21. W. Pfeiler, JOM **52**, 14 (2000).
22. W. Pfeiler and B. Sprušil, *Mater. Sci. Eng. A* **324**, 34 (2002).
23. H. Lang, K. Rohrhofer, P. Rosenkranz, R. Kozubski, W. Püschl, and W. Pfeiler, *Intermetallics* **10**, 283 (2002).
24. W. Pfeiler in: *Properties of Comple Inorganic Solids*, A. Gonis, A. Meike, and P.E.A. Turchi, editors. Plenum Press, New York 1997, p. 219.
25. G.S. Collins, L.S.-J. Peng, M. Wie, *Mater. Res. Soc. Symp. Proc.* **552**, KK4.2.1 (1999).
26. M. Kogachi and T. Haraguchi, *Mater. Sci. Eng. A* **230**, 124 (1997).
27. J. Wolff, M. Franz, A. Broska, R. Kerl, M. Weinhagen, B. Kohler, M. Brauer, F. Faupel, und T. Hehenkamp, *Intermetallics* **7**, 289 (1999).
28. R. Würschum and H.-E. Schaefer, *Mater. Sci. Forum* **81**, 255 (1997).
29. J.P. Rivière and J. Grilhe, *Acta Metall.* **20**, 1275 (1972).
30. M. Kogachi and T. Haraguchi, *Intermetallics* **7**, 981 (1999).
31. A.M. Mebed, T. Koyama, and T. Miyazaki in: *Solid-Solid Phase Transformations '99*, M. Koiwa, K. Otsuka, and T. Miyazaki, editors. The Japan Institute of Metals, Sendai 1999, p. 61.
32. S.M. Allen and J.W. Cahn, *Acta Metall.* **24**, 425 (1976).
33. J.R. Beeler, *Phys. Rev. A* **138**, 1259 (1965).
34. M. Athènes, P. Bellon, G. Martin, and F. Haider, *Acta mater.* **44**, 4739 (1996).
35. F. Soisson, A. Barbu, and G. Martin, *Acta Mater.* **44**, 3789 (1996).
36. P. Fratzl and O. Penrose, *Phys.Rev. B* **55**, R6101(1997).
37. C.Z. Wagner, *Elektrochem.* **65**, 581 (1961).
38. I.M. Lifshitz and V.V. Slyozov, *J. Phys. Chem. Solids* **19**, 35 (1961).

CALCULATION OF THE PHASE DIAGRAMS OF ALLOYS WITH NONPAIR ATOMIC INTERACTIONS WITHIN THE RING APPROXIMATION

R. V. Chepulskii[1,2]

[1]Faculty of Physics, Taras Shevchenko Kyiv National University
UA-03022 Kyiv, Ukraine
[2]Department of Solid State Theory, Institute for Metal Physics, N.A.S.U.
UA-03680 Kyiv-142, Ukraine

ABSTRACT

The elaborated in [R. V. Chepulskii, Analytical method for calculation of the phase diagram of a two-component lattice gas, *Solid State Commun.* 115:497 (2000)] analytical method for calculation of the phase diagrams of alloys with *pair* atomic interactions is generalized to the case of many-body atomic interactions of arbitrary orders and effective radii of action. The method is developed within the ring approximation in the context of a modified thermodynamic perturbation theory with the use of the inverse effective number of atoms interacting with one fixed atom as a small parameter of expansion. By a comparison with the results of the Monte Carlo simulation, the high numerical accuracy of the generalized method is demonstrated in a wide concentration interval.

INTRODUCTION

In Ref. 1 the new analytical method for calculation of the phase diagrams of alloys (or more generally of a two-component lattice gas[2]) with arbitrary complex crystal lattice and any long-range order in atomic distribution was developed. The method was elaborated within the ring approximation in the context of a modified thermodynamic perturbation theory with the use of the inverse effective number of atoms interacting with one fixed atom as a small parameter of expansion[3,4]. The numerical accuracy of the method proved to be high in a wide temperature-concentration interval and turns higher with increase of the effective radius of atomic interactions. The consideration of the lattice gas with arbitrarily long-range atomic interactions is possible within the method, because the interaction's parameters appear in the corresponding expressions only through the Fourier transforms of the interatomic potentials. It should be noted the much comparative generality and simplicity of the method in comparison with, *e.g.*, the most widely used Monte Carlo[5] and cluster-variation methods[6].

However, the developed method can be applied in case of alloys with only *pair* atomic interactions. The aim of the present letter is to generalize the method to the case of alloys with many-body atomic interactions of arbitrary orders and effective radii of action.

THEORY

Let us consider a two-component A-B alloy (within the lattice gas model) whose primitive unit cell consists of ν crystal lattice sites. Taking into account the many-body atomic interactions of arbitrary orders and radii of action, the Hamiltonian H of it can be written in the following form[7]

$$\begin{aligned} H &= N v_0 + \sum_{n=1}^{N\nu} \frac{1}{n!} \sum_{i_1, i_2, \ldots, i_n} \sum_{\mathbf{R}_1, \mathbf{R}_2, \ldots, \mathbf{R}_n} V^{(n)}_{i_1,\mathbf{R}_1; i_2,\mathbf{R}_2; \ldots; i_n,\mathbf{R}_n} C_{i_1,\mathbf{R}_1} C_{i_2,\mathbf{R}_2} \ldots C_{i_n,\mathbf{R}_n} \\ &= N v_0 + \sum_i V_i^{(1)} \sum_{\mathbf{R}} C_{i,\mathbf{R}} + \frac{1}{2} \sum_{i_1,i_2} \sum_{\mathbf{R}_1,\mathbf{R}_2} V^{(2)}_{i_1,\mathbf{R}_1; i_2,\mathbf{R}_2} C_{i_1,\mathbf{R}_1} C_{i_2,\mathbf{R}_2} \\ &+ \frac{1}{6} \sum_{i_1,i_2,i_3} \sum_{\mathbf{R}_1,\mathbf{R}_2,\mathbf{R}_3} V^{(3)}_{i_1,\mathbf{R}_1; i_2,\mathbf{R}_2; i_3,\mathbf{R}_3} C_{i_1,\mathbf{R}_1} C_{i_2,\mathbf{R}_2} C_{i_3,\mathbf{R}_3} + \ldots \, . \end{aligned} \quad (1)$$

In (1): N is the number of primitive unit cells of the crystal lattice, v_0 is the energy per unit cell of the lattice gas in which all $N\nu$ sites are occupied by B-type atoms, $V^{(n)}_{i_1,\mathbf{R}_1;i_2,\mathbf{R}_2;...;i_n,\mathbf{R}_n}$ is the mixing potential of n-th order (n=1,2,..., $N\nu$),

$$C_{i,\mathbf{R}} = \begin{cases} 1, \text{ if the site } (i,\mathbf{R}) \text{ is occupied by an A - type atom} \\ 0, \text{ otherwise} \end{cases}, \tag{2}$$

the summations on the indices $i_1,i_2,...,i_n$ and on the radius-vectors $\mathbf{R}_1,\mathbf{R}_2,...,\mathbf{R}_n$ are carried over all ν sublattices and N primitive unit cells of the crystal lattice, respectively.

Following to the procedure described in Refs. 3 and 4, the expression for the free energy f per one primitive unit cell of the system in question may be presented in such form[8]:

$$f = v_0 + \sum_i \left\{ P_i\left(V_i^{(1)} - \frac{\mu_i}{2}\right) + k_B T\left[P_i \ln P_i + (1-P_i)\ln(1-P_i)\right] + \Delta f \right\}, \tag{3}$$

where

$$\Delta f = -k_B T N^{-1} \times \ln\left\langle \exp\left[-\frac{1}{k_B T}\sum_{n=2}^{N\nu}\frac{1}{n!}\sum_{i_1,i_2,...,i_n}\sum_{\mathbf{R}_1,\mathbf{R}_2,...,\mathbf{R}_n} W^{(n)}_{i_1,\mathbf{R}_1;i_2,\mathbf{R}_2;...;i_n,\mathbf{R}_n} C_{i_1,\mathbf{R}_1}C_{i_2,\mathbf{R}_2}\cdots C_{i_n,\mathbf{R}_n}\right]\right\rangle, \tag{4}$$

$$W^{(n)}_{i_1,\mathbf{R}_1;i_2,\mathbf{R}_2;...;i_n,\mathbf{R}_n} = \begin{cases} V^{(2)}_{i_1,\mathbf{R}_1;i_2,\mathbf{R}_2} + \mu_{i_1}\delta_{i_1,i_2}\delta_{\mathbf{R}_1,\mathbf{R}_2}, \text{ if } n=2 \\ V^{(n)}_{i_1,\mathbf{R}_1;i_2,\mathbf{R}_2;...;i_n,\mathbf{R}_n}, \text{ if } n>2 \end{cases}, \tag{5}$$

$$\mu_i = 2\left(V_i^{(1)} - \mu_i^A + \mu_i^B\right), \tag{6}$$

μ_i^A and μ_i^B are the chemical potentials of A- and B-type atoms situated at the i-th sublattice, respectively, T is the absolute temperature, k_B is the Boltzmann constant, the sign $\langle ... \rangle$ means the statistical average over all states with given values of the LRO parameters, δ_{i_1,i_2} and $\delta_{\mathbf{R}_1,\mathbf{R}_2}$ are the Kronecker deltas,

$$P_i = \langle C_{i,\mathbf{R}} \rangle \tag{7}$$

and is equal to the probability of finding an A-type atom at the site belonging to the i-th sublattice. Note that the quantities P_i are independent from the radius-vector $\mathbf{R}$ of primitive unit cells due to the translational invariance of these cells. The values of the chemical potentials μ_i^{A}, μ_i^{B} and, therefore [see (6)], the values of μ_i (i=1,2,..., ν) must satisfy the following relationships

$$\partial f / \partial \mu_i = 0, \tag{8}$$

which follow from the general thermodynamic ones.

According to the general approach of the thermodynamic perturbation theory,[3,4,9-12] the expression (4) for Δf can be expanded in a cumulant series in powers of the inverse temperature. Following to the Brout's approach,[3,4,9] let us select the contributions to the cumulant expansion from the summands proportional, respectively, to zeroth and first powers of the quantity z^{-1} with z being equal to the effective number of atoms interacting with one fixed atom:

$$\begin{aligned}\Delta f = &\frac{1}{N}\sum_{n=2}^{N\nu}\frac{1}{n!}\sum_{i_1,i_2,\dots,i_n}\sum_{\mathbf{R}_1,\mathbf{R}_2,\dots,\mathbf{R}_n} W^{(n)}_{i_1,\mathbf{R}_1;i_2,\mathbf{R}_2;\dots;i_n,\mathbf{R}_n} P_{i_1}P_{i_2}\dots P_{i_n}\\ &-\frac{k_{\mathrm{B}}T}{2N}\sum_{n=1}^{\infty}\frac{1}{n}\sum_{i_1,i_2,\dots,i_n}\sum_{\mathbf{R}_1,\mathbf{R}_2,\dots,\mathbf{R}_n} f^{i_1,i_2}_{\mathbf{R}_1-\mathbf{R}_2} f^{i_2,i_3}_{\mathbf{R}_2-\mathbf{R}_3}\dots f^{i_n,i_1}_{\mathbf{R}_n-\mathbf{R}_1},\end{aligned} \tag{9}$$

where

$$f^{i_l,i_m}_{\mathbf{R}_l-\mathbf{R}_m} = -\frac{\sqrt{P_{i_l}\left(1-P_{i_l}\right)}\varphi^{i_l,i_m}_{\mathbf{R}_l-\mathbf{R}_m}\sqrt{P_{i_m}\left(1-P_{i_m}\right)}}{k_{\mathrm{B}}T}, \tag{10}$$

$$\varphi^{i_l,i_m}_{\mathbf{R}_l-\mathbf{R}_m} = \sum_{t=0}^{N\nu-2}\frac{1}{n!}\sum_{i'_1,i'_2,\dots,i'_t}\frac{P_{i'_1}P_{i'_2}\dots P_{i'_n}}{t!}\sum_{\mathbf{R}'_1,\mathbf{R}'_2,\dots,\mathbf{R}'_n} W^{(2+t)}_{i'_1,\mathbf{R}'_1;i'_2,\mathbf{R}'_2;\dots;i'_t,\mathbf{R}'_t;\ i_l,\mathbf{R}_l;i_m,\mathbf{R}_m}. \tag{11}$$

Introducing the Fourier transforms of the mixing potentials

$$\tilde{V}^{(n)}_{i_1,\mathbf{k}_1;i_2,\mathbf{k}_2;\dots;i_{n-1},\mathbf{k}_{n-1};i_n} = \sum_{\mathbf{R}_1,\mathbf{R}_2,\dots,\mathbf{R}_{n-1}} V^{(n)}_{i_1,\mathbf{R}_1;i_2,\mathbf{R}_2;\dots;i_{n-1},\mathbf{R}_{n-1};i_n,\mathbf{0}}\exp\left(-i\sum_{i=1}^{n-1}\mathbf{k}_i\mathbf{R}_i\right), \tag{12}$$

and performing a number of matrix transformations, one can obtain[8]

$$\Delta f = \frac{1}{2}\sum_i \mu_i\left(P_i\right)^2 + \sum_{n=2}^{N\nu}\frac{1}{n!}\sum_{i_1,i_2,\dots,i_n}\tilde{V}^{(n)}_{i_1,\mathbf{0};i_2,\mathbf{0};\dots;i_{n-1},\mathbf{0};i_n} P_{i_1}P_{i_2}\dots P_{i_n} + \frac{k_{\mathrm{B}}T}{2N}\sum_{\mathbf{k}}\ln\det\mathbf{A}_{\mathbf{k}}, \tag{13}$$

where the summation on **k** is carried over all the points specified by the cyclic boundary conditions in the corresponding first Brillouin zone and the designation det$\mathbf{A}_{\mathbf{k}}$ means the determinant of the matrix $\mathbf{A}_{\mathbf{k}}$ with the following elements

$$\mathbf{A}_{\mathbf{k}}^{i_l,i_m} = \delta_{i_l,i_m} - \frac{\sqrt{P_{i_l}\left(1-P_{i_l}\right)}\widetilde{\varphi}_{\mathbf{k}}^{i_l,i_m}\sqrt{P_{i_m}\left(1-P_{i_m}\right)}}{k_{\mathrm{B}}T}, \tag{14}$$

$$\begin{aligned}\widetilde{\varphi}_{\mathbf{k}}^{i_l,i_m} &= \sum_{t=0}^{N\nu-2}\frac{1}{n!}\sum_{i'_1,i'_2,\ldots,i'_t}\frac{P_{i'_1}P_{i'_2}\ldots P_{i'_n}}{t!}\widetilde{W}^{(2+t)}_{i'_1,0;i'_2,0;\ldots;i'_t,0;\,i_l,\mathbf{k};i_m}. \\ &= \mu_{i_l}\delta_{i_l,i_m} + \widetilde{V}^{(2)}_{i_l,\mathbf{k};i_m} + \sum_{i'_1}P_{i'_1}\widetilde{V}^{(3)}_{i'_1,0;\,i_l,\mathbf{k};i_m} + \sum_{i'_1,i'_2}\frac{P_{i'_1}P_{i'_2}}{2}\widetilde{V}^{(4)}_{i'_1,0;i'_2,0;\,i_l,\mathbf{k};i_m} + \ldots.\end{aligned} \tag{15}$$

Substituting (13) into (3), we obtain the following expression for the free energy

$$f_{\text{ring}} = f_{\text{MF}} - \frac{1}{2}\sum_i \mu_i P_i\left(1-P_i\right) + \frac{k_{\mathrm{B}}T}{2N}\sum_{\mathbf{k}} \ln\det\mathbf{A}_{\mathbf{k}}, \tag{16}$$

where f_{MF} corresponds to the mean-field approximation

$$\begin{aligned}f_{\text{MF}} &= v_0 + \sum_i V_i^{(1)}P_i + \sum_{n=2}^{N\nu}\frac{1}{n!}\sum_{i_1,i_2,\ldots,i_n}\widetilde{V}^{(n)}_{i_1,0;i_2,0;\ldots;i_{n-1},0;i_n}P_{i_1}P_{i_2}\ldots P_{i_n} \\ &+k_{\mathrm{B}}T\sum_i\left[P_i\ln P_i + \left(1-P_i\right)\ln\left(1-P_i\right)\right]\end{aligned} \tag{17}$$

or writing the series explicitly

$$\begin{aligned}f_{\text{MF}} &= v_0 + \sum_i V_i^{(1)}P_i + \frac{1}{2}\sum_{i_1,i_2}\widetilde{V}^{(2)}_{i_1,0;i_2}P_{i_1}P_{i_2} + \frac{1}{6}\sum_{i_1,i_2,i_3}\widetilde{V}^{(3)}_{i_1,0;i_2,0;i_3}P_{i_1}P_{i_2}P_{i_3} \\ &+\frac{1}{24}\sum_{i_1,i_2,i_3,i_4}\widetilde{V}^{(4)}_{i_1,0;i_2,0;i_3,0;i_4}P_{i_1}P_{i_2}P_{i_3}P_{i_4} + \ldots \\ &+k_{\mathrm{B}}T\sum_i\left[P_i\ln P_i + \left(1-P_i\right)\ln\left(1-P_i\right)\right].\end{aligned} \tag{18}$$

Using (8), we derive the following equations for determination of the quantities μ_i (i=1,2,...,ν)

$$N^{-1}\sum_{\mathbf{k}}\left[\mathbf{A}_{\mathbf{k}}\right]_{ii}^{-1} = 1, \tag{19}$$

where the designation $\left[\mathbf{A}_{\mathbf{k}}\right]_{ii}^{-1}$ means the i-th diagonal element of the matrix inverse to $\mathbf{A}_{\mathbf{k}}$.

Note that in a particular case of the disordered alloy with a Bravais crystal lattice, when $P_i = c = N_A/(\nu N)$ for any I=1,2,...,ν (c is the concentration of the A-type atoms, N_{A} is the total number of these atoms in the considered system), the obtained expression (16) for the free energy correspond to those for the grand thermodynamic potential derived within the ring approximation in Ref. 4. We shall also call the approximation obtained above as the ring one. Note that such name of the approximation is in accordance with the topology of the diagrams being taken into account within the ring approximation in the context of the corresponding diagram technique[8]. Neglecting the nonpair atomic interactions in the expression (16), we arrive to those obtained in Ref. 1.

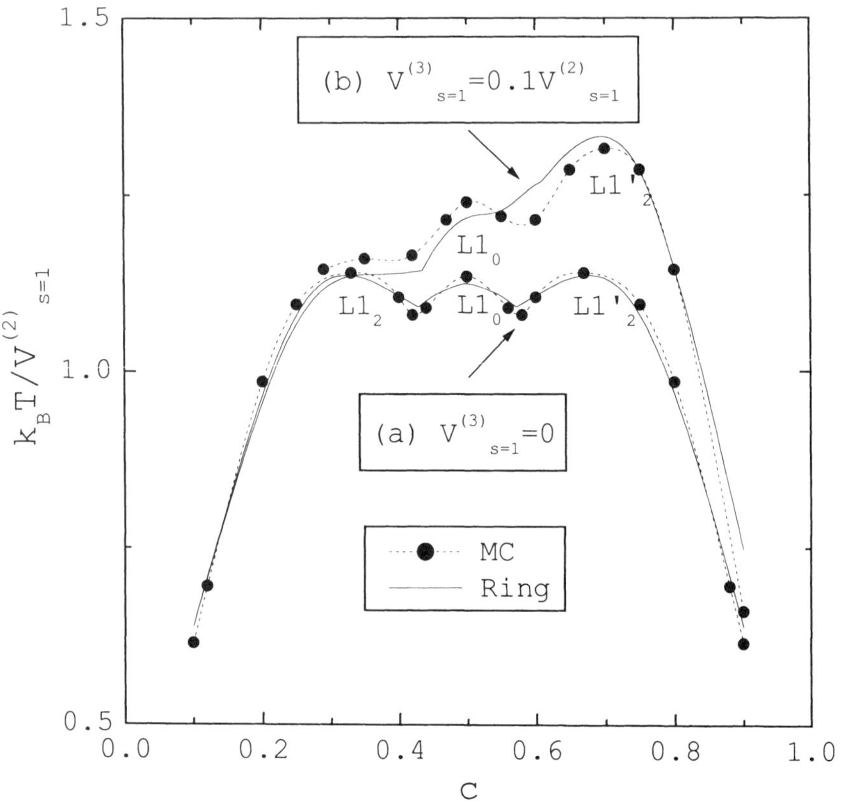

Figure 1. The values of the order-disorder phase transformation temperatures calculated within the ring approximation (16) (Ring) as well as by the Monte Carlo simulation (MC) in case of the f.c.c. crystal lattice with $V_i^{(1)} = 0$, $V_{s=1}^{(2)} > 0$, $V_{s=2}^{(2)} = -0.5V_{s=1}^{(2)}$ and (a) $V_{s=1}^{(3)} = 0$ (lower case), (b) $V_{s=1}^{(3)} = 0.1V_{s=1}^{(2)}$ (upper case). $V_s^{(n)}$ is the value of the n-th order mixing potential for the s-th coordination shell[4] of the f.c.c. crystal lattice. All the other mixing potentials except the denoted ones are equal to zero. The MC data were obtained in accordance with the procedure described in Ref. 3. The superstructures designations see *e.g.* in Ref. 13.

NUMERICAL RESULTS AND CONCLUSIONS

Using the expressions derived above, one can calculate the complete phase diagram of a two-component alloy with an arbitrary crystal structure and with many-body atomic interactions of arbitrary orders and effective radii of action[1,8]. As an example and with the aim to study the numerical accuracy of the ring approximation, we considered the model case appropriate to the face-centred cubic (f.c.c.) crystal lattice with $V_i^{(1)} = 0$, $V_{s=1}^{(2)} > 0$, $V_{s=2}^{(2)} = -0.5V_{s=1}^{(2)}$, $V_{s=1}^{(3)} = 0.1V_{s=1}^{(2)}$ ($V_s^{(n)}$ is the value of the n-th order mixing potential for the s-th coordination shell[4] of the f.c.c. crystal lattice; all the other mixing potentials except the denoted ones are equal to zero). In Fig. 1, we present the values of the order-disorder phase transformation temperatures calculated by the Monte Carlo simulation as well as within the ring approximation (from the condition of the equality of the free energies of the disordered and corresponding ordered states). In this figure, with the aim of comparison, we also included the data from Ref. 1 corresponding to the same case but without triplet atomic interactions. Note that in Fig. 1, we do not include the two-phase regions. This will be done elsewhere[8].

Accepting the results of the Monte Carlo simulation as a standard, on the basis of the data presented in Fig. 1, one may conclude the following. In the both considered cases, the ring approximation yields the adequate results at the entire concentration interval. There is some decrease of the numerical accuracy of the ring approximation when the triplet atomic interactions are taken into account.

ACKNOWLEDGEMENTS

The author thanks Dr. I.A. Abrikosov, Prof. E. Bruno, Prof. B.L. Gyorffy, Dr. L.V. Pourovskii, Dr. A.V. Ruban, Dr. S. Shallcross and Dr. V.A. Tatarenko for stimulating discussions.

REFERENCES

1. R. V. Chepulskii, Analytical method for calculation of the phase diagram of a two-component lattice gas, *Solid State Commun.* 115:497 (2000).
2. T. D. Lee and C. N. Yang, Statistical theory of equations of state and phase transitions. II. Lattice gas Izing model, *Phys. Rev.* 87:410 (1952).
3. R. V. Chepulskii and V. N. Bugaev, Analytical methods for calculation of the short-range order in crystal compounds. I. General theory, *J. Phys.: Condens. Matter* 10:7309 (1998); R. V. Chepulskii and V. N. Bugaev, Analytical methods for calculation of the short-range order in crystal compounds. II. Numerical accuracy study, *J. Phys.: Condens. Matter* 10:7327 (1998).
4. R. V. Chepulskii, Analytical description of the short-range order in crystal compounds with many-body atomic interactions II, *J. Phys.: Condens. Matter* 11:8645 (1999); R. V. Chepulskii, Effect of nonpair atomic interactions on the short-range order in disordered crystal compounds, *J. Phys.: Condens. Matter* 11:8661 (1999).
5. K. Binder and D. W. Heermann. Monte Carlo Simulation in Statistical Physics: an Introduction, Springer, Berlin (1988).
6. D. de Fontaine, Configurational thermodynamics of solid solutions, *Solid State Physics* 34:73 (1979).
7. V. N. Bugaev and R. V. Chepulskii, The symmetry of interatomic lattice potentials in general crystal structures. I. Basic theory, *Acta Cryst.* A 51:456 (1995).
8. R. V. Chepulskii, to be published.
9. R. Brout. Phase Transitions, Benjamin, New York (1965).
10. A. G. Khachaturyan, Ordering in substitutional and interstitial solid solutions, *J. Prog. Mat. Sci.* 22:1 (1978).
11. J. G. Kirkwood, Order and disorder in binary solid solutions, *J. Chem. Phys.* 6:70 (1938).
12. R. Kubo, Generalized cumulant expansion method, *J. Phys. Soc. Japan* 17:1100 (1962).
13. A. Finel and F. Ducastelle, On the phase diagram of the FCC Ising model with antiferromagnetic first-neighbour interactions, *Europhys. Lett.* 1:135 (1986).

MODELLING OF PHASE SEPARATION IN IRON-BASED TERNARY ALLOYS

Yoshiyuki Saito[1]

1 INTRODUCTION

Control of atomic scale modulated structures associated with phase separation is considered to be one of the most important factors for the design of nano materials. Phase separation behavior of ternary alloys [4, 5] is different from that of binary alloys [1, 2, 3]. According to a numerical simulation of phase separation in Fe-Cr-Mo ternary alloys given by Honjo and Saito [5], a periodic microstructure including high Cr and Mo was formed by phase separation in an Fe-40at.%Cr-3at.%Mo alloy and the Mo rich regions were formed inside the Cr rich region. However a little decrease in the amplitude of Mo concentration at the peak positions of Cr.

This paper deals with kinetics of phase separation in ternary alloys, especially asymptotic behavior of a minor element such as Mo in the above alloy associated with the decomposition of a major element. A theory on asymptotic behavior of the minor element in Fe-based ternary alloy was proposed. And then Numerical simulation models based on the Cahn-Hilliard equation[1, 2] have been applied to the investigation of phase separation in Fe-Cr-Mo ternary alloys. Simulated asymptotic behavior of Mo or Cr asscociated with decomposition of Cr or Mo is compared with that predicted by the theory.

2 MODEL

2.1 The Cahn-Hilliard Equation for Multicomponent System

The free energy of inhomegenous system of N-component system for a cubic lattice, F, is given by

$$F = \int_V f dv$$

$$(2.1) \qquad = \int_V \left[f_1(c_1, \cdots, c_N) + \frac{1}{2}\sum_{i=1}^{N}\sum_{j=1}^{N}[K_i\delta_{ij} + (1-\delta_{ij})L_{ij}]\nabla c_i \nabla c_j \right] dV$$

where f is the local free energy per unit volume, f_1 is the local free energy per unit volume of a solute of uniform composition $(c_1, \cdots, c_N)$, $c_i(\mathbf{x}, t)$ is the time-dependent concentration. δ_{ij} is the Kronecker's delta and K_i and L_{ij} are given

[1]Y.Saito, Department of Materials Science and Engineering, Waseda University, Okubo Shinjuku-ku, Tokyo 169-8555

by

$$K_i = \left[-2\frac{\partial^2 f}{\partial c_i \partial(\nabla^2 c_i)} + \frac{\partial^2 f}{\partial(\nabla c_i)^2} \right]_0 \tag{2.2}$$

$$L_{ij} = \left[\frac{\partial^2 f}{\partial(\nabla c_i)\partial(\nabla c_j)} \right]_0 \tag{2.3}$$

The subscript zero indicates the value of the parameter in a solution of uniform composition. It is well known that the chemical potenitial in inhomogeneous system is proportional to the functional derivative of the free energy.

$$\mu_i(\mathbf{x}, t) = \frac{\delta f}{\delta c_i} \tag{2.4}$$

Accoording to Onsager the current density of the i element, $\mathbf{J}_i(\mathbf{x}, t)$, is proportional to the gradient of the chemical potential.

$$\mathbf{J}_i(\mathbf{x}, t) = M_i \nabla \mu_i(\mathbf{x}, t) \tag{2.5}$$

where M_i is the moblity of the i element. Insertion of eqs.(2.4) and (2.5) into a continuity equation

$$\frac{\partial c_i(\mathbf{x}, t)}{\partial t} + \nabla \cdot \mathbf{J}_i(\mathbf{x}, t) = 0 \tag{2.6}$$

we obtain the Cahn-Hilliard equation [1].

$$\frac{\partial c_i}{\partial t} = \nabla \left[M_i \nabla \left(\frac{\partial f_1}{\partial c_i} - K_i \nabla^2 c_i - \sum_{j \neq i} L_{ij} \nabla^2 c_j \right) \right] \tag{2.7}$$

2.2 The Cahn-Hilliard Equation for a Ternary Alloy

Hereafter we consider one-dimentional case for simplicity. If the mobility of elements are not dependent on their positions in the space, the Cahn-Hillirad equation for a Fe-X-Y ternary alloy is given by

$$\frac{\partial c_X}{\partial t} = M_X \left[\frac{\partial^2}{\partial x^2} \left(\frac{\partial f_1}{\partial c_X} - K_X \frac{\partial^2 c_X}{\partial x^2} - L_{XY} \frac{\partial^2 c_Y}{\partial x^2} \right) \right] \tag{2.8}$$

$$\frac{\partial c_Y}{\partial t} = M_Y \left[\frac{\partial^2}{\partial x^2} \left(\frac{\partial f_1}{\partial c_Y} - L_{YX} \frac{\partial^2 c_X}{\partial x^2} - K_Y \frac{\partial^2 c_Y}{\partial x^2} \right) \right] \tag{2.9}$$

where $c_X(x, t)$ and $c_Y(x, t)$ are concentration fields of Y and Y elements, respectively. Equations (2.8) and (2.9) yield

$$\begin{aligned} \frac{\partial c_X}{\partial t} = M_X \Bigg[& \frac{\partial^2 f_1}{\partial c_X^2} \frac{\partial^2 c_X}{\partial x^2} + \frac{\partial^2 f_1}{\partial c_X \partial c_Y} \frac{\partial^2 c_Y}{\partial x^2} + 2\frac{\partial^3 f_1}{\partial c_X^2 \partial c_Y} \frac{\partial c_X}{\partial x} \frac{\partial c_Y}{\partial x} + \frac{\partial^3 f_1}{\partial^3 c_X} \left(\frac{\partial c_X}{\partial x} \right)^2 \\ & + \frac{\partial^3 f_1}{\partial c_X \partial^2 c_Y} \left(\frac{\partial c_Y}{\partial x} \right)^2 - K_X \frac{\partial^4 c_X}{\partial x^4} - L_{XY} \frac{\partial^4 c_Y}{\partial x^4} \Bigg] \end{aligned} \tag{2.10}$$

$$\frac{\partial c_Y}{\partial t} = M_Y \left[\frac{\partial^2 f_1}{\partial c_X \partial c_Y} \frac{\partial^2 c_X}{\partial x^2} + \frac{\partial^2 f_1}{\partial c_Y^2} \frac{\partial^2 c_Y}{\partial x^2} + 2\frac{\partial^3 f_1}{\partial c_X \partial c_Y^2} \frac{\partial c_X}{\partial x} \frac{\partial c_Y}{\partial x} + \frac{\partial^3 f_1}{\partial c_X^2 \partial c_Y} \left(\frac{\partial c_X}{\partial x} \right)^2 \right.$$
$$\left. + \frac{\partial^3 f_1}{\partial c_Y^3} \left(\frac{\partial c_Y}{\partial x} \right)^2 - L_{YX} \frac{\partial^4 c_X}{\partial x^4} - K_Y \frac{\partial^4 c_Y}{\partial x^4} \right] \tag{2.11}$$

With use of the regular solution model, the local free energy is written as

$$\begin{aligned} f_1 &= f_X(1 - c_X - c_Y) + f_X c_X + f_Y c_Y + \Omega_{FeX} c_X (1 - c_X - c_Y) \\ &+ \Omega_{FeY} c_Y (1 - c_X - c_Y) + \Omega_{XY} c_X c_Y \\ &+ RT[(1 - c_X - c_Y)\ln(1 - c_X - c_Y) + c_X \ln c_X + c_Y \ln c_Y] \end{aligned} \tag{2.12}$$

where R is the gas constant, T is the absolute temperature, Ω_{FeX}, Ω_{FeY} and Ω_{XY} are interaction parameters and f_X, f_Y and f_Z are the molar free energies of pure X, pure Y and pure Y, respectively. From eq.(2.12) it follows that

$$\frac{\partial^2 f_1}{\partial c_X^2} = -2\Omega_{FeX} + RT \left(\frac{1}{c_X} + \frac{1}{1 - c_X - c_Y} \right) \tag{2.13}$$

$$\frac{\partial^2 f_1}{\partial c_Y^2} = -2\Omega_{FeY} + RT \left(\frac{1}{c_Y} + \frac{1}{1 - c_X - c_Y} \right) \tag{2.14}$$

$$\frac{\partial^2 f_1}{\partial c_X \partial c_Y} = \Omega_{XY} - \Omega_{FeX} - \Omega_{FeY} + RT \frac{1}{1 - c_X - c_Y} \tag{2.15}$$

Functions G_X and G_Y are difined as

$$G_X \left(t, x, c_X, c_X, \frac{\partial c_X}{\partial x}, \frac{\partial c_Y}{\partial x}, \frac{\partial c_X^2}{\partial x^2}, \frac{\partial c_Y^2}{\partial x^2}, \frac{\partial^4 c_X}{\partial x^4}, \frac{\partial^4 c_Y}{\partial x^4} \right) \equiv M_X \left[\frac{\partial^2 f_1}{\partial c_X^2} \frac{\partial^2 c_X}{\partial x^2} + \frac{\partial^2 f_1}{\partial c_X \partial c_Y} \frac{\partial^2 c_Y}{\partial x^2} \right.$$
$$\left. +2\frac{\partial^3 f_1}{\partial c_X^2 \partial c_Y} \frac{\partial c_X}{\partial x} \frac{\partial c_Y}{\partial x} + \frac{\partial^3 f_1}{\partial c_X^3} \left(\frac{\partial c_X}{\partial x} \right)^2 + \frac{\partial^3 f_1}{\partial c_X \partial^2 c_Y} \left(\frac{\partial c_Y}{\partial x} \right)^2 - K_X \frac{\partial^4 c_X}{\partial x^4} - L_{XY} \frac{\partial^4 c_Y}{\partial x^4} \right] \tag{2.16}$$

$$G_Y \left(t, x, c_X, c_X, \frac{\partial c_X}{\partial x}, \frac{\partial c_Y}{\partial x}, \frac{\partial^2 c_X}{\partial x^2}, \frac{\partial^2 c_Y}{\partial x^2}, \frac{\partial^4 c_X}{\partial x^4}, \frac{\partial^4 c_Y}{\partial x^4} \right) \equiv M_Y \left[\frac{\partial^2 f_1}{\partial c_X \partial c_Y} \frac{\partial^2 c_X}{\partial x^2} + \frac{\partial^2 f_1}{\partial c_Y^2} \frac{\partial^2 c_Y}{\partial x^2} \right.$$
$$\left. +2\frac{\partial^3 f_1}{\partial c_X \partial c_Y^2} \frac{\partial c_X}{\partial x} \frac{\partial c_Y}{\partial x} + \frac{\partial^3 f_1}{\partial c_X^2 \partial c_Y} \left(\frac{\partial c_X}{\partial x} \right)^2 + \frac{\partial^3 f_1}{\partial c_Y^3} \left(\frac{\partial c_Y}{\partial x} \right)^2 - L_{YX} \frac{\partial^4 c_X}{\partial x^4} - K_Y \frac{\partial^4 c_Y}{\partial x^4} \right] \tag{2.17}$$

These function satisfy the following conditions:

$$G_X \left(t, x_p, c_X, c_X, 0, \frac{\partial c_Y}{\partial x}, 0, \frac{\partial^2 c_Y}{\partial x^2}, \frac{\partial^4 c_X}{\partial x^4}, \frac{\partial^4 c_Y}{\partial x^4} \right) =$$
$$M_Y \left[\frac{\partial^2 f_1}{\partial c_X \partial c_Y} \frac{\partial^2 c_Y}{\partial x^2} + \frac{\partial^3 f_1}{\partial c_X \partial c_Y^2} \left(\frac{\partial c_Y}{\partial x} \right)^2 - K_X \frac{\partial^4 c_X}{\partial x^4} - L_{XY} \frac{\partial^4 c_Y}{\partial x^4} \right] \tag{2.18}$$

$$G_Y\left(t, x_p, c_X, c_X, 0, \frac{\partial c_Y}{\partial x}, 0, \frac{\partial^2 c_Y}{\partial x^2}, \frac{\partial^4 c_X}{\partial x^4}, \frac{\partial^4 c_Y}{\partial x^4}\right) =$$

$$M_Y\left[\frac{\partial^2 f_1}{\partial c_Y^2}\frac{\partial^2 c_Y}{\partial x^2} + \frac{\partial^3 f_1}{\partial c_Y^3}\left(\frac{\partial c_Y}{\partial x}\right)^2 - L_{YX}\frac{\partial^4 c_X}{\partial x^4} - K_Y\frac{\partial^4 c_Y}{\partial x^4}\right]$$

(2.19)

At a peak position $p(x_p, t)$

$$\frac{\partial c_X}{\partial x} = 0 \quad \frac{\partial^2 c_X}{\partial x^2} < 0 \tag{2.20}$$

Applying the mean value theorem of differential calculus for compound fuction [6, 7], we obtain the following equation for an intermediate value ζ in the open interval $(\partial^2 c_X/\partial x^2, 0)$

$$\begin{aligned}
\frac{\mathrm{d}c_X}{\mathrm{d}t}(x_p,t) &= G_X\left(t, x_p, c_X, c_X, 0, \frac{\partial c_Y}{\partial x}, \frac{\partial^2 c_X}{\partial x^2}, \frac{\partial^2 c_Y}{\partial x^2}, \frac{\partial^4 c_X}{\partial x^4}, \frac{\partial^4 c_Y}{\partial x^4}\right) \\
&- G_X\left(t, x_p, c_X, c_X, 0, \frac{\partial c_Y}{\partial x}, 0, \frac{\partial^2 c_Y}{\partial x^2}, \frac{\partial^4 c_X}{\partial x^4}, \frac{\partial^4 c_Y}{\partial x^4}\right) \\
&+ M_X\left[\frac{\partial^2 f_1}{\partial c_X \partial c_Y}\frac{\partial^2 c_Y}{\partial x^2} + \frac{\partial^3 f_1}{\partial c_X \partial c_Y^2}\left(\frac{\partial c_Y}{\partial x}\right)^2 - K_X\frac{\partial^4 c_X}{\partial x^4} - L_{XY}\frac{\partial^4 c_Y}{\partial x^4}\right] \\
&= \frac{\partial^2 c_X}{\partial x^2}(\zeta,t)\frac{\partial G_X(t, x_p, c_X, c_X, 0, \partial c_Y/\partial x, \zeta, \partial^2 c_Y/\partial x^2, \partial^4 c_X/\partial x^4, \partial^4 c_Y/\partial x^4)}{\partial(\partial^2 c_X/\partial x^2)} \\
&+ M_X\left[\frac{\partial^2 f_1}{\partial c_X \partial c_Y}\frac{\partial^2 c_Y}{\partial x^2} + \frac{\partial^3 f_1}{\partial c_X \partial c_Y^2}\left(\frac{\partial c_Y}{\partial x}\right)^2 - K_X\frac{\partial^4 c_X}{\partial x^4} - L_{XY}\frac{\partial^4 c_Y}{\partial x^4}\right] \\
&= M_X\left[\frac{\partial^2 f_1}{\partial c_X^2}\frac{\partial^2 c_X}{\partial x^2}(\zeta,t) + \frac{\partial^2 f_1}{\partial c_X \partial c_Y}\frac{\partial^2 c_Y}{\partial x^2} + \frac{\partial^3 f_1}{\partial c_X \partial c_Y^2}\left(\frac{\partial c_Y}{\partial x}\right)^2\right. \\
&\left. - K_X\frac{\partial^4 c_X}{\partial x^4} - L_{XY}\frac{\partial^4 c_Y}{\partial x^4}\right]
\end{aligned}$$

(2.21)

$$\begin{aligned}
\frac{\mathrm{d}c_Y}{\mathrm{d}t}(x_p,t) &= \frac{\partial^2 c_X}{\partial x^2}(\zeta,t)\frac{\partial G_Y(t, x_p, c_X, c_X, 0, \partial c_Y/\partial x, \zeta, \partial^2 c_Y/\partial x^2, \partial^4 c_X/\partial x^4, \partial^4 c_Y/\partial x^4)}{\partial(\partial^2 c_X/\partial x^2)} \\
&+ M_Y\left[\frac{\partial^2 f_1}{\partial c_Y^2}\frac{\partial^2 c_Y}{\partial x^2} + \frac{\partial^3 f_1}{\partial c_Y^3}\left(\frac{\partial c_Y}{\partial x}\right)^2 - L_{YX}\frac{\partial^4 c_X}{\partial x^4} - K_Y\frac{\partial^4 c_Y}{\partial x^4}\right] \\
&= M_Y\left[\frac{\partial^2 f_1}{\partial c_X \partial c_Y}\frac{\partial^2 c_X}{\partial x^2}(\zeta,t) + \frac{\partial^2 f_1}{\partial c_Y^2}\frac{\partial^2 c_Y}{\partial x^2} + \frac{\partial^3 f_1}{\partial c_Y^3}\left(\frac{\partial c_Y}{\partial x}\right)^2\right. \\
&\left. - L_{YX}\frac{\partial^4 c_X}{\partial x^4} - K_Y\frac{\partial^4 c_Y}{\partial x^4}\right]
\end{aligned}$$

(2.22)

2.3 Asymptotic Behavior of the Y element Associated with Decomposition of the X elements in a Fe-X-Y Ternary Alloy

We deal with the case in which the concentration of the X element in a Fe-X-Y ternary alloy, c_X and the concentration of the Y element, c_Y satisfy the following conditions

$$c_X > c_X, \quad \frac{\partial^2 f_1}{\partial c_X^2} < 0, \quad \frac{\partial^2 f_1}{\partial c_Y^2} > 0 \tag{2.23}$$

Let us consider the variation in the concentration of the Y element when phase decomposition of the X element proceeds. During the phase decomposition of the X element, the amplitude of c_X increases until it reaches the equilibruim concentration. The second term on the right hand side of eq.(2.22), $M_Y(\partial^2 f_1/\partial c_Y^2)(\partial^2 c_Y/\partial x^2)$, acts for averaging the value of c_Y because a diffusion constant $D_Y \equiv M_Y(\partial^2 f_1/\partial c_Y^2)$ is positive. At the intial stage the value of $(\partial c_Y/\partial x)^2$ is much smaller than that of $|\partial^2 c_X/\partial x^2|$ near the peaks of c_X. Once a peak or a bottom of c_Y is formed due to the contribution of the first term, $(\partial^2 f_1/\partial c_X \partial c_Y)(\partial^2 c_X/\partial x^2)$, then we obtain $\partial c_Y/\partial x = 0$. So the contributions of the second and third terms to the value of c_Y is very small. The fourth order derivative terms are attributed to the interfacial energies of the Y or Z rich regions in the X matrix. The contribution of these terms to the value of c_Y at the peak top of the X element is assumed to be smaller than that of the first term. Then at a peak of c_X, eq.(2.22) can be approximated by

$$\frac{\partial c_Y}{\partial t}(x_p, t) \simeq M_Y \frac{\partial^2 f_1}{\partial c_X \partial c_Y} \frac{\partial^2 c_X}{\partial x^2} \tag{2.24}$$

From eq.(2.24) it is indicated that the behavior of the Y elements at a peak position of the X element depends on the sign of $\partial^2 f_1/\partial c_X \partial c_Y$. If $\partial^2 f_1/\partial c_X \partial c_Y > 0$, then we have

$$\frac{\mathrm{d}c_Y(x_p, t)}{\mathrm{d}t} < 0 \tag{2.25}$$

From eq.(2.25) it follows that the concentration profile of the Z at a peak position of c_X decreases with time. While $\partial^2 f_1/\partial c_X \partial c_Y < 0$, then we have

$$\frac{\mathrm{d}c_Y(x_p, t)}{\mathrm{d}t} > 0 \tag{2.26}$$

Equation (2.26) indicates that peaks of c_Y will be formed at the same positions of the peak tops of c_X. The wavelength of the concentration profile of the Y element will be equal to that of the X element. This is a phase separation induced modulated structure. From eq.(2.15) it follows that

$$\frac{\mathrm{d}}{\mathrm{d}t}\left(\frac{\partial f_1}{\partial c_X \partial c_Y}\right) = \frac{RT}{(1 - c_X - c_Y)^2}\left(\frac{\mathrm{d}c_X}{\mathrm{d}t} + \frac{\mathrm{d}c_Y}{\mathrm{d}t}\right) \tag{2.27}$$

If $\mathrm{d}c_X/\mathrm{d}t > 0$ and $\mathrm{d}c_Y/\mathrm{d}t > 0$ then

$$\frac{\mathrm{d}}{\mathrm{d}t}\left(\frac{\partial f_1}{\partial c_X \partial c_Y}\right) > 0 \tag{2.28}$$

The sign of $\partial f_1/\partial c_X \partial c_Y$ may change from negative to positive in lower tempeartures at which the equilibrium concenration of the X element is high. This indicates that bifurcation of peaks will occur at the later stage of phase decomposition.

From the analyses described above the aysmptoptic behavior of the minor element, Y, in a Fe-X-Y ternary alloy along a trajectory of a peak top of the major element, X, formed by phase separation is classified into three groups.

Group1($\partial^2 f_1/\partial c_X \partial c_Y > 0$): The concetration of the Y element along the trajectory of the peak top of the Y decreases with time.

Group2($\partial^2 f_1/\partial c_X \partial c_Y < 0, \quad 0 \le t < \infty$): Peaks of the concentration of the Y element formed at the same positions of the peaks tops of the Y. Once a peak is formed the amplitute of the concentration increases with time.

Group3($\partial^2 f_1/\partial c_X \partial c_Y < 0, \quad 0 \le t < t_0, \quad \partial^2 f_1/\partial c_X \partial c_Y > 0, \quad t_0 < t < \infty$): At the initial stage, peaks of the concentration of the Y element formed at the same positions of the peaks tops of the Y. Bifurcation of peaks occurs at the later stage.

Although experimental verifications are left as future problem, numerical simulations of phase separation in Fe-Cr-Cu and Fe-Cr-Mo ternary alloys based on the Cahn-Hilliard equation demonstrated the validity of the present analyses [8, 9].

The present method of analyzing a peak top behavior of the solution of the Cahn-Hilliard equation for a Fe-X-Y ternary alloy can be extended to n-dimentional case. Suppose that, at a potion $\mathbf{x} = \mathbf{x}^0$ and a time $t_1 > 0$, a function $\alpha(\mathbf{x}) = c_X(\mathbf{x}, t_1)$ has a peak top which is characterized by $\partial c_X/\partial x_i(\mathbf{x}^0, t_1)$, $i = 1, 2, \cdots, n$ and the negative definiteness of the Hessian($\partial^2 u_X(\mathbf{x}^0, t_1)/\partial x_i \partial x_j$). According to the general discussions on the solutions of general nonlinear equations[10, 11], there exist an implicit function $\mathbf{g}(t)$, $t \in [t_1, t_f)$, $\mathbf{x}^0 = \mathbf{g}(t_1)$ such that $\partial c_X(\mathbf{g}(t), t)/\partial x_i = 0$, $\partial^2 c_X(\mathbf{g}(t), t)/\partial x_i^2 < 0$, $i = 1, 2, \cdots, n$. Using the mean value theorem, we can estimate the value of $c_Y(\mathbf{x}, t)$ along this trajectory $\mathbf{g}(t)$ of a peak top. Thus we can show that the present classification of asymptotic behavior of the minor element, Y, in a Fe-X-Y ternary alloy along the trajectory of a peak top of X, $\mathbf{g}(t)$, formally valid in n-dimentional case. However in 3 or higher dimentions, it is difficult to visualize a trajectory of a peak top in the space-time system. Application of the present method to three dimentional case is left to the future problem.

3 COMPUTER SIMULATION OF PHASE SEPARATION IN FE-CR-MO TERNARY ALLOYS

3.1 Conditions of numerical simulations

Numerical simulations based on the Cahn-Hilliard Equation were performed for Fe-Cr-Mo ternary alloys. Table 1 shows the conditions used for simulation.

Table 1 Conditions of simulations

number	alloy	temp.(K)
1	Fe-40at%Cr-5at%Mo	800
2	Fe-5at%Cr-40at%Mo	800
3	Fe-40at%Cr-5at%Mo	1025
4	Fe-30at%Cr-30at%Mo	750

The mobilities and the gradient energy coefficients for Fe-Cr-Mo ternary alloys are

$$M_{Cr} = \frac{c_1 \cdot D_{Cr}}{2\Omega_{FeCr} - 4RT}, \quad M_{Mo} = \frac{c_1 \cdot D_{Mo}}{2\Omega_{FeMo} - 4RT} \tag{3.1}$$

$$K_{cr} = \frac{1}{2} \cdot a_0^2 \cdot \Omega_{FeCr}, \quad K_{Mo} = \frac{1}{2} \cdot a_0^2 \cdot \Omega_{FeMo} \tag{3.2}$$

$$L_{CrMo} = L_{MoCr} = \frac{1}{2} \cdot a_0^2(\Omega_{CrMo} - \Omega_{FeCr} - \Omega_{FeMo}) \tag{3.3}$$

where a_0 is the lattice constant, D_{Cr} is the diffusion coefficient of Cr in Fe-50at%Cr steel and D_{Mo} is the diffusion coefficient of Mo in Fe-50at%Mo steel. The following values for D_{Cr} and D_{Mo} were used for simulation [12].

$$D_{Cr} = 0.19 \exp\left(-\frac{246000}{RT}\right) \tag{3.4}$$

$$D_{Mo} = 0.29 \exp\left(-\frac{264000}{RT}\right) \tag{3.5}$$

The constant c_1 is an adjustable parameter which modifies time scale in order that good agreement between calculated and observed phase separation kinetics is obtained[5]. The constant 0.01 for c_1 was used in the present simulation. The interaction parameters are [13]

$$\Omega_{FeCr} = 18.6kJ/mol, \quad \Omega_{FeMo} = 18.2kJ/mol, \quad \Omega_{CrMo} = 8.0kJ/mol \tag{3.6}$$

3.2 Bifurcation of peaks of the minor element associated with decomposition of the major element in Fe-based ternary alloys

Figure 1 shows the variation in the concentration profiles of Cr in an Fe-40at%Cr-5at%Mo alloy at 800K together with that of Mo. The formation of Cr rich regions by phase separation is clearly seen in this figure. A modulated structure of Mo with similar wavelength to that of Cr is observed. However the concentration of Mo at the peak of Cr concentration is found to be decreasing with time. The variation in the concentration profiles of Mo and Cr in an Fe-40at%Mo-5at%Cr alloy at 800K is shown in Fig.2. The formations of Mo rich regions and a modulated structure of Cr are observed in this figure. The bifurcation of peaks of Cr occurs also in this

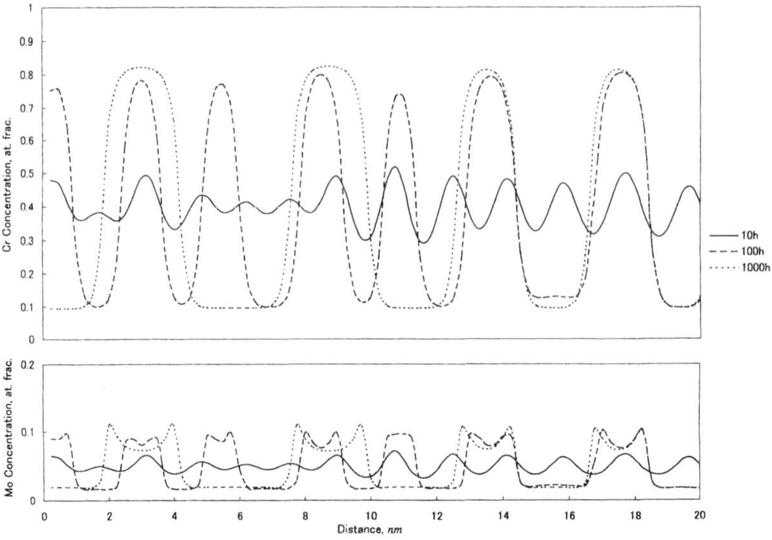

Figure 1: Variation of Cr and Mo profiles in an Fe-40at%Cr-5at%Mo alloy at 800K

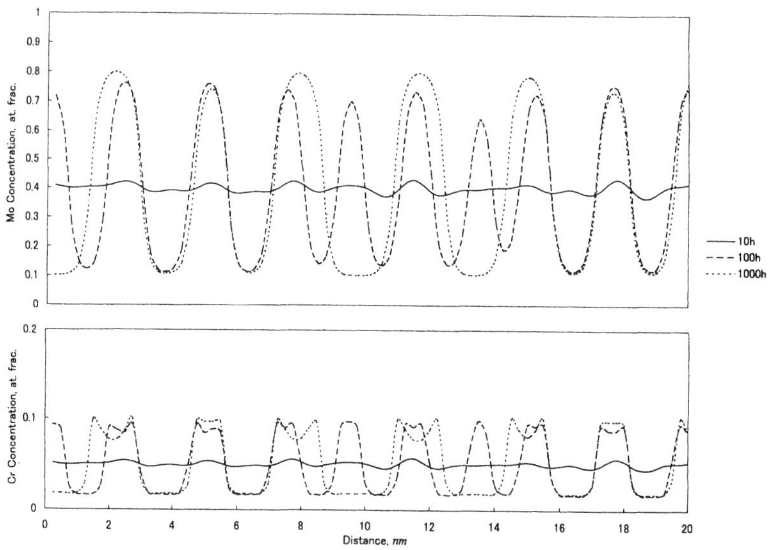

Figure 2: Variation of Mo and Cr profiles in an Fe-40at%Mo-5at%Cr alloy at 800K

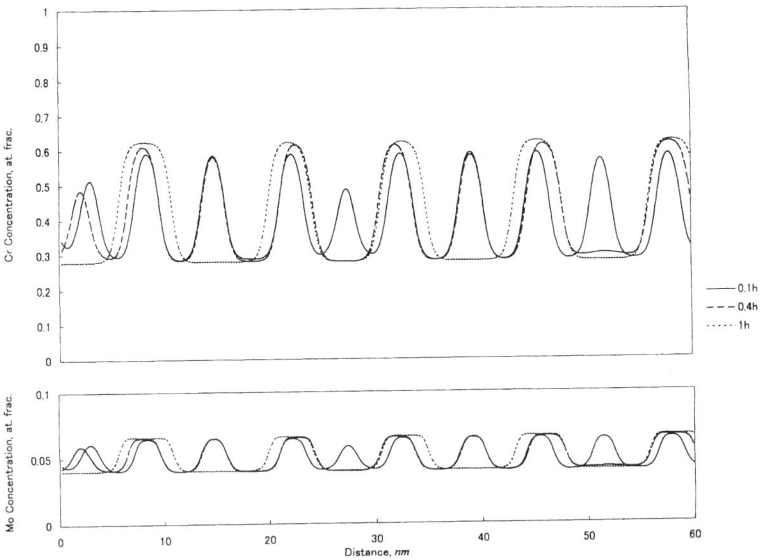

Figure 3: Variation of Cr and Mo profiles in an Fe-40at%Cr-5at%Mo alloy at 1025K

alloy at the later stage. The behavior of Cr in this alloy is quite similar to that in the Fe-40at%Cr-5at%Mo alloy.

The mechanism of bifurcation formation of peaks can be explained by the theory described in the previous section. The sign of $\partial^2 f_1/\partial c_{Cr}\partial c_{Mo}$ changes from negative to positive at 800K at which the equilibrium concentration of Cr is high. This indicates that bifurcation of peaks of the concentration of Mo will occur at the later stage of phase decomposition.

The variation in concentration profiles of Cr and Mo in an Fe-40at%Cr-5at%Mo alloy at 1025K is shown in Fig.3. The concentration of Mo at a peak position of Cr increases with time. The equilibrium concentration of Cr at a high temperature is smaller than that at a lower temperature. In this condition the value of $RT/(1 - c_{Cr} - c_{Mo})$ is smaller than the absolute value of $\Omega_{CrMo} - \Omega_{FeCr} - \Omega_{FeMo}$. So the value of $\partial^2 f_1/\partial c_X \partial c_Y$ is negative at the later stage. Thus the bifurcation of peaks will not occur in this case.

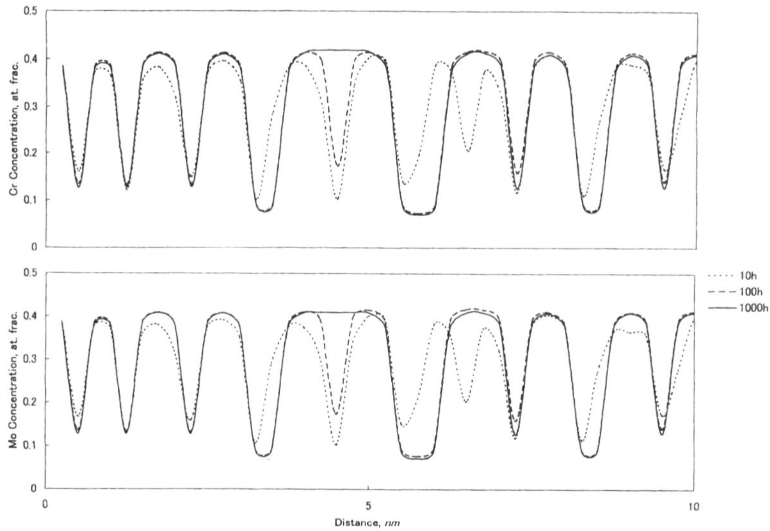

Figure 4: Variation of Cr and Mo profiles in an Fe-30at%Cr-30at%Mo alloy at 750K

3.3 Phase separation in an Fe-30at%Cr-30at%Mo alloy

Next let us consider the case in which both c_X and c_Y are within the spinodal region of an Fe-X-Y ternary phase diagram. In this case

$$c_X \simeq c_Y, \quad \frac{\partial^2 f_1}{\partial c_X^2} < 0, \quad \frac{\partial^2 f_1}{\partial c_Y^2} < 0 \tag{3.7}$$

It is expected that X and Y elements decompose separately. By repeating the above analyses, we may predict that the amplitude of c_Y increases at a peak position of the X element and that the amplitude of X element also increases at the peak position of the Y element. The modulated structure of the Y element is expected to be similar to that of the X element. As a result the wavelength of the modulated structure is predicted to be shorter than those of Fe-X and Fe-Y binary alloys.

Figure 4 shows the variation in concentration profiles of Cr and Mo in an Fe-30at%Cr-30at%Mo alloy. The concentration profile of Cr is quite similar to that of Mo. The wavelength of the modulated structure is smaller than those of Fe-Cr and Fe-Mo binary alloys.

3.4 Concentration dependence of atomic mobilities

In our analyses of asymptotic behavior of a minor element in an Fe-X-Y ternary alloy, the atomic mobilities M_X and M_Y are assumed to be not dependent on their position in the space. We neglected the concentration dependence of atomic mobilities. Now let us consider the case that the mobilities M_X and M_Y are functions of c_X and c_Y. In this case the Cahn-Hilliard equation for an Fe-X-Y ternary alloy is given by

$$\frac{\partial c_X}{\partial t} = M_X \left[\frac{\partial^2}{\partial x^2} \left(\frac{\partial f_1}{\partial c_X} - K_X \frac{\partial^2 c_X}{\partial x^2} - L_{XY} \frac{\partial^2 c_Y}{\partial x^2} \right) \right] + \frac{\partial M_x}{\partial x} \left[\frac{\partial}{\partial x} \left(\frac{\partial f_1}{\partial c_X} - K_X \frac{\partial^2 c_X}{\partial x^2} - L_{XY} \frac{\partial^2 c_Y}{\partial x^2} \right) \right] \tag{3.8}$$

$$\frac{\partial c_Y}{\partial t} = M_Y \left[\frac{\partial^2}{\partial x^2} \left(\frac{\partial f_1}{\partial c_Y} - L_{YX} \frac{\partial^2 c_X}{\partial x^2} - K_Y \frac{\partial^2 c_Y}{\partial x^2} \right) \right] + \frac{\partial M_Y}{\partial x} \left[\frac{\partial}{\partial x} \left(\frac{\partial f_1}{\partial c_Y} - L_{YX} \frac{\partial^2 c_X}{\partial x^2} - K_Y \frac{\partial^2 c_Y}{\partial x^2} \right) \right] \tag{3.9}$$

The terms $\partial M_X / \partial x$ and $\partial M_Y / \partial x$ are rewritten as

$$\frac{\partial M_X}{\partial x} = \frac{\partial M_X}{\partial c_X} \frac{\partial c_X}{\partial x} + \frac{\partial M_X}{\partial c_Y} \frac{\partial c_Y}{\partial x} \tag{3.10}$$

$$\frac{\partial M_Y}{\partial x} = \frac{\partial M_Y}{\partial c_X} \frac{\partial c_X}{\partial x} + \frac{\partial M_Y}{\partial c_Y} \frac{\partial c_Y}{\partial x} \tag{3.11}$$

At a peak position of the X element, the above equations yield

$$\frac{\partial M_X}{\partial x} = \frac{\partial M_X}{\partial c_Y} \frac{\partial c_Y}{\partial x} \tag{3.12}$$

$$\frac{\partial M_Y}{\partial x} = \frac{\partial M_Y}{\partial c_Y} \frac{\partial c_Y}{\partial x} \tag{3.13}$$

At the initial stage of phase separation $\partial c_Y / \partial x$ is very small. Once a peak or a bottom of Y element is formed the term $\partial c_Y / \partial x$ becomes zero. It follows that the asymptotic behavior of the Y element at a peak top of the X element can be approximated by eq.(2.24) also in the case that the mobilities of X and Y elements are concentration dependent.

4 SUMMARY

Asymptotic behavior of a minor element Y in a Fe-X-Y ternary system associated with phase decomposition of the major element, X, was investigated by using a model based on the Cahn Hillirad equation for multicomponent systems. Numerical simulations of phase separation in Fe-Cr-Mo ternary alloys were performed with use of the Cahn-Hilliard equation. The following results are obtained.

(1) The aysmptoptic behavior of the minor element, Y, in a Fe-X-Y ternary alloy along a trajectory of a peak top of the major element, X, is classified into three groups according to the sign of the second derivative of the chemical free energy f_1 with respect to the concentration of X, c_X and the concentration of Y, c_Y.

(2) If $\partial^2 f_1/\partial c_X \partial c_Y > 0$, the value of c_Y along the trajectory of the peak top of Y decreases with time.

(3) When $\partial^2 f_1/\partial c_X \partial c_Y < 0$ at $0 \le t < \infty$, peaks of c_Y formed at the peak tops of Y. Once a peak is formed, the amplitute of the peak increases with time.

(4) For the case that $\partial^2 f_1/\partial c_X \partial c_Y < 0$ at $0 \le t < t_0$ and $\partial^2 f_1/\partial c_X \partial c_Y > 0$ at $t_0 < t < \infty$, peaks of c_Y formed at the peak tops of Y at the initial stage. Bifurcation of peaks occurs at the later stage.

(5) Bifurcation of peaks of Mo along the peak tops of Cr concentration occurs in an Fe-40at%Cr-5at%Mo alloy at 800K. Bifurcation of peaks of Cr is also shown in an Fe-40at%Mo-5at%Cr alloy at 800K. In these cases, the sign of the second derivative of the chemical free energy f_1 with respect to the concentration of Cr, c_{Cr}, and the concentration of Mo, c_{Mo}, $\partial^2 f_1/\partial c_{Cr} \partial c_{Mo}$, changes from negative to positive at the later stage.

(6) In an Fe-40at%Cr-5at%Mo alloy, peaks of the concentration of Mo form along the peak tops of Cr at 1025K. Once a peak is formed the amplitude of the concentration increases with time.

(7) For the case in which both c_{Cr} and c_{Mo} are within the spinodal region of an Fe-Cr-Mo ternary phase diagram, the amplitude of Mo increases at the peak positions of Cr and the amplitude of Cr also increases at the peak positions of Mo. The wavelength of the modulated structure is shorter than those of Fe-Cr and Fe-Mo binary alloys.

(8) Simulated asymptotic behavior of Mo and Cr in Fe-Cr-Mo ternary alloys is in good agreement with that predicted by the theory.

Acknowledgements

The authors would like to express their appreciation to Professor Akihiko Kitada for drawing their attention to this problem.

References

[1] J.W. Cahn and J.E. Hilliard: J.Chem. Phys. **28**, (1958)258

[2] J.W. Cahn :Acta Metall., **9**(1961)795

[3] J.M. Hyde, A.Cerezo, M.K. Killer and G.U.W. Smith, Appl. Surf. Sci., **76/7**(1994), 233

[4] T. Koyama, T. Kozakai and T.Miyazaki: Materia Jpn, **38**(1999), 624

[5] M. Honjo and Y. Saito: ISIJ International, **40**(2000), 914

[6] A. Kitada: Jitsuyo Kaiseki Nyumon(Introduction to Practical Analysis), Yachiyo-Shuppan, Tokyo, (1985), 55[in Japanese]

[7] R. Courant and F. John: Introduction to Calculas and Analysis I, Springer-Verlag, New York, (1989), 222-223

[8] K. Ochi, Y.Suwa and Y.Saito: CAMP-ISIJ,**14**(2001)1207

[9] Y. Suwa, Y. Saito, K. Ochi, T. Aoki, K. Goto and K. Abe: submitted to Mater. Trans., (2001)

[10] A. Kitada: Ohyohbutsuri, **62**(1993) 322[in Japanese]

[11] A. Kitada, S. Nakamura and K.Ito: J. Phys. Soc. Jpn., **67**(1998)2149

[12] H. Bakker, H.P. Bonzel, C.M. Burff, M.A. Dayananda, W. Gust, J. Horvåth, I. Kaur, G.V. Kidson, A.D. LeClaire, H. Mehrer, G.E. Murch, G. Neumann, N.A. Stolwijk: Diffusion in Solid Metals and Alloys, Vol.26, ed. by H. Mehrer, (Springer-Verlag, Berlin, 1990)pp.125-125

[13] K. Hashiguchi and J.S. Kirkaldy, T. Fukuzumi and V. Pavaskar: Calphad **8**(1984)173-186

SHORT-RANGE ORDER PARAMETERS IN FCC BINARY ALLOYS

J. S. Faulkner, Silvia Pella, Aurelian Rusanu, and Yevgeniy Puzyrev

Alloy Research Center, Department of Physics
Florida Atlantic University
Boca Raton, FL 33431

INTRODUCTION

Monte-Carlo calculations on simple Ising models using the Metropolis algorithm are known to produce thermodynamic equilibrium states with a minimum of effort.[1] They have been used for many years to study the order-disorder transformations, phase boundaries, and the nature of critical points in substitutional solid-solution alloys. Although these calculations produce a supercell with the A and B atoms distributed on the sites of a Bravais lattice, this information has rarely been used to calculate the Warren-Cowley short-range order parameters (SROP). The SROP are interesting because they give a detailed picture of the short-range ordered state. We use periodically reproduced supercells from Monte Carlo calculations to study temperature and concentration dependence of SROP in Ising models with first and second nearest-neighbor interactions.

The SROP are defined by[2] $\alpha(\mathbf{r}_n) = 1 - x(\mathbf{r}_n)$ with

$$x(\mathbf{r}_n) = \frac{p_A(\mathbf{r}_n)}{c_A} = \frac{p_B(\mathbf{r}_n)}{c_B}$$

The probability that an A atom will be found on a site separated by the vector $\mathbf{r}_n$ from a site that definitely has a B atom on it can be calculated from the distribution of atoms in a supercell using the equation

$$p_A(\mathbf{r}_n) = \frac{N_{AB}(\mathbf{r}_n)}{N_{BB}(\mathbf{r}_n) + N_{AB}(\mathbf{r}_n)} = \frac{N_{AB}(\mathbf{r}_n)}{c_B N},$$

where $N_{AB}(\mathbf{r}_n)$ is the number of AB pairs in the supercell separated by $\mathbf{r}_n$ and $N_{BB}(\mathbf{r}_n)$ is the number of BB pairs. The probability that a B atom will be found on a site separated by the vector $\mathbf{r}_n$ from a site that definitely has an A atom on it is calculated similarly

$$p_B(\mathbf{r}_n) = \frac{N_{AB}(\mathbf{r}_n)}{N_{AA}(\mathbf{r}_n) + N_{AB}(\mathbf{r}_n)} = \frac{N_{AB}(\mathbf{r}_n)}{c_A N}.$$

Therefore, the most convenient definition to use for calculating the SROP in a supercell is

$$x(\mathbf{r}_n) = \frac{N_{AB}(\mathbf{r}_n)}{c_A c_B N}.$$

The probabilities obtained from experiments on samples containing 10^{23} atoms have the symmetry property $p_A(\mathbf{r}_n) = p_A(\mathbf{r}_m)$ when the vectors $\mathbf{r}_n$ and $\mathbf{r}_m$ connect to the same nearest neighbor shell. This will not be true for SROP calculated from a supercell of finite size because of statistical fluctuations, but it can be achieved by averaging

$$x(i) = \frac{1}{c_A c_B N n_{nn}^i} \sum_{|\mathbf{r}_n| = r_{nn}^i} N_{AB}(\mathbf{r}_n)$$

where r_{nn}^i and n_{nn}^i are the radius of the nth nn shell and the number of sites in it.

If the number of AB pairs connected by the vector $\mathbf{r}_n$ is zero, $\alpha(\mathbf{r}_n)$ is equal to 1, and this corresponds to clustering. Negative values of $\alpha(\mathbf{r}_n)$ correspond to ordering. If the number of AB pairs is equal to the average value in a random alloy, $c_A c_B N$, then $\alpha(\mathbf{r}_n) = 0$. The SROP are of interest to physicists for a number of reasons. We now consider some of them.

USES FOR SROP

Diffuse Scattering From Supercells

The diffraction of a beam of x-rays by a perfect crystal produces outgoing beams only in a discrete set of Bragg peaks. Any imperfection of the crystal leads to diffuse scattering away from the Bragg peaks. Warren and Cowley first defined the SROP to describe the diffuse scattering caused by a lack of long-range order in a binary substitutional alloy made up of A and B atoms. Their definition applies to infinite crystals, and must be modified for supercells. Consider a supercell containing N_A A atoms and N_B B atoms, with $N_A + N_B = N$, distributed on the sites of a periodic Bravais lattice. It is periodically reproduced to fill all space. The edges of the supercell are the vectors $La\mathbf{e}_1$, $La\mathbf{e}_2$, and $La\mathbf{e}_3$, where the $\mathbf{e}_i$ are a set of orthogonal unit vectors, a is a lattice constant, and L is a large integer. It is assumed, as in Warren's book,[2] that the sample is large enough to contain an infinite number of supercells but small enough to be described by small crystal theory. The direction of the incoming wave is described by the unit vector $\mathbf{s}_0$, the outgoing wave is in the direction $\mathbf{s}$, and the scattering vector is $\mathbf{k} = \frac{2\pi}{\lambda}(\mathbf{s} - \mathbf{s}_0)$, where λ is the wavelength of the radiation. Scattering from a supercell is just like scattering from a unit cell that contains N atoms. The only allowed outgoing waves have k-vectors

$$\mathbf{k} = \frac{2\pi n_1}{La}\mathbf{e}_1 + \frac{2\pi n_2}{La}\mathbf{e}_2 + \frac{2\pi n_3}{La}\mathbf{e}_3,$$

where the n_i are integers that range from 0 to L-1.

The scattering factor for the supercell can be written

$$I(\mathbf{k}) = \frac{1}{N}\sum_{\substack{m=1\\n=1}}^{N} f_m f_n e^{-\mathbf{k}\cdot(\mathbf{r}_m - \mathbf{r}_n)} = \frac{\Phi(\mathbf{k}) * \Phi(\mathbf{k})}{N},$$

where

$$\Phi(\mathbf{k}) = \sum_{i=1}^{N} f_i e^{i\mathbf{k}\cdot\mathbf{r}_i} = f_A \sum_{i=1}^{N} n_A(\mathbf{r}_i) e^{i\mathbf{k}\cdot\mathbf{r}_i} + f_B \sum_{i=1}^{N} n_B(\mathbf{r}_i) e^{i\mathbf{k}\cdot\mathbf{r}_i}.$$

In this expression, f_A and f_B are the scattering parameters for A and B atoms, $n_A(\mathbf{r}_i)$ is 1 if there is an A atom on site i and 0 otherwise, and $n_B(\mathbf{r}_i) = 1 - n_A(\mathbf{r}_i)$. Another way to write $\Phi(\mathbf{k})$ is

$$\Phi(\mathbf{k}) = f_A' \sum_{i=1}^{N} n_A(\mathbf{r}_i) e^{i\mathbf{k}\cdot\mathbf{r}_i} + f_B' \sum_{i=1}^{N} n_B(\mathbf{r}_i) e^{i\mathbf{k}\cdot\mathbf{r}_i} + \langle f \rangle \sum_{i=1}^{N} e^{i\mathbf{k}\cdot\mathbf{r}_i},$$

where $\langle f\rangle = c_A f_A + c_B f_B$, $\quad f_A' = f_A - \langle f\rangle = c_B(f_A - f_B)$, $\quad$ and $\quad f_B' = f_B - \langle f\rangle = -c_A(f_A - f_B)$. Manipulating the equations leads to $I(\mathbf{k}) = I(\mathbf{k})_{Bragg} + I(\mathbf{k})_{sro}$, in which $I(\mathbf{k})_{Bragg}$ is the scattering from the average atoms.

The positions of the atoms relative to the origin of the supercell are $\mathbf{r}_i = p_1^i\mathbf{a}_1 + p_2^i\mathbf{a}_2 + p_3^i\mathbf{a}_3$, where p_j^i are integers and $\mathbf{a}_j$ are the basis vectors of the Bravais lattice. It can be shown that $I(\mathbf{k})_{Bragg}$ is nonzero only at Bragg peaks for which the integers in the k-vector satisfy

$$\mathbf{k}\cdot\mathbf{a}_j = \frac{2\pi n_1}{La}\mathbf{e}_1\cdot\mathbf{a}_j + \frac{2\pi n_2}{La}\mathbf{e}_2\cdot\mathbf{a}_j + \frac{2\pi n_3}{La}\mathbf{e}_3\cdot\mathbf{a}_j = 2\pi\times(\text{integer})$$

for j equal 1, 2, and 3. The diffuse scattering term is given by

$$I(\mathbf{k})_{sro} = c_A c_B (f_A - f_B)^2 \sum_{n=1}^{N} \alpha(\mathbf{r}_n) e^{i\mathbf{k}\cdot\mathbf{r}_n},$$

where the $\alpha(\mathbf{r}_n)$ are the Warren-Cowley SROP defined for a supercell above.

Another quantity of considerable interest is the diffuse scattering parameter

$$\alpha(\mathbf{k}) = \frac{I(\mathbf{k})_{sro}}{c_A c_B (f_A - f_B)^2} = \sum_{n=1}^{N} \alpha(\mathbf{r}_n) e^{i\mathbf{k}\cdot\mathbf{r}_n},$$

which is the Fourier transform of the SROP. It can be shown that $\alpha(\mathbf{k})$ can be calculated very easily for a supercell using the functions

$$\phi(\mathbf{k}) = c_B \sum_{i=1}^{N} n_A(\mathbf{r}_i) e^{i\mathbf{k}\cdot\mathbf{r}_i} - c_A \sum_{i=1}^{N} n_B(\mathbf{r}_i) e^{i\mathbf{k}\cdot\mathbf{r}_i},$$

and the equation

$$\alpha(\mathbf{k}) = \frac{\phi(\mathbf{k})^* \phi(\mathbf{k})}{N c_A c_B}.$$

This equation can be checked by showing that it leads to the above expression containing the SROP.

The diffuse scattering parameter $\alpha(\mathbf{k})$ contains the information included in an infinite number of SROP. These equations will be used to calculate the diffraction from supercells generated by Monte Carlo calculations in the following.

Coulomb Energy

The Coulomb energy in alloys has been discussed in considerable detail.[3] The SROP appear in the definition of the Coulomb energy of isomorphous alloys, which are alloy models for which every A atom has the same charge $\bar{q}_A$ and every B atom has the charge $\bar{q}_B$. The charges in the supercell may be written $\bar{q}_A = c_B\Delta$ and $\bar{q}_B = -c_A\Delta$. These definitions guarantee that $q_A - q_B = \Delta$, a quantity that calculations show is approximately independent of concentration. It also ensures charge neutrality, $c_A\bar{q}_A + c_B\bar{q}_B = 0$.

The Coulomb energy per atom in a supercell is

$$U_C = \frac{1}{N}\sum_{\substack{i=1\\j=1}}^{N} M(|\mathbf{r}_j - \mathbf{r}_i|) q_i q_j,$$

where $M(|\mathbf{r}_j - \mathbf{r}_i|)$ is the Madelung matrix[4] that includes the effect of the charges in the infinity of periodically reproduced supercells and satisfies the conditions

$M(|r_i - r_i|) = M(0) = 0$. As the size of the supercell approaches infinity,

$M(|r_i - r_i|) \to \frac{1}{|r_i - r_i|}$.

Using $q_i = n_A(\mathbf{r}_i)\bar{q}_A + n_B(\mathbf{r}_i)\bar{q}_B$, it can be shown that

$$U_C = c_A c_B \Delta^2 \sum_{n=1}^{N} \alpha(\mathbf{r}_n) M(|\mathbf{r}_n|).$$

The Coulomb potential at a site i is

$$V_i = \sum_{j=1}^{N} M(|r_j - r_i|) q_j .$$

The average potential for an A site is

$$\bar{V}_{i \subset A} = \frac{1}{Nc_A} \sum_{i \subset A}^{N} V_i = c_B \Delta \sum_{n=1}^{N} M(|r_n|)\alpha(n),$$

while for a B site,

$$\bar{V}_{i \subset B} = \frac{1}{Nc_B} \sum_{i \subset A}^{N} V_i = -c_A \Delta \sum_{n=1}^{N} M(|r_n|)\alpha(n).$$

It follows that the Coulomb energy can be written

$$U_C = c_A \bar{q}_A V_{i \subset A} + c_B \bar{q}_B V_{i \subset B}$$

These expressions for the Coulomb energy and the potentials are exact.

As will be discussed in more detail below, the $\alpha(\mathbf{r}_n)$ are all zero for an infinitely large random supercell. For this case, the average potentials on a site and the Coulomb energy are zero. This is another proof that the Coulomb energy and Madelung potentials for an isomorphous model of a random alloy are zero.[5]

Parameterized Energy

The Hamiltonian for an Ising Model of an alloy may be written

$$\mathbf{H} = \sum_{\substack{i=1 \\ j=1}}^{N} V(|\mathbf{r}_j - \mathbf{r}_i|) S_i S_j - v \sum_{i=1}^{N} S_i ,$$

where $V(|\mathbf{r}_i - \mathbf{r}_i|) = V(0) = 0$. The "spins" S_i are 1 if there is an A atom on site i, and -1 if there is a B atom there, and v is the chemical potential. When this Hamiltonian is used, positive values of the potentials $V(r)$ correspond to antiferromagnetic or ordering interactions. It follows that the total energy for a given configuration of atoms is

$$E = \sum_{n=1}^{N} V(|\mathbf{r}_n|)[N - 4N_{AB}(\mathbf{r}_n)] - v(N_A - N_B).$$

Using the definitions in the introduction, the energy per atom may be written,

$$\frac{E}{N} = 4c_A c_B \sum_{n=1}^{N} V(|\mathbf{r}_n|)\alpha(\mathbf{r}_n) + (c_A - c_B)^2 \sum_{n=1}^{N} V(|\mathbf{r}_n|) - (c_A - c_B)v.$$

Since $V(|\mathbf{r}_n|)$ is the same for all $\mathbf{r}_n$ with magnitudes equal to the radius of a given nearest-neighbor shell r_{nn}^i, this expression can be changed to a sum over nearest-neighbor shells,

$$\frac{E}{N} = 4c_A c_B \sum_{i=1}^{N_{nn}} n_{nn}^i V(r_{nn}^i)\alpha(r_{nn}^i) + (c_A - c_B)^2 \sum_{i=1}^{N_{nn}} n_{nn}^i V(r_{nn}^i) - (c_A - c_B)v.$$

The energy may be manipulated into another form,

$$E = N\left(\sum_{n=1}^{N} V(|\mathbf{r}_n|) + \iota\right) - N_A 2\left(2\sum_{n=1}^{N} V(|\mathbf{r}_n|) + v\right) + 4\sum_{n=1}^{N} V(|\mathbf{r}_n|) N_{AA}(\mathbf{r}_n),$$

or,

$$E = N\left(\sum_{i=1}^{N_{nn}} n_{nn}^{i} V(r_{nn}^{i}) + v\right) - N_A 2\left(2\sum_{i=1}^{N_{nn}} n_{nn}^{i} V(r_{nn}^{i}) + v\right) + 4\sum_{i=1}^{N_{nn}} n_{AA}^{i} V(r_{nn}^{i}),$$

where

$$n_{AA}^{i} = \sum_{|\mathbf{r}_n| = r_{nn}^{i}} N_{AA}(\mathbf{r}_n)$$

This is the version used by Kanamori in his studies of the ground state configurations of alloys at T=0.[6] The first term is the energy of a pure metal with all B atoms. The second term is the number of A atoms times the energy that is used to insert one A atom. The last term is the interaction of pairs of A atoms.

The Meaning of Randomness in Terms of SROP

An ideal random substitutional alloy is made up of an infinite number of atoms distributed on the sites of a periodic Bravais lattice. The probability of finding an A atom on a lattice site is defined to be c_A and the probability of finding a B atom is $c_B = 1 - c_A$. A finite subset of the ideal lattice containing N atoms will be called R_N. The probability for finding a given number n_A of A atoms in R_N is given by the binomial distribution,

$$P(n_A) = \frac{N!}{n_A!(N - n_A)} c_A^{\,n_A} c_B^{\,(N-n_A)}$$

For large N, the binomial distribution is approximated by the normal distribution,

$$P(x) = \frac{1}{\sqrt{2\pi}\sigma} e^{\frac{(x-\mu)^2}{\sigma^2}}$$

It follows that the probability of finding x in the range $-0.6745\sigma \le (x - \mu) \le 0.6745\sigma$ is 0.50, while the probability for finding it in the range $-3.0\sigma \le (x - \mu) \le 3.0\sigma$ is 0.9973. In the calculation of $P(n_A)$, the mean is $\mu = Nc_A$ and the standard deviation (STD) is $\sigma = \sqrt{c_A c_B N}$. It should be noted that the probability c_A is the concentration only when the crystal is infinite in size. If we sample a region R_N containing a million atoms, the probability for measuring c_A in the range $0.4985 \le c_A \le 0.5015$ is 0.9973. In probability theory,[7] a quantity like c_A is theoretical, not experimental.

From the introduction it is seen that the SROP for the ith nearest neighbor shell can be written

$$\alpha(i) = 1 - \frac{n_{AB}^{i}}{c_A c_B n_p^{i}},$$

where $n_p^{i} = N n_{nn}^{i}$ is the number of pairs in the region R_N, and n_{AB}^{i} is the number of AB pairs separated by r_{nn}^{i} counted in R_N. The number of AB pairs is also described by a Binomial distribution or, for large N, a normal distribution with mean $\mu = c_A c_B n_p^{i}$ and the STD $\sigma = c_A c_B \sqrt{n_p^{i}}$.[8] It follows that $\alpha(i)$ is described by a normal distribution with mean $\mu = 0$ and the STD $\sigma = 1/\sqrt{n_p^{i}}$. For cosmetic reasons, supercells S_N are often used that differ from the regions R_N in that the fraction of A atoms n_A/N is forced to be precisely equal to c_A. This can be achieved by invoking additional interchanges of A and B atoms on random sites after a random number generator has been used to distribute the atoms on the sites. It has been shown[8] that the statistics of the SROP discussed in the preceding paragraph is not significantly affected by this process.

Leaving aside the unimportant dependence of n_{nn}^{i} on the size of the region, the STD for all the $\alpha(i)$ will approach zero uniformly like $1/\sqrt{N}$. It is in this sense, and only in this sense, that the statement can be made that the $\alpha(i)$ are zero for a region of a random alloy when $N \to \infty$. The reason that we emphasize this fact is that some confusion about this point has been expressed in the literature.

It has been proposed[9] that one can create a supercell that is more random than random by arranging the atoms of the first few nearest-neighbor shells so that $n^{i}_{AB} = c_A c_B n^{i}_{p}$ and thus the corresponding $\alpha(i)$ are zero. Supercells S_N with $8 \leq N \leq 16$ and the first few $\alpha(i)$ zero, called special quasi-random structures (SQS), have been used in quantum mechanical calculations on "random" alloys using the density functional theory and the local density approximation (DFT-LDA). The attraction of the idea stems from the fact that DFT-LDA calculations use an amount of computer time that scales as a power of N. There are a number of difficulties with the SQS concept. The probability of finding a region R_N in a random alloy that fits the SQS description is $P = c_A^{n_A} c_B^{(N-n_A)}$, the same as for any other completely specified arrangement containing n_A A atoms. The chance of finding such a structure periodically reproduced to fill all space is zero. Even the claim that the atomic arrangements in the first few nearest neighbor shells are similar to the ones in a random alloy is incorrect. Not to belabor the point, the argument that one can create a random distribution by rearranging atoms so that the $\alpha(i)$ are zero is equivalent, according to probability theory, to the following. Suppose a random sequence of N 1's and 0's are needed. The sequence can be prepared by flipping a coin or using a (pseudo-) random number generator. The test of the fairness of the coin flips or the quality of the random number generator is that the ratio $\rho(N) = \left(\sum \text{integers}\right)/N$ must approach 1/2 as N approaches infinity. A helpful person can point out that $\rho(N)$ can be made exactly equal to 1/2 for any even N by intervening in the random process. According to this view, it would follow that the most efficient random sequence is 0101010101... because, for any region of length 2, the ratio $\rho(2)$ has the "correct" value.

An argument for the SQS idea is based on the expression for the parameterized energy in terms of the SROP in the previous subsection.[9] If it is assumed that the Ising model expression for the energy can be used and that only the first three interaction parameters V_i are necessary, then the energy of a completely random alloy can be calculated using a supercell in which the magnitudes of the $\alpha(i)$ for i less than 4 are zero and the others are irrelevant. In the first place, people who do Monte Carlo calculations do not claim that the Ising model is anything more than a model. First principles theory provides some guidance for the choice of the V_i, but fundamentally they are fitting parameters that are used to approximate very complicated interactions. This causes no problem in Monte Carlo applications, but the argument cannot be translated into a first-principles environment. Secondly, there is charge transfer in metallic alloys that leads to Coulomb energies and Madelung potentials similar to the ones discussed in the subsection Coulomb Energy. In quantum mechanical calculations on condensed matter, sophisticated techniques such as the Ewald method[10] and fast multipole methods[11] have been developed to deal with them. It has been argued that the first few V_i are capable of modeling these long-range Coulomb effects,[9] but it is clear that this can only be done very approximately. Even if this argument is accepted, the $\alpha(i)$ in Table I illustrate another difficulty. These SROP are calculated for arrangements of 16 A and 16 B atoms on the sites of an fcc Bravais lattice in a 32 atom supercell. It is obvious from this table that the price one must pay for the zero values of $\alpha(1)$, $\alpha(2)$, and $\alpha(3)$ is an unnaturally large value for $\alpha(4)$ and other SROP. Of course, $\alpha(8)$ must be 1 because the atom sees its periodically reproduced self on the eighth nn shell. The quantity that appears in the sum for the Ising energy in the subsection Parameterized Energy is $V_i\alpha(i)$. Even if the V_i become smaller as i increases, they will become important when they are multiplied by the large $\alpha(i)$.

Table I. The SROP for some 32 atom supercells based on the fcc Bravais lattice. The 16 A and B sites are arranged to minimize the first three SROP.

0	0	0	-1/4	0	-1/4	0	1
0	0	0	-1/3	0	0	0	1
0	0	0	-1/6	0	-1/2	0	1
0	0	0	-1/12	0	-3/4	0	1

Useful information about alloys can be found using supercells containing as few as two atoms,[12] but the only standard that can be used for the correctness of a calculation on a random alloy is the degree to which it approximates the result that would be obtained from an infinitely large supercell. Finite-size effects should be given the same serious consideration in alloy calculations as they are in any other statistical study.[1]

CALCULATION OF SROP AND $\alpha(\mathbf{k})$

We use the formulas from the preceding sections to study the short-range ordered states in some typical disordered alloys. Our Monte Carlo calculations start from a code written for the simple cubic lattice by Loren P. Meissner[13] to illustrate the use of array intrinsic functions in fortran 90 and 95. It treats the 2 interpenetrating cubic lattices in bcc or the 4 interpenetrating lattices in fcc as blocks. We use the Ising Hamiltonian,

$$H = -V_1 \sum_{\langle i,j \rangle} s_i s_j - V_2 \sum_{[i,j]} s_i s_j - v \sum_{i=1}^{N} s_i$$

where $s_i = \pm 1/2$ if there is an A or B atom on the site. The notation $\langle i,j \rangle$ means the sum is over nearest neighbors, while $[i,j]$ indicates next-nearest neighbors. The chemical potential v plays the role of an applied field in the magnetic interpretation of this Hamiltonian. It should be noted that this is a different Hamiltonian from the one used in the subsection Parameterized Energy. For this Hamiltonian, an interaction is antiferromagnetic or clustering when $V_i \leq 0$. The expression

$$\alpha(i) = 1 - \frac{1}{c_A c_B N n_{nn}^i} \sum_{|\mathbf{r}_n| = r_{nn}^i} N_{AB}(\mathbf{r}_n)$$

is used to calculate the SROP for the supercell. The diffuse scattering intensity function $\alpha(\mathbf{k})$ is calculated by the method described in the introduction.

The ordering of binary alloys on an fcc Bravais lattice is particularly interesting because of the geometrical frustration of the Hamiltonian. For the bcc case, the transition temperatures T_c for ordering (antiferro-magnetism) or clustering (ferromagnetism) are essentially the same, while for the fcc case they differ by a factor of 5. We focus on fcc alloys with $V_1 = -1.0$ in the following.

The concentrations of a series of alloys based on the fcc Bravais lattice are shown as a function of temperature for a series of chemical potentials v in Fig. 1. For all these alloys, $V_1 = -1.0$ and $V_2 = 0.0$. It can be seen that any concentration can be obtained with a suitable choice of v as long as the temperatures are above the order-disorder phase boundaries. There are only three ground state ordered structures for the case of nearest-neighbor anti-ferromagnetic interactions; the $L1_0$ or CuAuI structure for the 50% alloy, the $L1_2$ or Cu_3Au structure at 75%, and the pure metal at 100%.[6] As the temperature approaches zero, the Monte Carlo program will search out one of these structures because they are the only thermodynamically stable ones. To agree with those that appear in other publications,[14] the transition temperatures must be multiplied by four and the chemical potentials by two. The reason for this is that most authors use $s_i = \pm 1$ rather than $s_i = \pm 1/2$.

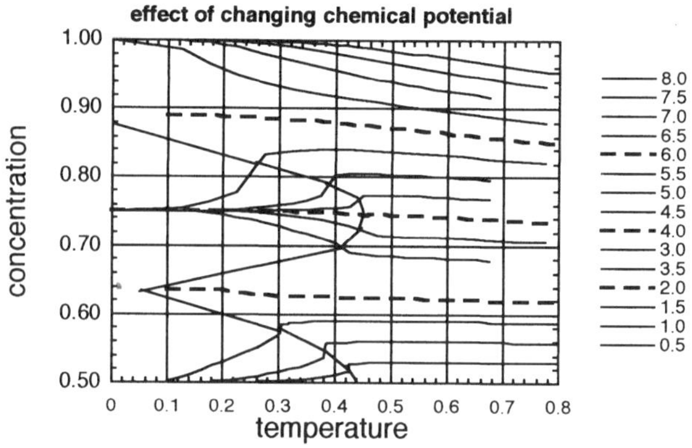

Figure 1. The concentrations as a function of temperature for various values of the chemical potential ν for AB alloys and the case that $V_1 = -1.0$ with $V_2 = 0.0$. The legend explaining the values of the chemical potentials is shown to the right of the drawing. The parabolas sketched in to the plot indicate the order-disorder transition temperatures.

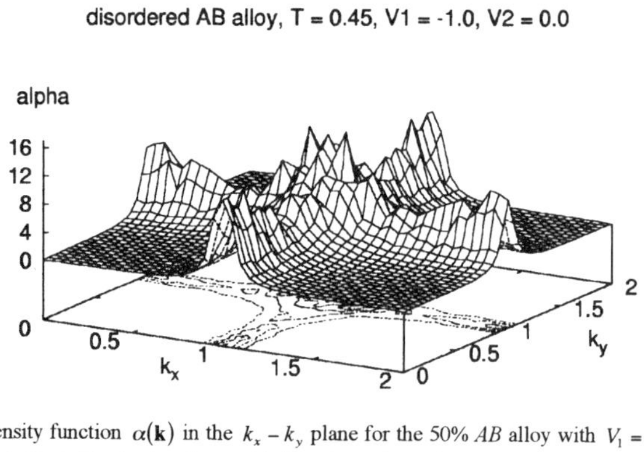

Figure 2. The intensity function $\alpha(\mathbf{k})$ in the $k_x - k_y$ plane for the 50% AB alloy with $V_1 = -1.0$ and $V_2 = 0.0$. The temperature is indicated in the drawing, and is just above the order-disorder transition temperature.

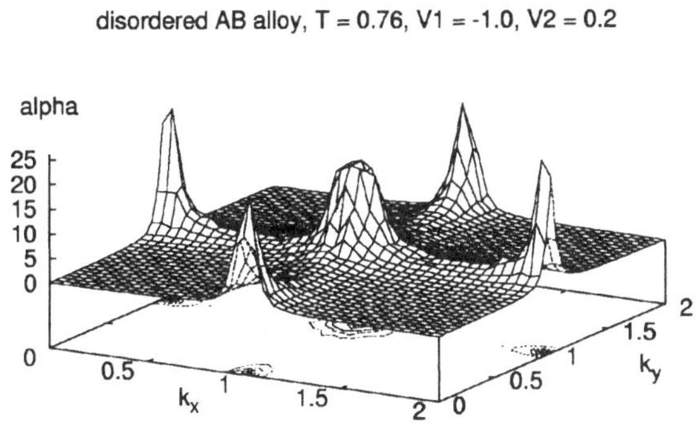

Figure 3. The intensity function $\alpha(\mathbf{k})$ in the $k_x - k_y$ plane for the 50% *AB* alloy with $V_1 = -1.0$ and $V_2 = 0.2$. The temperature is indicated in the drawing, and is just above the order-disorder transition temperature.

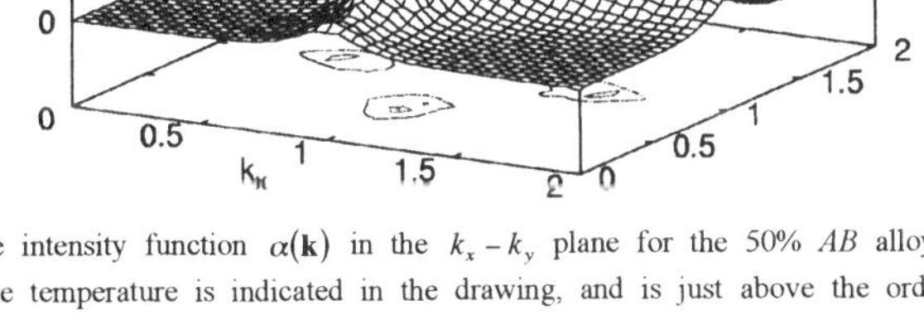

Figure 4. The intensity function $\alpha(\mathbf{k})$ in the $k_x - k_y$ plane for the 50% *AB* alloy with $V_1 = -1.0$ and $V_2 = -0.2$. The temperature is indicated in the drawing, and is just above the order-disorder transition temperature.

In Figs. 2, 3, and 4 we show the diffuse scattering intensity $\alpha(\mathbf{k})$ in the k_x, k_y plane for 50% AB alloys for the cases that $V_1 = -1.0$ with $V_2 = 0.0$, $V_2 = 0.2$, and $V_2 = -0.2$. The chemical potential is zero. Since the atoms are distributed on the sites of an fcc Bravais lattice, the peaks in $I(\mathbf{k})_{Bragg}$ are at the $\mathbf{k}$-points (0,0,0), (0,2,0), (2,2,0), etc. The temperatures in all cases are just above the order-disorder transition temperatures, which are approximately 0.44 for $V_2 = 0.0$, 0.75 for $V_2 = 0.2$, and 0.35 for $V_2 = -0.2$. For the cases where the second nearest-neighbor interactions are 0.0 or 0.2, the diffuse scattering peaks are at (0,1,0), (1,1,0), etc. The diffuse scattering intensity for $V_1 = -1.0$ and $V_2 = -0.2$ has peaks at the points (1,1/2,0), (1/2,1,0). All of these patterns are in keeping with the predictions for $\alpha(\mathbf{k})$ from the Clapp-Moss diagram.[15]

Ground state configurations for alloys based on the fcc lattice were predicted by Kanamori[6] and Allen and Cahn,[16] and the corresponding SROP are given in a table by Hata.[17] From Figs. 5 and 6, the first 8 SROP for the cases that V_2 is 0.0 or 0.2 have the values -1/3, 1, -1/3, 1, -1/3, 1, -1/3, and 1 at low temperatures. This pattern corresponds to the ordered $L1_0$ structure. The SROP for $V_2 = -0.2$ are shown in Fig. 7. For technical reasons the Monte Carlo calculations were initiated in the LI_0 state, but, as soon as the temperature reaches a point at which the system can find thermodynamic equilibrium after a reasonable number of Monte Carlo sweeps, the first 8 SROP become -1/3, 1/3, 1/3, -1/3, -1/3, -1, 1/3, and 1. This pattern is consistent with the N_2M_2 structure.

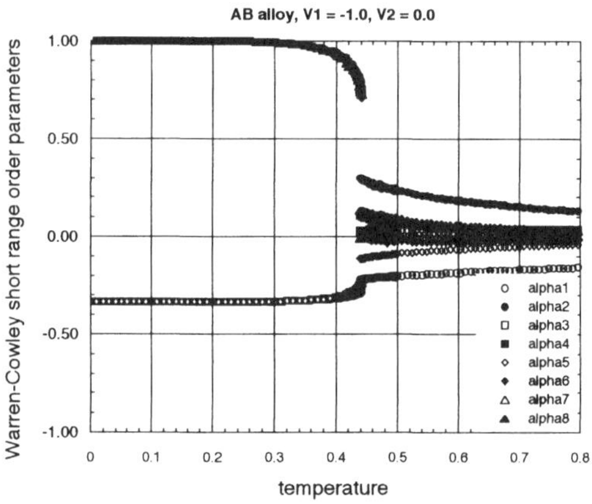

Figure 5. The first 8 SROP as a function of temperature for the AB alloy with $V_1 = -1.0$ and $V_2 = 0.0$. The concentration is 50%.

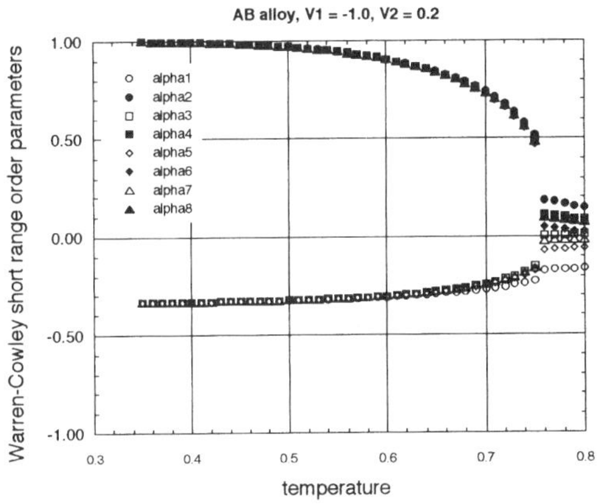

Figure 6. The first 8 SROP as a function of temperature for the *AB* alloy with $V_1 = -1.0$ and $V_2 = 0.2$. The concentration is 50%.

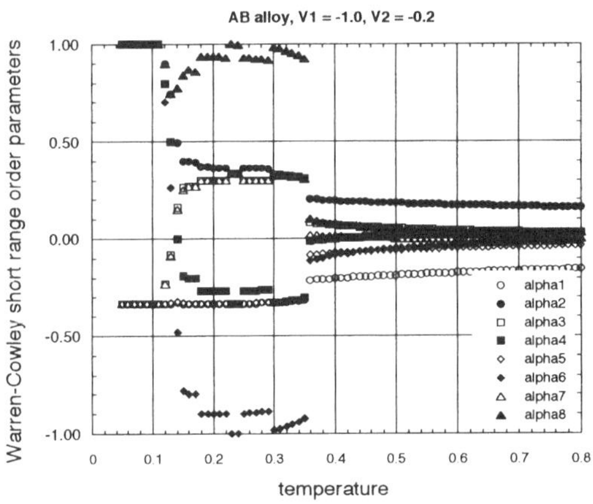

Figure 7. The first 8 SROP as a function of temperature for the *AB* alloy with $V_1 = -1.0$ and $V_2 = -0.2$. The concentration is 50%.

Calculations have been carried out for the chemical potential $\nu = 4.0$, so that the concentration is 75%. The results are similar to the 50% alloy described above. For $V_1 = -1.0$ and $V_2 = 0.0$ or $V_2 = 0.2$, the peaks in $\alpha(\mathbf{k})$ are at (0,1,0), (1,1,0), etc. From the pattern of SROP coefficients, it is seen that the ground state structure is $L1_2$. The diffuse scattering intensity for $V_1 = -1.0$ and $V_2 = -0.2$ has peaks at the points (1,1/2,0), (1/2,1,0),

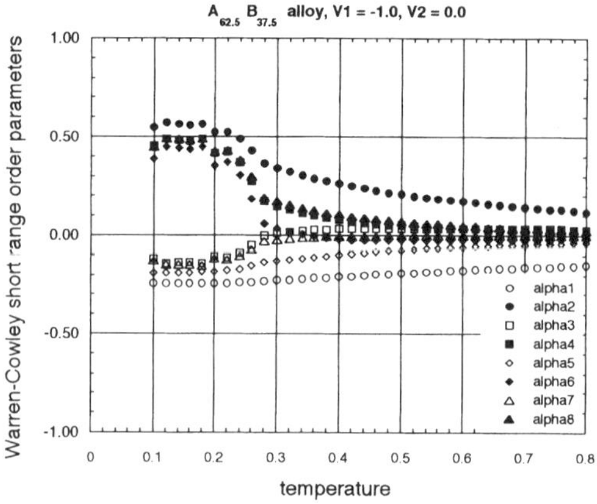

Figure 8. The first 8 SROP as a function of temperature for the *AB* alloy with $V_1 = -1.0$ and $V_2 = 0.0$. The concentration is 62.5%.

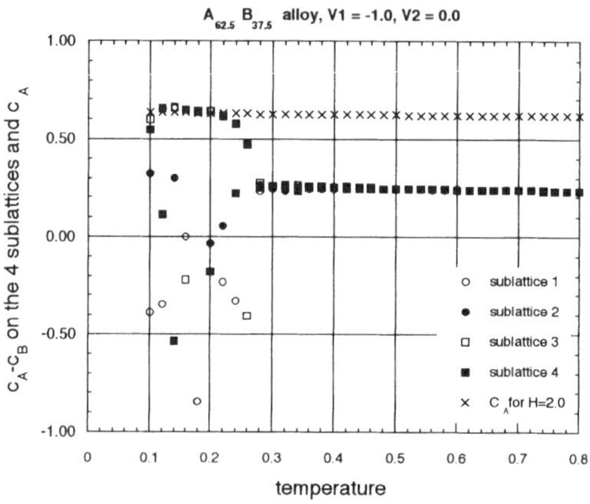

Figure 9. The difference $c_A - c_B$ for the four interpenetrating cubic sublattices for the 62.5% *AB* alloy is shown as a function of temperature for the case that $V_1 = -1.0$ and $V_2 = 0.0$. The concentration c_A is also shown.

etc., and it is seen from the pattern of the SROP that the ground state structure is DO_{22}. These ground state structures agree with the predictions.[6,16]

All of these calculations indicate that the short-range ordered state shown by the $\alpha(\mathbf{k})$ and the SROP for temperatures above the order-disorder transformation are closely related to the ground state structures. Remnants of the correlations that exist in the ground states will persist into the short-range ordered states as described, for example, by Hata.[17]

The SROP for concentrations in the neighborhood of 62.5%, which corresponds to the chemical potential $\nu = 2.0$, are shown in Fig. 8. The order parameter $c_A - c_B$ for the four interpenetrating sublattices are shown in Fig. 9, and it is clear that there is an order-disorder transformation with $T_C \approx 0.26$. There is little evidence for this transformation in the SROP. For temperatures as low as 0.1, they do not attain the values of 1 and -1/3 that they have for the 50% and 75% alloys. There is a triple point in the phase diagram near this concentration. More careful calculations[14] locate the triple point at $T_{tp} = 0.25$ at a chemical potential of $v_{tp} = 1.8$. The evidence is that below the transition temperature, the state is a mixture of $L1_0$ and $L1_2$. It would be expected that the short-range ordered state at a slightly higher temperature would also contain remnants of both of these ordered structures.

CONCLUSIONS

In the subsection The Meaning of Randomness in Terms of SROP, the Warren-Cowley analysis is used to rectify some misconceptions about disorder in alloys that have appeared in the literature.[9] The expressions for the Coulomb potentials and Coulomb energy in isomorphous alloys in terms of the SROP derived in the subsection Coulomb Energy provide another way to demonstrate that those functions are zero for an isomorphous random alloy, a question of some importance in alloy theory.[5] It is very easy to calculate the diffuse scattering intensity parameter $\alpha(\mathbf{k})$ using the method described here. The Krivoglaz-Clapp-Moss approximation,[15] the gamma expansion method,[18] the inverse Monte-Carlo method,[19] the Cowley method,[20] etc. are used to used to relate effective pair interactions such as V_1 and V_2 to $\alpha(\mathbf{k})$. Since excellent MC calculations can be done with an inexpensive PC, it seems to us that it is no longer necessary to employ these approximate methods.

The calculations in the preceding section indicate that the correlations in the low temperature ordered state are carried over into the short-range ordered state at higher temperature. This is similar to the conclusion reached from high-resolution electron diffraction studies.[17] It should be noted, however, that our Monte Carlo calculations generate thermodynamic equilibrium states. The electron diffraction experiments as well as the dynamic Monte Carlo method used to explain them focus on non-equilibrium states.[17]

REFERENCES

1. K. Binder and D. W. Heermann, *Monte Carlo Simulation in Statistical Physics* (Springer-Verlag, Berlin, 1992).
2. B. E. Warren, *X-ray Diffraction* (Addison-Wesley Book Co., New York, 1969).
3. J. S. Faulkner, Yang Wang, and G. M. Stocks, Phys. Rev. B **55**, 7 492 (1997).
4. M. P. Tosi, in *Solid State Physics, Vol. 16*, edited by F. Seitz and D. Turnbull (Academic Press, New York, 1964).
5. B. Ujfalussy, J. S. Faulkner, Nassrin Moghadam, G. M. Stocks and Yang Wang, Phys. Rev. B **61**, 12005 (2000); J. S. Faulkner, B. Ujfalussy, N. Y. Moghadam, G. M. Stocks and Yang Wang in *Properties of Complex Inorganic Solids, 2*, edited by A. Meike, A. Gonis, P. E. A. Turchi, and K. Rajan, (Kluwer Academic/Plenum Publishers, New York, 2000).
6. J. Kanamori and Y. Kakehashi, J. de Phys. (Colloque) **38-C7**, 274 (1977).
7. W. Feller, *An Introduction to Probability Theory and its Applications* (John Wiley & Sons; New York,1968).
8. J. S. Faulkner, Phys. Rev. B **64**, 233113 (2001).
9. S.-H. Wei, L. G. Ferreira, J. E. Bernard, and A. Zunger, Phys. Rev. B **42**, 9622 (1990); A. V. Ruban and H. L. Skriver, Phys. Rev. B **66**, 024201 (2002).
10. P. P. Ewald, Ann Physik 64, 253 (1921).

11. L. Greengard, The Rapid Evaluation of Potential Fields in Particle Systems, (MIT Press, Cambridge, 1988).
12. D. G. Pettifor, Phys. Rev. Letters **42**, 846 (1979).
13. L. P. Meissner, *Essential Fortran* (Unicomp, Inc., New York, 1997).
14. S. Kammerer, V. Dunweg, K. Binder, and M. d'O. de Meo, Phys. Rev. B **53**, 2345 (1996).
15. P. Clapp and S. C. Moss, Phys. Rev. **142**, 418 (1966); Phys. Rev. **171**, 754 (1968).
16. S. M. Allen and J. W. Cahn, Acta. Met. **20**, 423 (1972).
17. S. Hata, S. Matsumura, N. Kuwano, and K. Oki, Acta mater. **46**, 881 (1998).
18. V. I. Tokar, Phys. Lett. **110A**, 453 (1985).
19. V. Gerold and J. Kern, Acta Metall. **35**, 393 (1987).
20. J. M. Cowley, Phys. Rev. 77, 669 (1950).

DEPENDENCE OF ORDERING PROCESS IN Ni-BASED 1 1/2 0 ALLOY ON ALLOYING ELEMENTS

Satoshi Hata,[1] Makoto Inoue,[1] Noriyuki Kuwano,[2] and Yoshitsugu Tomokiyo[1]

[1] Department of Applied Science for Electronics and Materials
Kyushu University, Kasuga, Fukuoka 816-8580, Japan
[2] Advanced Science and Technology Center for Cooperative Research
Kyushu University, Kasuga, Fukuoka 816-8580, Japan

INTRODUCTION

Ni-based alloys such as Ni–V, Ni–Cr, Ni–Mo, Ni–W are called '1 1/2 0 alloys' because of their characteristic order-disorder transformation.[1–3] In the early stage of ordering, all the alloys form the short-range order (SRO) that is commonly identified by diffuse intensity maxima at *hkl* = 1 1/2 0 and equivalent positions, in addition to the fcc fundamental lattice reflections. In the later stage of ordering, on the other hand, various types of long-range order (LRO) structures develop from the 1 1/2 0 type SRO, depending on the alloy system and composition. Figure 1 illustrates the LRO structures of Ni-based 1 1/2 0 alloys: (a) A_4B type $D1_a$, (b) A_3B type $D0_{22}$ and (c) A_2B type Pt_2Mo structures and their schematic diffraction patterns (d). One can see the common features in the (420) stacking of the fcc lattices. Along the (420) stacking, certain numbers of Ni atom-planes are sandwiched between atom-planes composed of an alloying element. The numbers of the sandwiched Ni atom-planes are four for $D1_a$, three for $D0_{22}$ and two for Pt_2Mo structure, respectively. $D1_a$ is the LRO structure of Ni–Mo and Ni–W alloys, $D0_{22}$ is of Ni–V and Ni–Cr, and Pt_2Mo is of Ni–V, Ni–Cr and Ni–Mo alloys.

The ordering behavior in Ni-based 1 1/2 0 alloys was investigated by many research groups. However, a general understanding of the ordering mechanism in the alloys was not necessarily reached, since experimental methods and theoretical backgrounds are different among the research groups or the alloy systems to be investigated. For the general understanding of the ordering mechanism, the dependence of alloying elements on the ordering behavior in Ni-based 1 1/2 0 alloys was studied. Caudron *et al.*,[4] for example, studied the SRO in Ni–V and Ni–Cr alloys by *in-situ* neutron diffraction. They evaluated effective pairwise atomic interactions by the inverse cluster variation method and discussed the difference in phase stability using the evaluated effective interactions. Vlasova *et al.*,[5] Yamamoto *et al.*,[6] Martin and Williams[7, 8] and Arya *et al.*[9] studied the ordering in ternary

Ni–Mo–X (X = V, Ta, Al, W, Cr) alloys by transmission electron microscopy (TEM) and the first-principles calculations. They discussed influences of the alloying elements on the ordering behavior in terms of the effective atomic interactions, electronic structures of the ordered compounds and so on. In these previous studies, however, the ordering behavior was interpreted based on the mean-field approximation even for the SRO and imperfectly ordered states. Thus, the following points about the atomistic ordering process are not clear: (i) Do Ni-based 1 1/2 0 alloys form similar SRO structures in atomic level? (ii) How do the various LRO structures develop from the 1 1/2 0 type SRO state? (iii) What governs the SRO-LRO transition process in Ni-based 1 1/2 0 alloys?

The main purpose of the present study is to clarify these points (i), (ii) and (iii) described above. For attaining the purpose, the ordering processes in Ni–Mo and Ni–V alloys that show the different LRO structures with each other were investigated by TEM observation and Monte Carlo simulation using the kinetic Ising model. It has been demonstrated that the combination of TEM observation and the kinetic Monte Carlo simulation is a powerful method for studying the ordering processes in the Ni–Mo alloys.[10–13] The obtained results for Ni–Mo and Ni–V are compared with each other, and the atomistic ordering mechanism that can be commonly applicable for Ni-based 1 1/2 0 alloys is discussed. The physical meaning of the effective pairwise atomic interaction parameters for the Monte Carlo simulation is rationalized in terms of the concentration of free electrons in the alloys.

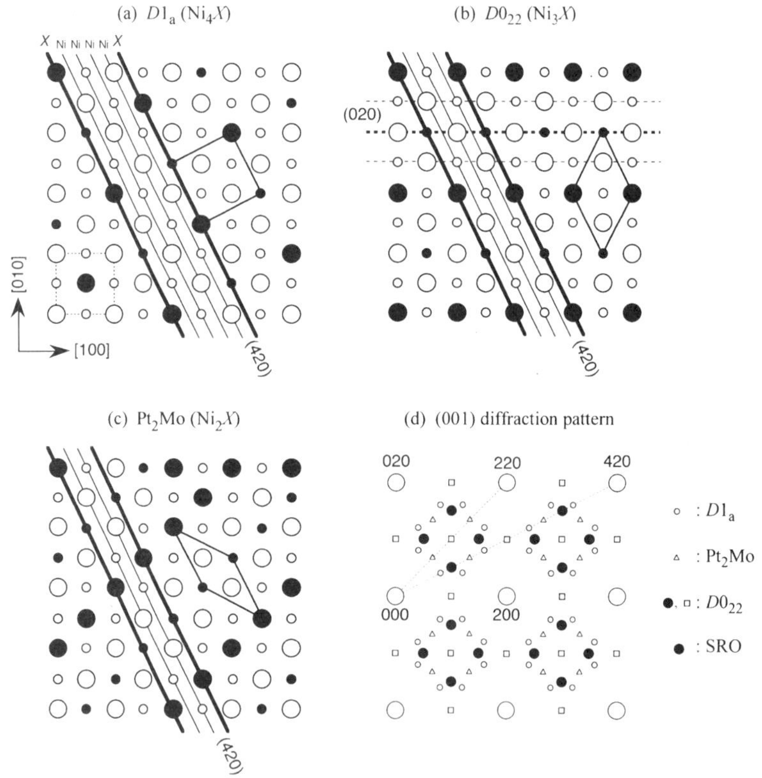

Figure 1. (001) cross-sectional views of $D1_a$ (a), $D0_{22}$ (b) and Pt_2Mo (c) structures and their schematic diffraction patterns (d).

TRANSMISSION ELECTRON MICROSCOPY OBSERVATION

Ordering Processes in Ni–Mo

Figure 2 shows (001) electron diffraction patterns of Ni_4Mo and Ni_3Mo alloys in the isothermal annealing at 873 K. The as-quenched specimens from the solid-solution

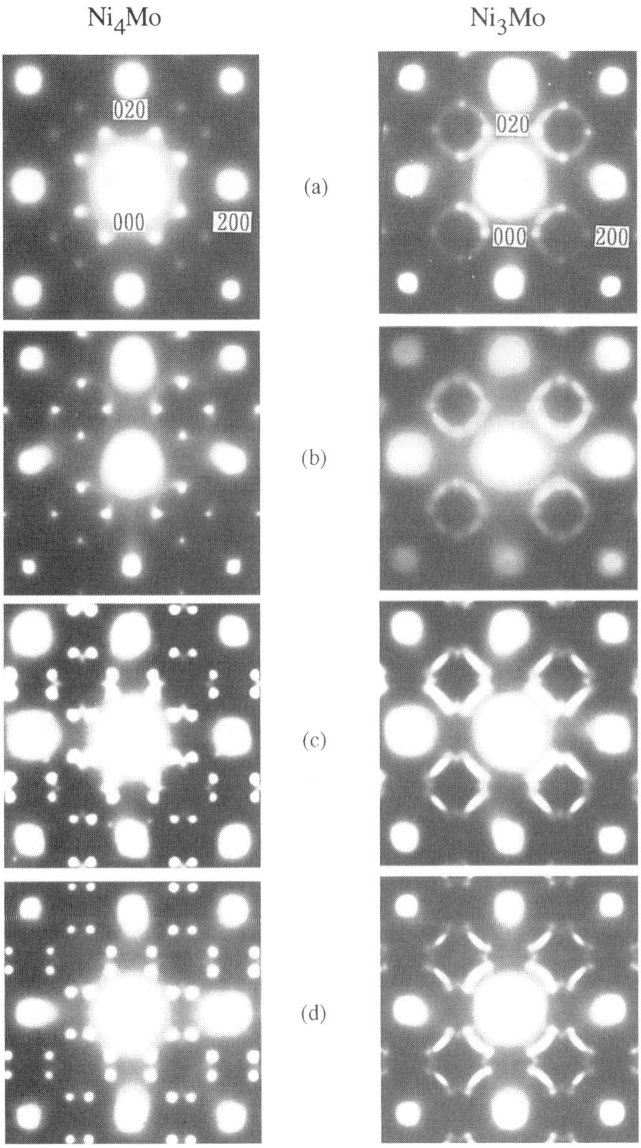

Figure 2. (001) electron diffraction patterns of Ni_4Mo (left column) and Ni_3Mo alloys (right column). As-quenched from 1273 K (1423 K for Ni_3Mo) (a), and annealed subsequently at 873 K for 7.2 (b), 86.4 (c) and 172.8 ks (d).

temperatures in (a) exhibit the intensity maxima at *hkl* = 1 1/2 0 and equivalent positions, as reported by many research groups previously. In addition, a ring-shaped diffuse scattering connecting the 1 1/2 0 type intensity maxima is observed for Ni_3Mo. As the long-range ordering proceeds from (b) to (d), the 1 1/2 0 type intensity maxima shift to the other positions, such as *hkl* = 4/5 2/5 0 for Ni_4Mo and *hkl* = 4/5 2/5 0 and 4/3 2/3 0 for Ni_3Mo. These intensity maxima correspond to the superlattice reflections of $D1_a$ and Pt_2Mo structures respectively, as illustrated schematically in Figure 1(d).

Kulkarni *et al.*[14] obtained a metastable fcc solid-solution with Ni_2Mo composition by the rapid solidification processing (RSP), and studied the ordering process in the Ni_2Mo solid-solution specimen. They reported that the 1 1/2 0 type intensity maxima and the ring-shaped diffuse scattering first appear, as observed for Ni_3Mo in Figure 1(a). The SRO state transforms into the Pt_2Mo type LRO state after subsequent annealing.

Ordering Processes in Ni–V

The SRO in fcc-based Ni–V alloy was extensively studied by neutron diffraction and the first-principles calculations rather than TEM. One of the main reasons is that $D0_{22}$ type LRO develops rapidly during quenching from the solid-solution temperatures.[15, 16] According to the *in-situ* neutron diffraction by Caudron *et al.*,[4] Ni_3V and Ni_2V alloys show the 1 1/2 0 type intensity maxima at the solid-solution temperatures. In the TEM observations of Ni_3V and Ni–29 at% V alloys[15, 16] quenched from the solid-solution temperatures, the $D0_{22}$ type ordering was recognized. The $D0_{22}$-ordered state in the Ni–29 at% V alloy[16] transformed into the two-phase LRO state of ($D0_{22}$+Pt_2Mo).

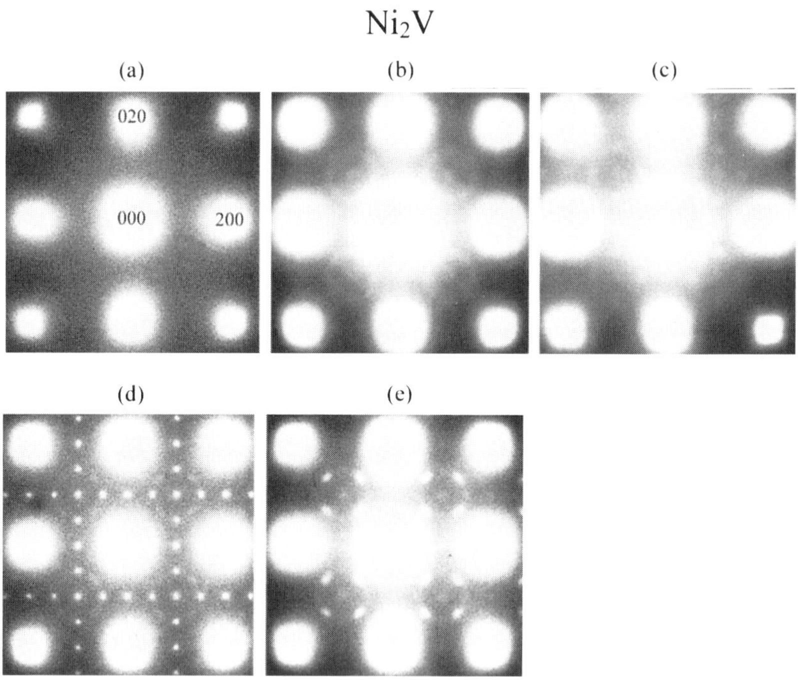

Figure 3. (001) electron diffraction patterns of Ni_2V alloy. As-quenched from 1373 K (a), and annealed subsequently at 843 K for 1.8 (b), 3.6 (c), 10.8 (d) and 43.2 ks (e).

In the present study, TEM observation of the ordering process in Ni_2V alloy was carried out, since the fcc solid-solution can be obtained by quenching.[17] Figure 3 shows (001) electron diffraction patterns in the isothermal annealing at 843 K, where the homogeneous ordering process through the 1 1/2 0 type SRO state was observed. It is confirmed from Figure 3(a) that the as-quenched specimen is of a disordered state. After annealing at 843 K, the ring-shaped diffuse scattering and the 1 1/2 0 type intensity maxima appear, as shown in (b) and (c). In addition, faint $D0_{22}$ type superlattice reflections at hkl = 110 and 100 are also seen. Another $D0_{22}$ type superlattice reflection at hkl = 1 1/2 0 may be superimposed on the SRO diffuse intensity maxima. The $D0_{22}$ type ordering develops from the SRO state, as seen in (d), although the matrix composition is the stoichiometric Ni_2V. As the Pt_2Mo type superlattice reflections become intense in the later stage of ordering in (e), the $D0_{22}$ type superlattice reflections turn to decrease in intensity. Sundararaman[18] also found such a peculiar ordering process in Ni_2V alloy.

MONTE CARLO SIMULATION

Simulation Procedure

In order to study the changes in local atomic arrangement during the ordering in Ni–Mo and Ni–V alloys, Monte Carlo simulation using the kinetic Ising model was carried out. An $A_{1-x}B_x$ alloy system consisting of $20_{[100]}\times20_{[010]}\times20_{[001]}$ fcc unit cells with the periodic boundary condition was considered. The internal energy of the system was given by

$$E = -\sum_{n} V(n)\left(\sum_{nNN} \sigma_i \sigma_j\right), \qquad (1)$$

where σ is the site-occupation operator defined by $\sigma_A = +1$ and $\sigma_B = -1$, and nNN denoting summation over the nth nearest neighbor pairs of i–j. $V(n)$ is the effective pairwise atomic interaction parameter for nth coordination shells defined in a conventional way as

$$V(n) = E_{AB}(n) - [E_{AA}(n) + E_{BB}(n)] / 2, \qquad (2)$$

where $E_{ij}(n)$ is the effective pairwise atomic interaction energy between i and j atoms. The system prefers unlike-atom pairs of the nth coordination shells if $V(n)$ is negative, and like-atom pairs if it is positive.

In the Monte Carlo simulation, the constituent A- and B-atoms were randomly

Table 1. Effective pairwise atomic interaction parameters $V(n)$ for Monte Carlo simulation.

$V(n)$	Ni_4Mo	$Ni_3Mo - Ni_2Mo$	$Ni_3V - Ni_2V$
$V(1)_{lmn=<110>}$	< 0 (ordering)	< 0 (ordering)	< 0 (ordering)
$V(2)_{<200>} / \lvert V(1)\rvert$	0.2	0.4	0.41
$V(3)_{<211>} / \lvert V(1)\rvert$	0.1	0.1	0.1
$V(4)_{<220>} / \lvert V(1)\rvert$	– 0.4	– 0.4	– 0.35
$V(5)_{<310>} / \lvert V(1)\rvert$	0.1	0.1	– 0.1
$V(6)_{<222>} / \lvert V(1)\rvert$	0	0.15	0.2
$V(7)_{<321>} / \lvert V(1)\rvert$	0	0.01	0.01
$V(8)_{<400>} / \lvert V(1)\rvert$	0	– 0.05	– 0.1
$V(9)_{<330>} / \lvert V(1)\rvert$	0	0	0.02
$V(9)_{<411>} / \lvert V(1)\rvert$	0	0	– 0.02
$k_BT / \lvert V(1)\rvert$	1.0	1.0	1.0

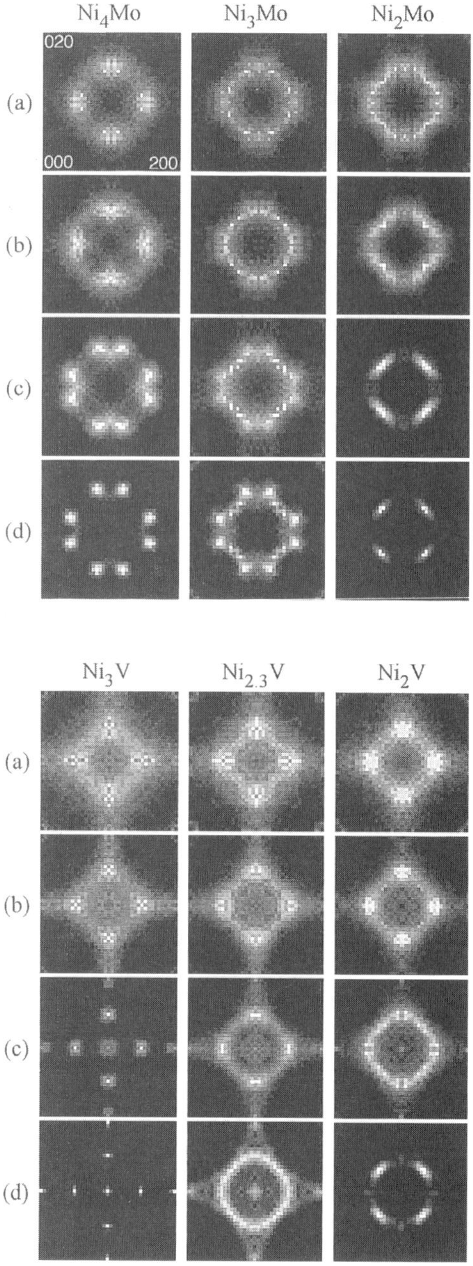

Figure 4. Temporal changes of (001) Fourier power spectra of atomic arrangements obtained by the kinetic Monte Carlo simulation. Monte Carlo steps (MCS) from (a) to (d) are 20, 50, 500, 3000 MCS for Ni_4Mo, 40, 200, 700, 3000 MCS for Ni_3Mo, 20, 100, 600, 3000 MCS for Ni_2Mo, 10, 100, 1500, 10000 MCS for Ni_3V, 10, 100, 1500, 30000 MCS for $Ni_{2.3}V$ (Ni–30 at% V), and 10, 100, 1 000, 10000 MCS for Ni_2V, respectively.

distributed on the fcc lattice sites. Two neighboring atoms were chosen at random, and attempted to be exchanged in position with each other, according to the Kawasaki spin-flip dynamics.[19] The exchange probability between the neighboring atoms was given by

$$W = \exp(-\Delta E / k_B T) / [1 + \exp(-\Delta E / k_B T)] = [1 - \tanh(\Delta E / 2k_B T)] / 2 \quad (3)$$

where ΔE is the energy gain caused by the exchange, and k_B and T are the Boltzmann's

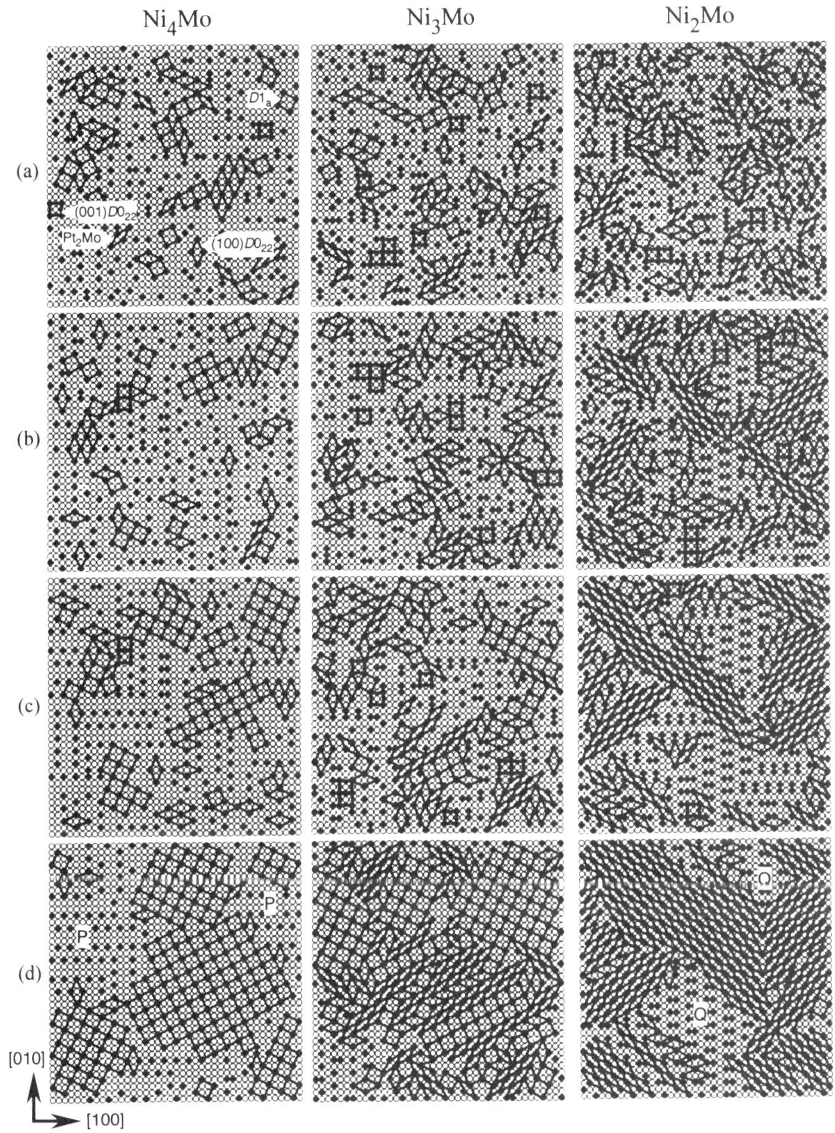

Figure 5. Temporal changes of (001) cross-sectional views of the simulated atomic arrangements for Ni–Mo alloys at 50 (a), 100 (b), 500 (c) and 3000 MCS (d).

constant and the absolute temperature, respectively. If W is larger than a random number R between 0 and 1, the neighboring atoms are exchanged with each other. The time for atomic exchange is measured in units of Monte Carlo step (MCS), one MCS corresponds to 32000 (20×20×20×4) attempts to exchange neighboring atoms, namely one attempt of exchanging per one atom.

In order to reproduce the ordering processes by the Monte Carlo simulation, the values of $V(n)$ should be optimized. The optimization procedure for $V(n)$ is based on the ground state analysis for fcc binary alloys as a function of the pairwise atomic interactions by Kanamori and Kakehashi.[20] According to their theory, $D1_a$, $D0_{22}$ and Pt_2Mo structures appear as the ground state structures when $V(n)$'s up to the fourth coordination shells ($n = 4$) are set at appropriate values. Thus, we first determined the ranges of $V(n)$ up to $n = 4$ that can produce $D1_a$, $D0_{22}$ and Pt_2Mo structures as the ground states in the composition range between A_4B and A_2B. Appropriate values of $V(n)$ with which the simulation could reproduce well the temporal changes in diffraction pattern of the Ni–Mo and Ni–V alloys were searched within the range of $V(n)$ determined above. However, it was found that $V(n)$ with $n > 4$ must be taken into account for reproducing the experimental results of the ordering processes in the wide composition range. Without setting $V(n)$ with $n > 4$, $D1_a$, $D0_{22}$ and Pt_2Mo type microdomains cannot grow into LRO domains at off-stoichiometric compositions. Then, the optimizing procedure was carried out again with taking account of $V(n)$ with $n > 4$. The values of $V(n)$ up to $n = 9$ were finally evaluated as listed in Table 1. The roles of $V(n)$'s with $n > 4$ will be discussed later. The value for T used in the Monte Carlo simulation was approximately to a half the critical temperature of Pt_2Mo structure estimated by the Bragg-Williams' approximation.

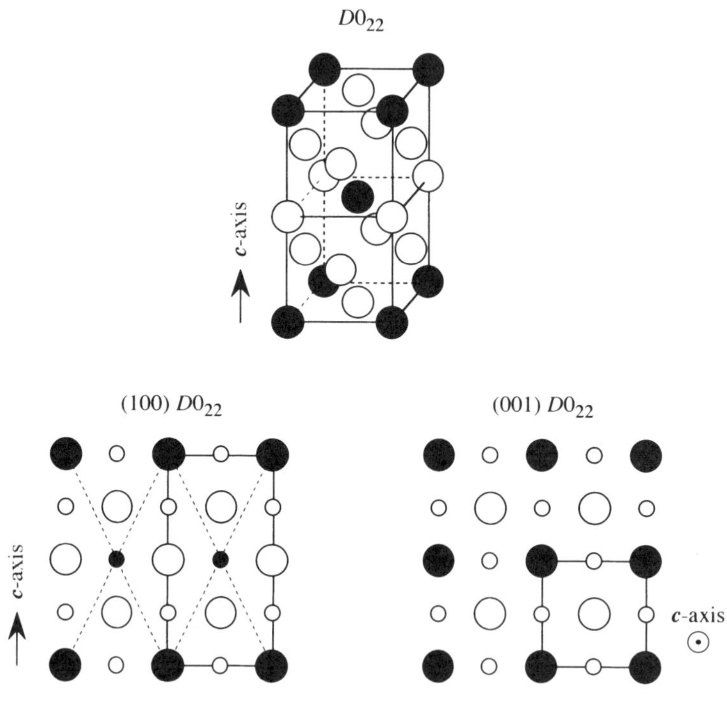

Figure 6. $D0_{22}$ structure and its derivatives of two-dimensional $(100)D0_{22}$ and $(001)D0_{22}$.

Simulation Results

Figure 4 shows temporal changes in (001) Fourier power spectra of atomic arrangements obtained by Monte Carlo simulation. Each spectrum is illustrated after averaging over three spectra of (100), (010) and (001) reciprocal lattice planes, and the fundamental 000, 200, 200 and 220 reflections located at the four corners in each spectrum are omitted. One can see that the simulated diffraction intensity distributions reproduce well the temporal changes in experimental diffraction pattern of the Ni–Mo and Ni–V alloys

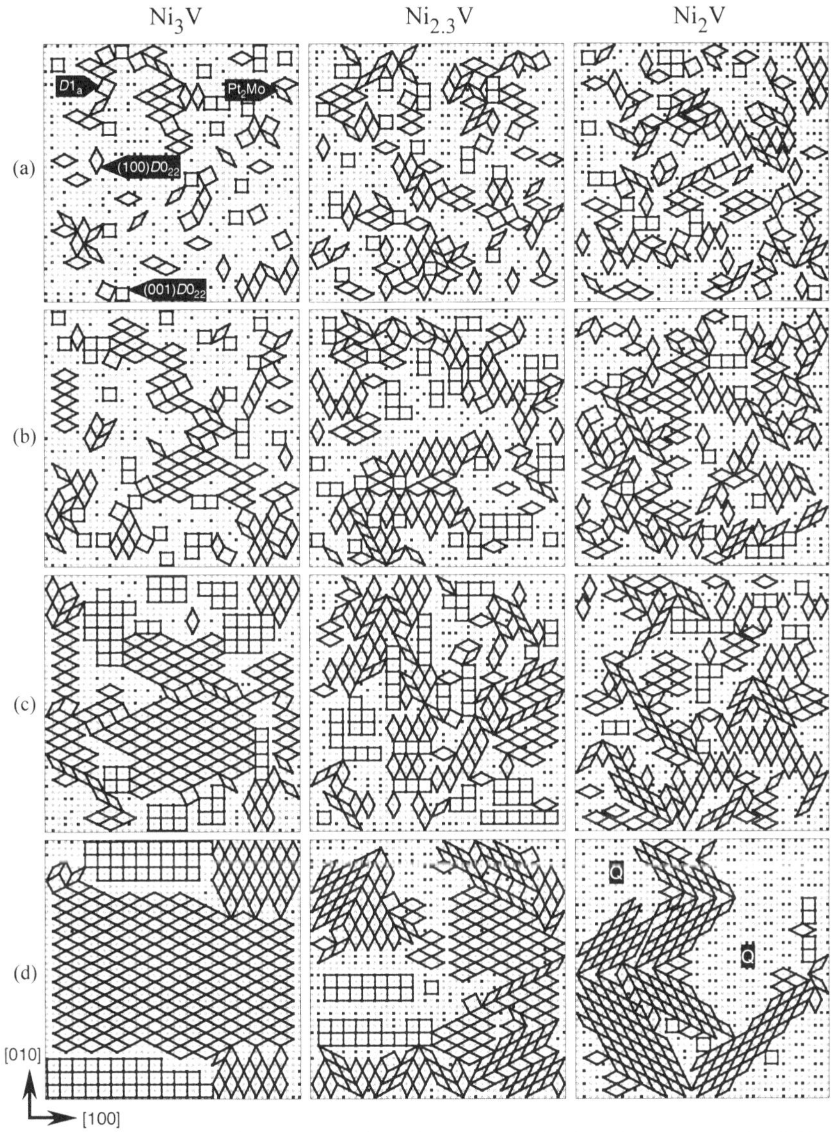

Figure 7. Temporal changes of (001) cross-sectional views of the simulated atomic arrangements for Ni–V alloys at 10 (a), 100 (b), 1000 (c) and 10000 MCS (30000 MCS for Ni_2V) (d).

described in the previous section (*e.g.* Figures 2 and 3).

Figure 5 demonstrates temporal changes in (001) cross-sectional views of the simulated atomic arrangements for the Ni–Mo alloys. In each cross-sectional view composed of two neighboring (002) atomic planes, black points correspond to Mo atoms. In order to help identification of ordered structures formed in the matrix, sub-unit cells of $D1_a$ (slanting large square), $D0_{22}$ (fat rhombus-shaped (100)$D0_{22}$ and small square-shaped (001)$D0_{22}$, illustrated in Figure 6) and Pt_2Mo (slanting lean rhombus) structures are depicted by solid lines.

For Ni_4Mo composition, microclusters of $D1_a$, $D0_{22}$ and Pt_2Mo types are formed together in the early stage of ordering in (a) and (b). It is known that such a mixed state of $D1_a$, $D0_{22}$ ad Pt_2Mo type clusters exhibit the diffuse intensity maxima at *hkl* = 1 1/2 0.[1, 10–13, 21] As the ordering proceeds, only $D1_a$ type clusters develop into LRO domains in (c) and (d). For Ni_3Mo composition, $D1_a$, $D0_{22}$ and Pt_2Mo type clusters are firstly formed in the same way as for Ni_4Mo composition, although the fractions of $D0_{22}$ and Pt_2Mo type clusters are larger than those for Ni_4Mo, as recognized in (a) and (b). It is noted that the number of (001)$D0_{22}$ (small square) type clusters is much smaller than that of (100)$D0_{22}$ (fat rhombus) type clusters. This suggests that the $D0_{22}$ type atomic arrangement tends to grow two-dimensionally to form the (100)$D0_{22}$ type clusters.[11] In the later stage of ordering, most $D0_{22}$ type clusters shrink, and $D1_a$ and Pt_2Mo type clusters grow into large domains to form the two-phase LRO state. In the case of Ni_2Mo composition, the fraction of $D1_a$ type clusters is small. Pt_2Mo type LRO domains develop monotonously with shrinking of $D1_a$ and $D0_{22}$ type clusters. Areas P and Q in (d) correspond to $D1_a$ and Pt_2Mo type domains with *c*-axes parallel to the (001) plane, respectively.

The mixed state of $D1_a$, $D0_{22}$ and Pt_2Mo type clusters in the SRO states of Ni–Mo and other 1 1/2 0 alloys has been experimentally observed by high-resolution transmission electron microscopy (HRTEM).[1, 12, 13, 22] Hata *et al.*[13] revealed that the fractions of $D1_a$, $D0_{22}$ and Pt_2Mo type clusters in Ni–Mo alloy depend on the alloy composition and heat treatment, in a good agreement with the present Monte Carlo simulation in Figure 5.

(001) cross-sectional views of the simulated atomic arrangements for the Ni–V alloys are demonstrated in Figure 7, hence black points in the cross-sectional views corresponding to V atoms. For Ni_3V composition, $D0_{22}$ type ordering is dominant during the ordering process. Nevertheless, $D1_a$ and Pt_2Mo type clusters are formed together with the $D0_{22}$ type clusters in the early stages of ordering in (a) and (b). It should be noted that the square-shaped (001)$D0_{22}$ type clusters are often formed with the (100)$D0_{22}$ clusters. This suggests that the $D0_{22}$ type atomic arrangements grow three-dimensionally for the Ni–V system. For $Ni_{2.3}V$ (Ni–30 at% V) composition, $D1_a$ type clusters are still generated with $D0_{22}$ and Pt_2Mo type clusters, and the Pt_2Mo type domains come to grow in the later stage of ordering with increasing V content. Even for Ni_2V composition, $D0_{22}$ type domains grow first from the mixed state of $D1_a$, $D0_{22}$ and Pt_2Mo type clusters in (a), (b) and (c), although Pt_2Mo type domains grow in the later stage with shrinking the $D0_{22}$ type domains. Areas Q in (d) correspond to Pt_2Mo type domains with *c*-axes parallel to the (001) plane.

From the results of Monte Carlo simulation, it is suggested that the 1 1/2 0 type SRO state is commonly described as a mixed state of $D1_a$, $D0_{22}$ and Pt_2Mo type microclusters,

Table 2. Warren-Cowley parameters of ordered structures in their perfectly ordered states.

Structure	$\alpha(1)$	$\alpha(2)$	$\alpha(3)$	$\alpha(4)$	$\alpha(5)$	$\alpha(6)$	$\alpha(7)$	$\alpha(8)$	$\alpha(9_{<330>})$	$\alpha(9_{<411>})$
$D1_a$	−0.250	0.167	0.167	−0.250	−0.042	−0.250	0.167	−0.167	−0.250	−0.250
Pt_2Mo	−0.250	0.000	0.250	−0.250	0.000	0.250	−0.250	0.000	1.000	0.250
$D0_{22}$	−0.333	0.556	0.111	0.111	−0.333	−0.333	0.111	1.000	−0.333	−0.333
(100)$D0_{22}$	−0.333	0.333	0.333	−0.333	−0.333	—	0.333	1.000	−0.333	−0.333
(001)$D0_{22}$	−0.333	1.000	−0.333	1.000	−0.333	—	0.333	1.000	−0.333	−0.333

and the SRO-LRO transition proceeds by the selected growth of the stable LRO structures from the mixed state of the microclusters.

DISCUSSION

Mechanism of Atomistic Ordering Processes

The ordering processes in the Ni–Mo and Ni–V alloys reproduced by Monte Carlo simulation are discussed based on the pairwise interaction model.[20] Table 2 shows Warren-Cowley parameters $\alpha(n)$ of the ordered structures in their perfectly ordered states. The parameter $\alpha(n)$ is defined by[23]

$$\alpha(n) = 1 - P_{AB} / x_{B}, \tag{5}$$

where P_{AB} is a conditional probability for finding a B-atom at a lattice point in the nth coordination shell of an A-atom, and x_B is the atomic fraction of B-atoms. (100)$D0_{22}$ and (001)$D0_{22}$ in Table 2 refer to two-dimensionally extending atomic arrangements of two adjoining (200) and (002) planes of $D0_{22}$ structure respectively, as illustrated in Figure 6. All the structures have the following tendencies of $\alpha(n)$ except for the three-dimensional $D0_{22}$ and (001)$D0_{22}$,

$$\alpha(1),\ \alpha(4),\ \alpha(5) \le 0, \tag{6a}$$

$$\alpha(2),\ \alpha(3) \ge 0. \tag{6b}$$

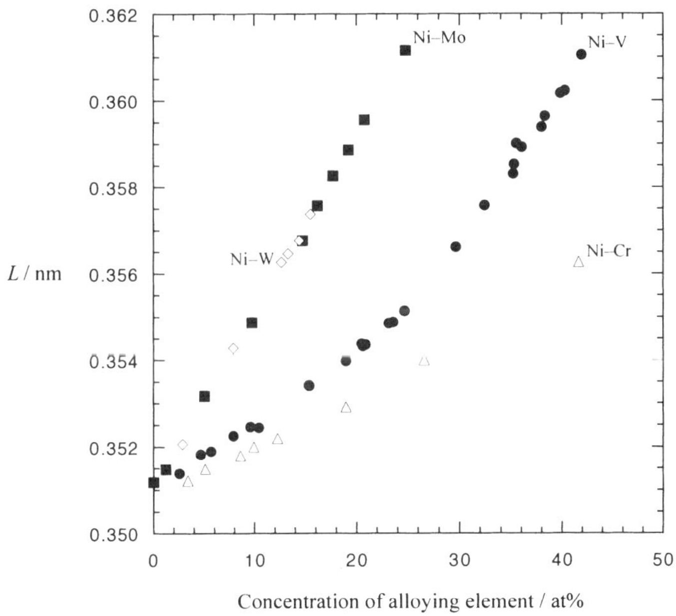

Figure 8. fcc lattice parameters of Ni solid-solutions.[26]

It is noted that the $D0_{22}$ type atomic configuration keeps a negative value of $\alpha(4)$ if the rhombus-shaped $(100)D0_{22}$ type clusters extend two-dimensionally. Such a two-dimensional growth of the $D0_{22}$ type atomic configuration may be one of the reasons for the fact that the 100_{D022} and 110_{D022} superlattice reflections are forbidden in the ordering process of Ni–Mo alloy. The common tendencies of $\alpha(n)$ suggest that the formation probabilities of $D1_a$, $D0_{22}$ and Pt_2Mo type atomic configurations in the SRO states are nearly even to each other. In other words, it can be rationalized that the formation of the mixed state of $D1_a$, $D0_{22}$ and Pt_2Mo type microclusters is thc fundamental features of the 1 1/2 0 type SRO.

The roles of $V(n)$ in the SRO-LRO transition process are also interpreted using $\alpha(n)$ in Table 2. In the case of the Ni–Mo type ordering, the positive $V(5)$ and negative $V(8)$ contribute to producing like- and unlike-atom pairs in the fifth and eighth coordination shells, respectively. As a result, $D0_{22}$ type ordering taking the negative $\alpha(5)$ and positive $\alpha(8)$ is prevented, while $D1_a$ or Pt_2Mo type ordering that takes $\alpha(5) \approx 0$ and $\alpha(8) \leq 0$ develops selectively. For the Ni–V type ordering, let us compare the values of $V(n)$ for the Ni–Mo ($D1_a$ + Pt_2Mo) alloy with those for the Ni–V ($D0_{22}$ + Pt_2Mo) alloy, listed in Table 1. If all the Mo atoms in the Ni–Mo alloy are substituted with V atoms, $V(4)/|V(1)|$ increases as $-0.4 \rightarrow -0.35$, and $V(5)/|V(1)|$ decreases as $+0.1 \rightarrow -0.1$. The changes of $V(4)/|V(1)|$ and $V(5)/|V(1)|$ contribute to $D0_{22}$ type ordering, since

$$\alpha(4)_{D1a}, \alpha(4)_{Pt2Mo} \ll \alpha(4)_{D022}, \tag{7a}$$

$$\alpha(5)_{D1a}, \alpha(5)_{Pt2Mo} \gg \alpha(5)_{D022}. \tag{7b}$$

On the other hand, the increases in $V(6)/|V(1)|$ ($+0.15 \rightarrow +0.2$) and $V(9_{<330>})/|V(1)|$ ($0 \rightarrow +0.1$) and the decrease in $V(8)/|V(1)|$ ($-0.05 \rightarrow -0.1$) contribute to Pt_2Mo type ordering, since

$$\alpha(6)_{D1a}, \alpha(6)_{D022} \ll \alpha(6)_{Pt2Mo}, \tag{8a}$$

$$\alpha(9_{<330>})_{D1a}, \alpha(9_{<330>})_{D022} \ll \alpha(9_{<330>})_{Pt2Mo}, \tag{8b}$$

$$\alpha(8)_{D022} \gg \alpha(8)_{Pt2Mo}, \alpha(8)_{D1a}. \tag{8c}$$

Caudron *et al.*[4] also reported that the role of $V(9_{<330>})$ is crucial for the Pt_2Mo type ordering. From the relationships between $V(n)$ and $\alpha(n)$ described above, it may be rationalized that the characteristic ordering sequence in Ni_2V, 1 1/2 0 type SRO → $D0_{22}$ type LRO → Pt_2Mo type LRO, is due to the different tendencies of the long-range pair correlations among $D1_a$, $D0_{22}$ and Pt_2Mo structures. As mentioned previously,[20] $D1_a$, $D0_{22}$ and Pt_2Mo structures appear as the ground state structures when the pairwise atomic interactions up to the fourth coordination shells are taken into account. The present Monte Carlo simulation, however,

Table 3. Fitting parameters for the Friedel oscillation expressed in Eq. (9).

Alloy	Lattice parameter [a] L / nm	Fermi wavelength k_F / nm^{-1}	Amplitude A	Phase factor ϕ / rad
Ni_3Mo	0.363	14.97	0.9553	– 2.589
Ni_2Mo	0.366	14.84	0.9556	– 2.588
Ni_3V	0.357	14.85	0.8577	– 2.351
Ni_2V	0.359	14.78	0.8585	– 2.356

[a] Estimated from Figure 8.[26]

suggests that the SRO-LRO transition process in 1 1/2 0 alloys depends strongly on the pairwise atomic interactions in larger distances than the fourth coordination distance. In this connection, Caudron *et al.*[4] suggested that the influence of the long-range pairwise interactions may be interpreted as the influence of many-body interactions. The influences of the many-body interactions on the order-disorder transformation in Ni–V alloy was recently studied by Chepulskii.[24, 25]

Dependence of $V(n)$ on Alloying Elements

The physical meaning of the effective pairwise interaction parameters $V(n)$ evaluated from the kinetic Monte Carlo simulation is discussed. Figure 8 shows the fcc lattice constants L of Ni solid-solutions.[26] The Ni–Mo and Ni–W alloys that form a $D1_a$ type LRO have larger lattice constants than those of Ni–V and Ni–Cr alloys without forming the $D1_a$ type LRO. This suggests that the lattice spacing is one of the important factors to determine the LRO structures in the Ni-based 1 1/2 0 alloys. In the periodic table, the four elements, V (Period 4; Group V_a; Outer electron configuration $3d^3 4s^2$), Cr (4; VI_a; $3d^5 4s$), Mo (5; VI_a; $4d^5 5s$) and W (6; VI_a; $5d^4 6s^2$), that form 1 1/2 0 alloys with Ni (4; VII; $3d^8 4s^2$), are adjacent to each other. Kulkarni and Banerjee[27] pointed out that the differences in outer electron configuration may give large influences on the phase stability in the Ni-based 1 1/2 0 alloys. They estimated the number of free electrons per atom e/a for various Ni-based alloys, in accordance with the scheme proposed by Sinha.[28] The values of e/a for pure metals were given as e/a = 10 for Ni, 5 for V, and 6 for Cr, Mo and W. If Ni forms an Ni_3X type alloy with V or Mo, e/a decreases to 8.75 (= 10 × 0.75 + 5 × 0.25) for Ni_3V or 9 (= 10 × 0.75 + 6 × 0.25) for Ni_3Mo. From the estimation of e/a, they assumed the condition, $e/a \geq 9$, for stabilizing a phase mixture of $D1_a$ and Pt_2Mo structures in Ni_3X alloy.

In order to take the influences of the lattice spacing and the number of free electrons into account, the values of $V(n)/V(1)$ were plotted as a function of interatomic distance r, and compared with a simple Friedel oscillation expressed by[29, 30]

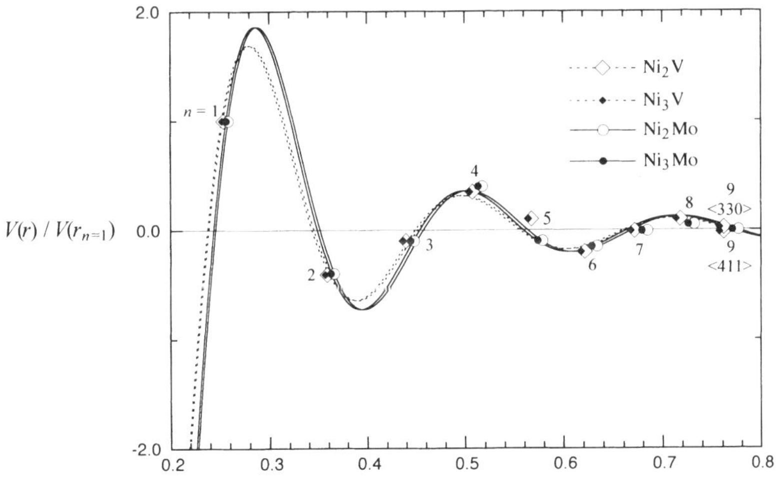

Figure 9. $V(r)/V(r_{n=1})$ as a function of interatomic distance (plots) and its fitting based on the Friedel oscillation (curves) expressed by Eq. (9). The fitting parameters used in Eq. (9) are listed in Table 3.

$$V(r) / V(r_{n=1}) = A \cos(2k_F + \phi) / (r / L)^3, \tag{9}$$

where A, k_F and ϕ are the amplitude, the Fermi wavenumber and the phase factor, respectively. As listed in Table 3, the values of these fitting parameters were determined for each alloy system, and the lattice parameters L were estimated from Figure 8.[26] The result in Figure 9 shows that $V(n)/V(1)$ is fitted well to the Friedel oscillation, although the values of $V(n)$ were evaluated by only reproducing the temporal changes in experimental diffraction pattern during the ordering processes. Such an applicability of the Friedel oscillation to the effective pairwise interactions in transition-metal alloys was also shown by Schweika and Haubold[31, 32] and Zou and Carlsson.[33, 34] Using the values of k_F estimated from the fitting of $V(n)/V(1)$, e/a can be calculated from the following equations assuming the spherical Fermi surface,[35]

$$k_F = (3\pi^2 N / V)^{1/3} = (3\pi^2 \cdot 4(e/a) / L^3)^{1/3}, \tag{10a}$$

$$(e/a) = L^3 k_F^3 / 12\pi^2, \tag{10b}$$

where N and V are the total number of free electrons and the volume of the unit cell, respectively. Figure 10 compares the values of e/a calculated from Eq. (10b) with the values of e/a reported by Kulkarni and Banerjee.[27] Although there are large differences in e/a due to the two different ways of estimating, the same tendency of e/a is shown; that is, larger for Ni–Mo than for Ni–V. This means that the different sets of $V(n)$ evaluated from the kinetic Monte Carlo simulation can derive qualitatively the differences in e/a between the Ni–Mo and Ni–V alloys. Therefore, it is suggested that for Ni-based 1 1/2 0 alloys, the dependence

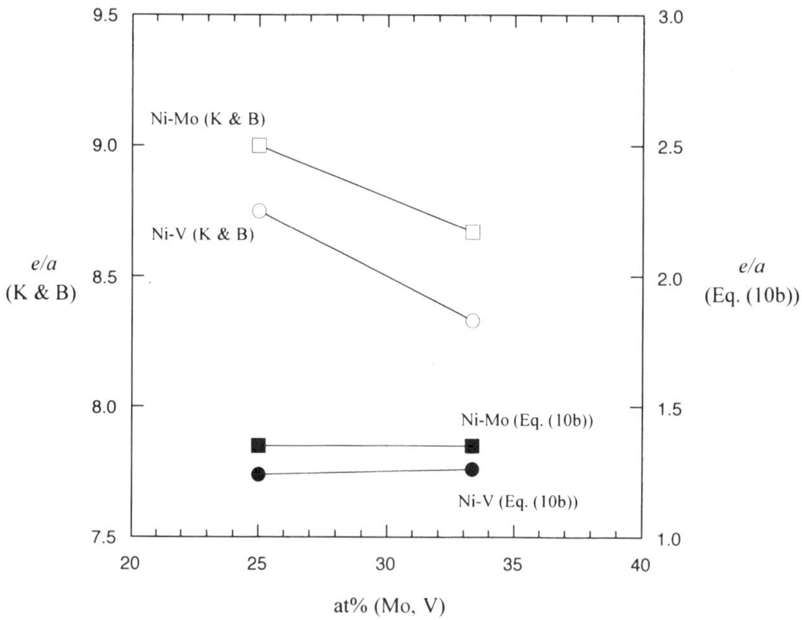

Figure 10. Numbers of free electrons per atom e/a for Ni–Mo and Ni–V alloys, estimated by Kulkarni and Banerjee (open symbols)[27] and Eq. (10b) (solid symbols).

of $V(n)$ on alloying elements can be rationalized in terms of the concentration of free electrons in the alloys.

CONCLUSIONS

In the present study, the ordering processes in fcc-based Ni–Mo and Ni–V alloys were investigated by TEM observation and the kinetic Monte Carlo simulation to clarify the atomistic ordering mechanism in Ni-based 1 1/2 0 alloys. The following conclusions were drawn.

1. It was experimentally confirmed that the ordering processes are described as 1 1/2 0 type SRO → $D1_a$ and/or Pt_2Mo type LRO for the Ni–Mo alloy, and 1 1/2 0 type SRO → $D0_{22}$ and/or Pt_2Mo type LRO for the Ni–V alloy. The $D0_{22}$ type ordering occurs before the development of Pt_2Mo type LRO even in the stoichiometric Ni_2V alloy.

2. The SRO structure in Ni-based 1 1/2 0 alloy is commonly described as a mixed state of $D1_a$, $D0_{22}$ and Pt_2Mo type microclusters. The formation of the mixed microclusters is due to the similar short-rage atomic arrangements of $D1_a$, $D0_{22}$ and Pt_2Mo structures. The fractions of $D1_a$, $D0_{22}$ and Pt_2Mo type microclusters change with the alloy system and composition.

3. The SRO-LRO transition proceeds by the selected growth of $D1_a$, $D0_{22}$ and Pt_2Mo type microclusters into LRO domains, depending on the alloy system and composition. The selection of the LRO structures is largely influenced by the effective pairwise atomic interactions in larger distances than the fourth coordination distance. The dependence of the effective pairwise interactions on alloying elements in Ni-based 1 1/2 0 alloys can be rationalized in terms of the concentration of free electrons in the alloys.

ACKNOWLEDGMENTS

This work was partly supported by Grant-in-Aid for Exploratory Research (No. 13875008) from the Japan Society for the Promotion of Science. This work was partly carried out at the Strategic Research Base 'Handai Frontier Research Center' supported by the Japanese Government's Special Coordination Fund for Promoting Science and Technology. The authors thank S. Matsumura, K. Oki, M. Sundararaman, J. S. Faulkner, A. G. Khachaturyan and K. Masuda-Jindo for valuable discussions.

REFERENCES

1. G. Van Tendeloo, S. Amelinckx, and D. de Fontaine, On the nature of the 'short-range order' in 1 1/2 0 alloys, *Acta Cryst.* B41:281 (1985).
2. W. M. Stobbs and S. J. Stobbs, Short-range order in <1 1/2 0> special-point alloys, *Phil. Mag. B* 53:537 (1986).
3. B. Schönfeld, Local atomic arrangements in binary alloys, *Prog. Mater. Sci.* 44:435 (1999).
4. R. Caudron, M. Sarfati, M. Barrachin, A. Finel, F. Ducastelle, and F. Solal, *In situ* diffuse scattering of neutrons in alloys and application to phase diagram determination, *J. Physique I France* 2:1145 (1992).
5. E. N. Vlasova, Y. D. Tyapkin, and V. D. Plakhtii, Ordering in Ni_4Mo and $Ni_{78.6}Mo_{19}V_{2.4}$ alloys, *Dokl. Akad. Nauk SSSR* 214:308 (1974).

6. M. Yamamoto, F. Shohno, and S. Nenno, The formation of $D0_{22}$-type ordered phase in the quenched $Ni_3(Mo,Ta)$ alloys, *Trans. Japan Inst. Metals* 19:475 (1978).
7. P. L. Martin and J. C. Williams, Long range order in Ni_3Mo based ternary alloys–I. Isothermal aging response, *Acta Metall.* 32:1681 (1984).
8. P. L. Martin and J. C. Williams, Long range order in Ni_3Mo based ternary alloys–II. Coherent phase solvii, *Acta Metall.* 32:1695 (1984).
9. A. Arya, G. K. Dey, V. K. Vasudevan, and S. Banerjee, Effect of chromium addition on the ordering behavior of Ni–Mo alloy: experimental results vs. electronic structure calculations, *Acta Mater.* 50:3301 (2002).
10. S. Hata, H. Fujita, C. G. Schlesier, S. Matsumura, N. Kuwano, and K. Oki, Monte Carlo study of ordering processes in fcc-based Ni–Mo alloys, *Mater. Trans., Japan Inst. Metals* 39:133 (1998).
11. S. Hata, S. Matsumura, N. Kuwano, and K. Oki, Short range order and its transformation to long range order in Ni_4Mo, *Acta Mater.* 46:881 (1998).
12. S. Hata, S. Matsumura, N. Kuwano, K. Oki, and D. Shindo, Short range order in Ni_4Mo and its high resolution electron microscope images, *Acta Mater.* 46:4955 (1998).
13. S. Hata, T. Mitate, N. Kuwano, S. Matsumura, D. Shindo, and K. Oki, Short range order structures in fcc-based Ni–Mo studied by high resolution transmission electron microscopy with image processing, *Mater. Sci. Eng.* A312:160 (2001).
14. U. D. Kulkarni, G. K. Dey and S. Banerjee, Ordering in rapidly solidified Ni_2Mo, *Scripta Metall.* 22:437 (1988).
15. L. E. Tanner, The ordering of Ni_3V, *Phys. Stat. Sol.* 30:685 (1968).
16. J. B. Singh, M. Sundararaman, and P. Mukhopadhyay, Evolution and thermal stability of Ni_3V and Ni_2V phases in a Ni-29 at. pct V alloy, *Metall. Mater. Trans.* 29A:1883 (1998).
17. L. E. Tanner, The ordering transformation in Ni_2V, *Acta Metall.* 20:1197 (1972).
18. M. Sundararaman, *private communication.*
19. K. Binder and D. W. Heermann, *Monte Carlo Simulation in Statistical Physics: An Introduction*, Springer-Verlag, Berlin (1988).
20. J. Kanamori and Y. Kakehashi, Conditions for the existence of ordered structure in binary alloy systems, *J. Physique* Colloque C7:274 (1977).
21. U. D. Kulkarni, S. Muralidhar, and S. Banerjee, Computer simulation of the early stages of ordering in Ni–Mo alloys, *Phys. Stat. Sol. (a)* 110:331 (1988).
22. P. De Meulenaere, G. Van Tendeloo, J. Van Landuyt, and D. Van Dyck, On the interpretation of HREM images of partially ordered alloys, *Ultramicroscopy* 60:265 (1995).
23. B. E. Warren, *X-ray Diffraction*, Addison-Wesley, Reading (1969).
24. R. V. Chepulskii, The effect of nonpair atomic interactions on the short-range order in disordered alloys, *J. Phys.: Condens. Matter* 11:8661 (1999).
25. R. V. Chepulskii, Analytical approximation for calculation of the short-range order in disordered alloys with nonpair atomic interactions, *Phys. Rev. B* 61:8606 (2000).
26. W. B. Pearson, *A Handbook of Lattice Spacings and Structures of Metals and Alloys, International Series of Monographs on Metal Physics and Physical Metallurgy Vol. 4*, G. V. Raynor, ed., Pergamon Press, London (1958).
27. U. D. Kulkarni and S. Banerjee, Phase separation during the early stages of ordering in Ni_3Mo, *Acta Metall.* 36:413 (1988).
28. A. K. Sinha, Close-packed ordered AB_3 structures in ternary alloys of certain transition metals, *Trans. Metall. Soc. AIME* 245:911 (1969).
29. T. Nakashima, S. Mizuno, T. Ido, K. Sato, S. Mitani, and K. Adachi, Atomic short range order and pair interaction of disordered Au_4Mn by neutron diffraction, *J. Phys. Soc. Japan* 43:1870 (1977).
30. M. Hirabayashi, M. Koiwa, S. Yamaguchi, and K. Kamata, Atomic short range order in Cu_3Mn studied by TOF neutron diffraction, *J. Phys. Soc. Japan* 45:1591 (1978).
31. W. Schweika and H.-F. Haubold, Neutron-scattering and Monte Carlo study of short-range order and atomic interaction in $Ni_{0.89}Cr_{0.11}$, *Phys. Rev. B* 37:9240 (1988).
32. W. Schweika, *Disordered Alloys: Diffuse Scattering and Monte Carlo Simulations*, Springer-Verlag, Berlin (1998).
33. J. Zou and A. E. Carlsson, Constant-volume pair potential for Al–transition-metal compounds, *Phys. Rev. B* 47:2961 (1993).
34. J. Zou and A. E. Carlsson, Preferred Mn Spacings in Al-Mn Compounds, *Phys. Rev. Lett.* 70:3748 (1993).
35. *e.g.* C. Kittel, *Introduction to Solid State Physics, Sixth Edition*, John Wiley & Sons, Inc., New York (1986).

CHANGES OF LRO IN ANISOTROPIC $L1_0$-ORDERED FePd

Andreas Kulovits,[1] William A. Soffa,[2] Wolfgang Püschl,[1] and Wolfgang Pfeiler[1]

[1]Institut für Materialphysik, University of Vienna, Strudlhofgasse 4 A-1090 Vienna, Austria
[2]Department of Materials Science and Engineering, University of Pittsburgh, 842 Benedum Hall, Pittsburgh, PA 15261, U.S.A.

INTRODUCTION

Long-range ordered intermetallic compounds have been in the centre of interest in materials science since decades due to their promising technological properties as, for example, high corrosion resistance and high-temperature mechanical strength.

Recently, the interest has turned to intermetallics with the tetragonal $L1_0$-ordered structure. Some of them are ferromagnetic at ambient temperature and display a very marked mechanical and magnetic anisotropy with the tetragonal c-axis being the easy axis of magnetization. This high anisotropy might be used as a way to stabilise a preferential orientation of magnetic domains with the easy axis of magnetization perpendicular to the surface. It is hoped that this way the storage density for magnetic media will be significanly increased opening a new dimension in high-density magnetic and magneto-optic recording. A further increase in storage density is expected if these materials can be stabilised in the form of magnetic nanostructures or thin films.

In spite of the great technical importance there is only very limited detailed knowledge on the basic physical properties which are essential for materials design and processing. These are especially kinetic properties such as the atomistic diffusion mechanism. It controls the state of order and its variation with temperature and at the same time yields criteria for the thermomechanical stability of the alloy. For this reason knowledge on atomic diffusion is vital when one ventures to design a high-performance material with specified technological properties.

FePd together with FePt and CoPt belongs to these magnetically and mechanically highly anisotropic intermetallics. The microstructural evolution during ordering in FePd has already been investigated intensively by TEM [1,2], although for very limited thermal treatment. We therefore decided to start a careful investigation of the evolution and dissolution of LRO during an isochronal annealing treatment.

To study the evolution of LRO for different thermo-mechanical starting conditions the method of monitoring *residual electrical resistivity* (REST) was used. This method because of its ultra high resolution for structural changes has proved to be a very advantageous tool for the analysis of order-disorder transformation, in particular the so called 'order-order' relaxations [3,4,5]. In this case, the relaxation to a new degree of LRO after a small change of temperature is observed, always remaining in a still highly ordered state. The present work reports on the establishment and dissolution of LRO in nearly stoichiometric FePd as measured by means of REST for two initial states, plastically deformed by cold-rolling and undeformed (completely recrystallized).

EXPERIMENTAL

The composition of the sample material alloyed by the Department of Materials Science and Engineering, University of Pittsburgh was determined by wet-chemical analysis (Institut für Geochemie, University of Vienna) as 52%±0.5% Fe and 48%±0.5% Pd, a composition marginally within the α-Fe/$L1_0$-FePd two-phase field (figure 1). Yet, neither during the detailed TEM investigations [1,2,6,7] nor during our resistivity measurements (total annealing time: several thousand hours) a beginning decomposition process never showed up. For details of preparation and measurement we refer to a forthcoming paper [8].

Measurements were carried out during a series of isochronal annealing treatments (ΔT=20K, Δt=20min) at rising and falling temperatures. A first run was made on the as-deformed samples cold rolled to a thicknes reduction of about 60% followed by a run after a recrystallization treatment of 2h at 1163K with subsequent slow cooling to 1060K and a

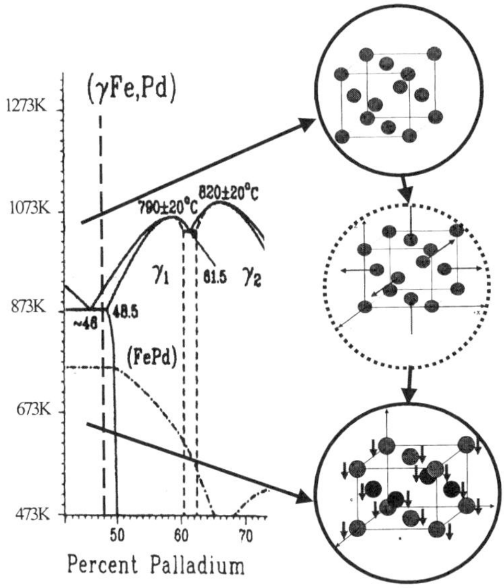

Figure 1. Corresponding part of the FePd phase diagram [10]. The alloy composition is given by dashed line. The inserts show the structural units for the disordered, the chemically and the magnetically ordered states, respectively.

water quench to room temperature, defining thus the undeformed state. For didactic reasons in the following the latter results will be reported first.

RESULTS

Undeformed state

Figure 2 gives the relative change of resistivity of the undeformed sample as a function of temperature.

Below 630K the resistivity during the isochronal step-heating procedure remains constant (range I in figure 2). Above 630K the resistivity decreases drastically by about 33% until a minimum is reached at 830K (range II). The resistivity then first increases slightly (range III, ≅10%/100K) and above 920K (starting temperature of range IV) drastically up to the order–disorder transition temperature at 963K (≅32%/100K). Above $T_{O/D}$ the resistivity continues to increase with about 10%/100K (range V).

Following the step-wise increase of temperature (forward isochrone) up to 1053K the temperature was then step-wise decreased (reverse isochrone). A decrease of resistivity with an almost constant slope (≅7%/100K) is followed at about 900K by a steep decrease down to 833K. A marked hysteresis is observed. At lower temperatures the resistivity changes with a decreasing rate (≤10%/100K).

In a subsequent second isochronal run (not shown here) the results of the first run were qualitatively reproduced.

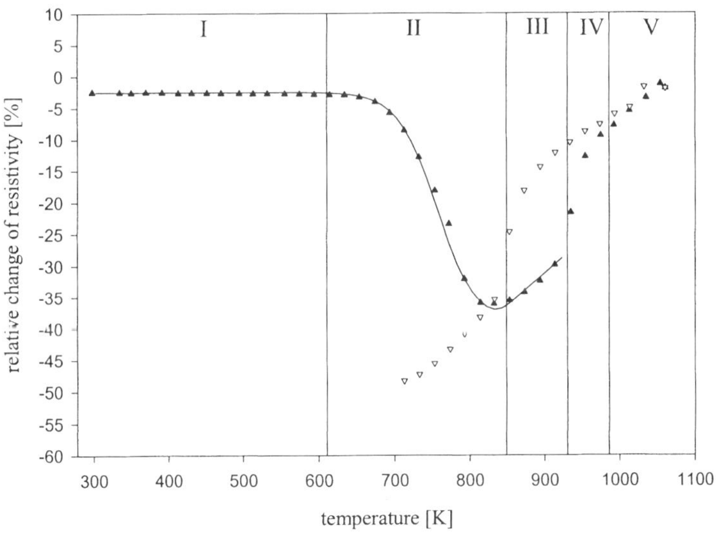

Figure 2. Relative change of resistivity during isochronal annealing (Δt=20min, ΔT=20K) as a function of temperature of an undeformed (recrystallized) sample after quenching from 1060K. (▲) increasing temperature, (▽) decreasing temperature; full line: calculated values, see text.

Deformed state

The isochronal annealing treatment of the deformed samples started at 473K just after sample preparation without any additional heat treatment (figure 3).

In contrast to the recrystallized sample the resistivity decrease starts at much lower temperatures, probably already slightly above room temperature (dashed line; no range I is observed). After a drastic decrease a minimum is reached at 790K (range II). Similar to the results of the recrystallized sample this steep decrease is followed by a slight and almost constant increase of resistivity with a slope of about 10%/100K in range III. Above 910K a steep increase is observed (range IV) which is followed by a linear increase of about 10%/100K at the beginning of range V. As a difference to the results of the recrystallized sample the relative change in resistivity in range V shows a very small increase with temperature above 990K.

During the subsequent cooling treatment a similar hysteresis is observed as in the recrystallized case. Below 740K the atomic mobity is essentially reduced and the changes of resistivity freeze at about 700K.

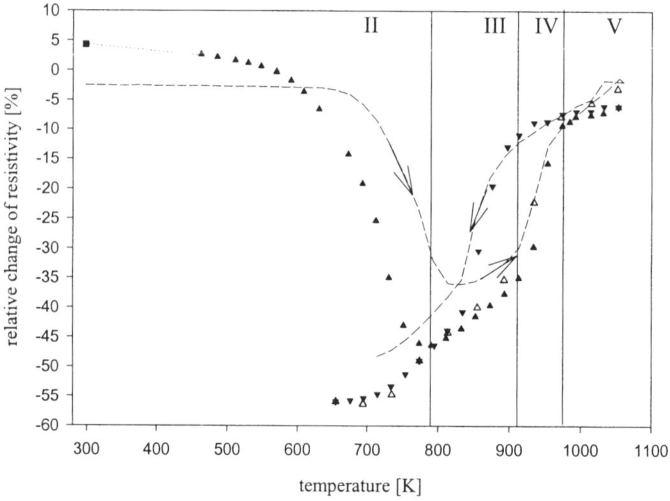

Figure 3. Relative change of resistivity during isochronal annealing (Δt=20min, ΔT=20K) as a function of temperature of a cold-rolled sample (≅60% thickness reduction). (▲) increasing temperature, (▼) decreasing temperature, (△) increasing temperature. The resistivity change of the undeformed sample is given for a comparison (dashed line).

In a second run the temperature was again step-wise increased. From about 660K up to $T_{O/D}$ the results of the previous run were qualitativley reproduced. In contrast to the first run the resistivity continued to increase above 990K almost similar to the recrystallized case (≅10%/100K).

DISCUSSION

Undeformed state

The initial state of the sample when starting isochronal annealing is the as-quenched disordered state. In temperature range I the concentration and/or mobility of vacancies are not sufficient to enable the establishment of LRO and resistivity therefore remains constant reflecting the value of the disordered state. With increasing temperature the increasing mobility and number of vacancies allows the ordering process to start and a corresponding decrease in electrical resistivity is observed above a temperature of 634K (beginning of range II).

In range II resistivity decreases steeply (establishment of LRO) until a temperature is reached where the isochronal annealing time (20 min) is long enough for the sample to reach the equilibrium state of order (minimum of resistivity change).

At still higher temperatures (range III) the degree of order that corresponds to the current annealing temperature as an equilibrium value is adjusted within the isochronal time interval. With increasing temperature, therefore, an increase of resistivity now reflects a decrease of LRO. This part of the curve is interpreted as an equilibrium curve.

In range IV disordered material starts to nucleate in the ordered matrix. The passing of the two phase region is observed as a steep increase in resistivity.

Above 975K (range V) the resistivity continues to increase with a constant slope of about 10%/100K. This may be an effect of remaining short-range order. A similar behaviour was found earlier for $L1_1$ ordering in off-stoichiometric $Cu_{30}Pt_{70}$ [9].

Stepwise decreasing the temperature a marked hysteresis of nearly 100K is observed. At the beginning of the first order order-disorder phase transition [10] ordered nuclei of a critical size have to be formed which can then grow and coarsen. Therefore a certain undercooling is necessary which is observed as a thermal hysteresis in the isochronal annealing curve.

This usual interpretation of the isochronal curve is corroborated by translating quantitatively resistivity changes into changes of LRO parameter η by applying the theory of Rossiter [11]. Taking a Rossiter parameter $A = 0.5$ the curves of figure 4 result. The

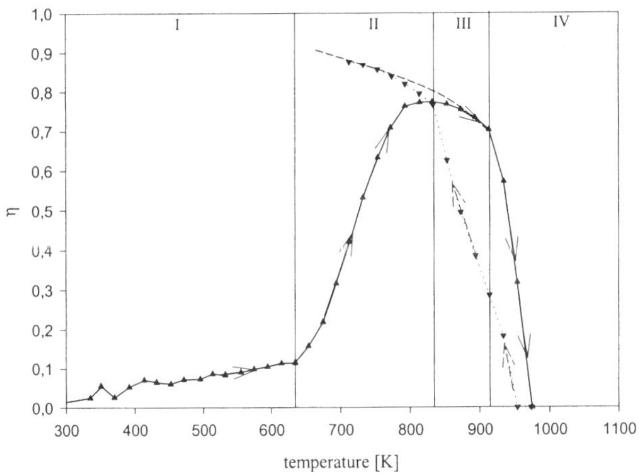

Figure 4. LRO-parameter η of the undeformed specimen as a function of temperature as calculated by using the theory of Rossiter [11] using A=0.5. (▲), full line: increasing temperature, (▼), dashed line: decreasing temperature.

LRO parameter η approaches an equilibrium curve in accordance with the Bragg-Williams model (dashed line).

We also tried to model the isochronal curve by just taking a single exponential relaxation process for isothermal ordering. Earlier investigations have shown that such a simple model of ordering kinetics originally developed for simulating the changes of SRO in concentrated solid solutions in certain cases can be applied to LRO systems (see [12,13] for more details). A total ordering activation energy of 3.1eV (H_f=1.5eV, H_m=1.6eV) as determined by a preliminary isothermal temperature treatment and a dislocation density (vacancy sinks) of $1x10^7$ cm^{-2} were used. This way the full line in figure 2 was calculated, being in very good correspondence with the measured points. Minor differences probably arise from a slight deviation of the order relaxation process from single-exponential kinetics.

Deformed state versus undeformed state

Several differences between the isochronal runs in the deformed state were found as compared to the recrystallized state.

The initial resistivity value in the as-deformed state is about 5% higher than that of the recrystallized sample. This difference in initial resistivity is mainly due to dislocations produced by the cold rolling deformation of about 60% reduction in thickness.

Differently from the undeformed state the resistivity in the initially cold-rolled state begins to decrease already slightly above room temperature. We ascribe this to the high concentration of vacancies produced during the process of rolling at room temperature.

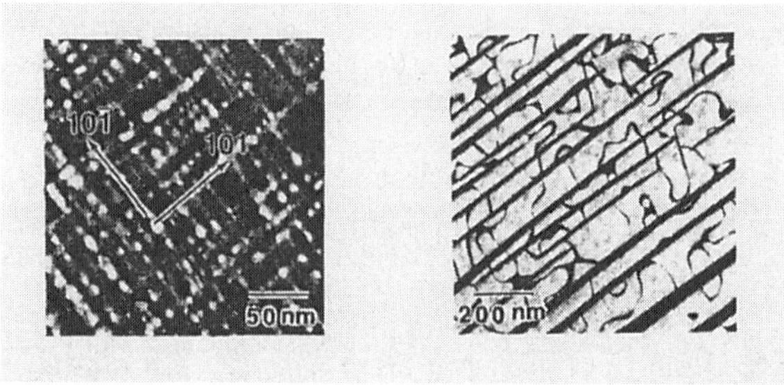

Figure 5. Dark-field TEM images of undeformed FePd aged at 773K for 3h (left) and for 61 h (right) [1].

The qualitative behaviour of resistivity in the ranges II, III, IV is very similar in both cases but the minimum resistivity value is both lower and is reached at lower temperatures (earlier in the isochronal annealing program) so that the equilibrium curve of the deformed samples is shifted by about 6-7% to lower resistivity values as compared to the recrystallized samples. As well known from TEM investigations [1,2] the resulting microstructures after ordering in the undeformed and in the deformed state, respectively, are considerably different. It was found that in the undeformed material the formation of order is correlated with a 'tweed' contrast for the early stages and the formation of a typical polytwinned structure for later stages (figure 5). Mean field computer simulations [14] meanwhile gave evidence that both effects are due to the transformation-induced elastic strain fields within the sample (figure 6). When ordering starts from a plastically deformed (cold-rolled) state, however, a completely different microstructure is found. In this case the polytwinned structure is missing and it seems that at least in certain parts of the sample one variant of ordered domain is predominantly formed by a process of 'massive ordering' (figure 7) by the moving grain boundary. Corresponding to the lower defect density and the higher degree of LRO lower values of resistivity are observed in the deformed case.

The influence of compressive strain and magnetic fields on ordering of FePd single crystals has been carefully investigated by K. Tanaka et al. [15]. Ordering was studied by X-ray diffraction during slow cooling and isothermal annealing at certain temperatures, respectively, under the influence of different values of external fields. A strong correlation

Figure 6. Mean-field model calculations of ordering kinetics taking into account transformation-induced elastic strain energy for different normalised times: t=1 (left) and t=50 (right) [14].

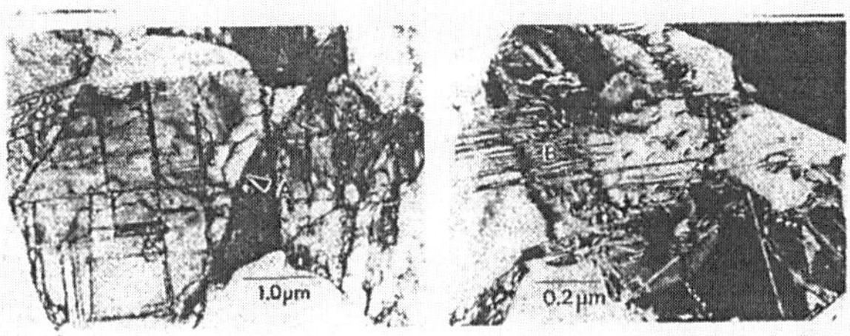

Figure 7. TEM images of deformed FePd (80% thickness reduction) aged at 798K for 50h (left) and at 698K for 17h (right) [2].

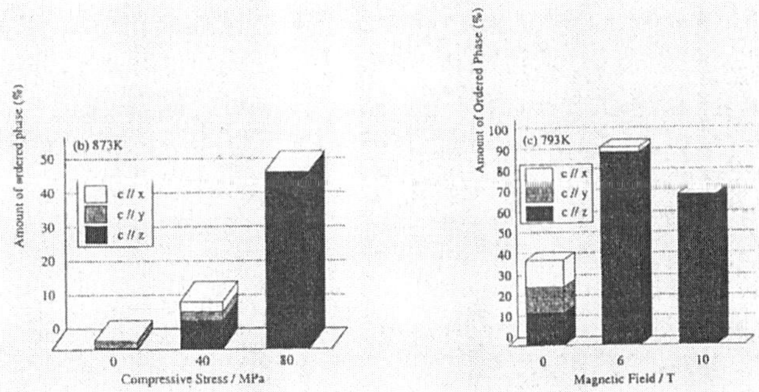

Figure 8. Volume fraction of different variants of ordered domains for increaing compressive stress (left) and increasing magnetic field (right) [15].

of the degree of LRO with the different thermal treatments and the value of the externalfield was found. As figure 8 shows, slow cooling (1K/min) under the influence of 10MPa compressive stress is enough to get a single variant of the three possible ordered domains which means 100% ordered phase. This means that the ordering process under the application of external fields, e.g. mechanical stress, leads to a prevalence of one favourably oriented variant of ordered domains over the other two possible variants. In our case effects of texture and internal stresses generated by plastic deformation may in a similar way result in a highly preferred variant of ordered domains. A detailed investigation by TEM is planned.

CONCLUSIONS

(i) The typical behavior of long-range ordered alloys with a first order LRO phase transition is found for both the deformed and the undeformed FePd alloys: a corresponding temperature hysteresis and the crossing of a two-phase region are definitely observed.

(ii) The undeformed and the deformed alloys differ by showing their corresponding equilibrium curves in different ranges of resistivity. This is in correspondence with the completely different microstructures when ordering takes place in the undeformed or the deformed samples, respectively.

(iii) It was shown that after a controlled plastic deformation a 'combined reaction' [16] of recrystallization and ordering takes place, leading to a marked preference of one variant of ordered domains. Thus a state with a considerably higher degree of order than in undeformed alloys is achieved. This is a good example that by proper choice of parameters it is possible to produce specific desired microstructures and to design attractive materials for proper technical application.

Acknowledgements

The financial support by the Austrian 'Fonds zur Förderung der wissenschaftlichen Forschung' is gratefully acknowledged.

REFERENCES

1. C. Yanar, J.M.K.Wiezorek, and W.A. Soffa in: *Phase Transformations and Evolution in Materials,* E.A. Turchi, A. Gonis, editors. The Minerals, Metals & Materials Society, Warrendale, 2000, p. 39.
2. T. Klemmer and W.A. Soffa in: *Solid-Solid Phase Transformations*, W.C. Johnson, J.M. Howe, D.E. Laughlin, and W.A. Soffa, editors. The Minerals, Metals & Materials Society, Warrendale 1994, p. 969.
3. W. Pfeiler, H. Lang, W. Püschl, and R. Kozubski in: *Solid-Solid Phase Transformations 1999 (PTM 99)*, M. Koiwa, K. Otsuka, T. Miyazaki, editors. The Japan Institute of Metals, Kyoto, Japan 1999, p. 457.
4. W. Pfeiler, Ordering Phenomena in Alloys: Access to kinetic parameters and atomic-jump processes, *JOM* **52**, 14 (2000).
5. H. Lang, K. Rohrhofer, P. Rosenkranz, R. Kozubski, W. Püschl, and W. Pfeiler, Long-range ordering in B2 FeAl: a resistometric study, *Intermetallics* **10,** 283 (2002).
6. B. Zhang and W.A. Soffa, Magnetic domains and coercitivity in polytwinned ferromagnets, *phys. stat. sol. (a)* **131**, 707 (1992).
7. T. Klemmer, D. Hoydick, H. Okumura, B. Zhang, and W.A. Soffa, Magnetic hardening and coercitivity mechanisms in $L1_0$ ordered FePd ferromagnets, *Scripta Met. Mater.* **33,** 1793 (1995).

8. A. Kulovits, W.A. Soffa, W. Püschl, and W. Pfeiler, Ordering and disordering in $L1_0$ FePd alloys as studied by residual resistivity measurement, submitted to *Intermetallics*, 2002
9. J. Banhart, W. Pfeiler, and J.Voitländer, Order-disorder transition in CuPt-alloys, investigated by resistivity measurement, *Phys. Rev.* **B37**, 6027 (1988).
10. O. Kubatschewski, *Iron-Binary Phase Diagrams,* Springer, Berlin 1982, p 88.
11. P.L. Rossiter, *The Electrical Resistivity of Metals and Alloys,* Cambridge University Press, Cambridge 1987.
12. D. Trattner and W. Pfeiler, Information about SRO-kinetics from isochronal annealing of electrical resistivity, *Scripta Metall.* **17**, 909 (1983).
13. R. Kozubski, M. Migschitz, and W. Pfeiler, Variations of long-range order in Ni_3Al+B after deformation by cold-rolling, *NATO-ASI Series* **B355**, 699 (1996).
14. L.-Q. Chen, Y. Wang and A.G. Khachaturyan, Kinetics of tweed and twin formation during ordering transition in a substitutional solid solution, *Phil. Mag. Letters* **63**, 15 (1992).
15. K. Tanaka, T. Ichitsubo, and M. Koiwa, Effect of external fields on ordering of FePd, *Mater. Sci. Eng.* **A312,** 118 (2001).
16. E. Hornbogen, Combined Reactions, *Met. Trans.* **10A**, 1979, 947.

KINETICS, DIFFUSION AND TRANSPORT

Ordering Process analyzed by Phase Field Method, CVM and PPM

M. Ohno and T. Mohri

Division of Materials Science and Engineering,
Graduate School of Engineering, Hokkaido University,
Kita-13 Nishi-8, Kita-ku, Sapporo, Japan

1. INTRODUCTION

Phase Field Method(hereafter PFM) has been attracting broad attentions as a powerful tool to describe inhomogeneous evolution process in microstructure. The PFM is a continuum model traced back to celebrated Cahn-Hilliard[1] and Allen-Cahn[2] equations. A key to the success of the PFM is the efficient parameterization of a microstructure through *field variable(s)* which constitute a free energy functional. Within the PFM, an interfacial boundary is not a specific entity to be separately described, but is merely an inhomogeneous localization of the field variables. The shape of the free energy determines final equilibrium state and the transition path. Therefore, by suitably defining both a free energy and field variable(s), one is capable of describing kinetic evolution of various microstructures, e.g., dendrite growth,[3] spinodal decomposition,[4] nucleation and growth,[5] crystal grain growth[6] and the evolution of anti-phase domain structure.[7] The latter is the main concern of this article.

The field variable in the PFM is a continuum quantity in the sense that atomic information is averaged out over discrete lattice points. Hence, most PFM calculations provide no direct information of the atomic configuration both in the equilibrium and non-equilibrium states. In reality, however, microstructural evolution process is driven by configurational kinetics through atomic movements, and detailed information fed from an atomistic scale is essential for a rigorous description of the time evolution of a microstructure. It is, therefore, desirable to combine PFM with an atomistic theory in a coherent manner.

Cluster Variation Method(CVM)[8-10] has been recognized as one of the most reliable theoretical tools to derive thermodynamic properties under a given set of atomic interaction energies on a discrete lattice. A key to the CVM is that the wide range of atomic correlations is explicitly taken into account in the free energy functional. A set of cluster probabilities or correlation functions in the free energy functional are employed as variational parameters by which the free energy is minimized to obtain the equilibrium state, while the optimized cluster probabilities describe the equilibrium atomic configuration on a discrete lattice.

Path Probability Method(PPM)[11] is a natural extension of the CVM to the time domain and, therefore, inherits various advantageous features of the CVM. Unique to the PPM is that the free energy and/or its derivative are not explicitly dealt in the constitutive equations, which enables one to extend the applicability to *far-from-equilibrium* regime. Furthermore, it has been amply demonstrated that the derived quantities by the PPM in the long time limit coincide exactly with the equilibrium ones independently obtained by the CVM. Hence, the combination of the CVM and PPM provides a unique theoretical tool in predicting and analyzing thermodynamic properties of a given alloy both in equilibrium and non-equilibrium states.[12-14]

In the CVM(PPM), the level of the approximation is determined by the largest cluster involved in the free energy(path probability function). Generally, the larger the basic cluster is, the better results one can expect. On the other hand, however, the number of variables becomes intractable with increasing the size of the basic cluster. Hence, even in the highest approximation employed so far, the extension of the basic cluster is limited only to a few atomic distances, and most of the CVM and PPM calculations assume the homogeneous distribution of the cluster probabilities in the microstructural scale.

In order to describe both the atomistic and microstructural processes, it is natural to hybridize PFM with CVM and PPM within a single theoretical framework. However, this is not a trivial task, since both time and spatial scales dealt within the two approaches are quite different. In particular, a rapid quantities derived from PPM are anticipated to be averaged out in the time scale of PFM. As a first step toward such a rigorous calculation, we attempted to hybridize PFM and CVM. The hybridized calculations were applied to disorder-B2[15] and disorder-$L1_0$[16] transitions at a fixed composition of 1:1 stoichiometry. Atomistic cooperative ordering processes through Long-Range-Order parameters(LRO) and Short-Range-Order parameters(SRO) and evolution/devolution process of anti-phase domain boundary in microstructural scale studied in the authors' group are reviewed in the present article.

The stability of the system against a small fluctuation of order parameters is determined by the second order derivative of the free energy functional. The spontaneous loss of the stability takes place at the temperature at which the second order derivative vanishes. The locus of this particular temperature(spinodal ordering temperature[17]) coincides with the phase boundary for the second order transition, while the deviation is manifested for the first order transition. Hence, for a disorder-B2 transition which is a typical example of the second order transition, the disordered phase is unstable against a configurational fluctuation below the transition temperature and the system spontaneously transforms from the disordered to B2 ordered phases. In microstructural scale, anti-phase domain structure results from gradual amplification of B2 ordering wave. On the other hand, for a disorder-$L1_0$ transition which is of the first order, the system requires the fluctuation at temperatures just below the transition temperature. Then anti-phase domain is formed by the nucleation and growth mechanism. These different ordering behavior were successfully described by the hybridized calculation of the CVM and PFM.

However, the results for these two types of transition were separately discussed and published.[15,16] The main objective of the present paper is, therefore, to provide a detailed comparison of ordering processes for a disorder-B2 transition and a disorder-$L1_0$ transition. Our focus is placed on both the atomistic and microstructural ordering behavior. The organization of the present paper is as follows. In the following two sections, the essential formulas of CVM, PPM and PFM are summarized. In the fourth section, time evolution of atomic configuration is focused, and we compare the PPM and PFM kinetics for disorder-B2 and disorder-$L1_0$ transitions. In the final section, the microstructural evolution processes are described by the hybridized calculation of the CVM and PFM, and characteristic microstructural features of B2 and $L1_0$ ordered phases are discussed.

2. ATOMISTIC THEORY

2.1 Cluster Variation Method(CVM)

As was described in the introduction, the key of the CVM is the fact that the wide range of atomic correlations is explicitly taken into account in the free energy functional. The level of the approximation constitutes a hierarchy structure and is determined by the basic cluster that is the largest one explicitly considered in the free energy functional. Generally, the choice of the basic clusters is made by the trade-off between the accuracy and computational burden. The tetrahedron approximation is known as a minimum meaningful approximation for a *fcc*-based system and the entropy formula for a $L1_0$ structure is given as

$$S = N \cdot k_B \cdot \left\{ \sum_{ij} \left(y_{ij}^{\alpha\alpha} \ln y_{ij}^{\alpha\alpha} + 4 y_{ij}^{\alpha\beta} \ln y_{ij}^{\alpha\beta} + y_{ij}^{\beta\beta} \ln y_{ij}^{\beta\beta} \right) - \frac{5}{2} \sum_{i} \left(x_i^{\alpha} \ln x_i^{\alpha} + x_i^{\beta} \ln x_i^{\beta} \right) - 2 \sum_{ijkl} w_{ijkl}^{\alpha\alpha\beta\beta} \ln w_{ijkl}^{\alpha\alpha\beta\beta} \right\}, \tag{1}$$

where N is the total number of lattice points, k_B the Boltzmann constant, x_i^{α}, $y_{ij}^{\alpha\beta}$ and $w_{ijkl}^{\alpha\alpha\beta\beta}$ are cluster probabilities of finding an atomic configuration specified by subscript(s) on point, pair and tetrahedron clusters, respectively, and α and β in the superscript distinguish the sub-lattices necessary for the description of the $L1_0$ ordered phase. In the conventional description, +1 and –1 are assigned to i, j, k and l to specify A and B atoms, respectively.

According to the ground state analysis,[18] the $L1_0$ ordered phase can be stabilized merely by the nearest neighbor pair interactions. Then, for the sake of simplicity, the internal energy, E, of the present study is limited to the nearest pair interaction and is written as

$$E = \frac{1}{2} \cdot \omega \cdot N \cdot \sum_{ij} e_{ij} \sum_{\gamma\delta} m^{\gamma\delta} \cdot y_{ij}^{\gamma\delta}, \tag{2}$$

where ω is the coordination number, e_{ij} the pair interaction energy between the nearest neighbor i-j pair, γ and δ indicate either α or β. The coefficient, $m^{\gamma\delta}$, is equivalent to the multiplicity and is given as $m^{\alpha\alpha} = m^{\beta\beta} = 1/6$ and $m^{\alpha\beta}$=4/6 for the $L1_0$ phase. Based on Eqs.(1) and (2), the Helmholtz free energy is symbolically written as

$$F_{L1_0} = E - T \cdot S = F_{L1_0} \left[T, \{e_{ij}\} \{x_i^{\gamma}\} \{y_{ij}^{\gamma\delta}\} \{w_{ijkl}^{\alpha\alpha\beta\beta}\} \right]. \tag{3}$$

The cluster probabilities employed in Eq.(3) are mutually dependent through the normalization conditions and reduction relationships. Hence, it is convenient to replace the cluster probabilities by correlation functions, $\{\xi_l\}$, that form a set of independent variables to describe atomic configurations and are defined as

$$\xi_l = \langle \sigma_1 \cdot \sigma_2 \cdots \sigma_p \cdots \rangle , \tag{4}$$

where the subscript, l, of the correlation function indicates a cluster and σ_p is a spin operator which takes +1 or −1 depending upon A and B atoms at a lattice point p, respectively, and $\langle\ \rangle$ denotes an averaging over all the lattice point. The relations between a cluster probability and correlation functions are readily grasped by the following examples,

$$x_i^{\gamma} = (1 + \xi_1^{\gamma})/2 , \tag{5}$$

$$y_{ij}^{\gamma\delta} = (1 + i \cdot \xi_1^{\gamma} + j \cdot \xi_1^{\delta} + ij \cdot \xi_2^{\gamma\delta})/2^2 , \tag{6}$$

and

$$w_{ijkl}^{\alpha\alpha\beta\beta} = \{1 + (i+j) \cdot \xi_1^{\alpha} + (k+l) \cdot \xi_1^{\beta} + ij \cdot \xi_2^{\alpha\alpha} + (ik + il + jk + jl) \cdot \xi_2^{\alpha\beta} + kl \cdot \xi_2^{\beta\beta} + (ijk + ijl) \cdot \xi_3^{\alpha\alpha\beta} + (ikl + jkl) \cdot \xi_3^{\alpha\beta\beta} + ijkl \cdot \xi_4^{\alpha\alpha\beta\beta}\}/2^4 . \tag{7}$$

Then the free energy given in Eq.(3) are rewritten in terms of correlation functions.

$$F_{L1_0} = F_{L1_0}\left[T, \{e_{ij}\} \xi_1^{\alpha}, \xi_1^{\beta}, \xi_2^{\alpha\alpha}, \xi_2^{\alpha\beta}, \xi_2^{\beta\beta}, \xi_3^{\alpha\alpha\beta}, \xi_3^{\alpha\beta\beta}, \xi_4^{\alpha\alpha\beta\beta}\right]. \tag{8}$$

Note that the distinction of the sub-lattices vanishes for a disordered phase and the number of correlation functions is reduced to four. One can determine an equilibrium state by minimizing the free energy of Eq.(8) with respect to correlation functions.

The phase diagram calculated within the tetrahedron approximation for $L1_0$ ordered system is demonstrated in Fig.1.[16] The temperature axis is normalized with respect to the nearest neighbor effective pair interaction energy, $v_2 = (e_{11} + e_{\bar{1}\bar{1}} - 2 \cdot e_{1\bar{1}})/4$. The transition temperature at 1:1 stoichiometry calculated by the present study is 1.893. The dashed line indicates the locus of the spinodal ordering temperature at which the second derivative of the free energy with respect to order parameters vanishes, and the system becomes unstable against the configurational fluctuation below this temperature. Throughout this

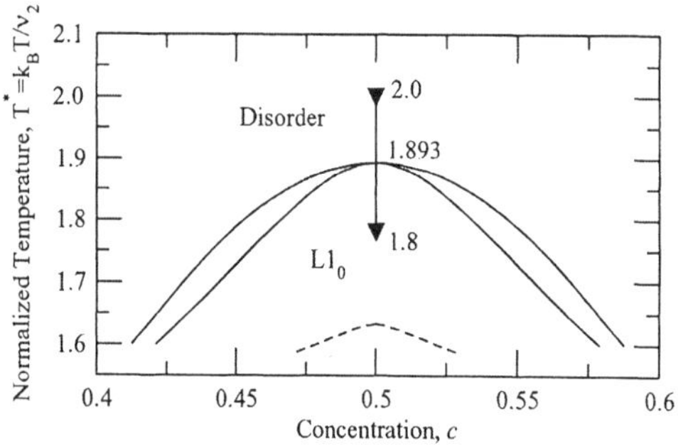

Figure 1. The phase diagram for disorder-$L1_0$ transition.[16] Temperature axis is normalized with respect to v_2. Solid lines indicate phase boundary and Dashed lines represents the locus of the spinodal ordering temperature.

study, the focus of the kinetic calculations for a disorder-$L1_0$ transition is placed on 1:1 stoichiometric composition at which the system is quenched from temperature 2.0 down to 1.8 followed by the annealing operation. It is noted that at $T^*=1.8$, the system is metastable state and the ordering process is expected to be nucleation-growth type. Due to the symmetry at 1:1 stoichiometry, the number of independent variables is reduced to five through the relations $\xi_1^{\beta} = -\xi_1^{\alpha}, \xi_2^{\alpha\alpha} = \xi_2^{\beta\beta}$ and $\xi_3^{\alpha\alpha\beta} = -\xi_3^{\alpha\beta\beta}$, and the free energy of $L1_0$ ordered phase is rewritten as

$$F_{L1_0} = F_{L1_0}\left[T, \{e_{ij}\} \xi_1^{\alpha}, \xi_2^{\alpha\alpha}, \xi_2^{\alpha\beta}, \xi_3^{\alpha\alpha\beta}, \xi_4^{\alpha\alpha\beta\beta}\right], \tag{9}$$

where ξ_1^{α} serves as the LRO and other correlation functions are Short-Range-Order parameter(SRO).

Contrary to the *fcc*-based system for which frustration is manifested, a simple pair approximation has been proved to provide fairly reasonable results for a *bcc*-based system. And we formulate the entropy of B2 ordered phase within the pair approximation and the internal energy of the system is limited to the nearest neighbor pair interaction. Also, by considering the symmetry of 1:1 stoichiometry, we reduce the number of independent variables of free energy and, then, the free energy of B2 phase is symbolically written as

$$F_{B2} = F_{B2}\left[T, \{e_{ij}\} \xi_1^{\alpha}, \xi_2^{\alpha\beta}\right], \tag{10}$$

where ξ_1^{α} serves as the LRO.

The phase diagram calculated within the pair approximation for B2 ordered system is shown in Fig.2.[15] The temperature axis is normalized with respect to the nearest neighbor effective pair interaction energy, $v_2 = e_{11} + e_{\bar{1}\bar{1}} - 2e_{1\bar{1}}$, and the transition temperature is 1.739 at 1:1 stoichiometry. The absence of two-phase field region indicates that the transition is of the second order. We focus on the ordering process for the disorder-B2 transition at 1:1 stoichiometric composition during the annealing operation at $T^*=1.0$ following the quenching from $T^*=2.0$.

It should be noted that the independent variables in Eqs.(9) and (10) are averaged

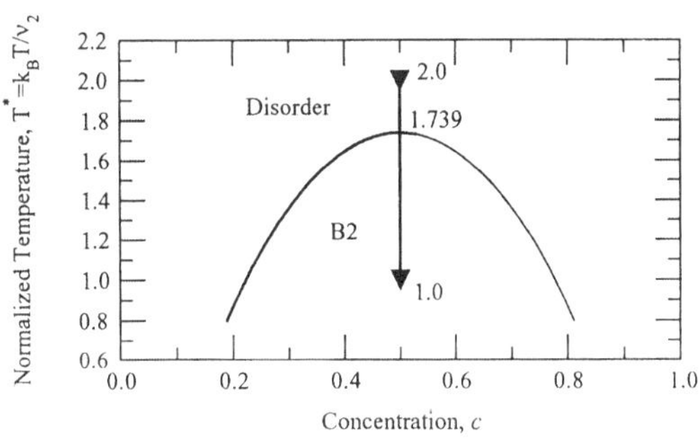

Figure 2. The phase diagram for disorder-B2 transition.[15] Temperature axis is normalized with respect to v_2. Solid line indicates phase boundary.

quantities over a space occupied by N's lattice points, hence the free energies given by Eqs.(9) and (10) are applicable only to a uniform system which is equivalent to a local system in the description of a microstructure within the PFM.

2.2 Path Probability Method(PPM)

As was already described, PPM inherits many features of the CVM. As a counter part of cluster probabilities of the CVM, the path variables are defined for the PPM. The path variable correlates the cluster probabilities at time t and $t+\Delta t$. Instead of providing a general formula, examples are given for point path variables as follows,

$$x_1(t) = X_{1,1}(t, t+\Delta t) + X_{1,\bar{1}}(t, t+\Delta t), \tag{11}$$

and

$$x_1(t+\Delta t) = X_{1,1}(t, t+\Delta t) + X_{\bar{1},1}(t, t+\Delta t), \tag{12}$$

where $X_{i,j}(t, t+\Delta t)$ is the point path variable of describing the configurational transition on a lattice point(a point cluster) from i at time t to j at $t+\Delta t$.

By employing path variables, a Path Probability Function, P, which corresponds to a free energy of the CVM, is written as the product of three terms, P_1, P_2 and P_3. Each term for disorder-$L1_0$ transition is given in the following logarithmic expression.

$$\begin{aligned} \ln P_1 = (N/2) \cdot \Big\{ & \left(X^{\alpha}_{1,\bar{1}} + X^{\alpha}_{\bar{1},1} + X^{\beta}_{1,\bar{1}} + X^{\beta}_{\bar{1},1} \right) \ln(\theta \cdot \Delta t) \\ & + \left(X^{\alpha}_{1,1} + X^{\alpha}_{\bar{1},\bar{1}} + X^{\beta}_{1,1} + X^{\beta}_{\bar{1},\bar{1}} \right) \ln(1 - \theta \cdot \Delta t) \Big\}, \end{aligned} \tag{13}$$

$$\ln P_2 = -\Delta E / (2 k_B T), \tag{14}$$

and

$$\begin{aligned} \ln P_3 = N \cdot \Bigg\{ & \sum_{ijkl} \left(Y^{\alpha\alpha}_{ij,kl} \ln Y^{\alpha\alpha}_{ij,kl} + 4 Y^{\alpha\beta}_{ij,kl} \ln Y^{\alpha\beta}_{ij,kl} + Y^{\beta\beta}_{ij,kl} \ln Y^{\beta\beta}_{ij,kl} \right) \\ & - \frac{5}{2} \sum_{ij} \left(X^{\alpha}_{i,j} \ln X^{\alpha}_{i,j} + X^{\beta}_{i,j} \ln X^{\beta}_{i,j} \right) - 2 \sum_{ijklmnop} W^{\alpha\alpha\beta\beta}_{ijkl,mnop} \ln W^{\alpha\alpha\beta\beta}_{ijkl,mnop} \Bigg\}, \end{aligned} \tag{15}$$

where $Y^{\gamma\delta}_{ij,kl}$ and $W^{\alpha\alpha\beta\beta}_{ijkl,mnop}$ are the path variables for a pair and tetrahedron clusters, respectively, θ the spin flipping probability per unit time which corresponds to the diffusivity in an alloy system and ΔE is the change of the internal energy during Δt.

For the disorder-B2 transition, in order to keep the consistency between the CVM and PPM calculations, the pair approximation is employed and, P_3, is given as

$$\ln P_3 = N \cdot \left\{ \frac{7}{2} \sum_{i} \left(X^{\alpha}_{i,j} \ln X^{\alpha}_{i,j} + X^{\beta}_{i,j} \ln X^{\beta}_{i,j} \right) - 4 \sum_{ijkl} Y^{\alpha\beta}_{ij,kl} \ln Y^{\alpha\beta}_{ij,kl} \right\}, \tag{16}$$

P_1 and P_2, on the other hand, are common irrespective of the approximation and given by Eqs.(13) and (14), respectively. It is worth repeating that the free energy and/or its derivative are not explicitly considered in the path probability function.

The formulation of the path probability function entirely depends on the kinetics assumed in the study. The vacancy-mediated kinetics or the exchange kinetics(Kawasaki dynamics[19]) requires a large number of path variables which make numerical operations intractable. The spin flipping kinetics(Glauber dynamics[20]) generally does not conserve the species with time, however the conservation is assured at 1:1 stoichiometric composition without imposing any additional constraints. In this regard, the spin system approximates an alloy system. Therefore, in order to avoid numerical complications with larger number of path variables resulted from the vacancy-mediated or exchange kinetics, the present study is confined to the spin flipping kinetics at 1:1 stoichiometry.

The most probable path of time evolution is determined by maximizing path probability function with respect to a set of path variables, which corresponds to the minimization condition of the free energy in the CVM. By optimizing the path variables for each time step, the cluster probability at time $t+\Delta t$ is uniquely determined with the knowledge of cluster probabilities at time t. Such a relation is exemplified for the point cluster in the following,

$$x_1(t+\Delta t) = x_1(t) + X_{\Gamma,1}(t, t+\Delta t) - X_{1,\Gamma}(t, t+\Delta t), \tag{17}$$

which is deduced from Eqs.(11) and (12). Hence, once the initial equilibrium state is obtained by the CVM calculation, non-equilibrium time evolution process is pursued by the PPM towards the new equilibrium state. Moreover, it has been demonstrated that the final equilibrium state predicted by the PPM at $t \to \infty$ agrees with the one independently calculated by the CVM.

3. HYBRIDIZED CALCULATION OF PHASE FIELD METHOD AND CLUSTER VARIATION METHOD

Within the PFM, microstructure is characterized by a spatial distribution of field variables such as concentration, order parameter etc., and the driving force for temporal evolution of microstructure is described as the gradient of the free energy with respect to field variables. The free energy of an inhomogeneous system is generally constituted by bulk free energy, interfacial energy and elastic-strain energy contributions. The present calculation neglects an elastic-strain energy contribution to microstructural evolution process, and the chemical energy of inhomogeneous system is given as,[1]

$$F_{chem}[\{\phi_i[\mathbf{r},t]\}] = \int \left\{ f_{local}[\{\phi_i[\mathbf{r},t]\}] + \sum_i \kappa_i (\nabla\phi_i[\mathbf{r},t])^2 \right\} dV, \tag{18}$$

where ϕ_i is a field variable which is a function of spatial coordinate, $\mathbf{r}$, and time, t, and V is a volume of the specimen, κ_i is the gradient energy coefficient and is assumed not to depend on an annealing temperature and the field variables. The first term of the integrand, f_{local}, is the local free energy density function for a uniform system and the second term corresponds to the interfacial energy.

The temporal and spatial variations of the field variable describe time evolution process of inhomogeneous system and the local free energy density determines the kinetic path. Hence, the key to the PFM calculation is the definitions of both the field variables and the local free energy density. In the present study, the CVM free energy(Eqs.9 and 10) is adopted as the local free energy density and the field variable $\{\phi_i\}$ in Eq.(18) is replaced by correlation functions defined in the previous section. Then, Eq.(18) is rewritten as

$$F_{chem}[\{\xi_i[\mathbf{r},t]\}] = \int \left\{ F_{CVM}[\{\xi_i[\mathbf{r},t]\}] + \sum_i \kappa_i (\nabla\xi_i[\mathbf{r},t])^2 \right\} dV, \tag{19}$$

where F_{CVM} denotes either F_{L1_0} of Eq.(9) or F_{B2} of Eq.(10). It should be stressed that in the CVM free energy, $\{\xi_i\}$ do not depend on position and F_{L1_0} in Eq.(9) and F_{B2} in

Eq.(10) are the free energies for a uniform system, while in Eq.(19) $F_{CVM}\left(F_{L1_0} \text{ or } F_{B2}\right)$ explicitly depends on $\mathbf{r}$ through the dependency of $\{\xi_i\}$.

Non-conserved correlation functions constitute field variables and their temporal evolutions are described by the following Time-Dependent Ginzburg-Landau(TDGL) equation,

$$\frac{\partial \xi_i[\mathbf{r},t]}{\partial t} = -L_i \frac{\delta F_{chem}[\{\xi_i[\mathbf{r},t]\}]}{\delta \xi_i[\mathbf{r},t]} + \xi_i^f[\mathbf{r},t], \tag{20}$$

where $\xi_i^f[\mathbf{r},t]$ is a noise term which is necessary to trigger the fist order transition and L_i is the relaxation coefficients for ξ_i. In the present calculation, L_i is assumed not to depend on annealing temperature and other field variables. It is noteworthy that kinetic evolution of TDGL Eq.(20) is based on the principle that the evolution/devolution rate of a field variable is proportional to the gradient of the free energy, which may be rationalized only in the near-equilibrium regime. The substitution of Eq.(19) into Eq.(20) yields

$$\frac{\partial \xi_i[\mathbf{r}',t']}{\partial t} = -L_i^{\bullet} \cdot \left(\frac{\partial F_{CVM}^{\bullet}[\{\xi_i[\mathbf{r}',t']\}]}{\partial \xi_i[\mathbf{r}',t']} - \kappa_i^{\bullet} \nabla^2 \xi_i[\mathbf{r}',t'] \right) + \xi_i^{f'}[\mathbf{r}',t'], \tag{21}$$

where, in order to generalize the treatment, dimensionless parameters are introduced by normalizing time and spatial coordinates and other parameters as follows,

$$t' = N \cdot v_2 \cdot L_b \cdot t, \quad \mathbf{r}' = \mathbf{r} \cdot \sqrt{(N \cdot v_2)/(2\kappa_b)}, \quad L_i^{\bullet} = L_i / L_b, \quad F_{CVM}^{\bullet} = F_{CVM} / (N \cdot v_2)$$

and

$$\kappa_i^{\bullet} = \kappa_i / \kappa_b .$$

Here, L_b and κ_b represent the relaxation coefficient and gradient energy coefficient for a basic cluster, respectively.

The PFM is formally constructed by both the Cahn-Hilliard equation for conserved variables such as a concentration and the TDGL equation for non-conserved variables, and both the evolution equations are coupled through the local free energy density function. Generally, the inhomogeneity of concentration is anticipated to influence the entire ordering kinetics through the coupling of $\{\xi_i\}$.[21] Our main interest in the present study, however, is configurational evolution manifested by LRO and SRO at a fixed composition of 1:1 stoichiometry. Therefore, in order to save computation time, Cahn-Hilliard's diffusion equation is not explicitly considered in the present study. This is regarded as the first approximation to more realistic alloy kinetics.

4. ORDERING PROCESSES DESCRIBED BY PPM AND CVM

As was described, the PPM does not explicitly deal with free energy or its derivative, which is in marked contrast with other kinetic theories including PFM. In order to compare the ordering process calculated by PFM with the one by PPM, we neglect interfacial energy terms in Eq.(21) in the PFM and focus on a uniform system.

4.1 B2 ordering process in a uniform system

Within the PPM, the kinetics is entirely dominated by spin flipping probability once the initial equilibrium state is specified by the CVM, and the relaxation behavior of

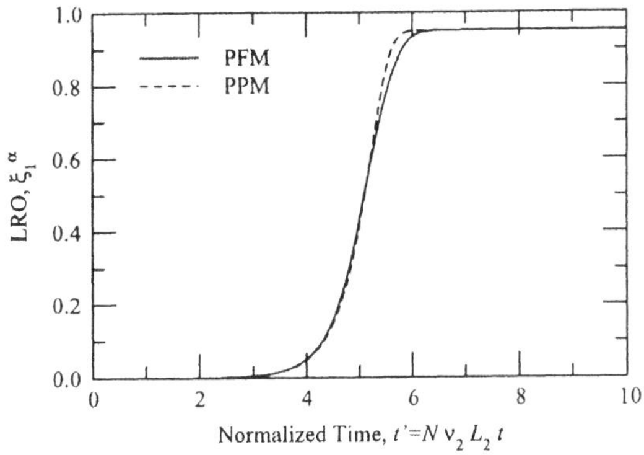

Figure 3. The relaxation curve for disorder-B2 transition. The solid and dashed lines indicate the results of the PFM and PPM, respectively.

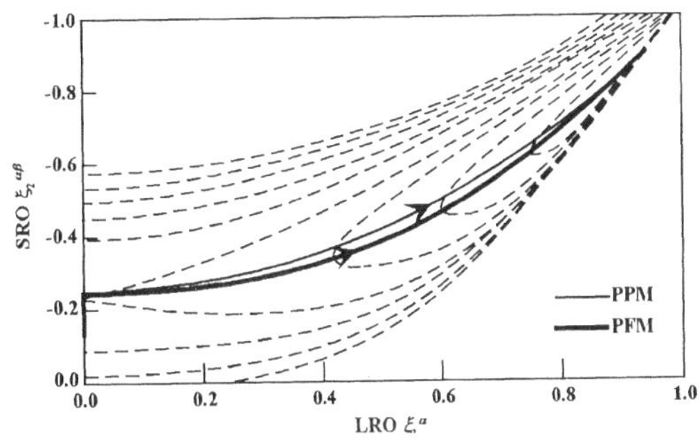

Figure 4. The kinetic path for disorder-B2 transition obtained by PFM and PPM. Dashed lines indicate the contour lines of the CVM free energy at $T^*=1.0$. The arrow designates the direction of transition.

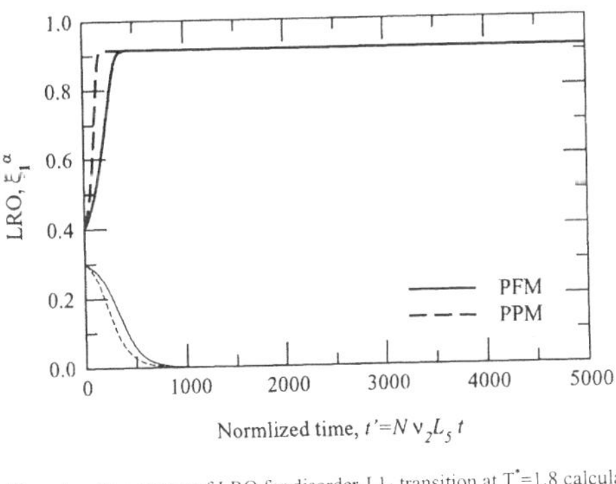

Figure 5. The relaxation curves of LRO for disorder-$L1_0$ transition at $T^*=1.8$ calculated by PPM with $\theta^*=0.162$ and PFM with $L_1^*=0.1$, $L_2^*=0.2$, $L_3^*=0.2$ and $L_4^*=0.2$.[16]

correlation functions are uniquely determined. In Fig.3, the relaxation curve of LRO during B2 ordering process at $T^*=1.0$ following quenching operation from $T^*=2.0$ is represented by the dashed line, which is calculated with $\theta^*=2.5$, the normalized spin flipping probability defined as $\theta^* = \theta/(N \cdot v_2 \cdot L_2)$. Within the PFM, the ordering behavior depends on the relaxation coefficient(s). When 1.2 is assigned to L_1^* which is the only kinetic factor of the PFM calculation for B2 ordering process, the relaxation curve indicated by the solid line is obtained. In this calculation, the noise term in Eq.(21) is omitted since the ordering process for second order transition spontaneously proceeds without fluctuation. One can confirm that, in both cases, the disordered phase transforms to the ordered phase without fluctuation and attains the same steady state value of LRO in the long time limit which is confirmed to be the equilibrium one at $T^*=1.0$ independently calculated by the CVM. Despite the difference in the underlying kinetics principles, both the PFM and PPM provide nearly isomorphic relaxation curve under well-tuned kinetic coefficient factors, θ^* and $\{L_i^*\}$. This fact implies the existence of a scaling property between two models.

In order to gain more detailed information into ordering process, a kinetic path is calculated. Shown in Fig.4 is the kinetic path traced in the thermodynamic configuration space spanned by LRO(ξ_1^α) and SRO($\xi_2^{\alpha\beta}$). Dashed lines represent the contour lines of the free energy of B2 phase at $T^*=1.0$ calculated by the CVM. Thick and thin solid lines indicate the kinetic paths obtained by the PFM and PPM, respectively, and arrows designate the direction of transition. One can see that the kinetic path of PFM exactly traces the steepest descent direction, which is the natural consequence of the fact that the PFM relaxation process is driven by the gradient of the free energy. However, a slight deviation from the steepest descent direction is observed in the kinetic path of the PPM, which may be ascribed to the different kinetic principle adopted in the PPM.

4.2 $L1_0$ ordering process in a uniform system

For $L1_0$ ordering process, the system requires the fluctuation to transform from the metastable disordered to the stable ordered phases. In the PFM, the fluctuations are represented by the noise term of Eq.(21) which generates triggered correlation functions in a local system, while the fluctuations are not explicitly incorporated in the conventional PPM formula. In order to introduce fluctuation, arbitrary values of LRO, ξ_1^α, are assigned in both models, and initial values of other correlation functions are determined so that the free energy is minimized, $\partial F_{L1_0} / \partial \xi_i \big|_{\xi_1^\alpha} = 0$, which is a constrained minimization. It is noted interfacial energy fluctuation is excluded in the present study.

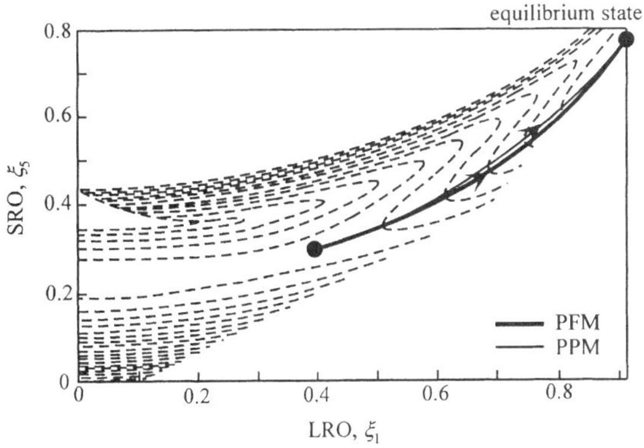

Figure 6. The kinetic path for disorder-$L1_0$ transition obtained by PFM and PPM. [16] Dashed lines indicate the contour lines of the CVM free energy at T^*=1.8. The arrow designates the direction of transition.

The relaxation curves of LRO calculated by the PFM and PPM are represented by solid and dashed lines in Fig.5,[16] respectively, when the system is quenched from T^*=2.0 to 1.8 and annealed at T^*=1.8. One can see that, in both PFM and PPM, with sufficient fluctuation(thick solid and thick dashed lines) which is represented by a large deviation of LRO from null value at t'=0, the system transforms to the ordered phase. The steady state values of LRO in both models are confirmed to be the equilibrium one at T^*=1.8 independently obtained by the CVM. The existence of a critical amount of fluctuation suggested by the present results implies a nucleation-growth type ordering process as opposed to B2 ordering process.

In Fig.6,[16] the kinetic paths which correspond to the thick solid and dashed lines in Fig.5 are traced in the thermodynamic configuration space spanned by LRO(ξ_1^{α}) and $\xi_4^{\alpha\alpha\beta\beta}$, one of SRO's. Dashed lines indicates the contour lines of the free energy of $L1_0$ phase at T^*=1.8 calculated by the CVM. One sees that there exists saddle point configuration in the vicinity of $\left(\xi_1^{\alpha},\xi_4^{\alpha\alpha\beta\beta}\right)\approx\left(0.3,0.25\right)$. The initial imposition of the fluctuation is indicated by the deviation of LRO from null value at the starting point of the kinetic path. It is seen that both the kinetic path nearly coincide.

As was amply described, the principle of the PFM is that the temporal evolution rate of a field variable is proportional to the gradient of the free energy and this may be rationalized only in the near-equilibrium regime. One the other hand, the free energy is not explicitly considered in the PPM formula, which enables one to extend the applicability to far-from-equilibrium regime. In the previous study,[16] it was shown that the discrepancy between ordering behavior of the PPM and PFM is manifested in far-from equilibrium regime. However, it is emphasized that both models yield nearly identical transition process when the transition slowly proceeds in near-equilibrium regime.

5. ORDERING PROCESS IN A NON-UNIFORM SYSTEM

The main advantage of the PFM is the capability of dealing with inhomogeneous system. We firstly focus on a simple one-dimensional system and gain basic insight into the evolution process both for the disoder-B2 and for the disorder-$L1_0$ transitions. Then, the microstructural features of B2 ordered and $L1_0$ ordered alloys are studied by a two-dimensional calculation.

5.1 Evolution process in one-dimensional inhomogeneous system

As was demonstrated in Figs.3 and 5, disorder-B2 ordering process spontaneously proceeds without fluctuation, while for disorder-$L1_0$ transition, the system requires fluctuation to transform to ordered state. Shown in Fig.7 is a one-dimensional profile of LRO at several times during the disorder-B2 ordering process. The vertical axis represents the square value of LRO, $\xi_1^{\alpha^2}$, and the horizontal axis is the normalized spatial coordinates, $x' = x \cdot \sqrt{(N \cdot v_2)/(2\kappa_2)}$. At t'=0, a positive(negative) small value is assigned as an initial value of LRO to each grid point for $x' \leq 10$ ($x' > 10$), while the value of SRO, $\xi_2^{\alpha\beta}$, at each point is the equilibrium value at T^*=2.0 calculated by the CVM. It is seen that the system spontaneously transforms from disordered to ordered phase, which is characterized by the gradual enhancement of LRO, and ordered domains are separated by an anti-phase boundary(APB) at t'=15.

For the disorder-$L1_0$ transition, the ordering reaction is triggered by the fluctuation, which implies that the ordering transition proceeds by the nucleation and growth mechanism. The evolution process of inhomogeneous system during annealing operation at T^*=1.8 is demonstrated in Fig.8. The initial system is constructed by uniformly assigning equilibrium values of LRO and SRO at T^*=2.0 calculated by the CVM to each point. Additionally, a nuclei of ordered particle is generated in the vicinity of x'=250. One sees that ordering process proceeds by the lateral motion of the ordered domain

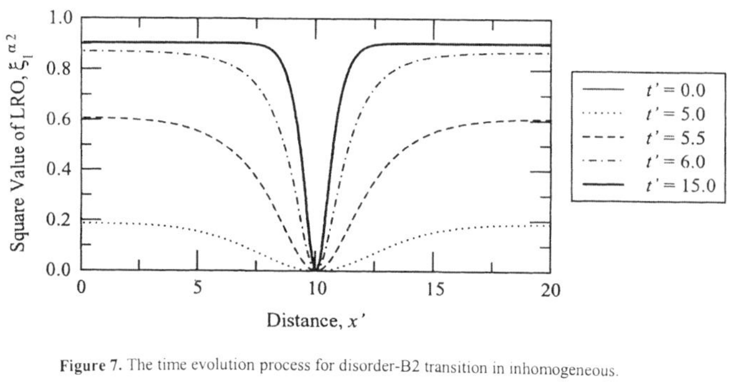

Figure 7. The time evolution process for disorder-B2 transition in inhomogeneous.

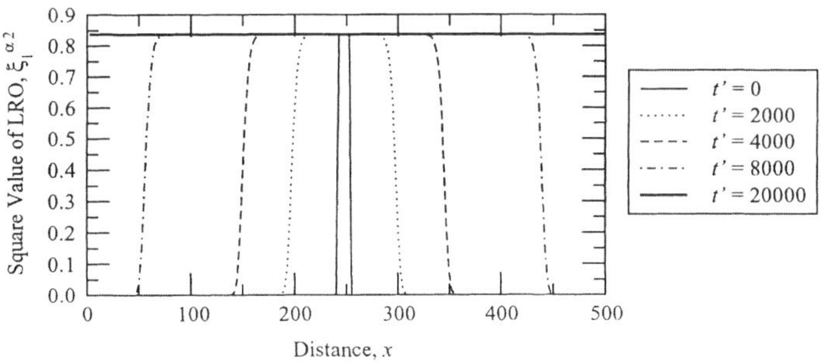

Figure 8. The time evolution process for disorder-$L1_0$ transition in inhomogeneous.

boundary. This evolution behavior is in marked difference with the one for disorder-B2 transition.

5.2 Microstructural features for B2 and $L1_0$ ordered systems

We employ a two-dimensional space divided by 100_100's grid points. In the calculation for disorder-B2 ordering process, the initial condition is given by assigning positive or negative small random values to LRO, ξ_1^{α} , at each grid point. Then, the initial value of SRO, $\xi_2^{\alpha\beta}$, is determined by a constrained optimization, $\partial F_{B2} / \partial \xi_2^{\alpha\beta} \big|_{\xi_1^{\alpha}} = 0$, for each grid point.

The microstructural evolution process for disorder-B2 transition during annealing operation at $T^{*}=1.0$ is shown in Fig.9. In these figures, the microstructure is visualized by gray levels representing different values of $\xi_1^{\alpha^2}$ which corresponds to a dark-field image of transmission electron micrograph. Dark and bright regions indicate a disordered phase or APB and an ordered phase, respectively. One observes that the small ordered domains are formed from the disordered matrix in the early period and an entire system transforms to the ordered phase separated by APB's. The resultant microstructural feature is in fairly good agreement with disorder-B2 transition observed, for instance, in α-Fe(bcc)_FeAl(B2)[2] for which anisotropy of an elastic energy is known as insignificant.

Shown in Fig.10 is a series of snapshots of microstructures during disorder-$L1_0$ ordering process. In this calculation, the noise term of Eq.(21) is operated to nucleate $L1_0$ ordered phase. Despite the different ordering modes, the microstructural feature in the later period shown in Fig.10(d) is quite akin to the one for the disorder-B2 transition shown in Fig.9(d). When an anisotropic elastic effect is neglected, a basic feature of morphology of anti phase boundaries is mainly determined by the number of variants of ordered domain.[22,23] Although there are six types of ordered variants for a $L1_0$ ordered phase, only two types of ordered domain are distinguished by the free energy of the

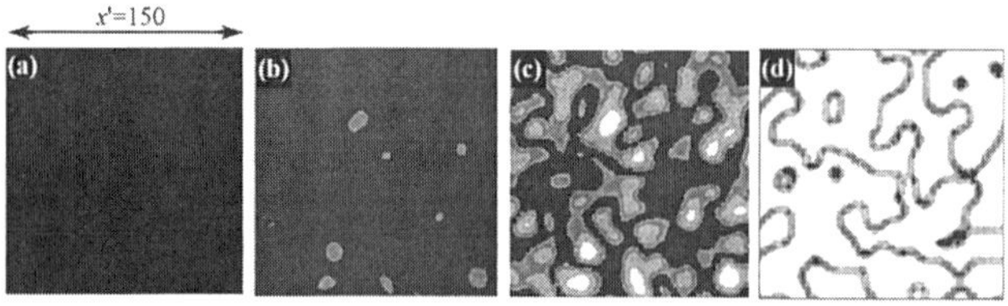

Figure 9. Microstructural evolution process for disorder-B2 transition at $T^*=1.0$. (a) $t'=0$, (b) $t'=11$, (c) $t'=13$, (d) $t'=25$. x' indicates the normalized spatial scale.

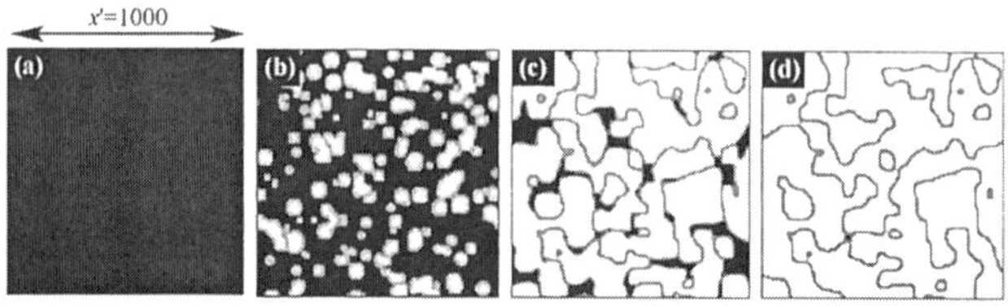

Figure 10. Microstructural evolution process for disorder-$L1_0$ transition at $T^*=1.8$. (a) $t'=0$, (b) $t'=3000$, (c) $t'=6000$, (d) $t'=10000$. x' indicates the normalized spatial scale.

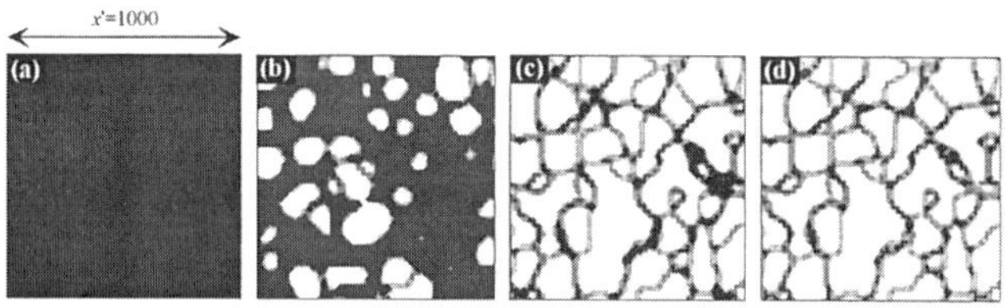

Figure 11. Competitive ordering process between six $L1_0$ ordered domains at $T^*=1.6$.[24] (a) $t'=0$, (b) $t'=10000$, (c) $t'=12000$, (d) $t'=20000$. x' indicates the normalized spatial scale.

present study given by Eq.(9). This is the reason that the microstructural feature shown in Fig.10(d) are similar to the one for disorder-B2 transition in which two types of ordered domain exist.

A competitive ordering process between six $L1_0$ ordered domains can be described by the present approach with a modification on the CVM free energy.[24] In order to specify six $L1_0$ ordered variants, the *fcc* lattice is divided into four simple cubic sub-lattices, and the resultant four lattice sites on a tetrahedron cluster are designated as α, β, γ and δ. Then the free energy of $L1_0$ ordered phase is symbolically written as

$$F_{L1_0} = F_{L1_0}\left[T, \{\varepsilon_{ij}\}, \xi_1^{\alpha}, \xi_1^{\beta}, \xi_1^{\gamma}, \xi_2^{\alpha\beta}, \xi_2^{\alpha\gamma}, \xi_2^{\alpha\delta}, \xi_3^{\alpha\beta\gamma}, \xi_3^{\alpha\beta\delta}, \xi_3^{\alpha\gamma\delta}, \xi_4^{\alpha\beta\gamma\delta}\right], \tag{22}$$

where three correlation functions for point cluster are correlated to LRO's for three types of orientational variant, η_i. The relations between correlation functions and LRO's are given as

$$\eta_1 = (\xi_1^{\alpha} + \xi_1^{\beta})/2, \tag{23}$$

$$\eta_2 = (\xi_1^{\beta} + \xi_1^{\gamma})/2, \tag{24}$$

$$\eta_3 = (\xi_1^{\gamma} + \xi_1^{\alpha})/2. \tag{25}$$

The microstructural evolution process is calculated by substituting Eq.(22) into TDGL Eq.(21) and the results are presented in Fig.11.[24] The microstructure is visualized by gray levels representing different values of $\eta = \eta_1^2 + \eta_2^2 + \eta_3^2$. In this calculation, the system is annealed at $T^* = 1.6$. One sees that a triple junction of APB's is formed in the later period. This is one of characteristic features of the anti-phase domain structure in $L1_0$ ordered alloy and is in marked difference with those shown in Figs.9 and 10.

The ordering kinetics depends crucially on the order of transition, i.e., the stability of the system against configurational fluctuation, as demonstrated in Figs.3, 5, 7 and 8. On the other hand, it is shown that the microstructural feature is mainly determined by the number of ordered variants rather than the stability of a system. It is noted that both the stability of a system and the number of the ordered variants originate from the atomistic nature of a given alloy such as atomic interaction energy, atomic correlation and the crystal symmetry. One realizes that the present result obtained by the hybridized calculation of the CVM and PPM explicitly reflects the atomistic nature of the system.

Within the present method, the atomistic information provided by the CVM is reflected in the evolution kinetics of microstructure. It has been also demonstrated that atomic configuration in a local system during microstructural evolution process can be synthesized.[16] However, atomistic kinetic information such as an elemental diffusion path and atomic mobility are not explicitly dealt in the present calculation. The significance of such information has been amply stressed in experimental studies for order-order relaxation behavior of LRO.[25,26] This problem is resolved only when the PPM is hybridized with PFM. We believe that this is one of the most important tasks for the further progress in a unified calculation from atomistic to microstructural levels.

REFERENCES

1. J. W. Cahn and J. E. Hilliard, On spinodal decomposition, *Acta Metal.*, **9**, 795-801(1961).
2. S. M. Allen and J. W. Cahn, A microscopic theory for antiphase boundary motion and its application to antiphase domain coarsening, *Acta Metall.*, **27**, 1085-1095(1979).
3. J. A. Warren and W. J. Boettinger, Prediction of dendritic growth and microsegregation patterns in a binary alloy using the phase-field method, *Acta mater.*, **43**, 689-703(1995).
4. T. Miyazaki, T. Koyama and T. Kozakai, Computer simulations of the phase transformation in real alloy systems based on the phase field method, *Mater. Sci. Eng., A***312**, 38-49(2001).
5. Y. Wang and A. G. Khachaturyan, Shape instability during precipitate growth in coherent solids, *Acta metall. mater.*, **43**, 1837-1857(1995).
6. D. Fan and L.-Q. Chen, Computer simulation of grain growth using a continuum field model, *Acta metal. mater.*, **45**, 611-622(1997).
7. L.-Q. Chen, Y. Wang and A. G. Khachaturyan, Kinetics of tweed and twin formation during an ordering transition a substitutional solid solution, *Phil. Mag. Let.*, **65**, 15-23(1992).
8. R. Kikuchi, A theory of cooperative phenomena, *Phys. Rev.*, **81**, 988-1003(1951).
9. J. M. Sanchez and D. de Fontaine, The fcc ising model in the cluster variation approximation, *Phys. Rev. B* **17**, 2926-2936(1978).
10. T. Mohri, J. M. Sanchez and D. de Fontaine, Short range order diffuse intensity calculations in the cluster variation method, *Acta Metall.*, **33**, 1463-1474(1985).
11. R. Kikuch, The path probability method, *Prog. Theor. Phys. Suppl.*, **35**, 1-64(1966).
12. T. Mohri, Atomic ordering process and a phase diagram, Solid-Solid transformations, ed. by W. C. Johnson,

J. M. Howe, D. E. Laughlin and W. A. Soffa(The Minerals, Metals & Materials Society, 1994), pp.53-74.

13. T. Mohri, Pseudo critical slowing down within the cluster variation method and the path probability method, *Modelling Simul. Mater. Sci. Eng.*, **8**, 239-249(2000).
14. K. Gschwend, H. Sato and R. Kikuchi, Kinetics of order-disorder transformation in alloys. II, *J. Chem. Phys.*, **69**, 5006-5019(1978).
15. M. Ohno and T. Mohri, Phase field calculation with CVM free energy for a disorder-B2 transition, *Mater. Sci. Eng., A***312**, 50-56(2001).
16. M. Ohno and T. Mohri, Disorder-$L1_0$ transition investigated by phase field method with CVM local free energy, *Mater. Trans.*, **42**, 2033-2041(2001).
17. D. de Fontaine, k-space symmetry rules for order-disorder reactions, *Acta Metall.*, **23**, 553-571(1975).
18. M. J. Richard and J. W. Cahn, Pairwise interactions and the ground state of ordered binary alloys, *Acta Metal.*, **19**, 1263-1277(1971).
19. K. Kawasaki, Diffusion constants near the critical point for time-dependent ising models. I, *Phys. Rev.* **145**, 224-230(1966).
20. R. J. Glauber, Time-dependent statistics of the ising model, *J. Math. Phys.*, **4**, 294-307(1953).
21. A. M. Gusak and A. O. Kovalchuk, Oscillatory regime of ordering interdiffusion, *Phys. Rev. B* **58**, 2551-2555(1998).
22. S. M. Allen and J. W. Cahn, Mechanisms of phase transformations within the miscibility gap of Fe-rich Fe-Al alloys, *Acta Metall.*, **24**, 425-437(1976).
23. A. Loiseau, C. Ricolleau, L. Potez and F. Ducastelle, Order and disorder at interfaces in alloys, Solid-Solid Phase Transition, edited by W. C. Johnson et al (The Mineral, Metals & Materials Society, 1994), p385-400.
24. M. Ohno and T. Mohri, Theoretical calculation of atomistic behavior and microstructural evolution process, *Bulletin of the Iron and Steel Institute of Japan*, **6**, 766-772(2001).
25. C. Dimitrov, X. Zhang and O. Dimitrov, Kinetics of long-range order relaxation in Ni_3Al: the effect of stoichiometry, *Acta mater.*, **44**, 1691-1699(1996).
26. R. Kozubski and W. Pfeiler, Kinetics of defect recovery and long-range ordering in Ni_3Al+B-II. atomic jump processes studied by "order-order" relaxation experiments, *Acta mater.*, **44**, 1573-1579(1996).

PHASE DISTRIBUTION AND TRANSFORMATION DYNAMICS USING IN-SITU SYNCHROTRON DIFFRACTION METHODS

Joe Wong

Lawrence Livermore National Laboratory, University of California,
PO Box 808, Livermore CA 94551, USA

INTRODUCTION

Novel site-specific and fast diffraction techniques utilizing intense synchrotron radiation have recently been developed and applied to map the phases and solid-state transformation in systems undergoing steep thermal gradients at high temperature. In the case of fusion welds, these in-situ probes yield kinetics and microstructural evolution data not previously obtained with conventional ex-situ, post-mortem methods. Thus far, the synchrotron data provide not only some of the crucial experimental information for understanding the phase dynamics and microstructural development in these technological systems, but also realistic inputs for simulations as well as validating various kinetic models of phase transformation in welding metallurgy. The experimental progress to date in these synchrotron studies will be reviewed in this paper. Recent time-resolved diffraction results on chemical dynamics of the ferrite $\rightarrow$ austenite transformation in carbon-manganese steel welds will also be presented.

SYNCHROTRON RADIATION

Synchrotron radiation is emitted as the major loss mechanism from charged particles such as electrons and positrons, in circular motion at relativistic energies. The properties of synchrotron light emitted from electrons with velocities near that of light are drastically different from the classical dipole radiation[1] and demonstrate its importance as a novel and powerful light source[2]. Synchrotron radiation was experimentally discovered by Elder et al.[3] using the General Electric 70 MeV betatron and studied theoretically by Schwinger[1]. The properties of synchrotron radiation may be summarized[4] as follows: (i) broad spectral distribution tunable from IR to the x-ray region; (ii) higher intensity, > 10^6 times that of an x-ray tube; (iii) plane polarized, with the electric vector in the orbital plane of the circulating particles; (iv) high natural collimation, which is important for small specimens and (v) sharply pulsed time structure. A detailed historical account of the discovery has been given by Pollock.[5]

The spectral distribution of synchrotron radiation from the Stanford Positron-Electron Accumulation Ring (SPEAR)[4] is shown in Fig. 1 with the electron beam energy E_e from 1.5 to 4.5 GeV as the parameter. As can be seen, the radiation is an intense continuous distribution extending from the infrared into the hard x-ray regime of the electromagnetic spectrum. Compared with the bremsstrahlung output of a 12kW standard x-ray tube, synchrotron radiation is higher in intensity by a factor of 10^6 or more. This reduces the measurement time for a typical diffraction measurement from hours to seconds or less.

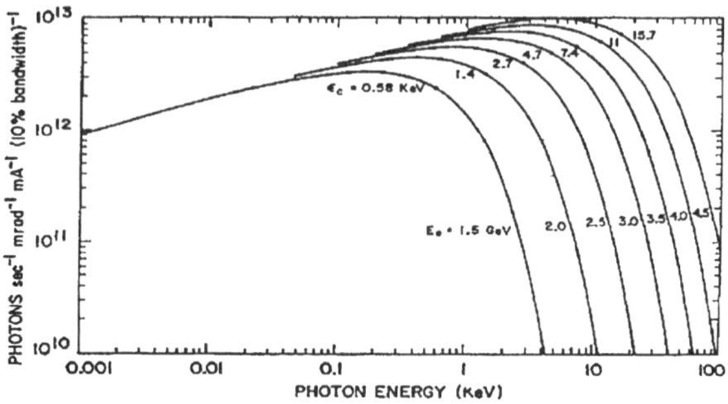

Figure 1. Spectral distribution of synchrotron radiation from the SPEAR storage ring with a radius of curvature of 12.7 m. After Doniach et al [4].

FUSION WELDS

During fusion welding, rapid thermal cycling induces solid state phase transformations both on heating and on cooling, and causes melting and solidification in those parts of the weld where the liquidus temperature has been exceeded. From a practical standpoint, solid state phase transformations play an important role in welding related problems such as sub-solidus cracking, cold cracking and distortion caused by residual stresses[6-8]. Solution and understanding of these problems will greatly be facilitated by the development of experimental methods capable of determining phase transformation behavior in steep thermal gradients and at the high cooling rates that occur *during* welding in the so-called heat-affected zone (HAZ) and fusion zone (FZ). Until a few years ago[9] no direct method existed for investigating solid state phase transformations that *are* taking place in fusion welds.

In this paper, we briefly describe a couple of novel synchrotron-based diffraction techniques that have been developed and applied successfully to investigate the phase distribution, transformation kinetics and chemical dynamics in the fusion welds of a number of metallurgical systems. These include a commercially pure titanium, *AISI* 1005 carbon-manganese steel and 2205 duplex stainless steel.

SYNCHROTRON DIFFRACTION METHODS

Space-Resolved X-ray Diffraction, SRXRD

The high intensity provided by synchrotron radiation emitted from a multi-pole wiggler insertion device was used to advantage to produce a sub-millimeter probe to record diffraction patterns as a function of position in the weld during welding to follow the phases and map their location in the HAZ or the FZ. The measurements were first performed on the 31-pole wiggler beam line, BL 10-2[10]
at Stanford Synchrotron Radiation Laboratory (SSRL) with SPEAR operating at an electron energy of 3.0 GeV and an injection current of ~100 mA. The synchrotron white beam was first focused by a toroidal mirror to the source size of ~1mm vertical x 2mm horizontal, and monochromatized downstream with a double Si(111) crystal. The focused monochromatic beam was then passed through a tungsten pinhole to render a sub-millimeter beam on the sample at an incident angle of ~25°. A photon energy of 12.0 keV (λ = 0.1033 nm) was chosen to maximize the number of Bragg peaks in a selected 2θ window to facilitate phase identification, and to be far enough in energy above the Ti and Fe-edges to minimize the background contribution due to Ti or Fe K-fluorescence from the sample (K-edge: Ti = 4.966 keV, Fe = 7.112 keV)[11].

SRXRD patterns were measured during welding by positioning the beam at a pre-determined location with respect to the welding electrode. A 50 mm, 2048-element position sensitive silicon photodiode array detector was used to record the diffraction patterns. The detector together with the associated ST121 data acquisition system was manufactured by

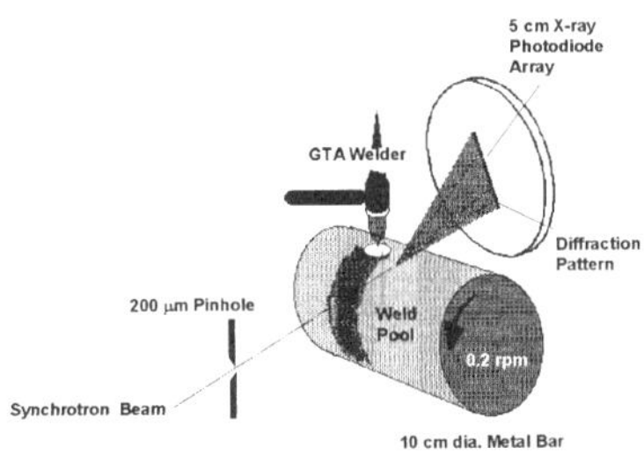

Figure 2. Schematics of the SRXRD setup for in-situ phase mapping and real time observation of phase transformation in fusion welds[13].

Princeton Instruments (now Roper Scientific)[12], and was used to store and display the x-ray diffraction data in real time. By incrementally jogging the weld to new locations in 200 μm intervals, a series of space-resolved x-ray diffraction patterns was collected along a lineal scan direction perpendicular to and away from the centerline of the weld. A schematic of the SRXRD setup is shown in Figure 2. Details of the apparatus has been described in detail elsewhere[13].

Fig.3 is a plot of an experimental SRXRD run of a carbon-manganese fusion weld at 1.9 kW. The starting position was 2.0 mm from the center of weld. The data were collected using a 10 keV photon beam, a pinhole of 180μm, jog size 200μm/step and an integration time of 20s/scan. It can be seen that in the first 8 jogs, no Bragg peaks were recorded indicative of the liquid state. The δ(110) reflection appeared at jog #9. This was followed by 11 frames of pure γ(fcc). Beyond which, γ and α coexisted for another 9 frames. At frame 29 and beyond, only α persisted as the x-ray now probed the base metal.

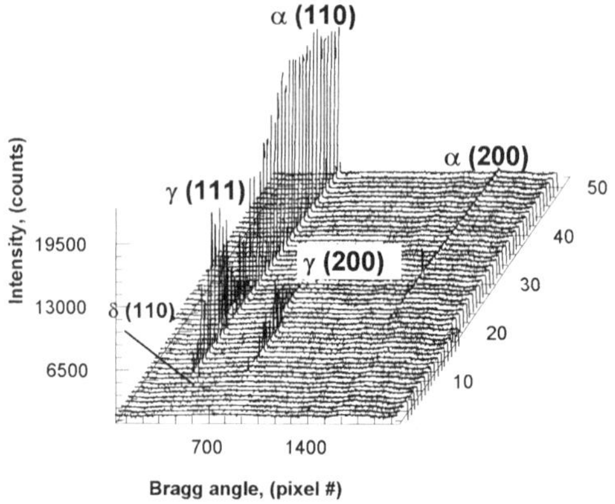

Figure 3. An experimental SRXRD scan recording the sequence of phase transformation from the liquid → δ(bcc) →γ(fcc) → α(bcc) as a function of position in a carbon-manganese fusion weld at 1.9 kW.

Time-Resolved X-ray Diffraction, TRXRD.

The TRXRD procedure consists of first positioning the x-ray beam at a pre-determined location in the fusion zone (FZ) or heat affected zone (HAZ) using the electrode position as center of the liquid pool. Experimentally, a time resolution of 100 ms in conjunction with a 260 μm pinhole used for phase was adequate to capture the α→γ phase transformation upon heating and the γ→α transformation upon cooling in the HAZ plain carbon steel welds. For phase transformations in the FZ, a 50 ms time resolution using a 730 μm pinhole was used to incorporate more grains into the diffraction beam[14]. The TRXRD technique was originally developed for chemical dynamics study of high temperature solid combustion reactions[15-16].

PHASE MAPPING

Commercially Pure Titanium.

In commercially pure grade 2 titanium, the allotropic transformation from a hcp α-phase to a bcc β-phase occurs at ~915°C. Transitions from a β-Ti bcc pattern at high temperature inside the HAZ, to a α+β-mixed zone, and eventually to the α-Ti hcp pattern have been recorded in real time. In addition to phase transformations, the material undergoes annealing and re-crystallization in the cooler region and outside of the HAZ. SRXRD data obtained from different starting positions with respect to the weld center may be classified into five principal diffraction patterns. Examples for each of the principal patterns are displayed in Fig.4 together with one for the $\alpha+\beta$-coexistence region.

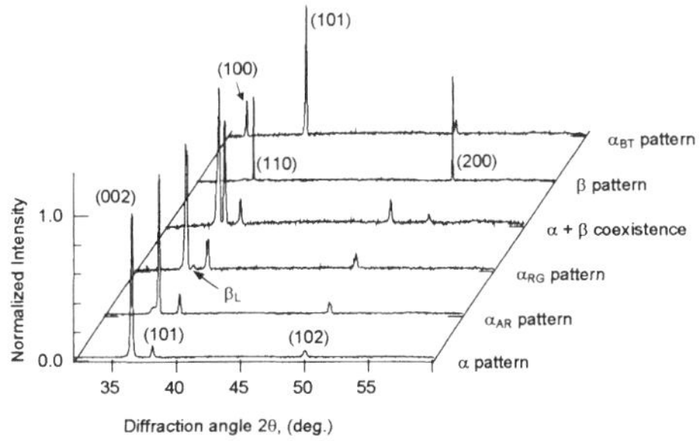

Figure 4. Principal diffraction patterns observed in various locations in the HAZ of commercially pure Ti fusion weld. Intensity is normalized to unity for the highest peak in each pattern[18].

In Fig.5 the completed phase map presenting the locations of four principal diffraction patterns plus β_L with respect to the HAZ of the weld is shown. Twenty one individual scans, each having 40 XSRXRD patterns, are displayed together with two calculated weld isotherms. For scans presented by two bars side-by-side, the left bar represents the presence

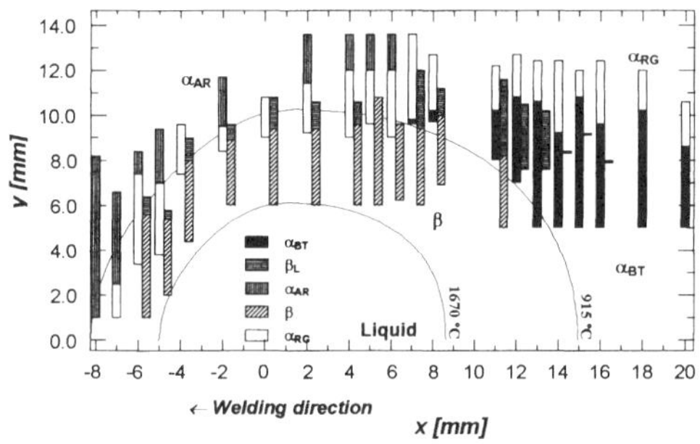

Figure 5. Complete phase map of a 1.9 kW fusion weld of commercially pure titanium and calculated transformation isotherms [18].

of α-Ti and the right bar for β-Ti. Each bar again is subdivided into regions with different shadings to indicate the different types of diffraction patterns that were present: (α_{AR}) is the annealed and recrystallized α-Ti; (α_{RG}) denotes a recrystallized α-Ti phase which exhibits large diffraction domains; (α_{BT}) is the back transformed α-Ti that forms from the region of the HAZ that once contained β-Ti; (β) is the β-Ti phase; (β_L) is the β-Ti that coexists with α-Ti in low amounts predominantly together with α_{BT}. The α pattern is not shown in this phase map since the SRXRD data were not recorded far enough in to the cold region of the HAZ. The principal diffraction patterns and associated microstructure have been described in detail elsewhere[17,18].

AISI 1005 Carbon-Manganese Steel.

In Fig. 6 the completed phase map and the distribution of the α-Fe, γ-Fe, δ-Fe, and liquid phases in the HAZ is shown. The coexistence of γ-Fe with α-Fe, or γ-Fe with δ-Fe, was identified at numerous SRXRD locations as evidenced by simultaneous recording of the fcc and bcc diffraction patterns. These regions indicate either a phase is in the process of transforming or that two phases are coexisting in a two-phase region of the HAZ. Superimposed on Fig. 6 are three major weld isothermal boundaries calculated using an analytical heat flow model described elsewhere[19]. The calculated isotherm at 1529°C represents the liquid weld pool boundary, which extends 4.4 mm from the weld centerline, and was made to equal to the actual weld pool width by adjusting the heat source distribution parameter[19]. The γ/(γ+α) boundary is represented by the 882°C isotherm, and the Fe_3C/(α+γ) eutectoid is represented by the 720°C isotherm. These calculated boundaries represent the locations where phase transformations would occur under equilibrium conditions.

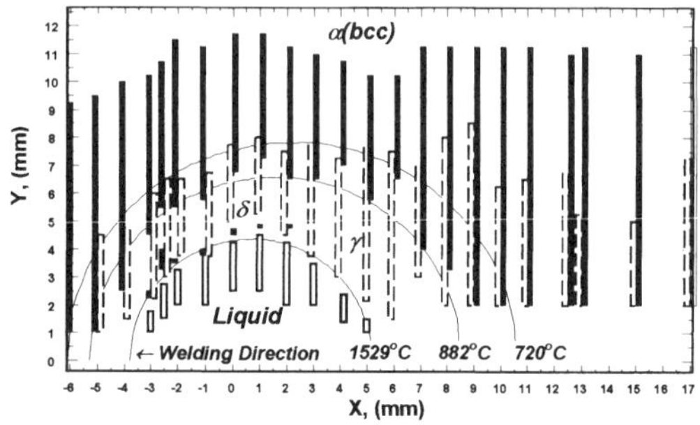

Figure 6. SRXRD map showing the locations of the α-Fe, γ-Fe, δ-Fe and liquid phases present in the AISI 1005 steel fusion weld at 1.9 kW. The calculated weld isotherms are superimposed on to the critical phase transformation temperatures. Notations - α: gray filled ; γ: dotted line; δ: black filled; liquid: thick line[19].

However, since the kinetics of the phase transformations require a finite time to take place, the location where the phase transformation is finally completed is displaced behind the calculated isotherms. Thus, the calculated isotherms represent the point where the phase transformations *can* begin to occur; the locations where the transformations are complete can be determined by SRXRD measurements. The difference between the calculated isotherms (start locations) and the SRXRD completion locations is related to the kinetics of a given phase transformation.

2205 Duplex Stainless Steel.

A detailed map of the locations of the austenite and ferrite phases in the HAZ of this duplex steel weld is shown in Fig. 7. The map is based on a combination of the SRXRD data and the calculated weld pool temperature profiles[20]. Three primary phase regions are observed: a liquid phase, a single phase δ-Fe (bcc), and a two phase δ-Fe (bcc) + γ-Fe (fcc) region. Superimposed on these phase regions are the calculated liquidus isotherm at

1443°C. The liquidus line corresponds closely with the locations where the liquid phase has been observed, thus validating the temperatures calculated using the turbulent heat transfer and fluid flow model. Adjacent to the liquid phase, a single- phase ferrite region is observed. This single

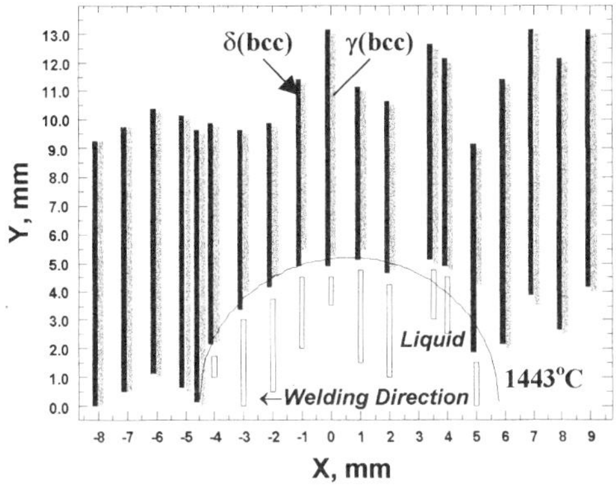

Figure 7. Complete SRXRD phase map of 2205 duplex stainless steel weld at 1.8 kW and calculated liquidus[20]

phase region denotes a complete transformation of austenite in the duplex ferrite/austenite microstructure to ferrite. As the distance from the weld centerline increases, two phase ferrite/austenite regions are observed adjacent to the single phase ferrite regions and extend in all directions to the base metal region. Details of this phase map has recently been described [20].

PHASE DYNAMICS

AISI 1005 carbon-manganese steel may be considered as a pseudo-binary Fe-C system containing 0.05 wt% of carbon. With increasing temperature, the system undergoes the following transformations:α(bcc)$\rightarrow\gamma$(fcc)$\rightarrow\delta$(bcc)$\rightarrow$liquid. The phase distribution and microstructural evolution in the vicinity of this fusion weld has studied using SRXRD[19]. Nature of the solidification product from the liquid pool, and chemical dynamics associated with the $\alpha\rightarrow\gamma$ transformation in a positive thermal gradient (heating) and the $\gamma\rightarrow\alpha$ transformation in negative thermal gradient (cooling) have also been recently elucidated in detail with a time resolution down to 50 ms[14]. With an experimental weld pool width of 8.5 mm, and by positioning the x-ray beam at a position 5 mm and another at 3 mm from the electrode, one can now probe the dynamics of phase transform-ation in the HAZ and FZ of the steel weld respectively. The results are summarized as follows.

Phase Transformation in the HAZ.

Fig. 8 shows a series of TRXRD patterns recorded in the HAZ at a location 5 mm away from the center of liquid pool. The data were recorded with a time resolution of 100ms. For clarity of presentation, only every 15th of the 600 TRXRD patterns were shown in the pseudo 3D plot given in Fig. 8(a). As the arc was turned on and off, diffraction patterns were continuously recorded to follow the annealing and phase transformation in real time in both the heating and cooling cycles. Frame-by-frame qualitative analysis of the TRXRD data yielded a real time sequence of events in the HAZ schematically given in Fig. 8(b). In this plot, t = 0 corresponds to the start of TRXRD measurement at room temperature. At t = 1.0s,

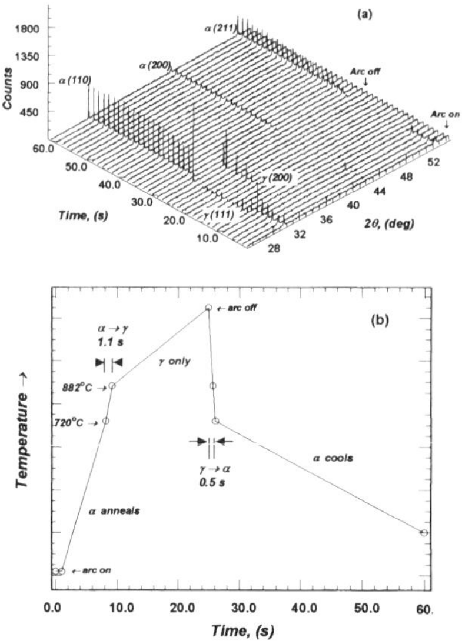

Figure 8. (a) TRXRD patterns recorded in the HAZ of a 1.9 kW carbon- manganese steel spot weld at a position 5 mm from the wel center with 100 ms time resolution. For clarity, only every 15th of the 600 frames is plotted. (b) Temperature-time schematics depicting the recorded TRXRD events shown in (a). After [14].

the arc was turned on and heating began. Annealing of the bcc α-phase took place in the next 7.1s. At t = 8.1 s, the fcc γ-phase first appeared, marking the start of the α→γ transformation. This transformation was completed within 1.1 s, and at t = 9.2s, only γ-Fe existed and persisted for the whole length of time the arc was on.

At t = 25.0s, the arc was turned off. The γ-phase cooled and at t = 25.7s, the first back transformed α-Fe was observed, signifying the start of the reverse γ→α transformation. At t = 26.1s, all γ-Fe disappeared, denoting completion of the γ→α. transformation. From then, only α-Fe existed and cooled with time. It is interesting to note that the α→γ transformation upon heating takes about twice as long as the γ→α transformation upon cooling.

Phase Transformation in the FZ

In Fig. 9(a), the TRXRD patterns recorded in the FZ at a position 3 mm away from the center of the liquid pool are shown. The data were recorded using a time resolution of 50 ms[14]. Again, for graphic clarity, only every 15th of the 600 TRXRD patterns were shown in the pseudo 3D plot given in Fig. 7(a). Frame-by-frame qualitative analysis of the TRXRD data yielded the following real time sequence of events in the FZ schematically given in Fig. 9(b).

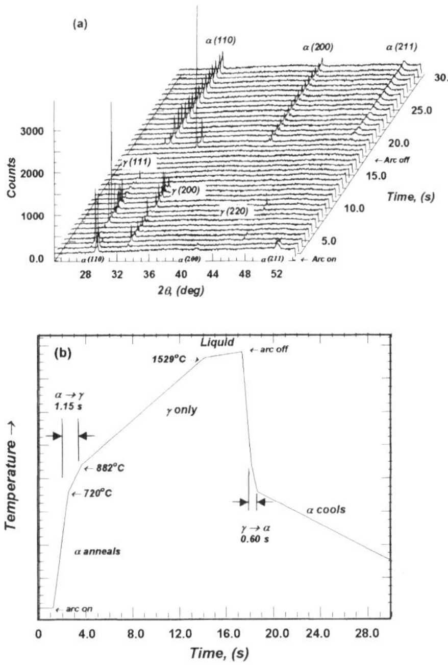

Figure 9. (a) TRXRD patterns recorded in the FZ of a 1.9kW carbon-manganese steel weld at a position 3 mm from the weld center with a 50 ms time resolution. For clarity, only every 15th of the 600 frames is plotted. (b) Temperature-time schematic depicting the recorded TRXRD events shown in (a).

Again, t = 0 corresponds to the start of TRXRD measurement at room temperature. At t = 1.20s, the arc was turned on and heating began. Annealing of the bcc α-phase took place in the next 1.30s, which was much shorter than that (7.1s) in the HAZ. At t = 2.50 s, the fcc γ-phase first appeared, marking the start of the α→γ transformation in the FZ on heating. Similar to that in the HAZ, this transformation was completed within 1.10 s, and at t = 3.60, only γ-Fe existed and persisted for an interval of 10.50s. At t = 14.10s, all γ disappeared and no diffraction peaks were observed signifying the system melted

At 17.35s, the arc was turned off. The γ-phase reappeared and cooled. At = 18.10s, the first back transformed α-Fe was observed, signifying the start of the γ→α transformation. At t = 18.65s, all γ-Fe disappeared, denoting completion of the γ→α transformation. From then, only α-Fe existed and cooled with time. It is interesting to note that, also in the FZ, the α→γ transformation upon heating takes about twice as long as the γ→α transformation upon cooling

Profile analysis[14] of the major Bragg reflections recorded in these TRXRD patterns reveals similarities and differences in the microstructural evolution with time in the HAZ and in the FZ. The latter undergoes melting and solidification in addition to solid state transformations. With increasing temperature, the cell constant of the α-phase prior to and during the α→γ transformation and that of the γ-phase just after the same transformation exhibit a decrease. This unusual lattice contraction with temperature rise may be attributed to dynamical processes involving diffusion of impurities or alloying elements. In the FZ, the γ-Fe that forms has a preferential (200)-texture upon solidification of the liquid, whereas upon cooling in the HAZ, the γ-Fe retains largely a (111)-texture induced in the α→γ transformation on heating. Upon cooling in the HAZ, the width of the γ(111) reflection increases initially, indicative of micro-strain developing in the fcc lattice, but decreases as expected with reduction of thermal disorder upon further cooling all the way to the completion of the γ→α transformation. In the FZ, however, the micro-strain in the γ-phase increases steadily upon solidification, and more rapidly in the entire duration of the γ→α transformation on further cooling. The final microstructure of the FZ is likely to consist of a single α-phase dispersed in two morphological entities, whereas in the HAZ, the α-phase persists in one morphological entity in the final microstructure[14].

CONCLUDING REMARKS

The synchrotron SRXRD and TRXRD methods have produced for the first time both real space and real time data on the phases, their location and transformation time in the HAZ and FZ of fusion welds. Profile analysis of the Bragg reflections associated with various occurring phases yielded detailed information on their microstructure in terms of annealing, re-crystallization, grain growth, texturing as well as micro-strain with the grains. These in-situ crystallographic and microstructural data must be taken into account for realistic and meaningful modeling of transformation kinetics in fusion welds[21] and serve as test-beds for evaluating various models for phase transformation in these technologically important systems.

ACKNOWLEDGMENTS

This work was performed under the auspices of the U. S. Department of Energy by the University of California, Lawrence Livermore National Laboratory under Contract No. W-7405-Eng-48. The author would like to thank J. W. Elmer, E.M. Larson, M. Froeba, T. Ressler, T. Palmer, P.A. Waide and A. Teruya for their collaboration and technical support in the course of developing these synchrotron tools for welding metallurgy. Synchrotron measurements were performed at Stanford Synchrotron Radiation Laboratory, supported by US Department of Energy, Office of Basic Energy Sciences.

REFERENCES

1. J. Schwinger, The properties of synchrotron radiation, Phys. Rev., **75**, 1912-1922 (1949)
2. H. Winick, in H. Winick and S. Doniach (eds), *Synchrotron Radiation Research*, Plenum Press, New York, **1980**, Chap. 2. p. 11.
3. F. R. Elder, R. V. Langmuir and H.C. Pollock, Observation of light emission from a 70

MeV betatron, Phys. Rev., **74**, 52 (1948).
4. S. Doniach, I. Lindau, W.R. Spicer and H. Winick, Synchrotron radiation: properties and applications, J. Vac. Sci. Technol., **12**, 1123-1129, (1975).
5. H. C. Pollock, History of the discovery of synchrotron radiation, Am. J. Phys., **51(3),** 278-285 (1983)
6. *International Trends in Welding Science and Technology*, ed. S.A David and J.M. Vitek, ASM Intern., **1993**.
7. B.K. Damkroger, G.R. Edwards and B.B. Rath, Crack growth in welded Ti alloys, Welding Journal, **68**, 290-296 (1989)
8. M.C. Mcquire, M.L. Santella and B.K Damkroger, Ductility in welded intermetallic alloys, in *The Science of Metal Joining*, ed. by M. H., Cieslak, J. Perepezko, S. Kang and M.E. Glicksman, TMS (1992), p.41.
9. Joe Wong, J. W. Elmer, P. Waide and E.M. Larson, In-situ X-ray Diffraction of an Arc weld showing the phase transformations of Ti and Fe as a Function of Position in the Weld, Adv. X-Ray Analysis, **37**, 479-485 (1994).
10. V. Karpenko, J.H. Kinney, S. Kulkarni, K. Neufeld, C. Poppe, K.G. Tirsell, Joe Wong, J. Cerino, T. Troxel, J. Yang, E. Hoyer, M. Green, D. Humpries, S. Marks, and D. Plate, A new wiggler beamline at SSRL, *Rev. Sci. Instrum.*, **60**, 1451-1460 (1989).
11. J. A. Bearden and A. F. Burr, X-ray Energies of the elements, Rev. Mod. Physics, **39**, 125-175 (1967)
12. ST121 Detector and Controller Manual, Princeton Instruments Rev. B. (1996).
13. Joe Wong, M. Fröba, J.W. Elmer, P.A. Waide, E. Larson, In-situ phase mapping and transformation study in fusion welds, J. Mater. Sc., **32**, 1493-1500 (1997)
14. Joe Wong, T. Ressler and J.W. Elmer, Dynamics of phase transformations and microstructure evolution in carbon-manganese steel arc welds using time-resolved synchrotron x-ray diffraction, *J. Synchrotron Rad.* (2002) *submitted*.
15. Joe Wong, E. Larson, J. Holt, P. Waide, B. Rupp and R. Frahm, Time-resolved diffraction study of solid combustion reactions, *Science*, **249**, 1406-1409 (1990).
16. E.M. Larson, Joe Wong, J. Holt, P. Waide, B. Rupp and L. Terminello, A time-resolved diffraction study of the Ta-C combustion synthesis, J. Mater. Res, **8(7)**, 1533-1541 (1993)
17. J. W. Elmer, Joe Wong and T. Ressler, SRXRD mapping in the HAZ of pure Ti arc welds, Met. Mat. Trans. **29A**, 2761-2773 (1998).
18. T. Ressler, Joe Wong and J.W. Elmer, Investigation of real-time microstructure evolution in steep thermal gradients using in-situ spatially resolved x-ray diffraction: A case study of Ti fusion welds, J. Phys. Chem. B, **102(52)**, 10724-10735 (1998)
19. J. W. Elmer, Joe Wong and T. Ressler, SRXRD mapping of phase transformations in the HAZ of carbon-manganese steel arc welds, Metal. Mater. Trans. **32A**, 1175-1187, (2001)
20. T.A. Palmer, J.W. Elmer and Joe Wong, In-situ observations of ferrite/austenite transformations in duplex stainless steel weldments using synchrotron radiation, Sci. Tech. Welding and Joining, (2002), *in press*
21. Z. Yang, J.W. Elmer, Joe Wong, and T. DebRoy, Modeling of phase transformation and grain growh in the HAZ of pure ti arc welds, Welding Journal, **79(4)**, 97-109 (2000)

MONTE CARLO STUDY OF THE PRECIPITATION KINETICS OF Al_3Zr IN Al-Zr

Emmanuel Clouet[1,2] and Maylise Nastar[2]

[1] Pechiney, Centre de Recherches de Voreppe
BP 27
38341 Voreppe cedex
France

[2] Service de Recherches de Métallurgie Physique
CEA/Saclay Bât. 520
91191 Gif-sur-Yvette
France

1. INTRODUCTION

Precipitation kinetics in alloys, like spinodal decomposition, nucleation and growth, or phase ordering, are now often studied at an atomistic scale using Monte Carlo simulations. So as to be able to reproduce the different kinetic behaviors during these transformations, one needs to adopt a realistic description of the diffusion. Therefore it is better to use a vacancy-diffusion mechanism than a direct atom exchange mechanism. It is then possible to explain why different kinetic pathways are observed. For instance, the vacancy diffusion mechanism can predict the importance of monomer diffusion with respect to the diffusion of small clusters [1–4]. This leads to a difference in the cluster size distribution during precipitation [2] and determines the coarsening mechanism (evaporation-condensation or coagulation) [3, 4]. One can predict too the slowdown of precipitation kinetics by vacancy trapping due to the addition of a third component impurity [1]. Finally, different interactions of solute atoms with vacancy lead to a difference of precipitate / matrix interface morphology during the kinetic pathway, the interface being diffuse for a repulsion between vacancy and solute atoms and sharp for an attraction [4].

One drawback of kinetic Monte Carlo simulations using vacancy-diffusion mechanism is that they limit themselves to pair interactions to describe configurational energy of alloys. Multisite interactions including more than two lattice sites are necessary if one wants to fully reproduce the thermodynamics of a system [5,6]. These interactions reflect dependence of bonds with their local environment and as a consequence break the symmetry imposed by pair interactions on phase diagram. It is interesting to notice that in Calphad approach [7] one naturally considers such interactions by describing formation energy of solid solutions with Redlich-Kister polynomials, and that coefficients of these polynomials can be mapped onto an Ising model to give effective interactions including more than two lattice sites [8,9]. Moreover these interactions allow one to understand shapes of precipitates in alloys [10] and can lead to a prediction of coherent interface energy in really good agreement with ab-initio calculations performed on supercells [11]. Nevertheless, to study kinetics in Monte Carlo simulations with such interactions, one usually uses a direct atom exchange mechanism [12], and thus looses all kinetic effects due to vacancy-diffusion mechanism.

We incorporate these multisite interactions in a kinetic model using a vacancy-diffusion mechanism to study precipitation kinetics of Al_3Zr in Al-Zr solid solution. For small supersaturation in zirconium of the aluminum solid solution, it has been experimentally observed that Al_3Zr precipitates are in the metastable $L1_2$ structure [13–15]. These precipitates are found to have mainly spherical shape (diameter $\sim 10-20$ nm), as well as rod-like shape [13]. For supersaturation higher than the peritectic concentration, nucleation is homogeneous and precipitates are coherent with the matrix [13,14]. With prolonged heat treatment, if the temperature is high enough, the metastable $L1_2$ structure can transform to the stable one DO_{23}. Using a phase field method, Proville and Finel [16] modelled these two steps of the precipitation, *i.e.* the transient nucleation of the $L1_2$ structure and the transformation to the DO_{23} structure. In this work, we only focus on the precipitation first stage, where Al_3Zr precipitates have the $L1_2$ structure and are coherent with the matrix.

We first use ab-initio calculations to fit a generalized Ising model describing thermodynamics of Al-Zr system. We then extend description of the configurational energy of the binary Al-Zr to the one of the ternary Al-Zr-Vacancy system and adopt a vacancy-atom exchange mechanism to describe kinetics. This atomistic model is used in Monte Carlo simulations to study diffusion in the Al-Zr solid solution as well as precipitation kinetics of Al_3Zr. We mainly focus our study on detecting any influence of multisite interactions on kinetics.

2. THERMODYNAMICS OF Al-Zr BINARY

2.1. Ab inito calculations

We use the full-potential linear-muffin-tin-orbital (FP-LMTO) method [17–19] to calculate formation energies of different compounds in the Al-Zr binary system, all based on a fcc lattice. Details of ab initio calculation can be found in appendix A. They are the same as in our previous work [20], except the fact that we use the generalized gradient approximation (GGA) instead of the local density approximation (LDA) for the exchange-correlation functional.

The use of GGA for the exchange correlation energy leads to a slightly better description of the Al-Zr system. The approximation does not fail to predict phase stability of pure Zr [21] as LDA does: if one does not include generalized-gradient corrections, the stable structure at 0 K for Zr is found to be the ω one (hexagonal with 3 atoms per unit cell) and not the hcp structure.

Another change depending on the approximation used for the exchange-correlation functional is that formation enthalpies obtained with GGA for the different Al-Zr compounds are a little bit lower (a few percent) than with LDA. For the DO_{23} structure of Al_3Zr (table 1), GGA predicts a formation energy which perfectly reproduces the one measured by calorimetry [22]: including generalized-gradient corrections has improved the agreement. The energy of transformation from the $L1_2$ to the DO_{23} structure, $\Delta E = -23$ meV/atom, agrees really well too with the experimental one measured by Desh *et al.* [23], but this was already true with LDA. Gradient corrections have improved the agreement for the equilibrium volumes too: with the LDA, they were too low compared to the available experimental ones. Considering the values of the relaxed equilibrium parameters, shape of the unit cell and atomic positions, no change is observed according to the approximation used, both LDA and GGA being in good agreement with measured parameters.

Table 1: Calculated equilibrium volumes V_0, c'/a ratios ($c' = c/2$ for the DO_{22} phase and $c' = c/4$ for the DO_{23} phase), atomic displacements (normalized by a), and energies of formation for Al_3Zr compared to experimental data.

		V_0 ($Å^3$/atom)	c'/a	Atomic displacements	ΔE (eV/atom)
$L1_2$	GGA[a]	16.89			−0.478
	LDA[b]	16.12			−0.524
DO_{22}	GGA[a]	17.40	1.138		−0.471
	LDA[b]	16.60	1.141		−0.525
DO_{23}	GGA[a]	17.16	1.080	$\delta_{Al} = +0.0013$	−0.502
				$\delta_{Zr} = -0.0239$	
	LDA[b]	16.35	1.087	$\delta_{Al} = -0.0021$	−0.548
				$\delta_{Zr} = -0.0273$	
	Exp.[c]	17.25	1.0775	$\delta_{Al} = +0.0004$	
				$\delta_{Zr} = -0.0272$	
	Exp.[d]				-0.502 ± 0.014

[a]FP-LMTO calculations (present work)
[b]FP-LMTO calculations [20]
[c]Neutron diffraction [24]
[d]Calorimetry [22]

2.2. Cluster expansion of the formation energy

In order to express the formation energy of any Al-Zr compound based on a perfect fcc lattice, we make a cluster expansion [25] of our FP-LMTO calculations to fit a generalized Ising model. This allows us to obtain the energy of any configuration of the fcc lattice.

Considering a binary alloy of N sites on a rigid lattice, its configuration can be described through an Ising model by the vector $\boldsymbol{\sigma} = \{\sigma_1, \sigma_2, \ldots, \sigma_N\}$ where the pseudo-spin configuration variable σ_i is equal to ± 1 if an A or B atom occupies the site i. Any structure is then defined by its density matrix ρ^s, $\rho^s(\boldsymbol{\sigma})$ being the probability of finding the structure s in the configuration $\boldsymbol{\sigma}$.

With any cluster of n lattice points $\alpha = \{i_1, i_2, \ldots, i_n\}$ we associate the multisite correlation function

$$\zeta_\alpha^s = \mathrm{Tr}\, \rho^s \prod_{i \in \alpha} \sigma_i = \frac{1}{2^N} \sum_{\boldsymbol{\sigma}} \rho^s(\boldsymbol{\sigma}) \prod_{i \in \alpha} \sigma_i, \tag{1}$$

where the sum has to be performed over the 2^N possible configurations of the lattice.

Clusters related by a translation or a symmetry operation of the point group of the structure have the same correlation functions. Denoting by D_α the number of such equivalent clusters per lattice site, or degeneracy, the energy, like any other configurational function, can be expanded in the form [25]

$$E = \sum_\alpha D_\alpha J_\alpha \zeta_\alpha^s, \tag{2}$$

where the sum has to be performed over all non equivalent clusters and the cluster interaction J_α is independent of the structure.

The cluster expansion as defined by equation 2 cannot be used directly: a truncated approximation of the sum has to be used. The truncation is made with respect to the number of points contained in a cluster, thus assuming that order effects on energy are limited to a small set of lattice points. It is truncated too with respect to distance between sites. Long range interactions are important mostly if one wants to fully reproduce elastic effects [26].

We use in the expansion of the energy six different clusters: the empty cluster {0}, the point cluster {1}, the pairs of first and second nearest neighbors {2,1} and {2,2}, the triangle of first nearest neighbors {3,1}, and the tetrahedron of first nearest neighbors {4,1}. The corresponding cluster interactions are obtained by making a least square fit of compound energies calculated with FP-LMTO. All 17 used compounds are lying on a perfect fcc lattice: energies are calculated without any relaxation of the volume, of the shape of the unit cell, or of the atomic positions. The lattice parameter used is the one which minimizes the cohesive energy of pure Al, $a = a_{Al} = 4.044$ Å. We choose to fit the cluster expansion for the equilibrium lattice parameter of Al because we are interested in describing thermodynamics of the Al rich solid solution as well as of Al_3Zr precipitates in the $L1_2$ structure. These precipitates have an equilibrium lattice parameter close to the one of pure Al, $a = 4.073$ Å as obtained from FP-LMTO calculations with GGA, and during the nucleation stage they are coherent with the Al matrix. Consequently, such an expansion should be able to give a reasonable thermodynamic description of the different configurations reached during this precipitation stage where precipitates are coherent.

Coefficients of the cluster expansion of the energy are given in table 3. Comparing the values of the many-body interactions, we see that the main contribution to the energy arises from the pair interactions and that the 3- and 4-point cluster contributions are only corrections. Signs of pair interactions reflect the tendency of Al and Zr atoms to

Table 2: Formation energy relative to pure fcc elements for Al-Zr compounds lying on a perfect fcc lattice ($a = a_{Al} = 4.044$ Å) obtained from a direct FP-LMTO calculations and from its cluster expansion.

	Pearson symbol	Structure type	E^{form} (eV/atom) FP-LMTO	CE
Al (fcc)	cF4	Cu	0.	0.
Al_4Zr ($D1_a$)	tI10	$MoNi_4$	−0.421	−0.491
Al_3Zr ($L1_2$)	cP4	Cu_3Au	−0.728	−0.671
Al_3Zr (DO_{22})	tI8	Al_3Ti	−0.617	−0.643
Al_3Zr (DO_{23})	tI16	Al_3Zr	−0.690	−0.657
Al_2Zr (β)	tI6	$MoSi_2$	−0.513	−0.482
AlZr ($L1_0$)	tP4	AuCu	−0.803	−0.780
AlZr ($L1_1$)	hR32	CuPt	−0.448	−0.466
AlZr (CH40)	tI8	NbP	−0.643	−0.723
AlZr (D4)	cF32	?[a]	−0.489	−0.414
AlZr (Z2)	tP8	?[a]	−0.345	−0.333
Zr_2Al (β)	tI6	$MoSi_2$	−0.443	−0.482
Zr_3Al ($L1_2$)	cP4	Cu_3Au	−0.640	−0.603
Zr_3Al (DO_{22})	tI8	Al_3Ti	−0.570	−0.574
Zr_3Al (DO_{23})	tI16	Al_3Zr	−0.603	−0.589
Zr_4Al ($D1_a$)	tI10	$MoNi_4$	−0.390	−0.437
Zr (fcc)	cF4	Cu	0.	0.

[a]Description of structures D4 and Z2 can be found in Ref. [6]

Table 3: Cluster expansion of the formation energy.

Cluster	D_α	J_α (eV/atom)
{0}	1	−4.853
{1}	1	0.933
{2,1}	6	97.5×10^{-3}
{2,2}	3	-28.4×10^{-3}
{3,1}	8	4.2×10^{-3}
{4,1}	2	13.1×10^{-3}

form heteroatomic first nearest neighbor pairs and homoatomic second nearest neighbor pairs.

In table 2, we compare the formation energies of the different compounds directly obtained from FP-LMTO calculations with the ones given by their cluster expansion. The standard deviation equals 41 meV/atom and the maximal difference is 79 meV/atom. This could have been improved by including more clusters in the expansion of the energy or by using a mixed-space cluster expansion [26]. Nevertheless, this would not have changed the main characteristics of the Al-Zr system, *i.e.* the short range order tendency given by pair interactions, as well as the dependence on local environment of the interactions given by 3- and 4-point cluster interactions. In order to be able to build a realistic kinetic model and to run Monte Carlo simulations in a reasonable amount of time, we have to keep the thermodynamic description of Al-Zr system as simple as it can be. Therefore we do not try to improve expansion convergence and we focus our work on the influence of the 3- and 4-point cluster interactions on the thermodynamic and kinetic properties.

2.3. Phase diagram

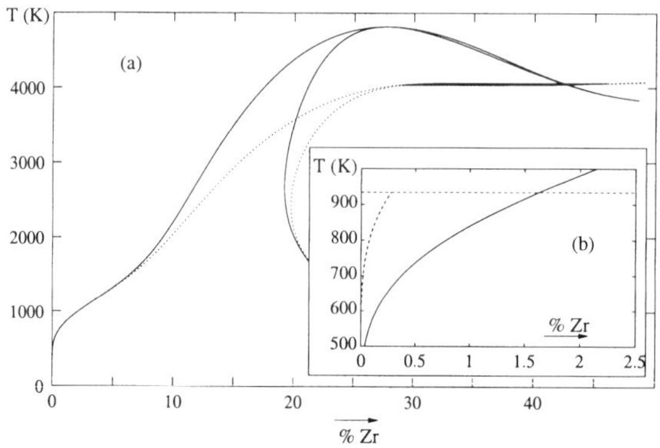

Figure 1: Al rich part of the phase diagram corresponding to the equilibrium between the fcc solid solution and the $L1_2$ structure given by our set of parameters (table 3). (a) Comparison of the phase diagrams obtained with pair, triangle, and tetrahedron interactions (solid line) and the one obtained with only pair interactions (dotted line). (b) Comparison with the predicted metastable solubility limit [20] (dashed line).

We use the cluster-variation method (CVM) [27] in the tetrahedron-octahedron (TO) approximation [28, 29] to study the equilibrium between the fcc Al-rich solid solution and the $L1_2$ structure (Fig. 1) corresponding to energy parameters of table 3. At low temperature, the 2 sublattices of the $L1_2$ structure remain highly ordered, as at the experimental peritectic melting temperature ($T \sim 934$ K) the Zr concentrations of the two sublattices are respectively 100. and 1.8 at.%. Turning out the energy coefficients of the first nearest neighbor triangle and tetrahedron ($J_3 = J_4 = 0$), we see that these many-body interactions have a thermodynamic influence only at high temperature (Fig. 1 (a)), as for temperatures below 1000 K the phase diagram remains unchanged with or without these interactions.

In figure 1 (b), we compare the Zr solubility limit in the fcc solid solution corresponding to the present work energy parameters with our previous estimation of this metastable solubility limit [20]. We should point out that the solubility limit obtained with the parameters given by table 3 corresponds to a coherent equilibrium between the fcc solid solution and the $L1_2$ structure as the energy coefficients of the expansion have been calculated for a perfect fcc lattice at the parameter of pure Al. This leads to a destabilization of the ordered phase and this is the main reason why we obtain a higher solubility than the estimated one corresponding to the equilibrium between incoherent phases. Another reason is that we use the cluster expansion to compute Al_3Zr cohesive energy, and thus get a small error on this energy, whereas in our previous study we directly used the value given by FP-LMTO calculation.

3. KINETIC MODEL

In order to be able to build an atomistic kinetic model, we have to generalize our thermodynamic description of the Al-Zr binary system to the one of the Al-Zr-Vacancy ternary system. To do so, we recast first the spin-like formalism of the cluster expansion into the more convenient one of the lattice gas formulation using occupation numbers [5]. This will allow us to obtain effective interactions for the different configurations of the tetrahedron of first nearest neighbors and of the pair of second nearest neighbors. Atom-vacancy interactions can then be introduced quite easily.

3.1. Effective interactions

Instead of using the pseudo-spin variables σ_n as we did in chap 2.2., this will be easier for the following to work with occupation numbers p_n^i, p_n^i being equal to 1 if an atom of type i occupies the site n and to 0 otherwise. In a binary alloy, occupation numbers and pseudo-spin variable at site n are related by

$$p_n^A = \frac{1+\sigma_n}{2}, \qquad \text{and} \qquad p_n^B = \frac{1-\sigma_n}{2}. \tag{3}$$

For the Al-Zr binary system, we included in our truncated cluster expansion of the energy first nearest neighbor interactions up to the pair, triangle, and tetrahedron clusters and a second nearest neighbor pair interaction. Thus, using the occupation numbers p_n^i, the expression of the energy becomes

$$E = \frac{1}{4N_s} \sum_{\substack{n,m,p,q \\ i,j,k,l}} \epsilon_{ijkl}^{(1)} p_n^i p_m^j p_p^k p_q^l + \frac{1}{2N_s} \sum_{\substack{r,s \\ i,j}} \epsilon_{ij}^{(2)} p_r^i p_s^j, \tag{4}$$

where the first sum runs over all sites (n, m, p, q) forming a first nearest neighbor tetrahedron and all their different configurations (i, j, k, l), and the second sum over all sites (r, s) forming a second nearest neighbor pair and all their different configurations (i, j). N_s is the number of lattice sites, $\epsilon_{ijkl}^{(1)}$ the effective energy of a first nearest neighbor tetrahedron in the configuration (i, j, k, l), and $\epsilon_{ij}^{(2)}$ the effective energy of a second nearest neighbor pair in the configuration (i, j).

Writing the energy with these effective interactions increases the number of dependent variables. Therefore several choices of these effective energies correspond to the same cluster expansion, then to the same thermodynamic and kinetic properties. If we make the assumption that second nearest neighbor interactions do not contribute to the cohesive energy of pure elements, *i.e.* $\epsilon_{AA}^{(2)} = 0$ and $\epsilon_{BB}^{(2)} = 0$, we obtain as many effective interactions as parameters in the truncated cluster expansion. Such an assumption does not have any physical influence and it just guarantees that homo-atomic effective interactions, $\epsilon_{AAAA}^{(1)}$ and $\epsilon_{AA}^{(2)}$, do not depend on the on the nature of B atom. Effective energies of the first nearest neighbor tetrahedron in its different configurations are then related to the cluster expansion coefficients by the equations

$$\begin{pmatrix} \epsilon^{(1)}_{AAAA} \\ \epsilon^{(1)}_{AAAB} \\ \epsilon^{(1)}_{AABB} \\ \epsilon^{(1)}_{ABBB} \\ \epsilon^{(1)}_{BBBB} \end{pmatrix} = \frac{1}{12} \begin{pmatrix} 6 & 6 & 6 & 6 & 6 \\ 6 & 3 & 0 & -3 & -6 \\ 6 & 0 & -2 & 0 & 6 \\ 6 & -3 & 0 & 3 & -6 \\ 6 & -6 & 6 & -6 & 6 \end{pmatrix} \begin{pmatrix} D_0 J_0 + D_{2,2} J_{2,2} \\ D_1 J_1 \\ D_{2,1} J_{2,1} \\ D_{3,1} J_{3,1} \\ D_{4,1} J_{4,1} \end{pmatrix}, \tag{5}$$

and the second nearest neighbor pair interaction by the equation

$$\epsilon^{(2)}_{AB} = -\frac{2}{3} D_{2,2} J_{2,2}. \tag{6}$$

For Al-Zr binary system, tetrahedron effective interactions corresponding to the cluster expansion of chap. 2.2. can be found in table 4: two sets are given depending if J_3 and J_4 are taken from the cluster expansion of table 3 or are supposed equal to zero. For both sets $\epsilon^{(2)}_{AB} = +0.057$ eV.

3.2. Decomposition of effective interactions

As we wrote before, several sets of effective interactions produce the same cluster expansion. In the following, we generate the set of interactions useful for our kinetic model for which we have to count bonds we break for vacancy-atom exchange.

Different contributions are included in the effective energy $\epsilon^{(1)}_{ijkl}$. One part of the energy is due to the bonding corresponding to the six different pairs of atoms contained in the tetrahedron, each of these pairs belonging to two different tetrahedrons. Then one has to add corrections due to order on the four triangles contained in the tetrahedron and another correction due to order on the tetrahedron itself. This decomposition leads to the relation

$$\begin{aligned} \epsilon^{(1)}_{ijkl} &= \frac{1}{2}\left(\epsilon^{(1)}_{ij} + \epsilon^{(1)}_{ik} + \epsilon^{(1)}_{il} + \epsilon^{(1)}_{jk} + \epsilon^{(1)}_{jl} + \epsilon^{(1)}_{kl}\right) \\ &+ \left(\tilde{\epsilon}^{(1)}_{ijk} + \tilde{\epsilon}^{(1)}_{ijl} + \tilde{\epsilon}^{(1)}_{ikl} + \tilde{\epsilon}^{(1)}_{jkl}\right) + \tilde{\epsilon}^{(1)}_{ijkl}, \end{aligned} \tag{7}$$

where $\epsilon^{(1)}_{ij}$ is the effective energy of the first nearest neighbor pair in the configuration (i, j) and $\tilde{\epsilon}^{(1)}_{ijk}$ and $\tilde{\epsilon}^{(1)}_{ijkl}$ the corrections to add to pair energy due to order on triangles and on the tetrahedron

Using the previous breakdown of the tetrahedron effective energy, the expression 4 of the energy becomes

$$\begin{aligned} E &= \frac{1}{2N_s} \sum_{\substack{n,m \\ i,j}} \epsilon^{(1)}_{ij} p^i_n p^j_m + \frac{1}{3N_s} \sum_{\substack{n,m,p \\ i,j,k}} \tilde{\epsilon}^{(1)}_{ijk} p^i_n p^j_m p^k_p \\ &+ \frac{1}{4N_s} \sum_{\substack{n,m,p,q \\ i,j,k,l}} \tilde{\epsilon}^{(1)}_{ijkl} p^i_n p^j_m p^k_p p^l_q + \frac{1}{2N_s} \sum_{\substack{r,s \\ i,j}} \epsilon^{(2)}_{ij} p^i_r p^j_s. \end{aligned} \tag{8}$$

As this is just another mathematical way to rewrite the cluster expansion 2 of the energy, the following relations holds :

$$D_0J_0 + D_{2,2}J_{2,2} = \frac{3}{2}\left(\epsilon^{(1)}_{AA} + 2\epsilon^{(1)}_{AB} + \epsilon^{(1)}_{BB}\right) + \tilde{\epsilon}^{(1)}_{AAA} + 3\tilde{\epsilon}^{(1)}_{AAB} + 3\tilde{\epsilon}^{(1)}_{ABB} + \tilde{\epsilon}^{(1)}_{BBB} + \frac{1}{8}\left(\tilde{\epsilon}^{(1)}_{AAAA} + 4\tilde{\epsilon}^{(1)}_{AAAB} + 6\tilde{\epsilon}^{(1)}_{AABB} + 4\tilde{\epsilon}^{(1)}_{ABBB} + \tilde{\epsilon}^{(1)}_{BBBB}\right) \quad (9a)$$

$$D_1J_1 = 3\left(\epsilon^{(1)}_{AA} - \epsilon^{(1)}_{BB}\right) + 3\left(\tilde{\epsilon}^{(1)}_{AAA} + \tilde{\epsilon}^{(1)}_{AAB} - \tilde{\epsilon}^{(1)}_{ABB} - \tilde{\epsilon}^{(1)}_{BBB}\right) + \frac{1}{2}\left(\tilde{\epsilon}^{(1)}_{AAAA} + 2\tilde{\epsilon}^{(1)}_{AAAB} - 2\tilde{\epsilon}^{(1)}_{ABBB} - \tilde{\epsilon}^{(1)}_{BBBB}\right) \quad (9b)$$

$$D_{2,1}J_{2,1} = \frac{3}{2}\left(\epsilon^{(1)}_{AA} - 2\epsilon^{(1)}_{AB} + \epsilon^{(1)}_{BB}\right) + 3\left(\tilde{\epsilon}^{(1)}_{AAA} - \tilde{\epsilon}^{(1)}_{AAB} - \tilde{\epsilon}^{(1)}_{ABB} + \tilde{\epsilon}^{(1)}_{BBB}\right) + \frac{3}{4}\left(\tilde{\epsilon}^{(1)}_{AAAA} - 2\tilde{\epsilon}^{(1)}_{AABB} + \tilde{\epsilon}^{(1)}_{BBBB}\right) \quad (9c)$$

$$D_{3,1}J_{3,1} = \tilde{\epsilon}^{(1)}_{AAA} - 3\tilde{\epsilon}^{(1)}_{AAB} + 3\tilde{\epsilon}^{(1)}_{ABB} - \tilde{\epsilon}^{(1)}_{BBB} + \frac{1}{2}\left(\tilde{\epsilon}^{(1)}_{AAAA} - 2\tilde{\epsilon}^{(1)}_{AAAB} + 2\tilde{\epsilon}^{(1)}_{ABBB} - \tilde{\epsilon}^{(1)}_{BBBB}\right) \quad (9d)$$

$$D_{4,1}J_{4,1} = \frac{1}{8}\left(\tilde{\epsilon}^{(1)}_{AAAA} - 4\tilde{\epsilon}^{(1)}_{AAAB} + 6\tilde{\epsilon}^{(1)}_{AABB} - 4\tilde{\epsilon}^{(1)}_{ABBB} + \tilde{\epsilon}^{(1)}_{BBBB}\right) \quad (9e)$$

As we want $\tilde{\epsilon}^{(1)}_{ijk}$ to be the energetic corrections due to order on triangles, the contribution to $J_{2,1}$ of $\tilde{\epsilon}^{(1)}_{AAA}$, $\tilde{\epsilon}^{(1)}_{AAB}$, ... must equal zero (second term in right hand side of eq. 9c). For the same reason, the contribution to $J_{2,1}$ and $J_{3,1}$ of $\tilde{\epsilon}^{(1)}_{AAAA}$, $\tilde{\epsilon}^{(1)}_{AAAB}$, ... must equal zero (last term in right hand side of eq. 9c and 9d). We require too that triangle and tetrahedron order corrections do not contribute to the cohesive energy of pure elements, as we did for second nearest neighbor pair interactions. Thus, $\tilde{\epsilon}^{(1)}_{AAA} = \tilde{\epsilon}^{(1)}_{BBB} = 0$ and $\tilde{\epsilon}^{(1)}_{AAAA} = \tilde{\epsilon}^{(1)}_{BBBB} = 0$. With these restrictions, all parameters entering in the expression 8 of the energy are well determined.

The first nearest neighbor pair effective energies are thus

$$\epsilon^{(1)}_{AA} = \frac{1}{6}\left(D_0J_0 + D_1J_1 + D_{2,1}J_{2,1} + D_{2,2}J_{2,2} + D_{3,1}J_{3,1} + D_{4,1}J_{4,1}\right) \quad (10a)$$

$$\epsilon^{(1)}_{AB} = \frac{1}{6}\left(D_0J_0 - D_{2,1}J_{2,1} + D_{2,2}J_{2,2} + D_{4,1}J_{4,1}\right) \quad (10b)$$

$$\epsilon^{(1)}_{BB} = \frac{1}{6}\left(D_0J_0 - D_1J_1 + D_{2,1}J_{2,1} + D_{2,2}J_{2,2} - D_{3,1}J_{3,1} + D_{4,1}J_{4,1}\right), \quad (10c)$$

the order corrections on first nearest neighbor triangle

$$\tilde{\epsilon}^{(1)}_{AAB} = -\tilde{\epsilon}^{(1)}_{ABB} = -\frac{1}{6}D_{3,1}J_{3,1}, \quad (11)$$

and the order corrections on first nearest neighbor tetrahedron

$$\tilde{\epsilon}^{(1)}_{AAAB} = \tilde{\epsilon}^{(1)}_{ABBB} = -D_{4,1}J_{4,1} \quad (12a)$$

$$\tilde{\epsilon}^{(1)}_{AABB} = 0. \quad (12b)$$

Inverting the system 5, one can easily express all these quantities from the effective tetrahedron energies ϵ_{ijkl} too.

3.3. Interactions with vacancy

Within the previous formalism, we can easily introduce atom-vacancy interactions. These interactions are a simple way to take into account the electronic relaxations around the vacancy. Without them, the vacancy formation energy E_V^{for} in a pure metal would necessarily equal the cohesive energy ($E_V^{for} = 0.69$ eV [30] and $E^{coh} = 3.36$ eV for fcc Al).

We only consider first-nearest neighbor interactions with vacancies and we do not include any order correction on triangle and tetrahedron configurations containing at least one vacancy, *i.e.* $\tilde{\epsilon}_{ijV}^{(1)} = \tilde{\epsilon}_{ijkV}^{(1)} = 0$ where i, j, and k are any of the species Al, Zr, and V. The vacancy formation energy in a pure metal A is then given by

$$E_V^{for} = 8\epsilon_{AAAV}^{(1)} - 6\epsilon_{AAAA}^{(1)} = 12\epsilon_{AV}^{(1)} - 6\epsilon_{AA}^{(1)} \tag{13}$$

The interaction $\epsilon_{AlV}^{(1)}$ is deduced from the experimental value of the vacancy formation energy in pure Al. For the interaction $\epsilon_{ZrV}^{(1)}$, we assume that the vacancy formation energy in the fcc structure is the same as in the hcp one, these two structures being quite similar. The only experimental information we have concerning this energy is $E_V^{for} > 1.5$ eV [30]. We thus use the ab-initio value calculated by Le Bacq *et al.* [31], $E_V^{for} = 2.07$ eV. This value is calculated at the equilibrium volume of Zr and cannot be used directly to obtain $\epsilon_{ZrV}^{(1)}$ as this interaction should correspond in our model to the equilibrium lattice parameter of pure Al. We have to use instead the vacancy formation enthalpy

$$H_V^{for} = E_V^{for} + P\delta\Omega_V^{for}, \tag{14}$$

where $\delta\Omega_V^{for} = -1.164$ Å^3 is the vacancy formation volume in pure Zr [30], and P is the pressure to impose to Zr to obtain a lattice parameter equal to the one of Al. P is calculated from the bulk modulus $B = 91$ GPa of fcc Zr and the equilibrium volumes of Al and Zr, $\Omega_{Al}^0 = 16.53$ Å^3 and $\Omega_{Zr}^0 = 23.36$ Å^3, these three quantities being obtained from our FP-LMTO calculations. This gives us the value $H_V^{for} = 1.88$ eV for the vacancy formation enthalpy in pure Zr at the lattice parameter of pure Al.

We use the experimental value of the divacancy binding energy $E_{2V}^{bin} = 0.2$ eV [30] in order to compute a vacancy-vacancy interaction, $\epsilon_{VV}^{(1)} = E_{2V}^{bin} - \epsilon_{AlAl}^{(1)} + 2\epsilon_{AlV}^{(1)}$. If we do not include this interaction and set it equal to zero instead, we obtain the wrong sign for the divacancy binding energy, divacancies being thus more stable than two monovacancies. This does not affect our Monte Carlo simulations as we only include one vacancy in the simulation box, but this will have an influence if we want to build a mean field approximation of our diffusion model.

We thus managed to add vacancy contributions to our thermodynamic description of the Al-Zr binary system. Using the breakdown 7 of the first nearest neighbor tetrahedron interaction, we can obtain the effective energies corresponding to the 15 different configurations a tetrahedron can have in the ternary system. These effective energies are presented in table 4 for the cases where energy correction due to order on first nearest neighbor triangle and tetrahedron are assumed different from zero or not (J_3 and J_4 given by table 3 or $J_3 = J_4 = 0$).

Table 4: Effective energies of the first nearest neighbor tetrahedron for Al-Zr-V ternary system. The set with order correction corresponds to the values J_3 and J_4 given by the cluster expansion of table 3 and the set without order correction assumes $J_3 = J_4 = 0$.

Configuration	Effective energy (eV)	
	with order correction	without order correction
Al Al Al Al	−1.680	−1.680
Al Al Al Zr	−2.257	−2.214
Al Al Zr Zr	−2.554	−2.554
Al Zr Zr Zr	−2.707	−2.698
Zr Zr Zr Zr	−2.647	−2.647
Al Al Al V	−1.174	−1.174
Al Al Zr V	−1.567	−1.561
Al Zr Zr V	−1.748	−1.754
Zr Zr Zr V	−1.751	−1.751
Al Al V V	−0.518	−0.518
Al Zr V V	−0.758	−0.758
Zr Zr V V	−0.804	−0.804
Al V V V	+0.288	+0.288
Zr V V V	+0.194	+0.194
V V V V	+1.243	+1.243

With this set of thermodynamic parameters, we calculate the binding energy between a Zr solute atom and a vacancy in Al,

$$E^{bin}_{ZrV} = 2\left(\epsilon^{(1)}_{AlAlAlAl} + \epsilon^{(1)}_{AlAlZrV} - \epsilon^{(1)}_{AlAlAlZr} - \epsilon^{(1)}_{AlAlAlV}\right). \tag{15}$$

The value obtained, $E^{bin}_{ZrV} = +0.369$ eV, agrees with the experimental observation that there is no attraction between Zr solute atoms and vacancies in Al [30, 32].

3.4. Migration barriers

Diffusion occurs via vacancy jumps towards one of its twelve first nearest neighbors. The vacancy exchange frequency with a neighbor of type A (A =Al or Zr) is given by

$$\Gamma_{A-V} = \nu_A \exp\left(-\frac{E^{act}_A}{k_B T}\right), \tag{16}$$

where ν_A is an attempt frequency and the activation energy E^{act}_A is the energy change required to move the A atom from its initial stable position to the saddle point position. It is computed as the difference between the contribution e^{sp}_A of the jumping atom to the saddle point energy and the contributions of the vacancy and of the jumping atom to the initial energy of the stable position. This last contribution is obtained by considering all bonds which are broken by the jump, *i.e.* all pair interactions the vacancy and the jumping atoms are forming as well as all order corrections on triangles and tetrahedrons containing the jumping atom,

$$E_A^{act} = e_A^{sp} - \sum_j \epsilon_{Vj}^{(1)} - \sum_{j \neq V} \epsilon_{Aj}^{(1)} - \sum_{jk} \tilde{\epsilon}_{Ajk}^{(1)} - \sum_{jkl} \tilde{\epsilon}_{Ajkl}^{(1)} - \sum_j \epsilon_{Aj}^{(2)}. \quad (17)$$

The attempt frequency ν_A and the contribution e_A^{sp} of the jumping atom to the saddle point energy can depend on the configuration [2]. Nevertheless, we do not have enough information to see if such a dependence holds in the case of Al-Zr alloys. We thus assume that these parameters depend only on the nature of the jumping atom, which gives us four purely kinetic parameters to adjust.

The contribution of Al to the saddle point energy, e_{Al}^{sp}, is deduced from the experimental value of the vacancy migration energy in pure Al, $E_V^{mig} = 0.61$ eV [30], and the attempt frequency ν_{Al} from the experimental Al self-diffusion coefficient, $D_{Al^*} = D_0 \exp(-Q/k_B T)$, the self-diffusion activation energy Q being the sum of the vacancy formation and migration energies in pure Al and $D_0 = 1.73 \times 10^{-5}$ m^2.s^{-1} [33].

To calculate ν_{Zr} and e_{Zr}^{sp}, we use the experimental value[1] of the diffusion coefficient of Zr impurity in Al, $D_{Zr^*} = 728 \times 10^{-4} \exp(-2.51 \text{ eV}/k_B T)$ m^2.s^{-1} [33,34]. The kinetic parameters can be deduced from this experimental data by using the five frequency model for solute diffusion in fcc lattices [35], if we make the assumption that there is no correlation effect. We check afterwards that such an assumption is valid: at $T = 500$ K the correlation factor is $f_{Zr^*} = 1$ and at $T = 1000$ K $f_{Zr^*} = 0.875$. Correlation effects are thus becoming more important at higher temperature but they can be neglected in the range of temperature used in the fitting procedure.

Table 5: Kinetic parameters for a thermodynamic description of Al-Zr binary with and without energy corrections due to order on first nearest neighbor triangle and tetrahedron.

	with order correction	without order correction
e_{Al}^{sp}	−8.219 eV	−8.219 eV
e_{Zr}^{sp}	−11.286 eV	−10.942 eV
ν_{Al}	1.36×10^{14} Hz	1.36×10^{14} Hz
ν_{Zr}	4.48×10^{17} Hz	4.48×10^{17} Hz

So as to study the influence on kinetics of energy corrections due to order on triangles and tetrahedrons, we fit another set of kinetic parameters corresponding to a thermodynamic description of Al-Zr binary with only pair interactions (*i.e.* $J_3 = J_4 = 0$, or equivalently $\tilde{\epsilon}_{ijk}^{(1)} = \tilde{\epsilon}_{ijkl}^{(1)} = 0$). This other set of kinetic parameters presented in table 5 reproduces as well coefficients for Al self-diffusion and for Zr impurity diffusion, the only difference being that these kinetic parameters correspond to a simpler thermodynamic description of Al-Zr binary.

[1]Diffusion coefficient measured in the temperature range between 800 and 910 K.

4. DIFFUSION IN SOLID SOLUTION

Diffusion in Al-Zr solid solutions can be fully characterized by the tracer correlation coefficients f_{Al} and f_{Zr} and by the phenomenological Onsager coefficients L_{AlAl}, L_{AlZr}, and L_{ZrZr}. These coefficients link fluxes of diffusing species, $\mathbf{J}_{Al}$ and $\mathbf{J}_{Zr}$, to their chemical potential gradients [36, 37] through the relations

$$\begin{aligned} \mathbf{J}_{Al} &= -L_{AlAl}\nabla\mu'_{Al}/k_BT - L_{AlZr}\nabla\mu'_{Zr}/k_BT \\ \mathbf{J}_{Zr} &= -L_{AlZr}\nabla\mu'_{Al}/k_BT - L_{ZrZr}\nabla\mu'_{Zr}/k_BT. \end{aligned} \quad (18)$$

Chemical potentials entering these equations are relative to the vacancy chemical potential, $\mu'_{Al} = \mu_{Al} - \mu_V$ and $\mu'_{Zr} = \mu_{Zr} - \mu_V$. We use to express diffusion fluxes the Onsager reciprocity condition, $L_{AlZr} = L_{ZrAl}$.

These coefficients can be used in finite-difference diffusion code so as to study "industrial" processes where diffusion is involved (precipitation, solidification, homogenization, ...) [38]. One way to obtain these coefficients is to adapt Calphad methodology to kinetics, *i.e.* to guess an expression for L_{AB} describing its variation with temperature and composition of the alloy and to adjust the model parameters on a large kinetic database [39, 40]. On the other hand one can use an atomistic model as the one described in chap. 3. to obtain the phenomenological coefficients [8, 9, 41]. If one carefully applies the same mean field approximation for thermodynamics and kinetics it is possible to get the whole Onsager matrix and not only diagonal terms and to catch all correlation effects [42]. Such an approach compared to the previous one does not need a huge experimental database. Moreover, as it is based on a realistic description of diffusion at the atomic scale, it appears safer to extrapolate kinetic quantities out of the range (composition or temperature) used in the fitting procedure.

In this study, we do not use any mean -field approximation to calculate phenomenological coefficients, but obtain them directly from kinetic Monte Carlo simulations by using generalization of the Einstein formula for tracer diffusion due to Allnatt [37, 43]

$$L_{AB} = \frac{\langle \Delta\mathbf{R}_A \cdot \Delta\mathbf{R}_B \rangle}{6\Delta t}, \qquad A, B = Al, Zr, \quad (19)$$

where the brackets indicate a thermodynamic ensemble average and $\Delta\mathbf{R}_A$ is the sum of total displacement $\Delta\mathbf{r}_i$ of all atoms i of type A during time Δt,

$$\Delta\mathbf{R}_A = \sum_{i \in A} \Delta\mathbf{r}_i. \quad (20)$$

We use residence time algorithm to run kinetic Monte Carlo calculations. The simulation box contains 125000 lattice sites, one of this site being occupied by a vacancy. Sum of total displacements $\Delta\mathbf{R}_{Al}$ and $\Delta\mathbf{R}_{Zr}$ in equation 19 are computed for a time interval corresponding to $\sim 10^6$ vacancy jumps, and their thermodynamic averages are obtained through simulations of 10^9 vacancy jumps. Such a big number of jumps is necessary to converge thermodynamic averages entering in the calculation of L_{AlZr} and L_{ZrZr}, whereas L_{AlAl} converges more quickly. This is due to the difference of diffusion coefficients between Al and Zr.

Results of calculations are presented in figure 2 for two different temperatures, $T = 1000$ K and $T = 900$ K, and different Zr concentration from 0 to 8 at.%. For the

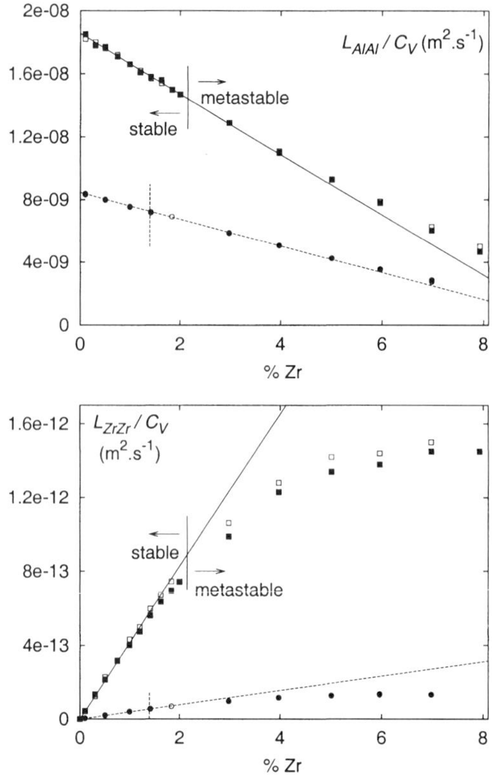

Figure 2: Onsager coefficients L_{AlAl} and L_{ZrZr}. Squares and solid lines correspond to $T = 1000$ K and circles and dashed lines to $T = 900$ K. The vertical lines indicate the corresponding solubility limit obtained from CVM calculations. Full symbols correspond to the set of parameters with order corrections on triangles and tetrahedrons and open symbols to the set without order corrections.

off-diagonal coefficient L_{AlZr} of Onsager matrix, dispersion is too important to get a precise value of thermodynamic average[2]. We interpret this as an indication that this coefficient can be neglected in this range of temperature and concentration.

Onsager coefficients are calculated for Zr concentration corresponding to the stable as well as to the metastable solid solution, the limit being given by the CVM calculations of chap. 2.2.. For calculations in the metastable solid solution, thermodynamic averages are computed during the incubation stage of precipitation kinetics when no stable precipitate is present in the simulation box (chap. 5.). L_{AlAl} behavior deviates only slightly from its linear extrapolation from the stable solid solution, but for L_{ZrZr} it

[2] $L_{AlZr} = 0 \pm 10^{-12}$ m^2.s^{-1} at $T = 1000$ K and $L_{AlZr} = 0 \pm 10^{-13}$ m^2.s^{-1} at $T = 900$ K

seems that no extrapolation from the stable to the metastable solid solution is possible.

So as to see the influence of triangles and tetrahedrons interactions, we ran simulations with only pair interactions considering the corresponding kinetic parameters of table 5. One can directly see on figure 2 that these two sets of parameters reproduce the same experimental data, *i.e.* the self-diffusion coefficient

$$D_{Al^*} = f_0 \lim_{C_{Zr} \to 0} L_{AlAl}, \tag{21}$$

where $f_0 = 0.78145$ for a fcc lattice, and the Zr impurity diffusion coefficient

$$D_{Zr^*} = \lim_{C_{Zr} \to 0} L_{ZrZr}/C_{Zr}. \tag{22}$$

Order corrections mainly affect L_{ZrZr}. This coefficient is slightly lower when one considers energy corrections due to order on triangles and tetrahedrons. The difference increases with Zr concentration and thus in the metastable solid solution: these order corrections lead to a slight slowdown of Zr diffusion. The two thermodynamic models are equivalent at these temperatures (*cf.* phase diagram on Fig. 1 (a)). As a consequence kinetic behaviors obtained from them are really close.

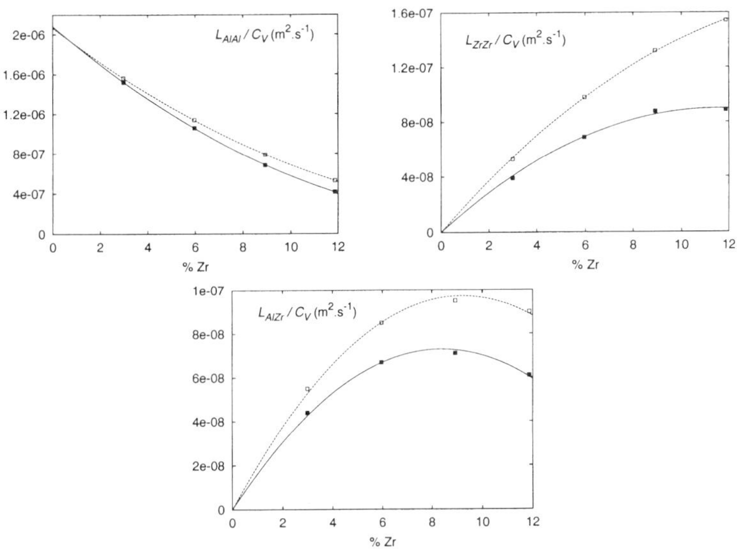

Figure 3: Onsager coefficients L_{AlAl}, L_{ZrZr}, and L_{AlZr} calculated at $T = 3000$ K. Full symbols and solid lines correspond to the set of parameters with order corrections on triangles and tetrahedrons and open symbols and dashed lines to the set without order corrections.

At higher temperatures, triangle and tetrahedron interactions change the phase diagram (Fig. 1 (a)). This thermodynamic influence leads to a kinetic change too: at

$T = 3000$ K, Onsager coefficients are lower when considering these multisite interactions (Fig. 3). One should notice that correlation effects cannot be neglected at this temperature as L_{AlZr} is far from being null. Thus one is not allowed anymore to assume Onsager matrix as diagonal. With triangle and tetrahedron interactions, Al_3Zr precipitate is more stable, which means that order effects are stronger. The kinetic corrolary of this thermodynamic influence is that they slow down diffusion.

5. KINETICS OF PRECIPITATION

Precipitation kinetics have been obtained by Monte Carlo simulations for four different supersaturations of the solid solution ($C^0_{Zr} = 5, 6, 7,$ and 8 at.%) at $T = 1000$ K. At this temperature, the equilibrium concentration is $C^{eq}_{Zr} = 2.1$ at.%. The simulation box contains 125000 lattice sites and its starting configuration is a completely disordered (random) solid solution.

5.1. Short range order parameters

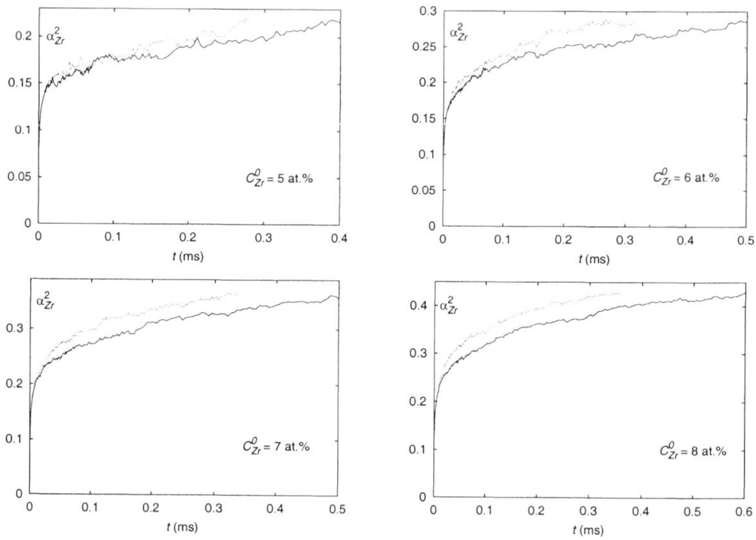

Figure 4: Evolution of second nearest neighbor short range order of Zr atoms, α^2_{Zr}, at $T = 1000$ K and four different nominal concentrations C^0_{Zr}. Full and dotted lines are respectively for the set of parameters with and without order corrections on triangles and tetrahedrons.

The quantities of interest to follow the global evolution of precipitation during the simulation are Warren-Cowley short range order (SRO) parameters [5]. SRO parameters for first-nearest neighbors evolve too quickly to give any really significant information on precipitation state. During simulation first steps, Zr atoms surround themselves with Al. Once this local equilibrium for first nearest neighborhood is reached, the corresponding SRO parameters do not evolve anymore. On the other hand, SRO parameters for second nearest neighbors slowly evolve until the end of the simulation. For Zr atoms, it is defined as

$$\alpha^2_{Zr} = \frac{\langle p_n^{Zr} \rangle_{Zr,2} - C^0_{Zr}}{1 - C^0_{Zr}}, \tag{23}$$

where $\langle p_n^{Zr} \rangle_{Zr,2}$ stands for the average of occupation numbers p_n^{Zr} on all second nearest neighbors of Zr atoms. For a randomly distributed configuration of the alloy (initial configuration) $\alpha^2_{Zr} = 0$, whereas for the $L1_2$ structure $\alpha^2_{Zr} = 1$. Looking at fig. 4, one sees that α^2_{Zr} evolves more quickly with the set of parameters with only pair interactions than with triangle and tetrahedron interactions. At first glance, this is in agreement with the slight difference on L_{ZrZr} measured in the metastable solid solution at this temperature (Fig. 2) for the two set of parameters. So as to see if the difference of precipitation kinetics can be understood only in terms of a difference of diffusion speed or is due to another factor, we measure the nucleation rate in our simulations and interpret it with classical theory of nucleation [1,44,45].

5.2. Precipitate critical size

We first need to give us a criterion to decide which atoms are belonging to $L1_2$ precipitates. As stable precipitates are almost perfectly stoichiometric at $T = 1000$ K (chap. 2.3.), we only look at Zr atoms and consider for each Zr atom in $L1_2$ precipitate that three Al atoms are belonging to the same precipitate. Zr atoms are counted as belonging to $L1_2$ precipitates if all their twelve first nearest neighbors are Al atoms and at least half of their six nearest neighbors are Zr atoms. Moreover, we impose that at least one Zr atom in a precipitate has its six second nearest neighbors being Zr, *i.e.* has a first and second nearest neighborhood in perfect agreement with the $L1_2$ structure.

Classical theory of nucleation predicts there is a critical radius, or equivalently a critical number i^* of atoms, below which precipitates are unstable and will re-disolve into the solid solution and above which precipitates will grow. i^* is obtained by considering the competition between the interface free energy σ and the nucleation free energy per atom ΔG^n,

$$i^* = \frac{2\pi}{3} \left(\frac{a^2 \sigma}{\Delta G^n} \right)^3. \tag{24}$$

Clusters of size $i < i^*$ are considered to be local variations of the solid solution composition and thus are not counted as $L1_2$ precipitates.

5.3. Nucleation free energy

The nucleation free energy per atom entering equation 24 is given by [44,45]

$$\Delta G^n = \frac{3}{4} \left(\mu_{Al}(C^{eq}_{Zr}) - \mu_{Al}(C^0_{Zr}) \right) + \frac{1}{4} \left(\mu_{Zr}(C^{eq}_{Zr}) - \mu_{Zr}(C^0_{Zr}) \right), \tag{25}$$

where $\mu_{Al}(C_{Zr})$ and $\mu_{Zr}(C_{Zr})$ are the chemical potentials of respectively Al and Zr components in the solid solution of concentration C_{Zr}, C^{eq}_{Zr} is the equilibrium concentration of the solid solution, and C^0_{Zr} the nominal concentration. The factors 3/4 and 1/4

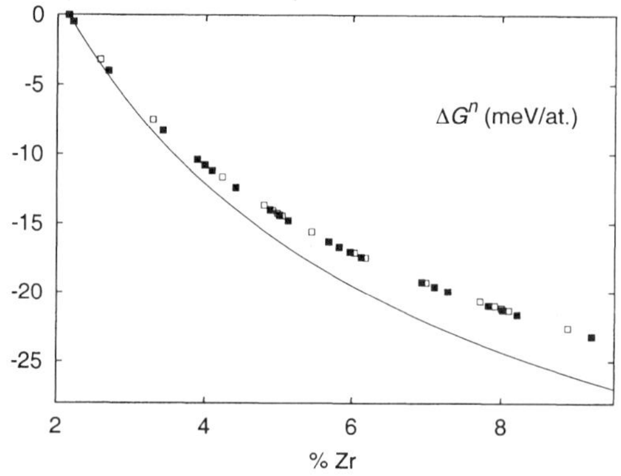

Figure 5: Nucleation free energy ΔG^n at $T = 1000$ K for different concentration of the solid solution. Square symbols correspond to CVM-TO calculation and the line to the ideal solution approximation (eq. 26). Full and open symbols are respectively for the set of parameters with and without order corrections on triangles and tetrahedrons.

arises from the stoichiometry of the precipitating phase Al_3Zr. Usually the nucleation free energy is approximated by

$$\Delta G^n = \frac{3}{4} k_B T \log \frac{1 - C_{Zr}^{eq}}{1 - C_{Zr}^{0}} + \frac{1}{4} k_B T \log \frac{C_{Zr}^{eq}}{C_{Zr}^{0}} \tag{26}$$

which is obtained by considering in equation 25 the expressions of the chemical potentials for an ideal solution. As at $T = 1000$ K we obtained the same solubility limit, $C_{Zr}^{eq} = 2.1$ at.%, with or without triangle and tetrahedron interactions (*cf.* phase diagram on fig. 1), the approximation 26 cannot be used to see if these interactions have any influence on the nucleation free energy. Therefore we use CVM-TO to calculate chemical potentials entering expression 25. Looking at figure 5, one should notice that the ideal solution approximation would have lead to an overestimation of ΔG^n, the error being $\sim 10\%$ for the maximal supersaturation considered. With CVM-TO, we do not obtain any change in the value of the nucleation free energy depending we are considering or not order corrections for first nearest neighbor triangle and tetrahedron. Thus slowdown of precipitation kinetics with these corrections cannot be explained by a decreasing of the nucleation free energy.

5.4. Interface free energy

To determine the precipitate critical size i^* using expression 24, we need to know the value of the interface free energy σ too. We calculate this energy at 0 K for different orientations of the interface. We therefore do not consider any configurational entropy and simply obtain the interface energy by counting the number by area unit of wrong "bonds" compared to pure Al and Al_3Zr in $L1_2$ structure. For (100) and (110) interfaces there is an ambiguity in calculating such an energy as two different planes, one pure Al and the other one of stoichiometry $Al_{1/2}Zr_{1/2}$, can be considered as interface. Considering $L1_2$ precipitates as stoichiometric will guarantee that to any type of the two possible interfaces is associated a parallel interface of the other type. Thus for (100) and (110) interfaces, we consider the average of these two different interface energies to be meaningful for the parameter σ entering in classical theory of nucleation. For (111) interface, as only one interface of stoichiometry $Al_{3/4}Zr_{1/4}$ is possible, we do not obtain such an ambiguity. The energies corresponding to these different interfaces are

$$\sigma_{100} = \frac{1}{\sqrt{2}}\sigma_{110} = \frac{1}{\sqrt{3}}\sigma_{111} = \frac{2\epsilon^{(2)}_{AB} - \epsilon^{(2)}_{AA} - \epsilon^{(2)}_{BB}}{2a^2}, \quad \text{with} \quad a^2\sigma_{100} = 57.0 \text{ meV}.$$

These interface energies only depend on second nearest neighbor interactions and therefore are the same with or without order corrections on first nearest neighbor triangle and tetrahedron. To determine the critical size of precipitates with equation 24 we use an interface free energy slightly higher than σ_{100}, $a^2\sigma = 64.1$ meV. With this interface free energy, nucleation rate obtained from Monte Carlo simulations are in better agreement than with σ_{100} (Fig. 7). As precipitates observed in Monte Carlo simulations do not exhibit sharp interfaces, this is quite natural to have to use an energy higher than the minimal calculated one.

5.5. Nucleation rate

Critical size for precipitates obtained from these nucleation and interface free energies are respectively $i^* = 187$, 104, 76, and 57 atoms for the different nominal concentrations $C^0_{Zr} = 5$, 6, 7, and 8 at.%. We use these critical sizes to determine the number N_p of supercritical precipitates contained in the simulation boxes, their average size $\langle i \rangle_p$, as well as the concentration of the solid solution C_{Zr}. The variation with time of these quantities are shown on fig. 6 for the simulation box of nominal concentration $C^0_{Zr} = 8$ at.%. After an incubation time, one observes a nucleation stage where the number of precipitates increases linearly until it reaches a maximum. We then enter into the growth stage: the number of precipitates does not vary and their size is increasing. At last, during the coarsening stage, precipitates are still growing but their number is decreasing. For this concentration, one clearly sees that precipitation kinetics is faster with only pair interactions as the number of precipitates is increasing more rapidly. Moreover precipitates have a bigger size than with triangle and tetrahedron order corrections.

The steady-state nucleation rate J^{st} is measured during the nucleation stage, when the number of precipitates is varying quite linearly with time. Slowdown of precipitation kinetics with triangle and tetrahedron order corrections can be seen on the steady-state nucleation rate (Fig. 7): without these corrections J^{st} is about two times higher than when these corrections are included.

In classical theory of nucleation, the steady-state nucleation rate is given by the expression [44],

$$J^{st} = N_0 Z \beta^* \exp -\frac{\Delta G^*}{kT}, \tag{27}$$

where N_0 is the number of nucleation sites, *i.e.* the number of lattice sites ($N_0 = 125000$ for Monte Carlo simulations), ΔG^* is the nucleation barrier and corresponds to the free

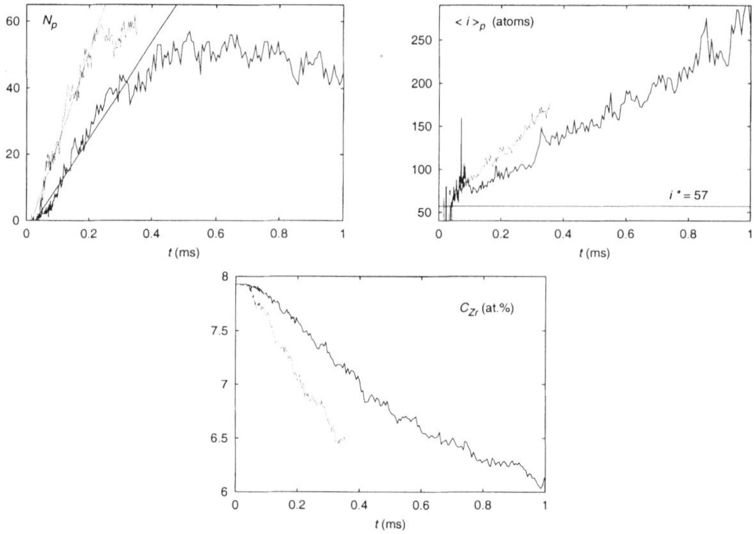

Figure 6: Kinetics of precipitation for a nominal Zr concentration C^0_{Zr} = 8 at.%: evolution with time of the number N_p of precipitates in the simulation box, of precipitates average size $< i >_p$, and of Zr concentration in the solid solution. Full and dotted lines are respectively for the set of parameters with and without order corrections on triangles and tetrahedrons.

energy of a precipitate of critical size i^*,

$$\Delta G^* = \frac{\pi}{3} \frac{(a^2\sigma)^3}{\Delta G^{n2}}, \tag{28}$$

Z is the Zeldovitch factor and describes size fluctuations of precipitates around i^*,

$$Z = \frac{1}{2\pi} \frac{\Delta G^{n2}}{(a^2\sigma)^{3/2}\sqrt{kT}}, \tag{29}$$

and β^* is the condensation rate for clusters of critical size i^*. Assuming the limiting step of the adsorption is the long range diffusion of Zr atoms in the solid solution, the condensation rate is [44]

$$\beta^* = 8\pi \frac{a^2\sigma}{\Delta G^n} \frac{D_{Zr}}{a^2} C^0_{Zr}. \tag{30}$$

Zr diffusion coefficient is obtained from our measure of Onsager coefficients in the metastable solid solution (chap. 4.). Assuming that vacancies are at equilibrium ($\mu_V = 0$), its expression is [35–37]

$$D_{Zr} = \left(L_{ZrZr} - \frac{C^0_{Zr}}{1 - C^0_{Zr}} L_{AlZr} \right) \frac{1}{k_B T} \frac{\partial \mu_{Zr}}{\partial C^0_{Zr}} \tag{31}$$

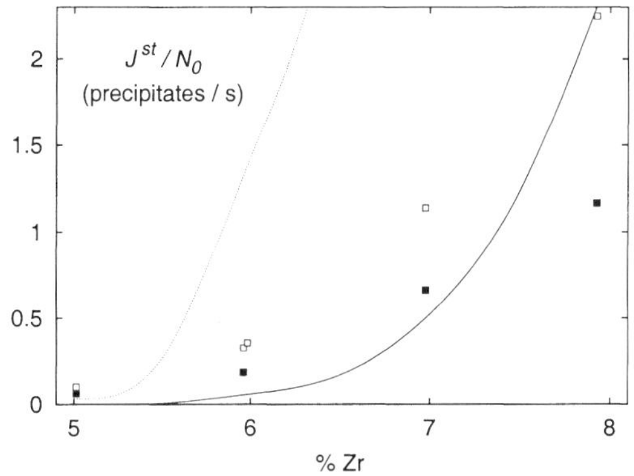

Figure 7: Evolution of the steady-state nucleation rate J^{st} with the nominal concentration for $T = 1000$ K. Full and open symbols are respectively for the set of parameters with and without order corrections on triangles and tetrahedrons. The full line corresponds to the nucleation rate predicted by classical theory of nucleation with $a^2\sigma = 64.1$ meV and the dotted line with $a^2\sigma = a^2\sigma_{100} = 57.0$ meV. J^{st} is normalized by the number of lattice sites in the simulation box, $N_0 = 125000$.

We obtain the thermodynamic factor $\partial\mu_{Zr}/\partial C^0_{Zr}$ using CVM-TO calculations. This factor is the same with or without order corrections on triangles and tetrahedrons. Therefore, the only difference these corrections induce on Zr diffusion arises from L_{ZrZr}. In classical theory of nucleation, the diffusion coefficient entering in the expression 30 of the condensation rate is only a scaling factor for time and does not have any other influence on kinetics. As a consequence the steady-state nucleation rate varies linearly with Zr diffusion coefficient as it clearly appears when combining equations 30 and 27. Thus small variations of L_{ZrZr} with the set of parameters used do not allow to explain the difference of the nucleation rate: with order corrections, L_{ZrZr} is far from being half the value it is with only pair interactions (Fig. 2). Thus slowdown of precipitation kinetics is not due to a slowdown of Zr diffusion.

One possible explanation would be a difference of the interface free energy σ. J^{st} is really sensitive to this parameter and one only needs a small decrease of σ to obtain a higher nucleation rate (see on fig. 7 the decrease of J^{st} when $a^2\sigma$ is going from 57.0 to 64.1 meV). Such a decrease would explain too why precipitates have a bigger size with only pair interactions. At $T = 0$ K, we obtain the same interface energy for all directions considered with the two sets of parameters, but at finite temperature the configurational entropy could lead to a difference of interface free energy. Nevertheless, this needs to be confirmed, by CVM calculations or using the Cahn-Hilliard method [46] for instance. Another possible explanation to understand the different kinetic pathways would be a different mobility of small clusters with the two sets of parameters. But this would be quite surprising, as we do not expect small clusters to be really mobile because of the repulsion between vacancy and Zr solute atoms.

6. CONCLUSIONS

We built an atomistic kinetic model for Al-Zr binary system using ab-initio calculations as well as experimental data. So as to be as realistic as it should be at this atomic scale, this model describes diffusion through vacancy jumps. Thanks to ab-initio calculations we could improve usual thermodynamic descriptions based on pair interactions and incorporate multisite interactions for clusters containing more than two lattice points so as to consider dependence of bonds with their local environment.

At temperatures lower than 1000 K, these energetic corrections due to local order do not modify thermodynamics: the phase diagram does not change when one does not consider these order corrections. For higher temperatures they lead to a stabilization of the ordered structure $L1_2$.

Concerning diffusion in the solid solution, these order corrections on first nearest neighbor triangle and tetrahedron do not really change the Onsager matrix, and thus diffusion characteristics. They just lead to a slight slowdown of Zr diffusion in the metastable solid solution. When looking at higher temperatures, the slowdown of Zr diffusion is more important.

For precipitation, kinetics are slower with these interactions. The slowdown is too important to be related to the small decrease of Zr diffusion in the metastable solid solution at the same temperature. One possibility would be a change of configurational entropy contribution to interface free energy.

ACKNOWLEDGMENTS

The authors are grateful to Prof. J. M. Sanchez for providing his CVM program and for his invaluable help and advices on the thermodynamics part, and to Dr F. Soisson for very useful discussions on Monte Carlo simulations, in particular for the comparison with the classical theory of nucleation. They would like to thank too Dr. B. Legrand, Dr. G. Martin, and Dr. C. Sigli for stimulating discussions. Financial support from Pechiney CRV (France) is acknowledged.

A DETAILS OF AB-INITIO CALCULATIONS

Ab initio calculations were carried out using a full-potential linear-muffin-tin-orbital (FP-LMTO) method [17–19] in the version developed by Methfessel and Van Schilfgaarde [47]. The basis used contained 22 energy independent muffin-tin-orbitals (MTO) per Al and Zr site: three κ values for the orbitals s and p and two κ values for the orbitals d where the corresponding kinetic energies were $\kappa^2 = 0.01$ Ry (spd), 1.0 Ry (spd), and 2.3 Ry (sp). A second panel with a basis composed of 22 energy independent MTO with the same kinetic energies was used to make a correct treatment of the 4p semicore states of Zr. The same uniform mesh of points was used to make the integrations in the Brillouin zone for valence and semicore states. The radii of the muffin-tin spheres were chosen to have a compactness of 47.6% for Al sites and 54.1% for Zr sites. Inside the muffin-tin spheres, the potential is expanded in spherical harmonics up to $l = 6$ and in the interstitial region spherical Hankel functions of kinetic energies $\kappa^2 = 1$ Ry and 3.0 Ry were fitted up to $l = 6$. The calculations were performed in the generalized gradient approximation (GGA) [48, 49] and the parameterization used was the one of Perdew *et al.* [50].

References

[1] F. Soisson and G. Martin, "Monte-Carlo simulations of the decomposition of metastable solid solutions: Transient and steady-state nucleation kinetics," *Phys. Rev. B*, vol. 62, pp. 203–214, 2000.

[2] Y. Le Bouar and F. Soisson, "Kinetics pathway from embedded-atom-method potential: Influence of the activation barriers," *Phys. Rev. B*, vol. 65, p. 0914103, 2002.

[3] M. Athènes, P. Bellon, and G. Martin, "Effects of atomic mobilities on phase separation kinetics: a Monte-Carlo study," *Acta mater.*, vol. 48, pp. 2675–2688, 2000.

[4] J. M. Roussel and P. Bellon, "Vacancy-assisted phase separation with asymmetric atomic mobility: Coarsening rates, precipitate composition, and morphology.," *Phys. Rev. B*, vol. 63, p. 184114, 2001.

[5] F. Ducastelle, *Order and Phase Stability in Alloys.* North-Holland, Amsterdam, 1991.

[6] Z. W. Lu, S.-H. Wei, A. Zunger, S. Frota-Pessoa, and L. G. Ferreira, "First-principles statistical mechanics of structural stability of intermetallic compounds," *Phys. Rev. B*, vol. 44, pp. 512–544, 1991.

[7] N. Saunders and A. P. Miodownik, *CALPHAD – Calculation of Phase Diagrams – A Comprehensive Guide.* Oxford: Pergamon, 1998.

[8] C. Desgranges, F. Deffort, S. Poissonet, and G. Martin, "Interdiffusion in concentrated quaternary Ag-In-Cd-Sn alloys: Modelling and measurements," in *Defect and Diffusion Science Forum*, vol. 143–147, pp. 603–608, 1997.

[9] C. Desgranges, *Compréhension et Prédiction du Comportement sous Irradiation Neutronique d'Alliages Absorbants à Base d'Argent.* PhD thesis, Université Paris XI Orsay, 1998.

[10] S. Müller, C. Wolverton, L.-W. Wang, and A. Zunger, "Prediction of alloy precipitate shapes from first principles," *Europhysics Letters*, vol. 55, pp. 33–39, 2001.

[11] M. Asta, S. M. Foiles, and A. A. Quong, "First-principles calculations of bulk and interfacial thermodynamic properties for fcc-based Al-Sc alloys," *Phys. Rev. B*, vol. 57, no. 18, pp. 11265–11275, 1998.

[12] S. Müller, L.-W. Wang, and A. Zunger, "First-principles kinetics theory of precipitate evolution in Al-Zn," *Modelling Simul. Mater. Sci. Eng.*, vol. 10, pp. 131–145, 2002.

[13] N. Ryum, "Precipitation and recrystallization in an Al-0.5 wt.% Zr alloy," *Acta Metall.*, vol. 17, pp. 269–278, 1969.

[14] J. D. Robson and P. B. Prangnell, "Dispersoid precipitation and process modelling in zirconium containing commercial aluminium alloys," *Acta Mater.*, vol. 49, pp. 599–613, 2001.

[15] E. Nes, "Precipitation of the metastable cubic Al_3Zr-phase in subperitectic Al-Zr alloys," *Acta Metall.*, vol. 20, pp. 499–506, 1972.

[16] L. Proville and A. Finel, "Kinetics of the coherent order-disorder transion in Al_3Zr," *Phys. Rev. B*, vol. 64, p. 054104, 2001.

[17] O. K. Andersen, "Linear methods in band theory," *Phys. Rev. B*, vol. 12, no. 8, pp. 3060–3083, 1975.

[18] M. Methfessel, "Elastic constants and phonon frequencies of Si calculated by a fast full-potential LMTO method," *Phys. Rev. B*, vol. 38, no. 2, pp. 1537–1540, 1988.

[19] M. Methfessel, C. O. Rodriguez, and O. K. Andersen, "Fast full-potential calculations with a converged basis of atom-centered linear muffin-tin orbitals: Structural and dynamic properties of silicon," *Phys. Rev. B*, vol. 40, no. 3, pp. 2009–2012, 1989.

[20] E. Clouet, J. M. Sanchez, and C. Sigli, "First-principles study of the solubility of Zr in Al," *Phys. Rev. B*, vol. 65, p. 094105, 2002.

[21] G. Jomard, L. Magaud, and A. Pasturel, "Full-potential calculations using the generalized-gradient corrections: Structural properties of Ti, Zr and Hf under compression," *Philos. Mag. B*, vol. 77, no. 1, pp. 67–74, 1998.

[22] S. V. Meshel and O. J. Kleppa, "Standard enthalpies of formation of 4d aluminides by direct synthesis calorimetry," *J. Alloys Compd.*, vol. 191, pp. 111–116, 1993.

[23] P. B. Desch, R. B. Schwarz, and P. Nash, "Formation of metastable $L1_2$ phases in Al_3Zr and Al-12.5%X-25%Zr (X≡Li, Cr, Fe, Ni, Cu)," *J. Less-Common Metals*, vol. 168, pp. 69–80, 1991.

[24] C. Amador, J. J. Hoyt, B. C. Chakoumakos, and D. de Fontaine, "Theoretical and experimental study of relaxation in Al_3Ti and Al_3Zr ordered phases," *Phys. Rev. Lett.*, vol. 74, no. 24, pp. 4955–4958, 1995.

[25] J. M. Sanchez, F. Ducastelle, and D. Gratias, "Generalized cluster description of multicomponent systems," *Physica*, vol. A 128, pp. 334–350, 1984.

[26] D. B. Laks, L. G. Ferreira, S. Froyen, and A. Zunger, "Efficient cluster expansion for substitutional systems," *Phys. Rev. B*, vol. 46, no. 19, pp. 12587–12605, 1992.

[27] R. Kikuchi, "A theory of cooperative phenomena," *Phys. Rev.*, vol. 81, no. 6, pp. 988–1003, 1951.

[28] J. M. Sanchez and D. de Fontaine, "Ordering in fcc lattices with first- and second-neighbor interactions," *Phys. Rev. B*, vol. 21, p. 216, 1980.

[29] T. Mohri, J. M. Sanchez, and D. de Fontaine, "Binary ordering prototype phase diagrams in the cluster variation approximation," *Acta Met.*, vol. 33, pp. 1171–85, 1985.

[30] P. Ehrhart, P. Jung, H. Schultz, and H. Ullmaier, "Atomic defects in metals," in *Landolt-Börnstein, New Series, Group III* (H. Ullmaier, ed.), vol. 25, Berlin: Springer-Verlag, 1991.

[31] O. Le Bacq, F. Willaime, and A. Pasturel, "Unrelaxed vacancy formation energies in group-IV elements calculated by the full-potential linear muffin-tin orbital method: Invariance with crystal structure," *Phys. Rev. B*, vol. 59, pp. 8508–8515, 1999.

[32] J. P. Simon, "Étude par trempe des interactions lacunes-impuretés dans les alliages dilués Al-Zr et Al-Cr," *Phys. Stat. Sol. (a)*, vol. 41, p. K107, 1977.

[33] H. Bakker, H. P. Bonzel, C. M. Bruff, M. A. Dayananda, W. Gust, J. Horvth, I. Kaur, G. Kidson, A. D. LeClaire, H. Mehrer, G. Murch, G. Neumann, N. Stolica, and N. A. Stolwijk, "Diffusion in solid metals and alloys," in *Landolt-Börnstein, New Series, Group III* (H. Mehrer, ed.), vol. 26, Berlin: Springer-Verlag, 1990.

[34] T. Marumo, S. Fujikawa, and K. Hirano, "Diffusion of zirconium in aluminum," *Keikinzoku - J. Jpn. Inst. Light Met.*, vol. 23, p. 17, 1973.

[35] J. L. Bocquet, G. Brebec, and Y. Limoge, "Diffusion in metals ans alloys," in *Physical Metallurgy* (R. W. Cahn and P. Haasen, eds.), ch. 7, pp. 536–668, Amsterdam: North-Holland, 1996.

[36] J. Philibert, *Atom Movements - Diffusion and Mass Transport in Solids*. Les Ulis, France: Les éditions de physique, 1991.

[37] A. R. Allnatt and A. B. Lidiard, *Atomic Transport in Solids*. Cambridge University Press, 1993.

[38] A. Borgenstam, A. Engström, L. Höglund, and J. Agren, "DICTRA, a tool for simulation of diffusional transformations in alloys," *J. Phase Equil.*, vol. 21, pp. 269–280, 2000.

[39] J.-O. Andersson and J. Ågren, "Models for numerical treatment of multicomponent diffusion in simple phases," *J. Appl. Phys.*, vol. 72, pp. 1350–1355, 1992.

[40] C. E. Campbell, W. J. Boettinger, and U. R. Kattner, "Development of a diffusion mobility database for Ni-base superalloys," *Acta Mat.*, vol. 50, pp. 775–792, 2002.

[41] G. Martin, "Atomic mobility in Cahn's diffusion model," *Phys. Rev. B*, vol. 41, p. 2279, 1990.

[42] M. Nastar, V. Y. Dobretsov, and G. Martin, "Self-consistent formulation of configurational kinetics close to equilibrium: the phenomenological coefficients for diffusion in crystalline solids," *Phil. Mag. A*, vol. 80, p. 155, 2000.

[43] A. R. Allnatt, "Einstein and linear response formulae for the phenomenological coefficients for isothermal matter transport in solids," *J. Phys. C: Solid State Phys.*, vol. 15, pp. 5605–5613, 1982.

[44] G. Martin, "The theories of unmixing kinetics of solid solutions," in *Solid State Phase Transformation in Metals and Alloys*, (Orsay, France), pp. 337–406, Les Éditions de Physique, 1978.

[45] D. A. Porter and K. E. Easterling, *Phase Transformations in Metals and Alloys*. London: Chapman & Hall, 1992.

[46] J. W. Cahn and J. Hilliard, "Free energy of a nonuniform system. interface free energy," *J. Chem. Phys.*, vol. 28, pp. 258–267, 1958.

[47] M. Methfessel and M. van Schilfgaarde, "Derivation of force theorem in density-functional theory: Application to the full-potential LMTO method," *Phys. Rev. B*, vol. 48, no. 7, pp. 4937–4940, 1993.

[48] P. Hohenberg and W. Kohn, "Inhomogeneous electron gas," *Phys. Rev.*, vol. 136, no. 3B, pp. B864–B871, 1964.

[49] W. Kohn and L. J. Sham, "Self-consistent equations including exchange and correlations effects," *Phys. Rev.*, vol. 140, no. 4A, pp. A1133–A1138, 1965.

[50] J. P. Perdew, K. Burke, and M. Ernzerhof, "Generalized gradient approximation made simple," *Phys. Rev. Lett.*, vol. 77, no. 18, pp. 3865–3868, 1996.

EXAMINATION OF MULTICOMPONENT DIFFUSION BETWEEN TWO NI-BASE SUPERALLOYS

C. E. Campbell*, W. J. Boettinger*, T. Hansen†, P. Merewether†, B. A. Mueller†

*Metallurgy Division, NIST, Gaithersburg, MD 20899-8555, USA.
†Howmet Corporation, Whitehall, MI, 49461, USA

ABSTRACT

The interdiffusion at 1293 °C between two multicomponent Ni-base superalloys, René-N4 and René-N5, was assessed by measuring the composition vs. distance curves and by comparing the measured curves to predictions obtained using a diffusion mobility database recently published by Campbell et al. (Acta Mat.50 (2002) 775-792). Although the diffusion database was constructed primarily from binary diffusion data, the extrapolation to the multicomponent systems gave good results in the prediction of the measured composition vs. distance curves. In addition, the location of the Kirkendall porosity on the René-N4 side of the diffusion couple was successfully predicted. This initial success points to the suitability of the general approach to the development of diffusion databases and to the desirability for additional database refinements including possible efforts from the first principles community.

INTRODUCTION

Diffusion data is needed for quantitative predictions of many materials processing and phase transformation models. In the area of superalloys alone, examples are abundant. The calculation of solid diffusion during solidification is required to predict microsegregation and the amount and type of second phase particles. Diffusion data is also required to predict incipient melting temperatures during reheating for solution treatment and to determine γ' size distributions during aging or cooling. The prediction of the phase sequence during transient liquid phase bonding and during the processing and exposure of thermal barrier coatings also requires multicomponent diffusion calculations.

Recently, there has been some effort to develop multicomponent diffusion mobility databases for the FCC phase of Ni based alloys[1-3] using the formalism put forth by Ågren[4,5] and Andersson and Ågren[6]. These databases are developed to reproduce measured tracer, intrinsic and chemical diffusion data in binary and ternary systems and to permit extrapolation to higher order systems. The development and utilization of a mobility database relies on the pre-existence of a separate thermodynamic database for the phases of interest. The thermodynamic database provides the necessary thermodynamic factors to convert from chemical potential gradients to concentration gradients.

A brief summary of the diffusion formalism of Ågren[4,5] and Andersson and Ågren[6] as applied to a disordered substitutional solid solution is useful. Diffusion is assumed to occur by a vacancy exchange mechanism, in which the equilibrium vacancy concentration is maintained. The partial molar volumes of the substitutional species are assumed to be equal. For a given phase, the flux of species i in the z-direction in volume-fixed frame of reference is given by

$$J_i = \sum_{j=1}^{n-1} \tilde{D}_{ij}^n \frac{\partial c_j}{\partial z} \tag{1}$$

where c_j is the concentration of species j, species n is the dependent (solvent) species, and $\tilde{D}_{ij}^n$ is the interdiffusion coefficient. The interdiffusion coefficient, also known as the chemical diffusivity, can be expressed as the difference between the intrinsic diffusivities.

$$\tilde{D}_{ij}^n = D_{ij} - D_{in} . \tag{2}$$

The intrinsic diffusivities, D_{ij}, are defined in terms of the atomic mobility, M_p, as

$$D_{ij} = \sum_{p=1}^{n} (\delta_{pi} - x_i) x_p M_p \frac{\partial \mu_p}{\partial x_j} \tag{3}$$

where the partial derivative of the chemical potential, μ_p, with respect to the mole fraction, x_i, defines the thermodynamic factors, which can be calculated using an appropriate multicomponent thermodynamic database, such as those by Saunders[7] and Kattner [8]. It should be noted that the thermodynamic factors must be evaluated in the form $\mu_k(x_1, x_2, \ldots x_n)$ where n is the dependent species.

The atomic mobility, M_p, of species p in a given phase is both composition and temperature dependent.

$$M_p = \Theta_p \frac{1}{RT} \exp\left(\frac{\Delta Q_p^{\bullet}}{RT} \right) \tag{4}$$

where Θ_p represents the effects of the atomic jump distance (squared) and the jump frequency and has the units of m^2/s. The variable $\Delta Q_p^{\bullet}$ is the diffusion activation energy of specie p in a given phase with units of (J/mol). The variable R is the gas constant is and the temperature, T, is in Kelvin. As $\Delta Q_p^{\bullet}$ and Θ_p can be combined into one parameter, it is customary[1] to let Θ_p equal 1 and only treat the temperature and composition dependence of $\Delta Q_p^{\bullet}$. Ågren and co-workers[3,6,9-12] expressed the composition and temperature dependence of each $\Delta Q_p^{\bullet}$ in terms of a Redlich-Kister[13] polynomial, as seen in eqn (5),

$$\Delta Q_p^{\bullet} = \sum_j x_j Q_i^j + \sum_q \sum_{j>q} x_q x_j \sum_k {}^k A_i^{qj} (x_q - x_j)^k , \tag{5}$$

where the Q_i^j and the ${}^kA_i^{qj}$ are linear functions of temperature. The expansion of the composition dependence in terms of a Redlich-Kister[13] polynomial is similar to the CALPHAD approach[14,15] used in the development of the thermodynamic databases. Note for a given diffusing species, i, that if all Q_i^j s are equal and A_i^{qj} equals zero, then $\Delta Q_p^{\bullet}$ and the corresponding M_p are not concentration dependent.

Optimized mobility functions[2] were obtained using the PARROT[16] optimization code to evaluate the composition and temperature dependence of $\Delta Q_p^{\bullet}$ using the available experimental diffusion tracer, intrinsic, and chemical diffusivity data. The experimental data were weighted giving preference to the tracer diffusivity, $D_p^{\bullet}$, data, which are independent of concentration, as seen by equation (6)

$$D_p^{\bullet} = RTM_p \ . \tag{6}$$

The current approach requires defining metastable end-member diffusion mobilities, such as the self-diffusion in fcc-W. Determination of these end-member quantities follows approaches similar to those used to determine the lattice stabilities of the metastable thermodynamic quantities of the elements[14,15]. This determination of diffusion activation energies for metastable end-member phases enables the extrapolation to higher order systems where diffusion data may be limited. Presently, the only available check on these assessed values is the application of a diffusion correlation, which states the following for pure FCC metals[17]

$$\frac{-Q}{RT_M} \approx 17 \tag{7}$$

where Q is the diffusion activation energy, T_M is the melting temperature. For Ni-W the activation energy, Q, for fcc-W was optimized to be –311420 J/mol and the metastable fcc melting temperature is 2229 K. This gives an acceptable ratio equal to 16.8. First principles calculations of these metastable quantities would be useful.

To examine the validity of this diffusion mobility database, diffusion simulations are compared to results from an experimental multicomponent diffusion couple profile.

EXPERIMENTAL AND COMPUTATIONAL PROCEDURES

René-N4 and René-N5, first-generation and second-generation superalloys respectively, were chosen as the two sides of the diffusion couple. The sample geometry of each half of the diffusion couple consisted of a 2.25 cm square with a thickness of 6.35 mm. The diffusion couples were pre-bonded at 1277 °C under a load of 180 N for 2 h. After bonding, the diffusion couples were diffusion heat-treated at 1293 °C for 10 h and 100 h in a vacuum furnace and gas-cooled. The 1293 °C temperature was chosen to ensure that both superalloys would be in the single phase γ region. The diffusion couples were characterized using scanning electron microscopy and microprobe analysis using standard ZAF correction and elemental standards. Gas cooling of the diffusion couples from 1293 °C produced coarse γ' precipitates (see Figure 1.) which resulted in significant scatter in the experimental composition profiles. This scatter is due to the chance of measuring the composition of either the γ or γ' phase, while performing the line scan. Therefore, the diffusion couples were re-heated for 1 hr at 1293 °C and water quenched to reduce the experimental scatter. The compositions of initial René-N4 and René-N5 were determined by an average of ten microprobe measurements obtained near the ends of the diffusion couples. The values are given in Table 1.

Table 1. Alloy Percent Mass Fraction Compositions. Balance Ni.

Alloy	Al	Co	Cr	Hf	Mo	Nb	Ta	Ti	Re	W
René-N4	4.23	7.79	10.29	--	1.48	0.47	4.64	3.46	--	6.38
René-N5	6.18	7.72	7.48	0.15	1.4	--	7.13	--	2.85	6.38

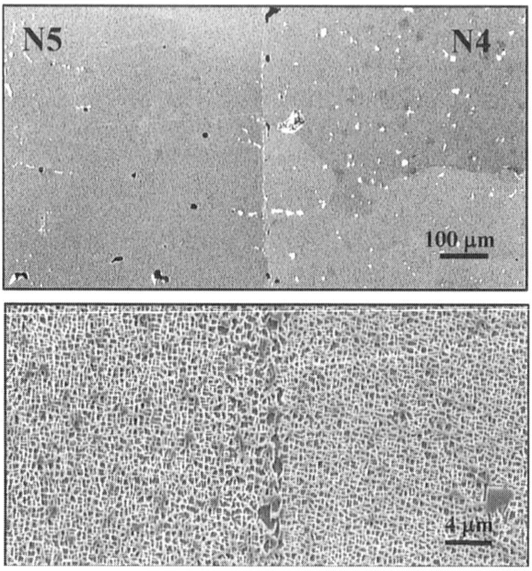

Figure 1. High and low magnification back scatter images of 10 h René-N5/René-N4 diffusion couple gas cooled from 1293 °C showing the precipitation of γ'. This coarse structure was eliminated with a post diffusion heat-treatment that included a rapid quench.

The experimental diffusion couples were simulated using the 1-D finite-difference diffusion code, DICTRA*[18], in conjunction with the thermodynamic database, Ni-Data[7], and the Ni diffusion mobility database of Campbell et al[2]. The simulation considered only single phase γ with a planar interface between René-N4 and René-N5. A 200-point geometric grid, which consisted of a higher density of grid points at the center, was used to describe the 6.35 mm couple. The calculations were done with concentration dependent diffusion coefficient matrices as described in eqns. (1-5). The matrices $\tilde{D}_{ij}^{Ni}$ for the initial compositions are shown in Tables 2 and 3.

* The use of any commercial product does not constitute an endorsement by the National Institute Standards and Technology.

Table 2. Interdiffusion Coefficients for René-N4 ($x10^{-14} m^2/s$) at 1293 °C

	Al	Co	Cr	Mo	Nb	Ta	Ti	W
Al	+119.5	+13.93	+34.83	+34.34	+42.43	+51.50	+49.51	+53.22
Co	-11.37	+17.00	-8.25	-5.67	-5.55	-1.83	-7.10	-9.69
Cr	-4.26	-5.37	+13.67	-3.21	+8.93	+9.91	+8.25	+2.49
Mo	-8.33	-0.280	-0.426	+7.57	-0.55	-0.36	-0.17	-0.45
Nb	+0.31	+0.25	+0.66	+0.27	+24.05	+0.74	+0.85	+0.31
Ta	-0.68	+0.33	+0.53	0.24	+0.26	+0.76	+0.50	+0.23
Ti	+1.63	+1.35	+4.94	+4.94	+6.25	+6.57	+23.62	+5.41
W	-1.81	-0.62	-0.55	-0.60	-1.22	-0.83	-0.70	+3.40

Table 3. Interdiffusion Coefficients for René-N5 ($x10^{-14} m^2/s$) at 1293 °C

	Al	Co	Cr	Hf	Mo	Re	Ta	W
Al	+93.16	+13.92	+33.46	-6.51	+33.42	+25.44	+48.63	+50.87
Co	-6.51	+27.22	-8.56	-27.64	+4.95	-5.11	+3.87	-9.21
Cr	+4.51	-4.23	+21.02	-6.25	-0.22	-0.78	+13.81	+6.89
Hf	+0.86	+0.07	+1.70	+262.1	+1.52	+0.87	+2.37	+1.84
Mo	-0.35	-0.30	-0.30	-1.91	+7.71	-0.25	-0.13	-0.19
Re	-0.75	-0.32	-0.36	-2.59	-0.25	+0.08	-0.51	-0.32
Ta	-0.03	+0.33	+0.98	-4.17	+0.64	+0.86	+7.75	+0.87
W	-1.18	-0.57	-0.54	-4.51	-0.39	-0.11	-0.76	+0.59

RESULTS

The experimental diffusion couples were compared to the diffusion simulations using both the composition profiles and the location of the porosity.

To compare the experimental composition profiles with the calculated profiles, the experimental Matano interface was determined by an averaging process. For three of the elements with large concentration differences (Cr, Re, and Ti), the Matano interface was independently located. With respect to the average position, the coordinates for Cr, Re, and Ti profiles for the 100 h treatment were –15 μm, -1 μm and +17 μm, respectively. These variations are small compared to the large diffusion distances.

Figure 2 shows the agreement between the experimental and calculated diffusion profiles after 10 h and 100 h at 1293 °C. In general, the calculations are in good agreement with the experiments; however, there are some discrepancies especially in the Cr and Re profiles. In the René-N4, the Cr diffuses more slowly than predicted and the Re diffuses more rapidly than predicted. To further evaluate the results, individual profiles are plotted as a function of distance over the square root of time ($z/\sqrt{t}$), as seen in Figure 3. (Note the composition differences in the Co, Mo, and Hf profiles are within the experimental error of the measurements and thus, the profiles are not evaluated.) For the given length scale and times of 10 h and 100 h for the Al, Cr, Ta, Ti, Re and W profiles the simulated profiles are independent of time.

The overall agreement is confirmed by a simple error analysis for the six profiles in Figure 3 at 100 h. This simple error analysis, shown in Table 4, consists of calculating the difference between the simulated and experimental value at each grid point, summing the differences, and then averaging over the sum of the differences by the total number of grid points. The W profile for 100 h had the smallest average error (1.8 %) and the Re profile for 100 h had the largest average (5.8 %). These errors are within the experimental scatter. Both 10 h and 100 h experimental Ta profiles show non-monotonic behavior, which is also observed in the calculated profiles but not to the same magnitude.

Table 4. Error Analysis of 100 h composition profiles

Al	Cr	Ta	Ti	Re	W
3.1 %	2.0 %	2.7 %	2.4 %	5.8 %	1.8 %

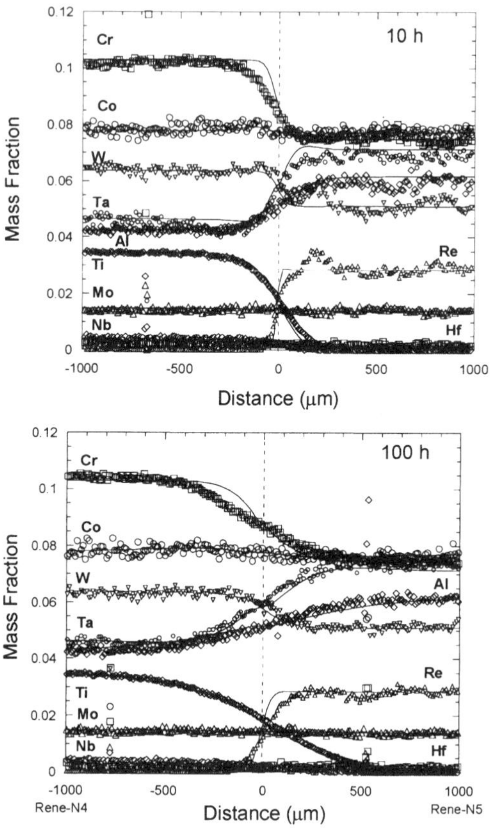

Figure 2. Calculated and experimental composition profiles for René-N4/René-N5 diffusion couples after 10 h and 100 h at 1293 °C.

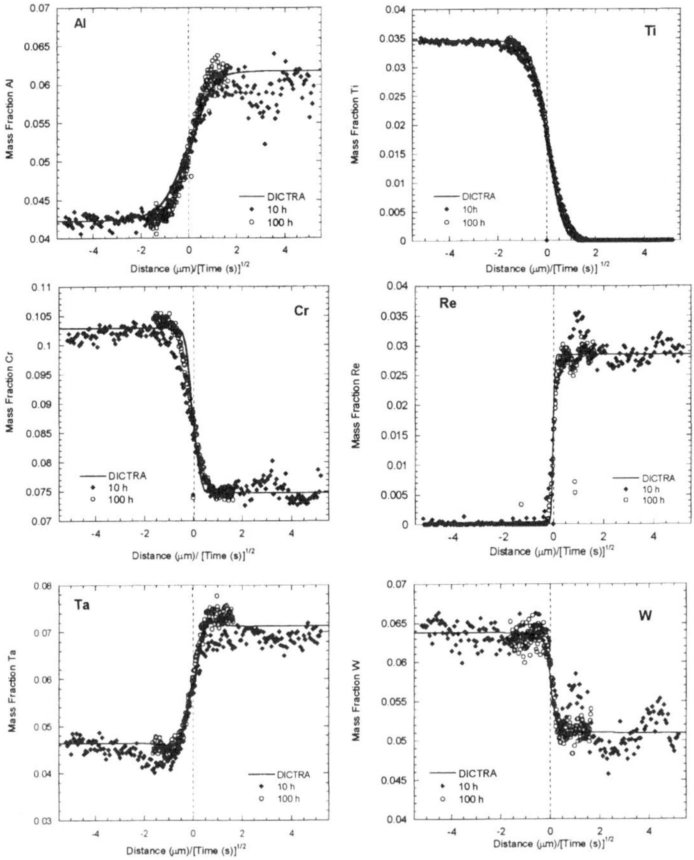

Figure 3. Composition profiles plotted as function of distance (μm) divided by the square root of time (s).

The location of porosity formed in the experimental couples can also be compared to diffusion simulations. In the lattice-fixed frame of reference, the sum of the net atom fluxes equals the vacancy flux. The negative gradient of the vacancy flux give the number of sinks (sources) necessary to maintain local equilibrium. The expected location of Kirkendal porosity is given by the position of the largest negative value of the gradient, as demonstrated by Höglund and Ågren[19]. Figure 4(a) shows this calculation for the René-N4/René-N5 couple. The location of maximum porosity is predicted to be at 65 μm to the left of the Matano interface on the René-N4 side of the couple. Porosity is observed on the René-N4 side of the interface of the experimental couple as shown in Figure 4(b). Thus, the diffusion simulation and experiment are in qualitative agreement.

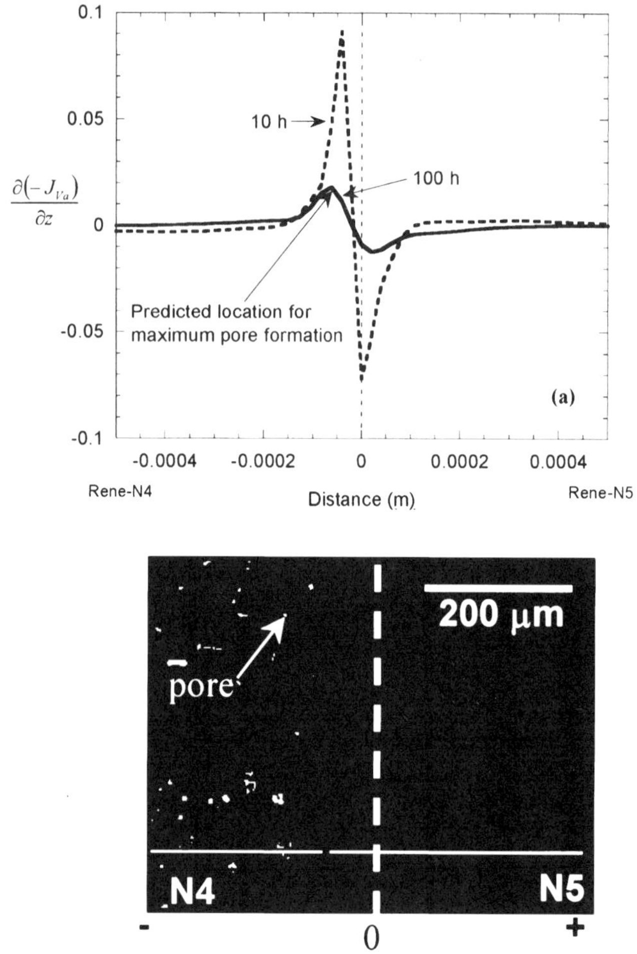

Figure 4. (a) Predicted location of where maximum pore density is expected for René-N4/René-N5 at 1293 °C. (b) Back scatter image of René-N4/René-N5 diffusion couple after 100 h at 1293 °C. Thin white line indicates position of microprobe scan. The dashed white line corresponds to the Matano interface. The dashed black line is location of the predicted maximum porosity.

CONCLUSIONS

Experimental diffusion data obtained by isothermal anneals of diffusion couples made of René-N4 and René-N5 in the single phase γ region were compared to calculations using a multicomponent diffusion mobility database. The results compare favorably. The calculations also correctly predict the location of Kirkendall porosity. This initial success points to the suitability of the general approach and the desirability of further database refinements.

ACKNOWLEGEMENTS

This work was partially supported under the GE/DARPA/AIM program. The authors would also like to thank Nigel Saunders for permission to use the Ni-Data thermodynamic database.

REFERENCES

1. M. S. A. Karunaratne, D. C. Cox, P. Carter, and R. C. Reed, Modelling of the Microsegregation in CMSX-4 Superalloy and its Homogenisation During Heat Treatment, in: *Superalloys 2000*, T. M. Pollock et al., ed., TMS, Warrendale, PA, (2000).
2. C. E. Campbell, W. J. Boettinger, U. R. Kattner, Development of a diffusion mobility database for Ni-base superalloys, *Acta Mater.* 50:775 (2002).
3. A. Engström, and J. Ågren, Assessment of diffusional mobilities in face-centered cubic Ni-Cr-Al Alloys, *Z. Metallkd.*, 87:92 (1996).
4. J. Ågren, Numerical treatment of diffusional reactions in multicomponent alloys, *J. Phys. Chem. Solids,* 43:385 (1982).
5. J. Ågren, Diffusion in phases with several components and sublattices, *J. Phys. Chem. Solids,* 43:421 (1982).
6. J.-O. Andersson, and J. Ågren, Models for numerical treatment of multicomponent diffusion in simple phases *J. Appl. Phys.*, 72:1350 (1992).
7. N. Saunders, *Superalloys 1996*, R. Kissinger, et al., eds., TMS, Warrendale, PA, (1996).
8. U. R. Kattner., Construction of a thermodynamic database for Ni-base superalloys: A case study, *CALPHAD and Alloy Thermodynamics,* P. E. A. Turchi, et al., eds., TMS, Warrendale, PA, (2002).
9. B. Jönsson, Assessment of the mobility of carbon in fcc C-Cr-Fe-Ni alloys *Z. Metallkde.*, 85:502 (1994).
10. B. Jönsson, Assessment of the mobilities of Cr, Fe, and Ni in fcc Cr-Fe-Ni alloys *Z. Metallkd,* 86:686 (1995).
11. T. Helander, and J. Ågren, A phenomenological treatment of diffusion in Al-Fe and Al-Ni alloys having B2-BCC ordered structure *Acta mater.*, 47:1141 (1999).
12. T. Helander, and J. Ågren, Diffusion in the B2-BCC phase of the Al-Fe-Ni system - Application of a phenomenological model, *Acta mater.*, 47:3291 (1999).
13. O. Redlich, and A. Kister, *Ind. Eng. Chem.*, 40:345 (1948).
14. N. Saunders, and P. Miodownik, *CALPHAD: A Comprehensive Guide,* Elsevier Science Inc., New York, (1998).
15. L. Kaufman, and H. Bernstein, *Computer Calculation of Phase Diagrams*, Academic Press, London, (1970).
16. B. Jansson, Internal Report Trita-Mac-234, Royal Institute of Technology, Stockholm, Sweden, (1984).
17. A. M. Brown, and M. F.Ashby, Correlations for diffusion constants, *Acta metal.*, 28:1085 (1980).
18. A. Borgenstam, A. Engtröm, L. Höglund, and J. Ågren, DICTRA, a tool for simulation of diffusion transformations in alloys, *J. Phase Equilibria,* 21:269 (2000).
19. L. Höglund, and J. Ågren, J., Analysis of the Kirkendall effect, marker migration and pore formation, *Acta Mater.*, 49:1311 (2001).

Curvature and basis function effects on electronic and transport properties of carbon nanotubes

Antonis N. Andriotis
Institute of Electronic Structure and Laser
Foundation for Research and Technology-Hellas
P.O. Box 1527, 71110 Heraklio, Crete
Greece

Madhu Menon
Center for Computational Sciences
325 McVey Hall
University of Kentucky
Lexington, KY 40506
USA

Abstract

Curvature effects are shown to distinguish carbon nanotubes from graphitic carbon in qualitative ways. In particular, bonding geometries and magnetic moments are found to be sensitively dependent on curvature. Furthermore, our work also reveals that use of full orbital basis set is necessary for realistic calculations of quantum conductance of carbon nanotubes.

1 Introduction

Since the discovery of carbon nanotubes by Iijima,[1] a rich variety of carbon nanotube morphologies have been experimentally observed. The carbon nanotubes consist of rolled up graphene sheet with various chiralities. The unusual electronic properties of single wall carbon nanotubes (SWCN) show great promise in their potential for use in molecular electronic devices. Drawing on the similarities between graphite and SWCN, many authors have attempted to extrapolate known results for graphite for use in the case of SWCNs. In this work we investigate the usefulness and limitations of such extrapolation. We also show that use of full basis set enables obtaining structural relaxation and quantum conductivity calculations using the *same* Hamiltonian, thereby providing consistency in predicting physical properties of real nano-systems using theoretical simulations.

One of the most efficient and, consequently, very popular calculational scheme for studying the electronic and transport properties of SWCNs is the use of tight-binding (TB) Hamiltonian with only one π-electron orbital (or p_z) per atom. This approximation has been shown to accurately reproduce the band structure of the graphene layer and/or that of graphite near the Fermi energy, E_F. This approximation also successfully reproduces the band structure of the infinite SWCNs near E_F. Following this success, it was argued that the effect of the σ-bond and, therefore, the use of an enlarged basis functions set is of no practical significance when calculating physical quantities for SWCNs. However, careful *ab initio* calculations and semi-empirical ones employing the TB approximation have revealed that the one π-orbital approximation is not sufficient enough if accurate results are desired. This is because the curvature of the carbon nanotube wall affects the C-C hopping integral. As a result, curvature effects are found to be responsible for gap opening at E_F in achiral metallic SWCNs.[2] The importance of the s and p re-hybridization was also emphasized by Choi *et al*,[3] who attributed the splitting of the π and π^* bands to this re-hybridization. Furthermore, it was also found that the $\sigma - \pi$ representation is very important for describing SWCNs with point defects (i.e. SWCNs exhibiting bent parts, vacancies, adsorbed gases on their walls etc).[3, 4]

In addition to the basis-set problem, it should be noted that band structure results (as well as results derived from these, e.g., transport data) are usually obtained by considering tubes of infinite length. The infinite-tube results, however, are quite different from those obtained for tubes of finite length.[5, 6, 7] Taking into account that realistic calculations have to be performed for tubes of finite length and that *ab initio* methods are not easily applicable to systems consisting of a moderate number of atoms ($\approx$ 1000 atoms are necessary for simulations of finite tubes), it can be seen that a practical calculation would require use of semi-empirical methods such as the TB.

It should also be noted that full symmetry unconstrained structural relaxation is essential before the theoretical determination of any physical properties. In quantum mechanical simulations, use of a Hamiltonian with only one π-electron orbital is untenable for dynamical relaxations even for graphite. Most authors attempt to circumvent this problem by using classical many body potentials for obtaining relaxation while still making use of the π-electron orbital approximation for conductivity calculations. Use of two completely different methods for the same system, can introduce inconsistency in the prediction of physical properties.

In this report we show that graphene (or graphite) results cannot be simply extrapolated and applied to curved systems such as SWCNs, C_{60}, other fullerenes and fullerites etc. Furthermore, it will be shown that the tight-binding Hamiltonian using the full basis set can be efficiently incorporated into computational schemes like the surface Green's function matching (SGFM) method to obtain transport properties of ideal and defected SWCNs while also providing accurate structural relaxation for the same systems.

2 Tight-Binding Molecular Dynamics Methodology

In this section we give brief overviews of our theoretical simulations methods.

The details of our tight-binding molecular dynamics scheme can be found in Refs. [8]. Here we give a brief overview.

The total energy U is written in its general form as a sum of several terms,[8]

$$U = U_{\rm el} + U_{\rm rep} + U_0, \tag{1}$$

where $U_{\rm el}$ is the sum of the one-electron energies E_n for the occupied states:

$$U_{\rm el} = \sum_n^{occ} E_n. \tag{2}$$

In the tight-binding scheme E_n is obtained by solving the characteristic equation:

$$(\mathbf{H} - E_n \mathbf{1})\mathbf{C}^n = 0, \tag{3}$$

where $\mathbf{H}$ is the Hamiltonian of the system.

The Hellmann-Feynman theorem for obtaining the electronic part of the force is given by[8],

$$\frac{\partial E_n}{\partial x} = \mathbf{C}^{n\dagger} \frac{\partial \mathbf{H}}{\partial x} \mathbf{C}^n. \tag{4}$$

The total energy expression also derives contributions from ion-ion repulsion interactions. This is approximated by a sum of pairwise repulsive terms and included in $U_{\rm rep}$. This sum also contains the corrections arising from the double counting of electron-electron interactions in $U_{\rm el}$[8]. U_0 is a constant that merely shifts the zero of energy. The contribution to the total force from $U_{\rm rep}$ is rather straightforward. One can then easily do molecular dynamics simulations by numerically solving Newton's equation,

$$m\frac{d^2x}{dt^2} = F_x = \frac{\partial U}{\partial x} \tag{5}$$

to obtain x as a function of time.

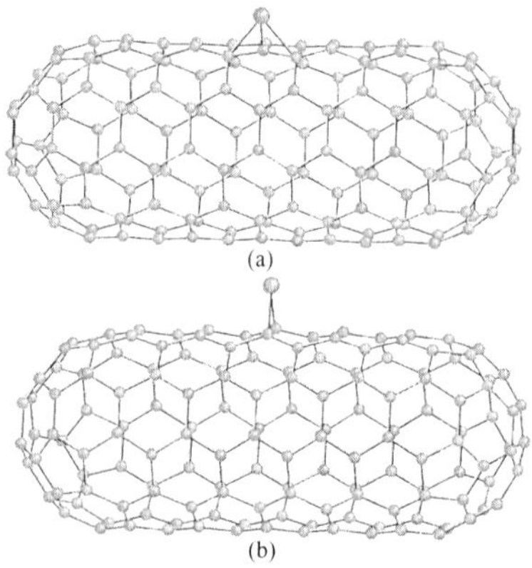

Figure 1: The two stable binding sites for a single Ni on graphite; (a) hole site and (b) atop site.

3 Interaction of Transition Metal Atoms with graphite, C_{60} and SWCNs

In this section we will demonstrate how curvature affects the bonding properties of TMAs on low dimension carbon surfaces.

3.1 Interaction of Ni and V with graphite

In the present work, the graphite is simulated by a portion of a graphene sheet consisting of 128 carbon atoms. This size was found to be sufficient for ensuring convergence of the results with the cluster size. The relaxation of the transition metal atom on the graphene layer is simulated with our TBMD computational method. Three distinct sites were considered for a single transition metal atom (Ni or V) on graphite, nanotube wall and on a C_{60} molecule. These consist of a Ni atom (i) directly above a C atom (atop site), (ii) above the center of an hexagon (hole site) and, (iii) over a C-C bond (bridge site).

At the atop site, the Ni atom induces a significant relaxation in the graphene layer that moves the C-atom (lying just beneath the Ni atom) below the graphene

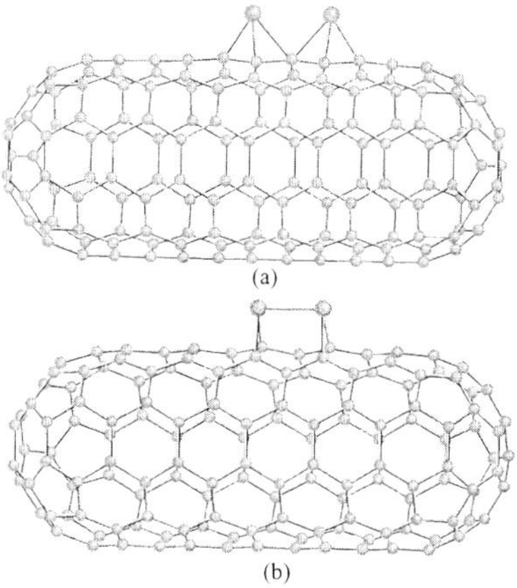

Figure 2: The two stable binding sites for a single Ni on carbon nanotube wall; (a) atop site and (b) bridge site.

plane. The Ni atom bonds in addition to the C atom beneath it and to the three next nearest C-atoms (Fig. 1a). At this position, the Ni atom looses 0.98 e of charge to the carbon atoms and also exhibits a magnetic moment of 0.3 μ_B.

At the hole site (Fig. 1b), the Ni forms six bonds with the six nearest carbon atoms on relaxation. The strong Ni-C interactions results in considerable distortions in the planar structure beneath the Ni atom. The Ni atom exhibits a magnetic moment of $0.29\mu_B$ and looses charge to the carbon atoms (approximately 0.44 e). The hole site was found energetically more stable than the atop site.

Similarly, it was found that the adsorption of V was accompanied by a considerable distortion of the graphene sheet, especially for C atoms in the neighborhood of the V atom. As in the case of the Ni atom, the hole site was found the most stable. The total energies for the fully relaxed geometries were obtained using *ab initio* calculations and show the ordering; $E_{hole} < E_{atop}$. At each adsorption site, the V atom exhibits a magnetic moment and there is appreciable charge transfer to or from the graphite atoms. However, our results do not support the high spin states for V on graphite reported recently.[9, 10] This disagreement may be attributed to surface relaxation effects which have been completely omitted in references [9] and [10].

There was no bonding found for both Ni and V on the C-C bridge site of graphite.

3.2 Interaction of Ni and V with C_{60} and SWCNs

In the case of interaction of Ni with the C_{60} and the SWCN, only the atop and bridge sites were found to be stable on molecular dynamics relaxation. The hole site (i.e. above the center of a hexagonal ring of carbon atoms) is found to be unstable for Ni on the nanotube wall. Instead, a Ni atom, initially placed at a hole site, moves and relaxes on an atop site. This is in striking contrast to the case of the interaction of Ni with graphite where the hole site was found to be the most stable.

At the atop site (Fig. 2a) the Ni atom forms three Ni-C bonds (1.79, 1.95, 1.95 $\mathring{A}$) and gains electronic charge (≈ 0.05 e) from the carbon atoms while displaying a magnetic moment of $0.15\mu_B$.

The Ni atom relaxed at the bridge site (Fig. 2b) forms bonds of length 1.76 $\mathring{A}$ each with the carbon atoms. There is a gain of electronic charge of magnitude 0.179 e. The magnetic moment on the Ni atom is $0.10\mu_B$. The atop site is energetically more favorable by 9.1 eV over the bridge site.

For V on a C_{60} or on a SWCN, we find that V binds at hole, atop, and bridge sites, while the total energies for these sites satisfy the relation $E_{hole} < E_{atop} < E_{bridge}$. Note that for Ni, the hole site was found to be unstable, while the atop site was the most stable on C_{60}.[11] The only similarity here is that the bridge site now becomes stable for both V and Ni as a result of re-hybridization due to the substrate-curvature. It is also worth noting that our results for V on both graphite and C_{60} indicate the preference for V to act as an η^6 ligand in contradistinction with Ni which acts as an η^2 or η^3 ligand[11, 12], in agreement with the experimental findings for both Ni and V interacting with C_{60}.[13, 14, 15, 16, 17] The V atom on the C_{60} exhibits a net charge and magnetic moment that depend on the adsorption site.

The qualitatively different behavior found in the case of V and Ni in their interactions with graphite and the C_{60} can be attributed mainly to the different occupancies of the adsorbate d-orbitals. Another factor contributing to this is the variation of the hybridization strength between the adsorbate d-orbitals and the p_z-orbitals of graphite (the z-axis is perpendicular to the surface). While the occupancy of the d-orbitals depends on the adsorbate atom and is affected by inter-atomic and intra-atomic charge transfer effects, the hybridization strength depends on the point group symmetry of the adsorption site (i.e,, C_{6v} for hole; C_{3v} for atop, and C_{2v} for bridge sites), the surface relaxation near the adsorbate, and the adsorbate-substrate distance. Presence of all these factors make a meaningful quantitative deduction of the contribution of each single factor to the interaction between a $3d$-element and the graphite (or the C_{60}) seem quite difficult.

The bonding differences found for Ni and V on a SWCN, lead us to believe that these may also manifest as the differences in the contact resistances when these transition metal atoms attach themselves to SWCN walls.[18]

4 Quantum Conductivity of SWCNs

In this section we describe a state-of-the-art formalism to efficiently calculate the quantum conductivity of SWCNs and present results using this formalism and compare them with available experiments. We also show that inclusion of full basis set enables structural relaxation and quantum conductivity calculations using the *same* Hamiltonian, providing a consistent approach for the treatment of SWCNs.

4.1 The Surface Green's Function Matching Method

This is an Embedding technique based on the work of Inglesfield[19] as extended by Fisher.[20] We consider the case of a finite length SWCN connected at both ends to semi-infinite metallic leads. Within an embedding scheme, these metallic leads take the place of a host lattice in which the nanotube is assumed to be embedded. We then construct a boundary surface **S** which separates the embedded system (tube) from the host lattice (leads). According to Inglesfield-Fisher scheme,[19, 20, 21] the effect of the host lattice can be efficiently incorporated into the bare tube Hamiltonian through the Green's function of the free host crystal Green's function satisfying Dirichlet's condition on the boundary surface **S**. In particular, if $G(\mathbf{r}_1, \mathbf{r}_2; E)$ is the Green's function of the host satisfying the Dirichlet's boundary condition on **S**, then the host-tube interaction, $\Sigma_S(\mathbf{r}_1, \mathbf{r}_2; E)$, ($\mathbf{r}_1, \mathbf{r}_2$, defined on **S**), can be evaluated from the following formula:

$$\Sigma_S(\mathbf{r}_1, \mathbf{r}_2; E) = -\frac{1}{4}\frac{\partial^2}{\partial n_1 \partial n_2} G(\mathbf{r}_1, \mathbf{r}_2; E) \tag{6}$$

where $\partial/\partial n$ denotes the derivative normal to **S**.

Here $\Sigma_S(\mathbf{r}_1, \mathbf{r}_2; E)$ is the electron self-energy (SE) term that describes the lead-tube interaction if $G(\mathbf{r}, \mathbf{r}'; E)$ is taken to be the Green's function for the unperturbed lead(s). The problem of constructing the lead-tube interaction, therefore, becomes a problem of constructing the Green function of a semi-infinite metal-lead satisfying the Dirichlet's condition on the lead-tube interface. This state-of-the-art embedding scheme, thus, allows a realistic description of the lead-tube interaction.

In order to utilize the Inglesfield-Fisher[19, 20, 21] embedding approach to calculate the tube-lead interaction in the present situation (SWCN of finite length

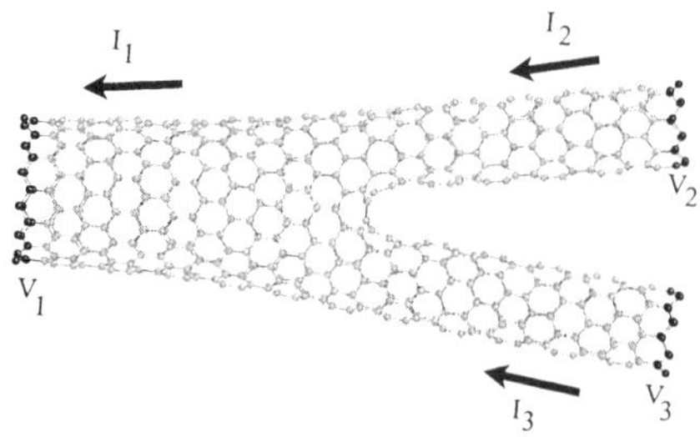

Figure 3: A carbon nanotube Y-junction consisting of a (14,0) tube branching into two (7,0) tubes with an acute angle between them. The structure contains 704 carbon atoms with six heptagons clustered in the middle and no pentagons. The dark-colored atoms are those participating in the metal-tube interaction.

connected at both ends to two semi-infinite metal leads), we take each metal lead as a semi-infinite metal and calculate its Green's function utilizing the recently proposed method of Sanvito *et al.*[22] From this Green's function, we then calculate the tube-lead interaction according to Eqn. 6. This is repeated for the other lead as well.

According to the embedding scheme employed here, the SE given is a local energy-dependent interaction acting only along the boundary surface **S**. This means that $\Sigma_S(E)$ will affect only the TB parameters of those tube atoms that lie on the boundary surface **S**. By repeating the same procedure, as described above, for every metal-lead, we can construct the SE's that will act on all tube-atoms that are in contact with all metal-leads.

We let $\Sigma_L = \Sigma_L(\mathbf{r}_S, \mathbf{r}'_S; E, V_b)$, and $\Sigma_R = \Sigma_R(\mathbf{r}_S, \mathbf{r}'_S; E, V_b)$, with $\mathbf{r}_S$, $\mathbf{r}'_S$ taken on the boundary surface(s), be the self-energies due to the left and right metal-tube interactions, respectively, under the bias-voltage V_b. In terms of Σ_L and Σ_R, the Green's function, G_C, of the conducting tube is obtained as follows[24, 25, 26]:

$$G_C = (E - H_C - \Sigma_L - \Sigma_R)^{-1} \quad , \tag{7}$$

where H_C is the Hamiltonian of the isolated tube.

Having evaluated G_C, the transmission function T(E) can now be obtained from the following equation [24] :

$$T(E, V_b) = tr[\Gamma_L G_C \Gamma_R G_C^\dagger] \ , \tag{8}$$

where

$$\Gamma_j(E; V_b) = i(\Sigma_j - \Sigma_j^\dagger) \ , j = L, R \ . \tag{9}$$

Having obtained T(E,V_b), we proceed with the evaluation of the current-voltage, I-V, characteristic of the tube utilizing the following formula for calculating the current:[24]

$$I(V_b) = \frac{2e}{h} \int_{-\infty}^{+\infty} T(E, V_b) \left[f_E(\mu_L) - f_E(\mu_R) \right] dE \ , \tag{10}$$

where $\mu_i = E_F - eV_i$; i=L,R, where V_L and V_R are applied voltages on the left and the right metal leads, respectively, e>0 is the electron charge, E_F the Fermi energy of the free tube and $f_E(\mu)$ the Fermi distribution. In the following we put V_b=V_L-V_R to be the bias-voltage.

Due to the lack of self-consistency, the bias-voltage V_b cannot be treated as an independent parameter of the system; the distribution of the voltage drop along the tube and the actual values of V_L and V_R relative to the Fermi energy are not strictly defined. Following Tian et al,[27] we introduce a division parameter η and set

$$V_L = E_F - \eta V_b \ , \tag{11}$$

$$V_R = E_F + (1 - \eta) V_b \ . \tag{12}$$

By putting $\eta = 0.5$ we achieve a symmetric division of V_b between the two contacts and we use this feature in all our subsequent applications in the present work.

In the tight-binding formulation used in the present work both the Hamiltonian and the Green's functions are each taken to be $N_{at}N_{orb} \times N_{at}N_{orb}$ matrices, where N_{at} is the number of atoms in the embedding subspace and N_{orb} is the number of orbitals on each atom. Contrary to most previous works on quantum transport which use only one π-electron orbital per atom, we use N_{orb}=4 for semiconductor atoms that includes 1-*s* and 3-*p* orbitals. Additionally, we use N_{orb}=9 for transition metal atoms (taken to be the material of the leads) that includes 1-*s*, 3-*p* and 5-*d* orbitals. The use of all these orbitals is necessary in order to allow for the correct description of the inter-atomic interactions in semiconductor, transition metal and their hetero-systems[11, 12, 28, 29]. Furthermore, the Hamiltonian used for performing conductivity calculations is the *same* Hamiltonian used for performing molecular dynamics relaxation of all systems on which conductivity calculations are to be performed.

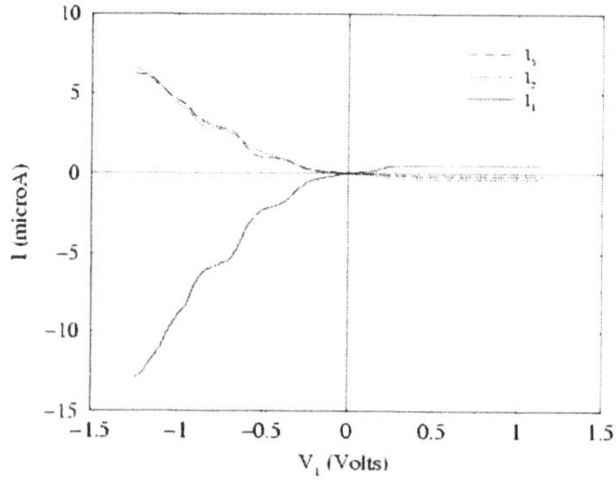

Figure 4: The calculated I-V curves for the Y-junction shown in Fig. 3 showing asymmetric behavior and rectification. The current directions and the voltages on the three arms are also indicated in Fig. 3. The voltage configuration for this plot has been set to V_2=V_3=0.0V, making it a two terminal device for enabling direct comparison with experiment.

4.2 Applications of the Quantum Conductivity Method

Very recently, experimentalists have succeeded in developing template-based chemical vapor deposition (CVD),[30] and pyrolysis of organometallic precursor with nickelocene and thiophene techniques[31] that allows for the reproducible and high-yield fabrication of carbon nanotube Y-junctions.[30, 31] The conductance measurements on these Y-junctions have shown intrinsic rectifying I-V behavior at room temperature.[31, 32]

Motivated by these experimental works we have applied our quantum conductivity formalism in Sec. 4.1 to a Y-junction consisting of a (14,0) tube branching into two (7,0) tubes with an acute angle between them (shown in Fig. 3), closely resembling the Y-junctions produced in experiments[30, 31, 32]. The Y-junction shown in Fig. 3 was fully relaxed using our tight-binding molecular dynamics scheme. The dark colored carbon atoms in Fig. 3 are in contact with the metal leads. All leads are taken to be from Ni metal and the Ni-C interaction is fully incorporated into our tight-binding Hamiltonian used in our previous works.[33]

The I-V characteristics for the Y-junction in Fig. 3 have been calculated using our formalism[18, 34]. In Fig. 4, we show the calculated currents in the three arms of the Y-junctions as a function of the bias voltage V_1. The current directions and the voltages on the three arms are indicated in Fig. 3. The current is taken to be positive when flowing towards the junction and negative otherwise. The voltage configuration for this plot has been set to $V_2=V_3=0.0V$. This setup makes the Y-junction a two terminal device for the investigation of rectifying behavior and allows direct comparison with the experimental results reported by Papadopoulos *et. al.*,[32]. As seen in the figure, there is a substantial increase in the current for negative values of the bias voltage V_1, while for positive values of V_1 the current is negligible. The I-V characteristics, thus, display a distinct asymmetry and rectifying behavior. This is in excellent agreement with the experimentally measured I-V curve of Papadopoulos *et. al.*,[32] for their CVD grown Y-junctions establishing the validity of our formalism..

5 Summary

In summary, using tight-binding molecular dynamics simulations, we have demonstrated qualitative differences in the physical properties of carbon nanotubes and graphitic carbon. Furthermore, we have presented an efficient Green's function formalism for calculating the quantum conductance of SWCNs. Our work reveals that use of full orbital basis set is necessary for realistic calculations of quantum conductance of carbon nanotubes. Furthermore, our approach allows us to use the *same* Hamiltonian to calculate quantum conductivity as well as to perform structural relaxation.

6 Ackowledgements

The present work is supported through grants by NSF Grant (NER-0165121, ITR-0266277), DOE Grant (00-63857), NASA Grant (02-465679), EU-GROWTH research project (G5RD-CT-2001-00478) and the University of Kentucky Center for Computational Sciences.

References

[1] S. Iijima, *Nature* **354**, 56 (1991).

[2] R.Saito, G.Dresselhaus and M.S.Dresselhaus, Phys. Rev. B**61**, 2981 (2000).

[3] H.J.Choi, J.Jhm, S.G.Louie and M.L.Cohen, Phys. Rev. Lett., **84**, 2917, (2000).

[4] A.Rochefort, Ph.Avouris, F.Lesage and D.R.Salahub, Phys. Rev. B**60**, 13824 (1999).

[5] D.Orlikowski, M.B.Nardelli, J.Bernholc and C.Roland, Phys. Rev. B**61**, 14194 (2000).

[6] A.N.Andriotis, M.Menon and G.E.Frudakis, Phys. Rev. Lett., **85**, 3193 (2000).

[7] A. N. Andriotis, M. Menon, D. Srivastava, and L. Chernozatonskii, Phys. Rev. Lett. **87**, 066802 (2001).

[8] A. N. Andriotis and M. Menon, Phys. Rev. B**57**, 10069 (1998).

[9] D.M.Duffy and J.A.Blackman, Phys. Rev. B**58**, 7443 (1998).

[10] P.Kruger, J.C.Parlebas and A.Kotani, Phys. Rev. B**59**, 15093 (1999).

[11] A.N.Andriotis, M.Menon and G.E.Froudakis, Phys. Rev. B**62**, 9867 (2000).

[12] A. N. Andriotis and M. Menon, Phys. Rev. B**60**, 4521 (1999).

[13] S.Nagao, T.Kurikawa, k.Miyajima, A.Nakajima and k.Kaya, J. Chem. Phys. A**102**, 4495 (1998).

[14] T.Kurikawa, S.Nagao, K.Miyajima, A.Nakajima and K.Kaya, J. Chem. Phys. A**102**, 1743 (1998).

[15] A.Nakajima, S.Nagao, H.Takeda, T.Kurikawa and K.Kaya, J. Chem. Phys. **107**, 6491 (1997).

[16] E.K. Parks, K.P. Kerns, S.J. Riley and B.J. Winter Phys. Rev. B**59**, 13431 (1999).

[17] P.Mathur, I.J.Mavunkal and S.B.Umbarkar, Journal of Clus. Sci., **9**, 393 (1998).

[18] A. N. Andriotis, M. Menon and G. E. Froudakis, Appl. Phys. Lett., **76**, 3890, (2000).

[19] J. E. Inglesfield, J. Phys. C **14**, 3795 (1981).

[20] A. J. Fisher, J. Phys.: Condens. Matter, **2**, 6079 (1990).

[21] A. N. Andriotis, Europhys. Lett. **17**, 349 (1992); A.N.Andriotis, J. Phys. Condens. Matter, **2**, 6079 (1990).

[22] S.Sanvito, C.J.Lambert, J.H.Jefferson and A.M.Bratkovsky, Phys. Rev. B**59**, 11936 (1999).

[23] A. N. Andriotis and M. Menon, J. Chem. Phys., **115**, 2737 (2001).

[24] S. Datta, in *Electronic Transport in Mesoscopic Systems*, Cambridge Univ. Press, Cambridge (1995).

[25] M.P.Samanta, W.Tian, S.Datta, J.I.Henderson and C.P.Kubiak, Phys. Rev. B**53**, R7626 (1996).

[26] M.B.Nardelli, Phys. Rev. B**60**, 7828 (1999).

[27] W. Tian, S. Datta, S. Hong, R. Reifenberger, J. I. Henderson and C. P. Kubiak, J. Chem. Phys. **109**, 2874 (1998).

[28] Antonis N. Andriotis, M. Menon, and George E. Froudakis, Phys. Rev. B**61**, R13393 (2000).

[29] A. N. Andriotis, M. Menon, and G. E. Froudakis, Phys. Rev. Lett., **85**, 3193 (2000).

[30] J.Li, C.Papadopoulos and J.Xu, *Nature*, **402**, 253 (1999).

[31] B. C. Satishkumar, P. J. Thomas, A. Govindraj, and C. N. R. Rao, Appl. Phys. Lett. **77** 2530 (2000).

[32] C.Papadopoulos, A.Rakitin, J.Li, A.S.Vedeneev and J.M.Xu, Phys. Rev. Lett. **85**, 3476 (2000).

[33] G.E.Froudakis, M.Muhlhauser, A.N.Andriotis and M.Menon, Phys. Rev. B**64**, 2421401(R) (2001).

[34] A. N. Andriotis, M. Menon, D. Srivastava and L. Chernozatonskii, Phys. Rev. Lett., **87**, 066802 (2001).

THE BEHAVIOR OF SOLID SOLUTIONS IN GEOLOGICAL TRANSPORT PROCESSES : THE QUANTIZATION OF ROCK COMPOSITIONS BY FLUID-ROCK INTERACTION

Bernard GUY[1]

[1]Ecole nationale supérieure des mines de Saint-Etienne
158 Cours Fauriel, 42023 Saint-Etienne cédex 2, France

INTRODUCTION: REACTIVE TRANSFER PROCESSES, ASSUMPTIONS FOR MODELLING

Most of the minerals found in rocks are solid solutions that are partly similar to alloys. In the present paper, we will not discuss the prevision of the thermodynamic properties of solid solutions by means of physical calculations at atomic scale. We will rather assume that these properties are known, and focus on their links with the behavior of solid solutions at the rock scale. We will particularly study the chemical transformation of rocks induced by aqueous fluids in disequilibrium with the rocks and flowing through their porosity. In geological literature, one speaks of *metasomatic* processes to characterize the evolution of open geological systems, in the thermodynamic sense, i.e. involving exchange of matter with an external medium: the composition of the rock is modified. In materials science, this would correspond to a coupling of corrosion with advection. An interesting phenomenon may be observed: the transformation may be organized in zones or steps separated by sharp contacts ("metasomatic zoning", Korzhinskii, 1970). Thereby, a quantization effect happens in the sense that all the compositions are not equally likely: some compositions may be selected by the process and others may be excluded. The overall transformation is due to the coupling between chemical exchange and matter transport. Through the transformation fronts, one observes sharp changes in the mineralogical and chemical composition of the rocks (Fig. 1). Within the various zones, one can often observe continuous variations of the composition of the minerals and the rocks. In the present paper, we will put these continuous and discontinuous variations of the composition of the rocks in relation to the "détente"- and shock- waves of the hyperbolic problems that are used to model the geological problems.

The model system will be composed of two phases: a solid phase, assimilated to one solid solution of specified type, with a given number of independent degrees of freedom; and an aqueous liquid phase containing various species. Under conditions of high temperature and slow fluid migration that we will consider here, the hypothesis of local equilibrium will be taken, so that transport becomes the limiting process and kinetics are not considered.

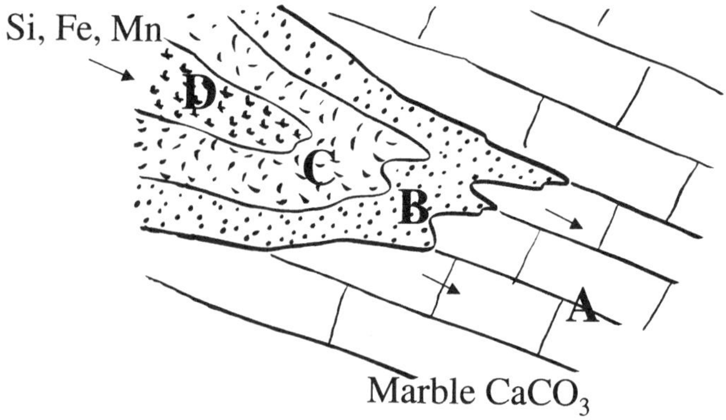

Figure 1. The transformation of starting rock A, here a limestone composed mostly of calcite $CaCO_3$, by the action of an aqueous fluid rich in Si, Fe, Mn and so on, often displays a series of zones B, C, D and so on, with different nature and chemical composition of solid solution minerals. Size of such systems is decimetric to hectometric.

CONSTRUCTION OF THE SOLID SOLUTION – AQUEOUS FLUID EQUILIBRIUM FUNCTION ("ISOTHERM")

The first step to model a problem is to derive the isotherm function ruling the equilibrium partition of the chemical components between the aqueous fluid and the solid. In our approach, the solid is made up either of a binary solid solution (one independent degree of freedom) or a three-component solid solution (two degrees of freedom). The isotherm function may be obtained from solid- and fluid- solution models: one writes the equality of the chemical potentials of each component in the fluid and in the solid. The thermodynamic properties of the solid and the liquid solutions are then needed. However, the equilibrium data are often known in terms of dissociation constants for end-members of solid solutions in equilibrium with bulk solution; we prefer to use these data rather than those on the properties of each species in solution separately. In that case, only the ratios of the dissociation constants matter. The thermodynamic validity of this approach is based on experimental works showing that, when writing the chemical exchange of two components between two minerals and one aqueous solution, the ratio of the activities of the species in solution is nearly equal to the ratio of the concentrations. This result makes it possible to get rid of the question of the activity coefficients of the species in the aqueous solution. It also allows to avoid the complete discussion of speciation, insofar as one can show that, under additional assumptions, the preceding ratios of concentrations can be written by making use of the dominant species (the chlorinated complexes $CaCl_2$, $MnCl_2$, and $FeCl_2$ if we discuss for example the case of a three-end-member garnet solid solution (Ca, Mn, Fe)$_3Al_2Si_3O_{12}$ reacting with aqueous so-called hydrothermal solutions under the conditions of high temperature (450°C- 700°C) and high pressure (1 to 2 Kbar)). This thus corresponds directly to the ratio of the total concentrations of the elements concerned with the exchanges with the solids, and which we need to compute for the isotherms. The isotherm function allows us to reduce the number of unknowns and to gather all the chemical equilibrium information. At high temperature, because of the properties of the solid solution parameters, the isotherms read like Langmuir isotherms and the exchange process becomes similar to the adsorption process. Such an approach has been followed for the three end-members grossular (Ca), spessartite (Mn), almandine (Fe) garnet solid solution. The following equations allow to compute the isotherms functions; one first considers the exchange equilibria:

$$\tfrac{1}{3}alm + MnCl_2 = \tfrac{1}{3}spe + FeCl_2 \qquad (K_{Fe/Mn}) \tag{1}$$

$$\tfrac{1}{3}gro + MnCl_2 = \tfrac{1}{3}spe + CaCl_2 \qquad (K_{Ca/Mn}) \tag{2}$$

$$\tfrac{1}{3}alm + CaCl_2 = \tfrac{1}{3}gro + FeCl_2 \qquad (K_{Fe/Ca}) \tag{3}$$

with alm = almandin, spe = spessartite, gro = grossular and K_{i/i_0} is the equilibrium constant for the exchange reaction between i et i_0. The equilibrium constants have been measured experimentally (Gavrieli et al., 1996; Bartholomew, 1989). They can be written as :

$$K_{Fe/Mn} = \frac{C^f_{Fe}.x^{gt}_{Mn}\gamma^{gt}_{Mn}}{C^f_{Mn}.x^{gt}_{Fe}\gamma^{gt}_{Fe}},\ K_{Ca/Mn} = \frac{C^f_{Ca}.x^{gt}_{Mn}\gamma^{gt}_{Mn}}{C^f_{Mn}.x^{gt}_{Ca}\gamma^{gt}_{Ca}} \text{ and } K_{Fe/Ca} = \frac{C^f_{Fe}.x^{gt}_{Ca}\gamma^{gt}_{Ca}}{C^f_{Ca}.x^{gt}_{Fe}\gamma^{gt}_{Fe}} \tag{4}$$

where γ_i and x_i are the activity coefficient and the molar fraction of component i in the solid phase respectively, and C^f_i is the concentration in the aqueous fluid, with gt = garnet. The equilibrium constants are linked by

$$K_{Fe/Ca}K_{Ca/Mn} = K_{Fe/Mn} \tag{5}$$

From equations (4) and (5), one deduces :

$$\frac{C^f_{Mn}}{x^{gt}_{Mn}\gamma^{gt}_{Mn}} = \frac{C^f_{Fe}}{K_{Fe/Mn}x^{gt}_{Fe}\gamma^{gt}_{Fe}} = \frac{C^f_{Ca}}{K_{Ca/Mn}x^{gt}_{Ca}\gamma^{gt}_{Ca}} = \frac{C^f_{Mn} + C^f_{Fe} + C^f_{Ca}}{x^{gt}_{Mn}\gamma^{gt}_{Mn} + K_{Fe/Ca}x^{gt}_{Fe}\gamma^{gt}_{Fe} + K_{Ca/Mn}x^{gt}_{Ca}\gamma^{gt}_{Ca}} \tag{6}$$

For sake of simplification, let us call K_{Mn}, K_{Ca} et K_{Fe} the equilibrium constants $K_{Mn/Mn} = 1$, $K_{Ca/Mn}$ et $K_{Fe/Mn}$ respectively. One then obtains the general form of the exchange isotherm written with molar fractions:

$$X^f_i \equiv \frac{C^f_i}{\sum_i C^f_i} = \frac{K_i x^{gt}_i \gamma^{gt}_i}{\sum_i K_i x^{gt}_i \gamma^{gt}_i} = F_i(x^s_i) \qquad i = Ca,\ Fe,\ Mn \tag{7}$$

If we assimilate the mineral to a regular solid solution and take for the excess energy that given by Grover (1977), the activity coefficients read as:

$$\gamma_i = e^{a\prod_{j\neq i} x_j} \tag{8}$$

with

$$a = \frac{w_{123}}{nRT} \tag{9}$$

where w_{123} is the principal parameter of the polynomial development of the solid solution and n is the the number of exchange sites.

TRANSPORT OF MATTER

The second step is the writing of the mass balance of the chemical components between the solid and the fluid during its movement within the porosity of the rock. The balance is given by equation

$$\partial_t[(1-\phi)C_i^s + \phi C_i^f] + Div(\phi v C_i^f) = 0 \qquad i = \text{Mn, Ca, Fe} \tag{10}$$

where

$$J_i = \phi v C_i^f \tag{11}$$

is the flux of chemical component i.. One will consider that the flux of the components is reduced to a convection flux due to the fluid movement (no diffusion); this is justified by the scale (metric to plurimetric) on which the phenomenon is displayed (Guy, 1988, 1993). The porosity of the rock is φ and the velocity of the fluid in the pores is v. Equation (10) can be simplified by making the following assumptions: monodirectional flow along the axis Ox; homogenized Darcy velocity φv constant; local chemical equilibrium reached in each point, due to high temperature of the system and slow percolation velocity; solid rock of constant and very low porosity because of the compaction.

One also makes the hypothesis that $\sum_i C_i^f = A = \text{Cst}$ because of the chromatographic exchange between the components fulfilling the electroneutrality. Under these assumptions and with the isotherm data obtained in the first step, we obtain an hyperbolic system of chromatographic type in the case of a three-component solid solution and with the change of variables $\tau = \phi v A.t$, the preceding system reads as (Sedqui, 1998; Sedqui and Guy, 2001):

$$\partial_\tau x_i^s + B\partial_x F_i(x_i^s) = 0 \tag{12}$$

with $B = V_0/n$ where V_0 is the average molar volume of the solid solution; the molar fraction and the concentration in the solid phase are related by $C^s_i = x^s_i/V_0$. $F_i(x_j^s)$ are the isotherm functions given by equations (7).

For our problem, for the relevant value of solid solution parameters, the eigenvalues λ_k of the hyperbolic system are real and distinct. The system obtained is then strictly hyperbolic. Moreover, the λ_k are positive. This means that the propagation of the reactional front is made in the same direction as the movement of the fluid, as it is rational from the physical point of view.

EVOLUTION WAVES AND NUMERICAL SIMULATIONS

The theory of hyperbolic systems makes it possible to solve the problem known as the Riemann Problem (R.P.) where one seeks to connect two constant states (see also Glueckauf, 1949). It corresponds to the transformation of an homogeneous rock which is the downstream medium (first constant state) by a fluid of constant composition (upstream medium, second constant state). Because the fluid circulates from the left to the right, the downstream medium is said on the right ("à droite", d) and the upstream medium is said on the left ("à gauche", g)). The theory shows that the solution of the R.P. is made of 3 constant states: in addition to the two states (g) and (d), another constant state (intermediate, i) appears, which is connected to (g) and (d) by shocks, "détentes" or contact discontinuities. The restriction of the evolution along curves in the composition space results from the condition known as the coherence condition; this one is a consequence of the mass balance governed by the equality of the ratios ΔC^f / ΔC^s for all the components (where C^f and C^s are the concentrations in fluid and solid phases and Δ is an infinitesimal or finite variation). The character given to a portion of curve ("détente" or shock) depends on the relative position in space (upstream or downstream) of the states to connect to the point pivot (d) or (g) and on the velocities of these states along the curves.

The first step to construct the solution of the R.P. consists in determining the curves of the k-waves (k = 1, 2) containing (g) and (d). The following step consists in connecting the states (g) and (d) by portions of wave curves; this allows to determine all the intermediate states; the various cases are determined according to the relative situation of (d) compared to (g). According to the rule: " wave-1 then wave-2 " and of the nature (shock or "détente") of the wave, one represents on figures 1, the possible evolution curves; they must respect the law of increasing velocity along the "détentes" or, in case of shock, the Lax condition (Lax, 1973). The intermediate constant state is determined by the intersection of the two portions of wave-curves. In the general case, one observes a separate plateau of composition between the two extreme states, connected to them either by two shocks, or by two "détentes", or a shock and a "détente", or a "détente" and a shock (Fig. 2, Fig. 3).

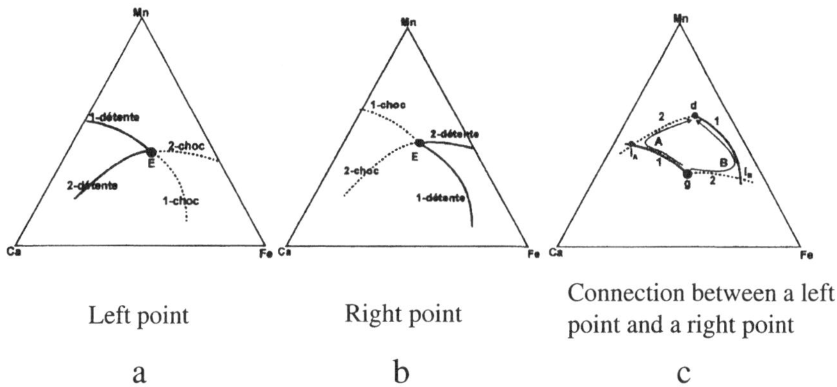

Figure 2. Rules for the use of the evolution curves in the composition triangles. For a given composition E in the triangle, and for k = 1, 2, one can determine the portions of curves which connect E to other compositions by drawing only the acceptable parts, either by a continuous evolution (relaxation or "détentes", full line) or by shocks (dotted line). **a.** Case where one connects point E to other points located on its right (downstream). Such a point is called a left point. **b.** Case when point E is connected to other points located on its left (upstream). Such a point is called a right point. **c.** When one wants to connect a left point to a right point, one uses the results given in (a) and (b) together with the condition that a 1-wave is followed by a 2-wave (path A and not path B). This makes an intermediate state appear (Sedqui and Guy, 2001; with permission from Elsevier).

All these results are based on the mass-balance equation. The study shows the influence of the problem parameters on the shape of the evolution path in the composition triangles. On the first hand, the solid solution parameters have an influence on the path curvature. Depending on the parameter values, the path is composed of straight line portions like in the case of Langmuir isotherm or is more or less curved. On the second hand, equilibrium constants influence the path orientation. Orientations themselves are responsible for the value of the intermediate composition plateaus. At last the hydrogeological parameters essentially modify the front velocity. Evolution paths in composition triangles are drawn depending on the initial conditions (starting rock and inlet fluid composition). The results of the numerical simulation may be compared with the data on natural spatial evolution of compositions for natural minerals (e.g. Einaudi and Burt, 1982; Guy, 1988; Le Guyader, 1982; Shimazaki, 1977). Other behaviours, such as a localised enrichment of one component with respect to the others, may be understood in the same frame and may provide a model for natural mineral concentrations. In the inverse problem, the hierarchy of the equilibrium constants may be predicted from observed compositional data. The method presented in this paper is general and may apply to many types of solid - fluid exchanges.

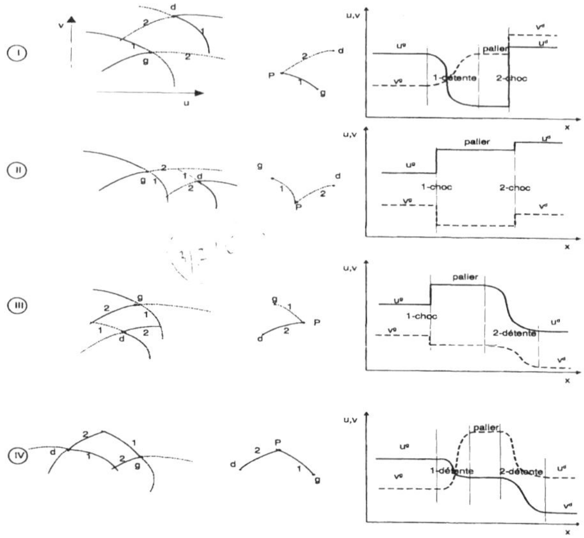

Figure 3. Representation in both the composition space (left) and the (composition, space) coordinates (right) of the evolution of solid solution composition for different types of initial and boundary conditions, giving rise to different types of Riemann problems. The two independent composition variables are termed u and v. The constant states of the Riemann problems are called u^g, v^g and u^d, v^d respectively for each case. Shock, détente waves appear, so as intermediate plateaus. The 1-waves and 2-waves are labelled by 1 and 2 respectively.

QUANTIZATION OF FLUID AND ROCK COMPOSITIONS

The theory allows to discuss more precisely why aspects of quantization (in the way we described them at the beginning of the paper) may occur. In the scalar case (Fig. 4) the theory (e.g. Guy, 1993) shows that the velocity of a given composition is proportional to the slope of isotherm for corresponding composition. The isotherm reads as $c_f = f(c_s)$·so that this velocity may be written as $v(c_0) = dx / dt_{\ c = c_0} = f'(c_0)$. Depending on the boundary conditions, there may be a problem if $v(c_1) > v(c_0)$, while c_0 is downstream to c_1. So a condition must rule the overall succession of the velocities, taking into account the boundary conditions: this condition is provided by the 2d principle of thermodynamics; it is a concavity condition $f^{*}{}'' < 0$ or > 0 where f^* is the envelope of the isotherm function containing the extreme (boundary points); the larger velocities go before the lower velocities so that shocks may appear. A quantization may be met with the possible appearance of a series of discrete zones of different compositions.In the case of systems (Fig. 5), the isotherm functions read as $c_{fi} = f_i(c_{sj})$; the jacobian matrix is $F_{ij} = \partial c_{fi}/\partial c_{sj}$.

The eigen vectors are noted as r_k. The theory shows that the system of compositions is in eigen states, i.e. the compositions that will be obtained (observed) along the waves are eigen states; along these waves, Riemann invariants (specific combinations of compositions) are also conserved. The λ_k are the eigen values of "operator" F. The theory shows that the velocity of the generalized compositions are given by $dx / dt_{(c1, c2...)} = \lambda_k$ along the eigen vector r_k . Similar to the scalar case, one must state that the velocities do not decrease from upstream to downstream: thus shocks may appear. As a general consequence, this gives rise to a quantization condition, that must take into account the boundary conditions: $\nabla\lambda_k . r_k$ has a constant sign along the path (generalization of concavity condition).

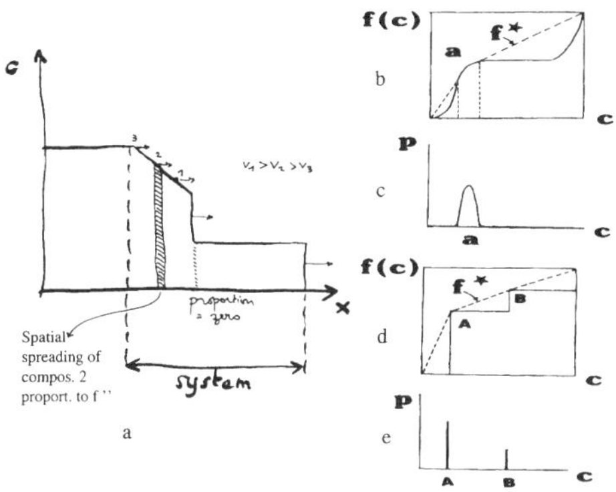

Figure 4. Quantization and probabilities of compositions, scalar case. In a) the composition profile c(x) (solid solution composition as a function of space at a given time) is represented. There is a sharp front (corresponding compositions have zero probability) and a continuous evolution, wherein the spatial spreading of a specific composition 2, lying between compositions 1 and 3 is represented; its probability p is proportional to f ''(c), i.e. the difference of the neighbouring velocities. In b) and c) the application of this rule is given for a continuous isotherm; the envelope f* between the extreme points is shown. The probability distribution is given in c). In d) and e) the same method is applied for a discontinuous isotherm (isotherm is given in d) and probability distribution in e)); Guy, 1993, with permission from Eur. J. Mineral.

FURTHER PROBABILISTIC ASPECTS

In our problem the probability aspects are related to the preceding spatial quantization (refer again to Fig. 4 and 5, and to Fig. 6). The probability to find a specific rock composition c_0 on the field is proportional to the ratio of the surface covered by c_0 to the total surface of outcrop of transformed rocks (Fig. 6). The spatial spreading of a composition (i.e. the proportion of the composition with respect to the whole of the transformed rocks) is proportional to the difference of the velocities of the compositions before and after it. This statement allows to compute the probability density p of c_0; in the scalar case: p is proportional to $f''(c_0)$. In the case of systems, p is proportional to $\nabla\lambda_k . r_k$ (scalar product with eigen vector; it is the co-ordinate of $\nabla\lambda_k$ along r_k).

As a summary, in our problem, there is a spatial quantization whenever there is a differential movement and a non-linear interaction between a moving entity (here fluid) and matter; this is not a property of solid (nor fluid) matter by itself: it is a collective or cooperative effect. The thermodynamic properties of the solid solution are most important to predict the overall structure. The quantization condition is derived from the second principle of thermodynamics. The overall condition combines the condition ruling the compositions that can be observed (use of eigen vectors and eigen values in order to compute the probability distributions) and the fitting within the general evolution taking into account the boundary conditions. Quantization is related to a choice of scale (no diffusion, chemical equilibrium); at smaller scale intermediate compositions might be observed.

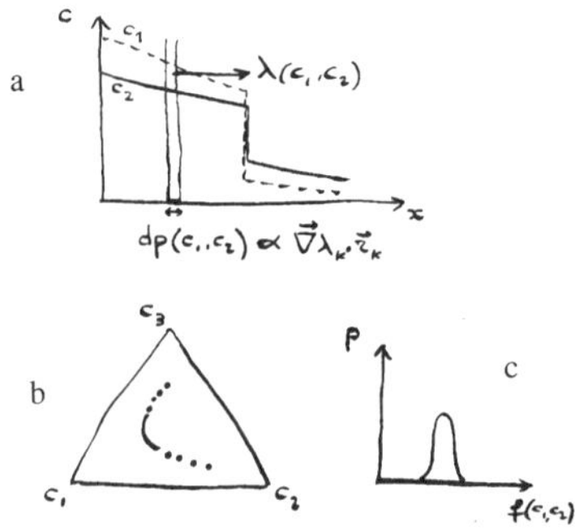

Figure 5. Quantization and probabilities of compositions, case of systems. In a) the composition evolution in space at a given time is shown for the two independent concentrations c_1 and c_2: $c_1(x)$ and $c_2(x)$ show a sharp change in composition and a continuous evolution. The reasoning is the same as in the scalar case: the probability of a composition pair (c_1, c_2) is proportional to the gradient of velocity taken along the evolution curve; this is proportional to $\nabla\lambda_k.r_k$ (scalar product with eigen vector; it is the co-ordinate of $\nabla\lambda_k$ along r_k). The corresponding evolution in composition triangle is given in b) and the probability distribution in c). Sharp changes or fronts correspond to zero probability.

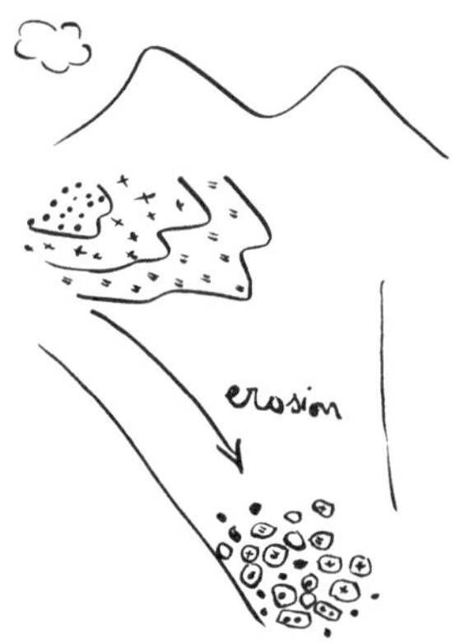

Figure 6. Quantization of the information on the system as seen from a given locality. We suppose that, because in our case of erosion, information comes from different points of the geological site with equal probability. The a priori average composition of eroded rocks is $c = p_1c_1 + p_2c_2 + \ldots$ where composition c_i has probability p_i. If one piece is collected, it will be but one of the possible states i; it will have one among the possible discrete compositions i with probability p_i..

APPLICATION TO MATERIALS SCIENCE?

In order the preceding theory would be applied to a problem in materials science, we would need to have in the same time low transport velocities (in order local equilibrium be achieved), large systems and a combination between kinetics factors, diffusion and fluid velocity parameters such that advection effects (sharp fronts) dominate over diffusion effects (smoothing). The following parameters would for instance meet these requirements: v of the order of cm/s, D = 10^{-6} cm^2/s; k (chemical kinetics parameter) = 10^{-6} $mol/m^2/s$, s (reactive surface) = 10^6 m^2/m^3, typical concentration c = 10^3 mol/m^3 so that x(Peclet number = 1) = 10^{-4} cm ; x(Damköhler number = 1) = 10 m. This may be met but seldom in some industrial systems; in earth sciences, one may think of the solidification of earth inner core where molten metal may migrate within solid alloy.

I thank the organizers, especially T. Goni, T. Mohri and P. Turchi for allowing me to participate to the conference. I thank A. Sedqui for his help.

REFERENCES

Bartholomew R. B. 1989. Interpretation of solution properties of Fe-Mg olivines and aqueous Fe-Mg chlorides from ion-exchange experiments; *American Mineralogist*, 74; 37-49.

Einaudi M. T. and Burt D. M. 1982. A special issue devoted to skarn deposits; introduction, terminology, classification and composition of skarn deposits, *Econ. Geol.*, 77, 4, 745-754.

Gavrieli I. T. , Matthews A. , Holland T. J. B. 1996. Ca-Mn exchange between grossular and $MnCl_2$ solution at 2 Kbar and 600°C : reaction mechanism end evidence for non ideal mixing in spessartine-grossular garnets, *Contrib. Mineral. Petrol.* 125; 251-262.

Glueckauf E. 1949. Theory of the chromatography : VII The general theory of two solutes following non-linear isotherms, *Discuss Faraday Soc.* 7, 12.

Grover J. 1977. Chemical mixing in multicomponent solutions : an introduction to the use of Margules and other thermodynamic excess functions to represent non ideal behaviour. D.G. Fraser ed.; D. Reidel, 67-97.

Guy B. 1988. *Contribution à l'étude des skarns de Costabone (Pyrénées Orientales, France) et à la théorie de la zonation métasomatique*. Thèse de Doctorat d'état, Université Paris VI, 928 p.

Guy B. 1993. Mathematical revision of Korzhinskii's theory of infiltration metasomatic zoning, *Eur. J. Mineral.* 5, 317-339.

Korzhinskii D. S. 1970. *Theory of metasomatic zoning*, Clarendon Press Oxford, 162 p.

Lax P. D. 1973. Hyperbolic systems of conservation laws and the mathematical theory of the shock waves, *SIAM regional conference series in applied mathematics*, N°11.

Le Guyader R. 1982. *Eléménts traces dans les skarns à scheelite et les roches associées à Costabonne (Pyrénées Orientales, France)*, Thèse Doct. 3° cycle, Université Paris VI, 169 p.

Sedqui A. 1998. *Modélisation et simulation numérique de l'interaction fluide-minéral. Application à l'échange chimique avec transport entre un fluide aqueux et une solution solide à trois pôles*. Thèse de Doctorat, Ecole nationale supérieure des mines et Université de Saint Etienne, 177 p.

Sedqui A. and Guy B. (2001) Chromatographic exchange of two independent chemical components between an aqueous fluid and a three-component solid solution : application to the evolution of skarn Ca-Mn-Fe garnet. Comptes-rendus de l'Académie des Sciences, Paris, 332, 227-234.

Shimazaki H. 1977. Grossular-spessartine-almandine garnets for some japanese scheelite skarns, *Can. Mineral.*, 15, 74-80.

MAGNETIC AND ELASTIC PROPERTIES

AB-INITIO STUDY OF DILUTED MAGNETIC SEMICONDUCTORS

Josef Kudrnovský[1,2], Václav Drchal[1], František Máca[1], Ilja Turek[3,4],

George Bouzerar[2,5], and Patrick Bruno[2]

[1] Institute of Physics, Academy of Sciences of the Czech Republic
Na Slovance 2, CZ-182 21 Praha 8, Czech Republic
[2] Max-Planck-Institut für Mikrostrukturphysik, Weinberg 2
D-06120 Halle, Germany
[3] Institute of Physics of Materials, Academy of Sciences of the Czech Republic, Žižkova 22, CZ-616 62 Brno, Czech Republic
[4] Department of Electronic Structures, Charles University
Ke Karlovu 5, CZ-121 16 Praha 2, Czech Republic
[5] Institut Laue Langevin, 6 Rue Jules Horowitz
38042 Grenoble, France

INTRODUCTION

The diluted III-V magnetic semiconductors (DMS) represent a new material with potential technological applications in the so-called spintronics by incorporating ferromagnetic elements into semiconductor devices[1]. The physics of the DMS is interesting because of the interplay between the hole-mediated ferromagnetism and disorder in the tetrahedrally bonded semiconductor compounds. It is now generally accepted that the ferromagnetic coupling in III-V DMS containing Mn impurities is mediated by holes in valence bands of the host semiconductor. Mn atoms substituting three-valent cations in the host act as acceptors which create holes in the valence band. The Fermi energy is then pinned inside the valence band for the finite concentration of Mn atoms. The ferromagnetic coupling of Mn atoms is explained by the fact that the period of RKKY oscillations exceeds the typical distance of Mn atoms in the low concentration case because of the small size of the corresponding hole Fermi surface. It was demonstrated recently[1] that the Curie temperature exceeding 100 K can be achieved in the zincblende-type (Ga,Mn)As alloy as well as in the diamond-based GeMn system[2]. The DMS with Curie temperatures of the order of room temperature are needed for practical applications and the theoretical study of the Curie temperature for realistic models is thus of a great importance.

Theoretical approaches to the DMS can be divided into two groups, namely the model studies and the studies based on the spin-density functional theory (DFT). The model studies mostly employ the kinetic-exchange (KE) model in the connection with a continuum approximation for the distribution of Mn atoms and other defects[3,4] that yields a disorder-free problem, although recently this model was refined to include the disorder via the supercell method in the framework of Monte Carlo simulations[5].

The DFT studies represent a natural step towards a more detailed, parameter-free understanding of the properties of the DMS. One possibility is to employ the supercell approach[6,7], in which big cells are needed to simulate experimentally observed low concentrations of magnetic atoms and other impurities. Alternatively, one can employ the Green function methods combined with the coherent potential approximation (CPA) in the framework of the Korringa-Kohn-Rostoker (KKR) method[8] or the tight-binding linear muffin-tin orbital method (TB-LMTO) as in the present case. The CPA-based approaches allow to treat any alloy composition which is advantageous for the DMS with low concentrations of various impurities.

The evaluation of the Curie temperature on *ab initio* level is still a challenging task, in particular for complex alloy systems like the DMS. Recently, we have calculated Heisenberg exchange parameters in a real space for bulk ferromagnets as well as for low-dimensional systems such as ultrathin films and employed them to estimate their magnon spectra and Curie temperatures in a good agreement with available experimental data[9,10]. The success of this two step process, which consists in a mapping of the complicated itinerant electron system onto an effective Heisenberg model (EHM) and consequent application of statistical mechanical methods to it[11,12], rely upon the fact that it provides an almost exact description of low-lying magnetic excitations which influence the Curie temperature significantly. The validity of this approach, based on the adiabatic approximation, is justified for magnetic atoms with large exchange splittings, like e.g. Mn-atoms. In the present paper we employ this approach for random system, in particular for DMS.

The DMS containing Mn atoms are highly compensated, i.e., the experimentally found number of holes in the valence band is strongly reduced as compared to the Mn-concentration thus suggesting the presence of other lattice defects. The As-antisites, namely the As atoms on the Ga-sublattice, which add two electrons into the valence band and compensate two holes, are found to be the most common defects. Alternatively, the interstitial Mn atoms acting as double donors have the same effect[7] as the As-antisites. The defect manipulation can influence the number of holes. It was shown recently[6] that the optically induced transition of As-antisites into an As interstitial-Ga vacancy pair (so-called EL2 defect) reduces the hole compensation thus strengthening the ferromagnetic coupling. We will demonstrate that a similar effect can be achieved also by a suitable doping, e.g., by adding Zn-impurities into GaMnAs.

A recent experiment[13] has demonstrated that the magnetization and the Curie temperature of the DMS depend on the temperature and duration of annealing. This effect can be understood by assuming that Mn moments are not completely aligned in the ground state. Recent theoretical studies confirm this assumption. The stability of the ferromagnetic state in the KE model was studied in[5] suggesting that the non-collinear ferromagnetism is common in the DMS. A new magnetic state stabilized by the As-antisites and characterized by a partial disorder in orientations of local Mn-magnetic moments was found recently also in first-principles studies of the GaMnAs alloys[14]. It is also interesting to investigate doping of GaAs with transition metals other than Mn. We will consider here the case of neighbors of Mn in the Periodic Table, namely Fe and Cr for which recent studies using supercell approach are available[16,28]. All this indicates a great importance of the influence of various defects on the electronic structure and Curie temperature of the DMS and calls for the first-principle, parameter-free study which is the subject of the present paper.

ELECTRONIC STRUCTURE

We have determined the electronic structure of the DMS in the framework of the first principles all-electron scalar-relativistic tight-binding linear muffin-tin orbital (TB-LMTO) method in the atomic-sphere approximation. We introduce empty spheres into interstitial positions of the zinc-blende GaAs semiconductor for a good space filling[8]. We have thus four fcc sublattices shifted along the [111]-direction by $a/4$, where a is the lattice constant and occupied in turn by Ga, As, and two (different) empty spheres. The substitutional disorder due to Mn-atoms, As-antisites, and other defects is treated in the CPA applied to a system consisting of several sublattices. The CPA correctly reproduces concentration trends and can also treat systems with small but finite concentrations of defects typical for the DMS, but it neglects the local environment effects and possible lattice relaxations. We have verified that we can neglect a weak dependence of the sample volume on defect concentrations and the lattice constant of the pure GaAs ($a = 5.653$ Å) was used in all calculations. We used equal Wigner-Seitz radii for all atoms and empty spheres. The charge selfconsistency is treated in the framework of the LSDA using Vosko-Wilk-Nusair parametrization for the exchange-correlation potential. The details of the method can be found in a book[17].

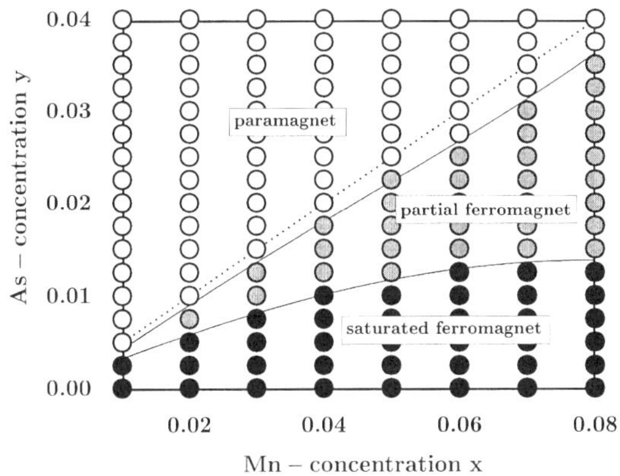

Figure 1: Three phases in the magnetic phase diagram of $(Ga_{1-x-y}Mn_xAs_y)As$. The dashed line separates the n- and p-type samples.

As explained in the Introduction, the DMS are characterized also by some degree of the magnetic disorder[5,14]. We treat this disorder in the framework of the disordered local moment (DLM) model[14,18,19] which is the simplest way of including the disorder in spin orientations. The DLM can be naturally treated in the framework of the CPA: the Mn atoms have collinear but random spin-up (Mn^+) and spin-down (Mn^-) orientations. Corresponding concentrations x^+ and x^- fulfill the condition $x = x^+ + x^-$, and the amount of the degree of magnetic disorder is characterized by the order parameter $r = x^-/x$. For example, the GaAs mixed crystal with prescribed concentrations x, y of Mn- and As-atoms on the Ga-sublattice is formally treated as a multicomponent $(Ga_{1-x-y}Mn^+_{(1-r)x}Mn^-_{rx}As_y)As$ alloy with varying order parameter r, $0 \leq r \leq 0.5$. In the saturated ferromagnetic (FM) state, $r = 0$, all magnetic moments are pointing in the direction of the global magnetization. The paramagnetic (PM) state, $r = 0.5$, is characterized by a complete disorder of spin-directions. To obtain the order parameter r corresponding to the ground state, we varied for each composition (x, y) the ratio r in steps $\Delta r = 0.025$ and evaluated the corresponding total energy. Strictly speaking, one should determine selfconsistently the orientation of each magnetic moment in the system, which in practice can be done using a supercell approach. This is possible for a simple KE model[5] but numerically prohibitive in the framework of the DFT. On the other hand, the basic physics is already present in the DLM model adopted here.

Magnetic phase diagram

The magnetic phase diagram (Fig. 1) at $T = 0$ is calculated for the whole range of concentrations (x, y) in steps of $\Delta x = 0.01$ and $\Delta y = 0.0025$. It should be noted that in addition to the FM and PM phases[8] mentioned above there is an additional phase, the partial ferromagnetic (pFM) state, in which some degree of disorder in the spin-orientations is present. The existence of such a new phase was predicted in[14] for a specific case of $x = 0.04$ and here we have extended this study to the whole plane of the physically relevant concentrations (x, y). In the region where the number of holes is reduced by As-antisites ($y > 0$) this phase separates the FM from the PM phase. On the p-type side of the phase diagram ($y \leq x/2$) the magnetization is zero in the PM region due to a complete orientational disorder of local moments. Above the diagonal $y = x/2$, even at the highest concentrations of donors, we did not find any indication for the electron-induced ferromagnetism. Further details can be read off from Fig. 2, in which for a fixed Mn concentration of $x = 0.05$ the total energy with respect to the

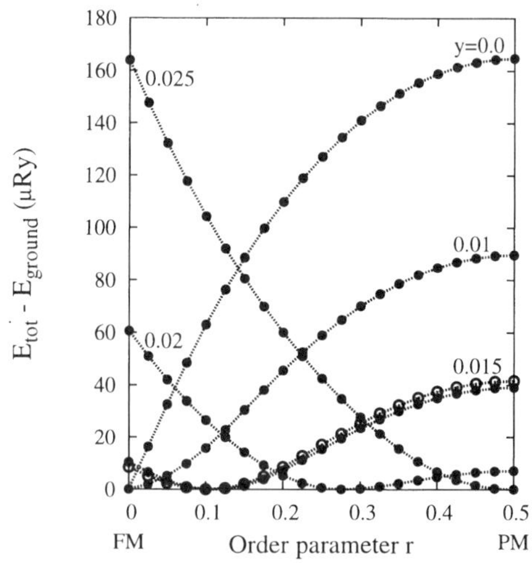

Figure 2: The differences $E_{\mathrm{tot}} - E_{\mathrm{ground}}$ are plotted as a function of the order parameter r for an alloy $(Ga_{0.95-y}Mn_{0.05}As_y)As$ and a set of concentrations y of As-antisites. The lines serve as a guide for eye.

ground state is displayed as a function of the order parameter r. For less than 1% of As-antisites the ground state is a saturated ferromagnet, for more than 2.25% of As-antisites a paramagnet, and a partial ferromagnet inbetween. We have also tested the robustness of this result with respect to the change of the sample volume. Empty symbols in Fig. 2 denote results obtained for y=0.015 assuming the experimental lattice constant of $(Ga_{0.95}Mn_{0.05})As$ alloy ($a = 5.672$ Å) and demonstrate that the ground state is not changed by this choice. This is important result as the total energy change due to the change of the sample volume is about 0.6 mRy, much higher as compared to the relative energy differences between various phases in Fig. 2.

Densities of states and Bloch spectral functions

In Fig. 3 we compare calculated total and local Mn-densities of states (DOS) for the case of a single Mn-impurity in GaAs (evaluated as a low concentration limit x=0.0001) and for the $(Ga_{0.95}Mn_{0.05})As$ alloy.

The total DOS for a single-impurity case is just the DOS of the pure GaAs host crystal. The local Mn-DOS exhibits, in agreement with the experiment[20], a majority t_2-bound state just above the top of the valence band. The effect of the finite concentration of Mn-impurities is: (i) the shift of the Fermi energy inside the valence band and creation of holes there, and (ii) the smearing out of the DOS features. The minority valence band is completely filled for $(Ga_{0.95}Mn_{0.05})As$ and this alloy is thus a semimetallic DMS. In random alloys is the bandstructure substituted by the notion of the Bloch spectral functions (BSF) which provide a detailed description of alloy electronic states. The BSF corresponding to a given sublattice B is defined as

$$A_B^\sigma(\mathbf{k}, E) = -\frac{1}{\pi}\,\mathrm{Im}\,\mathrm{tr}_L\,\bar{G}_{BB}^\sigma(\mathbf{k}, E + i\delta)\,, \qquad (1)$$

where $\bar{G}_{BB}^\sigma(\mathbf{k}, z)$ is the configurationally averaged Green function on a given sublattice B, tr_L denotes the trace over angular momenta $L = (\ell m)$, and $\mathbf{k}$ is the vector in the Brillouin zone.

The BSF for $(Ga_{0.95}Mn_{0.05})As$ alloys at $\mathbf{k} = \Gamma$ are presented in Fig. 4a and are

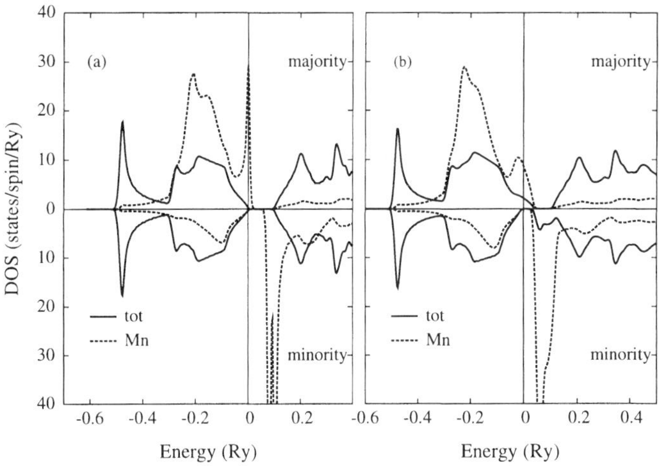

Figure 3: Spin-resolved total Mn-densities of states and local Mn-densities of states on the Ga-sublattice: (a) a single Mn-impurity in GaAs, and (b) $(Ga_{0.95}Mn_{0.05})As$ alloy. The Fermi level coincides with the energy zero.

compared with the BSF of the pure GaAs. The BSF of GaAs is a set of delta-peaks at corresponding band energies broadened by a small imaginary part δ. The Γ_{15}-peak at E=0 and the Γ_1-peak represent the top of the valence band and the bottom of the conduction band in GaAs, respectively. The effect of disorder is dramatic at the top of the valence band. It should be noted that the Mn-impurity in GaAs represents the so-called split-band limit of the alloy theory[17] as the d-levels of Ga and Mn are very different. We observe non-lorentzian behavior of the majority Γ_{15}-peak and also of the minority Γ_1-peak. There is also a non-zero spectral density in the energy region of majority Mn-states (see Fig. 3). On the contrary, the Γ_{15}-peak in the conduction band has a nearly lorentzian behavior indicating the weak influence of disorder in this energy region.

The calculated DOS in $(Ga_{0.94}Mn_{0.05}As_{0.01})As$ in the presence of As-antisites are shown in Fig. 5. The following conclusions can be drawn: (i) As-antisites reduce the number of holes in the GaMnAs valence band and shift the Fermi energy towards the top of the valence band. The minority valence bands remain fully occupied (semimetallic behavior); (ii) the unoccupied impurity s-states of the As-antisite are found in the gap while occupied impurity p-states form a broad resonance peak in the valence band (see Fig. 5b). Corresponding BSF are presented in Fig. 4b. Both the majority and minority conduction Γ_1-states are strongly modified by disorder (a bound state) resulting in the two-peak form of the BSF particularly well pronounced for majority states. Also majority valence Γ_{15}-states are strongly influenced by disorder due to Mn- and As-impurities, particularly in the energy region close to the top of the valence band. These results clearly demonstrate the fact that for a detailed study of electronic properties of the DMS are approximations adopted in a phenomenological KE model, in particular a complete neglect of the disorder and the use of the virtual-crystal approximation, not justified. On the other, some quantities like, e.g., the magnetic moment, which is obtained by integrating the BSF over the **k**-space and the energy, can be desribed reasonably well in the framework of the KE-model.

Magnetic moments

Calculated total and local Mn-magnetic moments are presented in Fig. 6 for $(Ga_{0.95-y}Mn_{0.05}As_y)As$ as a function of the concentration of As-antisites. The transition

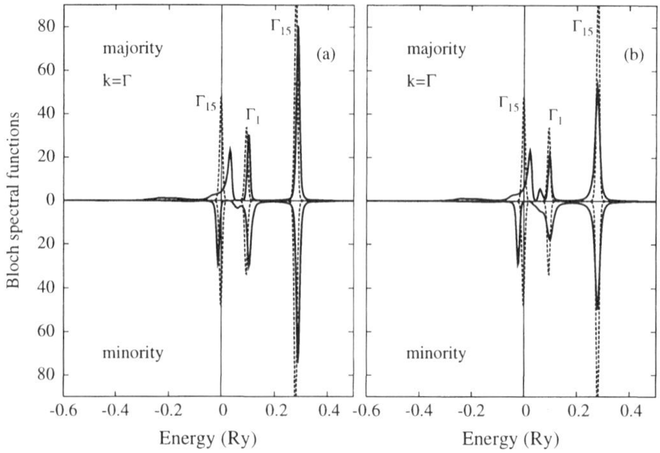

Figure 4: Spin-resolved Bloch spectral functions on the Ga-sublattice for (a) $(Ga_{0.95}Mn_{0.05})As$ and for (b) $(Ga_{0.94}Mn_{0.05}As_{0.01})As$ (full lines). Bloch spectral functions of the reference GaAs host (dashed lines) are labeled by corresponding group notations. The Fermi level coincides with the energy zero.

between the FM and pFM states is marked by a drop of both total and local Mn-moments for $y > 0.01$. The local moments m^{Mn^+} and m^{Mn^-} have opposite signs and nearly the same values of the order 4 μ_B only weakly depending on the composition thus justifying, *a posteriori*, the applicability of the DLM picture in the present case. As a result, the magnetization in the pFM state, $m \approx x^+ m^{\mathrm{Mn}^+} + x^- m^{\mathrm{Mn}^-}$, is strongly reduced as compared to the FM state[14].

HEISENBERG HAMILTONIAN

A particularly simple and yet accurate approach for the evaluation of the Curie temperature on an *ab initio* level consists in a mapping of the itinerant spin-polarized electron system onto an effective classical Heisenberg model (EHM)[9,11]

$$H = - \sum_{i \neq j} J_{ij} \, \mathbf{e}_i \cdot \mathbf{e}_j \,, \tag{2}$$

and subsequent application of statistical mechanics[9,11]. Here, $\mathbf{e}_i$ and $\mathbf{e}_j$ are unit vectors of the local magnetic moments at sites i and j, and the J_{ij} denote the effective exchange interactions between a respective pair of atoms carrying magnetic moments. For applications to the DMS, we have to generalize to disordered systems the theory developed in[9,11,12]. The Heisenberg parameters are obtained in terms of the magnetic force theorem[11,21] by (i) directly evaluating the change of energy associated with a constrained rotation of the spin-polarization axes in atomic cells i and j in the framework of the CPA, and (ii) using the vertex-cancellation theorem[22,23]. The summation in Eq. (2) is restricted to pairs of magnetic atoms M, M'. The results is

$$\bar{J}_{ij}^{M,M'} = \frac{1}{4\pi} \operatorname{Im} \int_C \operatorname{tr}_L \left[\delta_i^M(z) \, \bar{g}_{ij}^{M,M'\uparrow}(z) \, \delta_j^{M'}(E) \, \bar{g}_{ji}^{M',M\downarrow}(z) \right] \mathrm{d}z \,, \tag{3}$$

where tr_L denotes the trace over angular momenta $L = (\ell m)$, energy integration is performed in the upper half of the complex energy plane over a contour C starting below the bottom of the valence band and ending at the Fermi energy, $\delta_i^M(z) = P_i^{M,\uparrow}(z) - P_i^{M,\downarrow}(z)$, and $P_i^{M,\sigma}(z)$ are the L-diagonal potential functions ($\sigma = \uparrow, \downarrow$) corresponding to

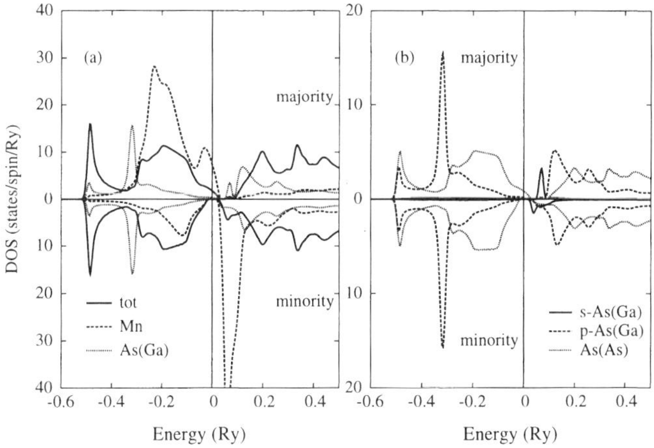

Figure 5: Spin-resolved densities of states for $(Ga_{0.94}Mn_{0.05}As_{0.01})As$: (a) total, local Mn, and local As densities of states on the Ga-sublattice; (b) The same but for s- and p-resolved local As-densities on the Ga sublattice and the total As-density on the As-sublattice. The Fermi level coincides with the energy zero.

the magnetic atom M. It should be noted that the quantity $\delta_i^M(z)$ is proportional to the exchange splitting of a given magnetic atom[12]. The quantities $\bar{g}_{ij}^{M,M'\uparrow}(z)$ and $\bar{g}_{ji}^{M',M\downarrow}(z)$ refer to site off-diagonal blocks of the conditionally averaged Green function, namely the average of the Green function over all configurations with an atom of the type M fixed at the site i and an atom of the type M' fixed at the site j. Such a quantity can be evaluated in the framework of the CPA[17]. We neglect for simplicity small induced moments on non-magnetic atoms and then the non-vanishing exchange interactions are only among substitutional Mn-atoms on the Ga-sublattice, namely M=M'=Mn. A detailed derivation will be the subject of another paper.

The low-concentration limit of Eq. (3) gives the RKKY-type expression for exchange interactions between two magnetic impurities in a non-magnetic host, namely

$$J_{ij}^{\rm RKKY} = \frac{1}{4\pi} \, {\rm Im} \int_C {\rm tr}_L \left\{ \Delta_i^{\rm imp}(z) \, g_{ij}^{\rm imp\uparrow}(z) \, \Delta_j^{\rm imp}(z) \, g_{ji}^{\rm imp\downarrow}(z) \right\} \, {\rm d}z \,. \qquad (4)$$

Here, $g_{ij}^{{\rm imp},\sigma}(z)$ is the Green function of two Mn-impurities embedded in the GaAs host and evaluated at sites i, j where impurities are located. The conventional RKKY-expression is obtained after substitution of the impurity Green function by the host GaAs Green function. The effect of neglected scatterings on impurities is two-fold: it introduces a phase factor and modifies the amplitude of oscillations as compared to the RKKY formula[24]. It should be noted that $\bar{J}_{ij}^{Mn,Mn}$ is closely related to the non-interacting local transversal susceptibility

$$\bar{\chi}_{ij}^{\rm Mn,Mn} \propto \frac{1}{\pi} \, {\rm Im} \int_C {\rm tr}_L \left[\bar{g}_{ij}^{\rm Mn,Mn\uparrow}(z) \, \bar{g}_{ji}^{\rm Mn,Mn\downarrow}(z) \right] {\rm d}z \qquad (5)$$

evaluated in the real space.

Curie temperature

For matters of simplicity we employ the mean-field approximation (MFA) to determine the Curie temperature from the EHM. The Curie temperature T_c in the FM

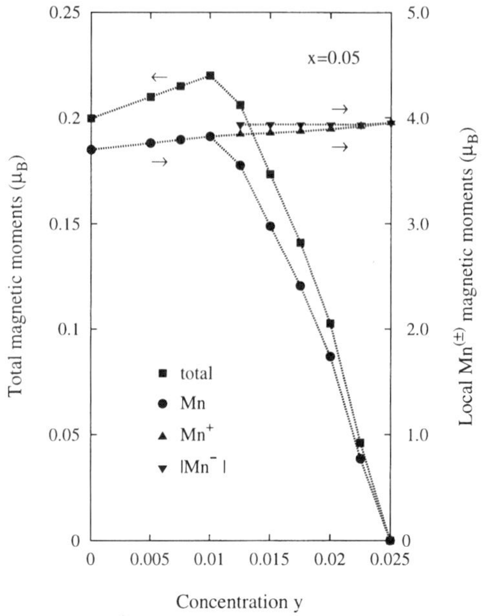

Figure 6: Total and local Mn-magnetic moments for $(Ga_{0.95-y}Mn_{0.05}As_y)As$ as a function of concentration of As-antisites y. Arrows indicate relation of a given curve to one of two different scales in the plot.

state is given by

$$k_B T_c^{MFA} = \frac{2x}{3} \sum_{i \neq 0} \bar{J}_{0i}^{\mathrm{Mn,Mn}} , \tag{6}$$

where x is the concentration of Mn atoms, and k_B is the Boltzmann constant. It should be noted that the MFA overestimates the Curie temperature as compared to more sophisticated approximations, like the random-phase approximation (RPA)[9,10]. The RPA theory of the Curie temperature of the DMS in the framework of the KE model has appeared recently[25] and the next step will be its first-principles implementation.

In practice, a more straightforward approach to determine the Curie temperature in the MFA is to evaluate the on-site effective exchange parameter for the magnetic atoms M, J_i^M. In the framework of the TB-LMTO approach[11,9]

$$\bar{J}_i^M = -\frac{1}{4\pi} \operatorname{Im} \operatorname{tr}_L \int_C \left[\delta_i^M(z) \, (\bar{g}_{ii}^{M,\uparrow}(z) - \bar{g}_{ii}^{M,\downarrow}(z)) + \delta_i^M(z) \, \bar{g}_{ii}^{M,\uparrow}(z) \, \delta_i^M(z) \, \bar{g}_{ii}^{M,\downarrow}(z) \right] \mathrm{d}z \,. \tag{7}$$

The quantities $\bar{g}_{ii}^{M,\sigma}(z)$ refer to site-diagonal blocks of the conditionally averaged Green function, namely, the average of the Green function over all configurations with the M-atom fixed at the site i[17], and other quantities were already defined. The mean-field value of the Curie temperature is (M=Mn here)

$$k_B T_c = \frac{2}{3} \bar{J}_i^{\mathrm{Mn}} \,. \tag{8}$$

A contour map of the MFA-Curie temperatures as a function of the chemical composition (x, y) as presented in Fig. 7 exhibits several important features: (i) without donors T_c increases with x and reaches room temperature at approximately $x = 0.05$. For higher Mn-concentrations T_c saturates which in turn indicates that the effective exchange interactions decrease for heavy Mn doping; (ii) the Curie temperature rapidly decreases with increasing concentration of donors. Detailed results are presented in Fig. 8 and show that for $y = 0.01$ is T_c reduced by approximately 100 K–150 K over the whole range of Mn-concentrations. For this concentration of As-antisites again a flat maximum develops at $x \approx 0.1$; the results for $x \leq 0.05$ are in a reasonable agreement with the experiment[1]. These results also agree qualitatively with the Zener model description of the DMS in the framework of the KE-model[3,4]; (iii) for a more quantitative analysis we have inserted the experimental points into Fig. 7. For each experimental point with a given Mn-content x we have estimated the concentration y of As-antisites necessary to produce a good fit of theoretical results, Eq. (6) to the experimental data[1]. Points obtained in this way follow approximately a straight line, i.e., the number of As-antisites increases proportionally with the concentration of Mn-atoms. We have calculated T_c assuming this kind of linear correlation between x and y, and the corresponding results[26] are presented in Fig. 8. As can be seen from this figure a pronounced maximum at approximately $x = 0.05$ is formed and T_c decreases for a higher Mn-doping.

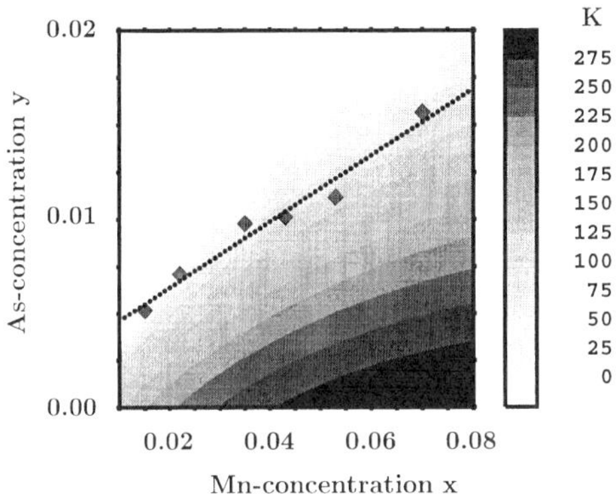

Figure 7: Contour plot of the Curie temperature as a function of the composition (x, y) assuming the ferromagnetic ground state. The symbols refer to experimental values[1], the dashed line represents a least-square fit to these data.

The concentration dependence of T_c results from an interplay of two effects: (i) an increase of T_c with increasing concentration of local Mn-moments (see Eq. (6)); and (ii) a non-trivial behavior of the leading $\bar{J}^{\mathrm{Mn,Mn}}$'s as a function of the chemical composition (x, y). Generally, we observe a decrease of the nearest-neighbor $\bar{J}_1^{\mathrm{Mn,Mn}}$ with increasing concentrations of both Mn- and As-defects, as illustrated in Fig. 9. The decrease of $\bar{J}^{\mathrm{Mn,Mn}}$ with the increasing concentration of Mn atoms can be understood from a simple RKKY model: the amplitude of oscillations is inversely proportional to the volume of the carrier Fermi sphere which increases with the Mn-content. The decrease of $\bar{J}_1^{\mathrm{Mn,Mn}}$ with the increasing concentration of Mn atoms was also obtained in[16] in the framework of the supercell approach. The leading $\bar{J}_1^{\mathrm{Mn,Mn}}$ are ferromagnetic in alloys without donors, but may change their sign for a highly compensated system. The dependence of $\bar{J}^{\mathrm{Mn,Mn}}$ on x and y cannot be scaled to a single function of $x - 2y$ as it should be in the simplest scheme of the hole-mediated magnetism. Also the dependence on the quantity $x + 2y$, which can be roughly related to the strength of the disorder scattering and, in turn, to the mean-free path is not universal. This is a strong indication that the degree of compensation, i.e., the number of holes available in the valence band, and the carrier mean-free path are not sufficient to describe the exchange mechanism. A material-specific reconstruction of the electronic structure around the gap region might thus have an important impact on the magnetic state of III-V DMS.

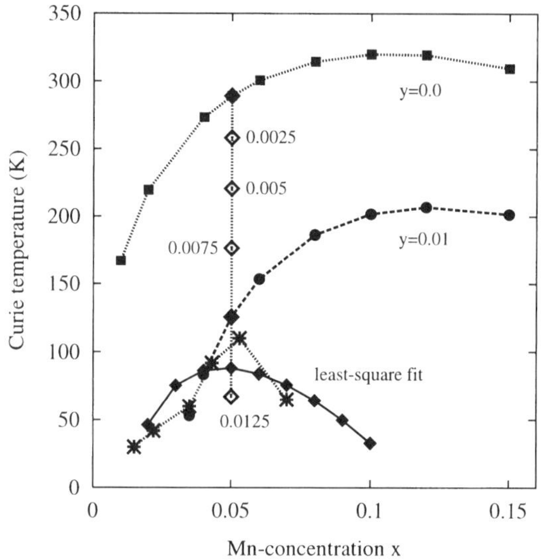

Figure 8: Curie temperatures of $(Ga_{1-x-y}Mn_xAs_y)As$ for various models as a function of x: (i) $y = 0$ (filled squares); (ii) $y = 0.01$ (filled circles); (iii) $x = 0.05$ with varying y (empty diamonds); (iv) the linear least-square fit to the experiment (filled diamonds, see Fig. 7); and (v) experiment (stars, Ref. [1]). The lines serve as a guide for eye.

Exchange interactions

Exchange interactions represent one of the basic characteristics of the magnetic state and therefore deserve a more detailed study. We have calculated exchange interactions $\bar{J}_s^{\mathrm{Mn,Mn}}$ for a given shell s as a function of the shell distance and studied the effect of various impurities on their values. In Fig. 10 we compare exchange interactions for the case of two-isolated Mn-impurities (RKKY-limit, Eq. (4)) and the case of the finite Mn-concentration $x = 0.03$ assuming the same number of holes in both cases. In the RKKY limit, Eq. (4), we used selfconsistent potentials for the single-impurity case (see Fig. 3) and just rigidly shifted the Fermi energy to accommodate the same number of holes as for the $(Ga_{0.97}Mn_{0.03})As$ alloy. The leading first nearest-neighbor interaction is ferromagnetic in the two-impurity case but for other neighbors the interaction changes its sign in an oscillatory manner. This picture is changed for a finite concentration of Mn-impurities: the couplings remain ferromagnetic up to large distances between impurities and this is a clear indication of the ferromagnetism in DMS. It should be noted that for small distances shown in the figure the effect of the alloy damping is weak, and, in addition, the disorder can change the phase of oscillations. A more detailed study of the dependence of exchange interactions on the distance between Mn-atoms is needed.

In Fig. 11 we demonstrate the dramatic effect of As-antisites on exchange interactions. The first nearest-neighbor interaction is reduced by nearly an order of magnitude as compared to the case without antisites while the reduction of other interactions is less pronounced. The Curie temperature drops from 289 K for the case without As-antisites to 126 K for the case with As-antisites. The reduction of $\bar{J}_s^{Mn,Mn}$ has few reasons as it can be seen from Eq. (3): (i) the most important is the change of the number of carriers which enters the expression (3) via the value of the Fermi energy (see Figs. 9 and 11); (ii) the disorder due to various impurities influences the quantity $\bar{g}_{ij}^{M,M'\sigma}$ and, in particular, its values close to the Fermi energy whose contribution to

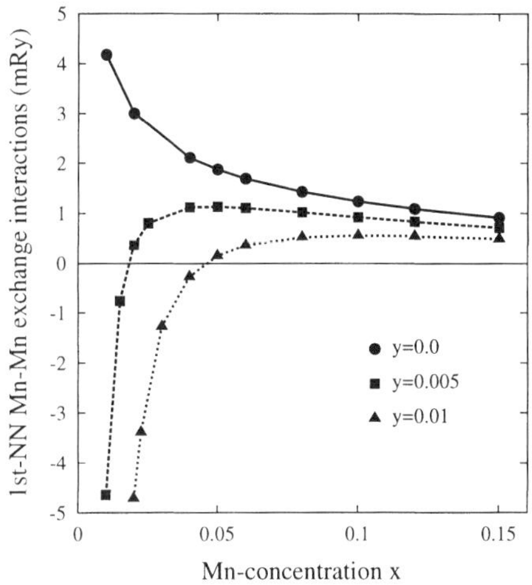

Figure 9: First nearest-neighbor exchange interaction $\bar{J}_1^{\mathrm{Mn,Mn}}$ for $(\mathrm{Ga}_{1-x-y}\mathrm{Mn}_x\mathrm{As}_y)\mathrm{As}$ as a function of x in the ferromagnetic state: $y = 0.0$ (filled circles), $y = 0.005$ (filled squares), and $y = 0.01$ (filled triangles). The lines serve as a guide for eye.

the integral (3) is largest. One can judge on changes of $\bar{g}_{ij}^{M,M'\sigma}$ only indirectly, e.g., via the values of the BSF (see Fig. 4) which characterize the strength of the disorder in an alloy; and (iii) the quantity δ_i^M, proportional to the exchange splitting, is generally only weakly influenced by alloying with non-magnetic impurities but the alloying with other magnetic elements and/or the inclusion of electron correlations beyond the LSDA can influence it non-negligibly. The values of $\bar{J}_{ij}^{\mathrm{Mn,Mn}}$ are thus the result of a complex interplay of above mechanisms.

Defect manipulation

In this section we consider the influence of EL2-defects and of the Zn-doping of the Curie temperature of the DMS.

Upon illumination of the sample, the As-antisite undergoes a structural transition into a pair formed by an interstitial As and a vacancy on the Ga-sublattice (photoquenched EL2-defect). This defect was studied by Sanvito[6] using the supercell approach. Because of technical reasons (the size of the supercell) he considered only the case of equal concentrations of Mn-atoms and As-antisites ($x = y \approx 0.03$) which corresponds to the less interesting case of n-doped sample. Here we present the CPA results for the case of p-doped sample with $x = 0.05$ and varying concentrations y of As-antisites assuming that the positions of As-interstials and vacancies are random. The calculated Curie temperature of $(\mathrm{Ga}_{0.95-y}\mathrm{Mn}_5\mathrm{As}_y)\mathrm{As}$ with $y = 0.005$ increases from 220 K for As-antisites to 238 K for the case of EL2-defects of the same concentration. The increase of the Curie temperature for $y = 0.01$ is larger, namely 126 K and 190 K for As-antisites and EL2-defects, respectively. We have also found enhancement of the ferromagnetic exchange couplings for the case of EL2-defects. The optically induced transitions of As-antisites into EL2-defects thus reduce the hole compensation and allow the tuning of the hole concentration without changing the chemical composition[6]. This mechanism, however, cannot be used to obtain higher Curie temperatures because the temperature of regeneration of antisites is of the order 100 K[6].

An alternative way how to reduce the hole compensation by As-antisites is changing of the chemical composition of the sample by doping. The zinc, which acts as an acceptor in GaAs, is one of possible candidates: it brings one hole into the valence

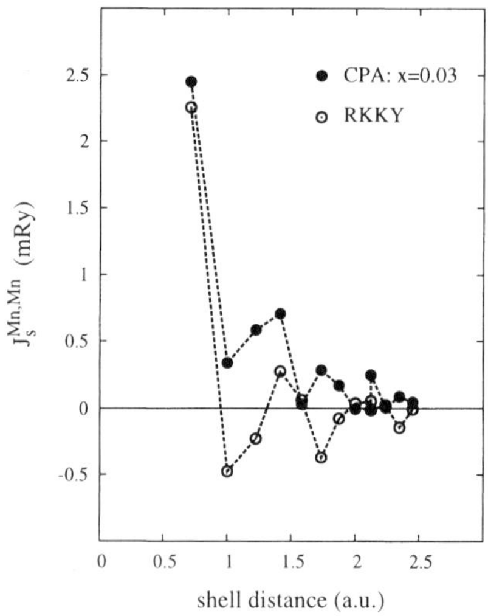

Figure 10: Exchange interaction $\bar{J}_s^{\mathrm{Mn,Mn}}$ as a function of the shell distance: the ferromagnetic $(Ga_{0.97}Mn_{0.03})As$ alloy (filled circles) and two Mn-impurities embedded in the GaAs host (empty circles) with the Fermi energy shifted inside the valence band to accommodate the same number of holes as in the alloy case.

band of GaAs when substituted for Ga and recently a high concentration of Zn-defects of the order $2 \times 10^{20}/\mathrm{cm}^3$ was prepared in thin layers of GaAs[27]. Zinc influences the low-lying part of the GaAs valence band and it has therefore only weak influence on Mn-states and As-states close to the Fermi level. Consequently, its main effect is to reduce the hole compensation due to As-antisites. Calculated Curie temperatures of $(Ga_{1-x-y-z}Mn_xAs_yZn_z)As$ (see Table 1) confirm this conclusion.

Table 1: Curie temperatures [K] of the ferromagnetic GaAs alloy with different concentrations of Mn atoms (x), As-antisites (y), and Zn-dopants (z).

$(Ga_{1-x-y-z}Mn_xAs_yZn_z)As$					
$x = 0.05$	$x = 0.05$	$x = 0.05$	$x = 0.03$	$x = 0.05$	$x = 0.05$
$y = 0.0$	$y = 0.005$	$y = 0.01$	$y = 0.0$	$y = 0.005$	$y = 0.01$
$z = 0.0$	$z = 0.01$	$z = 0.02$	$z = 0.02$	$z = 0.0$	$z = 0.01$
289	286	282	250	221	217

It should be noted that Zn-dopants fully compensate the effect of As-antisites if $z = 2y$. This is illustrated by first three columns in Table 1) corresponding to alloys with $x = 0.05$ which have the same number of holes $n_h = 0.05$ but different concentrations y and z, and which obey the above condition. Indeed, calculated Curie temperatures agree very well in all three cases. The effect is the same if the condition $z = 2y$ is not obeyed but the number of holes is still the same, namely $n_h = 0.04$ (see last two columns). Finally, the column four illustrates the case with $x = 0.05$, without antisites but doped with Zn atoms so that the hole concentration is $n_h = 0.05$ like for $(Ga_{0.95}Mn_{0.05})As$. The calculated Curie temperature is smaller in the case with the smaller Mn content.

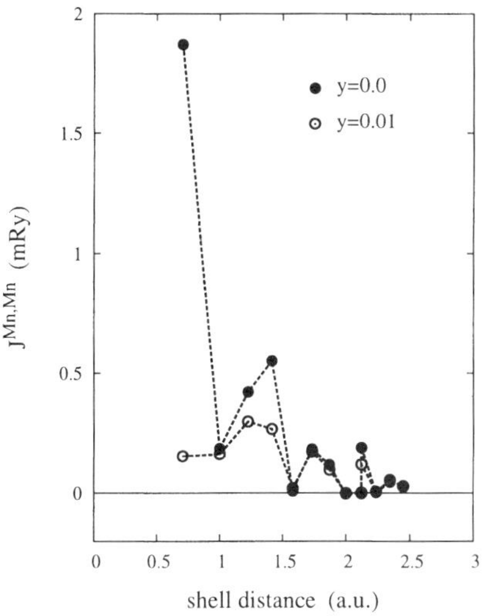

Figure 11: Exchange interactions $\bar{J}_s^{\rm Mn,Mn}$ as a function of the shell distance for the disordered ferromagnetic semiconductor alloys $(Ga_{0.95-y}Mn_{0.05}As_y)As$: $y = 0.0$ (filled circles) and $y = 0.01$ (empty circles). The number of holes is reduced in the case of As-antisites.

Other magnetic impurities

We shall study the effect of doping the GaAs host by magnetic atoms which are left (Cr) and right (Fe) neighbors of Mn in the Periodic Table. In particular, doping by Cr atoms is interesting as recently it was observed that thin films of CrAs grown on the GaAs substrate are ferromagnetic with the estimated Curie temperature higher than 400 K[28]. It was assumed that CrAs films adopt the zinc-blende (zb) structure and the full-potential *ab initio* calculations performed for the hypothetical zb-CrAs have shown that CrAs is a semimetallic ferromagnet with a total moment of 3 μ_B in agreement with the experiment. In Fig. 12a we present calculated total and local Cr-densities of states (DOS) for zb-CrAs assuming the lattice constant of GaAs. Our results for the DOS and the calculated total spin moment of 3 μ_B agree well with[28] as well as with recent full-potential KKR calculations of Galanakis[29]. This is an important test of our approach which adopts the atomic-sphere approximation.
The leading exchange interactions $J_1^{\rm Cr,Cr}$ are ferromagnetic and are smaller as compared to $J_1^{\rm Mn,Mn}$ without antisites (see Fig. 11). Exchange interactions $J_s^{\rm Cr,Cr}$ are quickly damped as a function of the shell distance, Fig. 12b, even if there is no disorder in the system and Cr atoms fully occupy one sublattices. The explanation for such an exponential damping of exchange interactions was given in[9] and the reason is the semimetallic character of the ordered CrAs, namely the fact that the Fermi level lies in the gap of minority states thus giving rise to the imaginary critical Fermi wave vector which causes exponential damping of exchange interactions. Cr atoms also induce non-negligible exchange interactions between Cr and As pairs and Cr and neighboring interstitial sites. The energy difference between the DLM state and the ground FM state is as large as 14 mRy per formula unit thus indicating a high Curie temperature in a qualitative agreement with the experiment[28].
Calculated DOSs for the GaAs with 5% of Cr impurities without and with As-antisites are presented in Fig. 13 and they should be compared with corresponding results for GaMnAs alloy, Figs. (3) and (5). The differences can be summarized as follows: (i) the Fermi energy is located within the impurity subband with a strong admixture of Cr-states. The Cr-impurity level is about 0.5 eV above the top of the valence band as

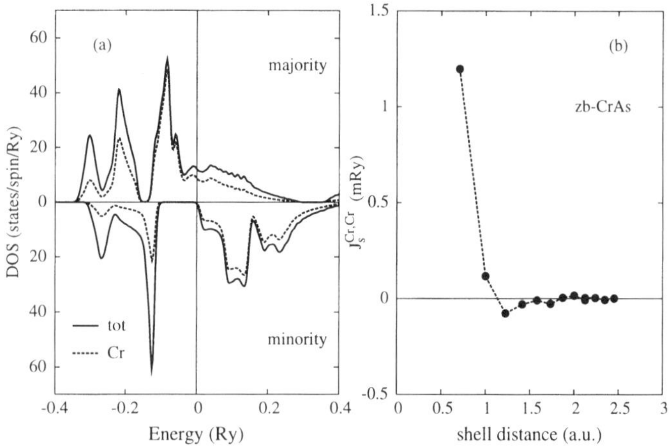

Figure 12: The zincblende CrAs with the GaAs lattice constant: (i) Spin-resolved total and local Cr-densities of states. The Fermi level coincides with the energy zero; (b) Exchange interactions $J_1^{Cr,Cr}$ as a function of the shell distance.

compared with the value of 0.1 eV for the Mn-impurity. The total spin moment per Cr atom is 3 μ_B; (ii) the difference between the DLM- and FM-states decreases with increasing As-concentration much more slowly as compared to the case of Mn-dopants; (iii) the effect of As-antisites is also different because the effect of compensation by As-antisites is weaker. The Fermi level is shifted towards the conduction band as expected but even for 3% of As-antisites, when it reaches the bottom of the minority conduction band, it is still located in the Cr-majority band peak as in the case without antisites. This is in an agreement with the point (ii) above; and (iv) calculated Curie temperatures in the MFA are 433 K and 454 K for the $(Ga_{0.95-y}Cr_{0.05}As_y)As$ alloy and $y = 0.0$ and $y = 0.01$, respectively. This is in agreement with behavior of the corresponding exchange interactions (see Fig. 14) which are quite similar. The Curie temperature decreases if the concentration of As-antisites further increases, namely the calculated values of T_c^{MFA} for $y = 0.02$ and $y = 0.03$ are 389 K and 214 K, respectively.

The system $(Ga_{0.95}Fe_{0.05})As$ is a metal (the Fermi level lies within the valence majority and minority bands) and its ground state is the DLM. This result is in an agreement with conclusions of the paper[15] that the zb-FeAs is an antiferromagnet. The difference between magnetic properties of GaCrAs and GaFeAs can be understood from the behavior of $\bar{J}_s^{Cr,Cr}$ and $\bar{J}_s^{Cr,Cr}$, Fig. 14. The leading nearest-neighbor exchange interactions in GaCrAs are ferromagnetic as contrasted with dominating antiferromagnetic nearest-neighbor exchange interactions found in GaFeAs. It should be noted that the leading $\bar{J}_1^{Cr,Cr}$ are much larger as compared to $\bar{J}_s^{Mn,Mn}$ (see Fig. 11).

CONCLUSIONS

We have investigated from first principles the effect of As-antisites and other defects on the electronic and magnetic properties of (Ga,Mn)As alloys. The Curie temperatures were estimated from the effective Heisenberg Hamiltonian constructed from the selfconsistent electronic structure calculations within the local density approximation and in the framework of the coherent potential approximation to treat the effect of disorder. The most important results are: (i) the existence of an additional phase with partially ordered local magnetic moments which separates the FM and PM states in the magnetic phase diagram. This phase is stabilized by As-antisites and has a reduced magnetization; (ii) a strong reduction of the Curie temperature with in-

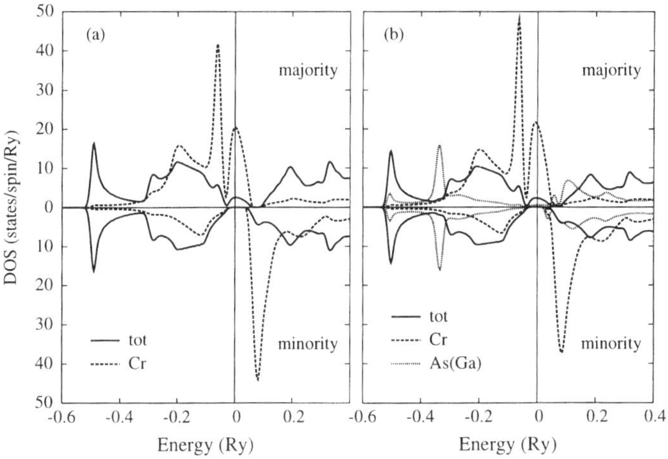

Figure 13: Spin-resolved total and local Cr-densities of states on the Ga-sublattice: (a) $(Ga_{0.95}Cr_{0.05})As$ alloy, and (b) $(Ga_{0.94}Cr_{0.05}As_{0.01})As$ alloy. In the latter case is also shown the local As-antisite density of states. The Fermi level coincides with the energy zero.

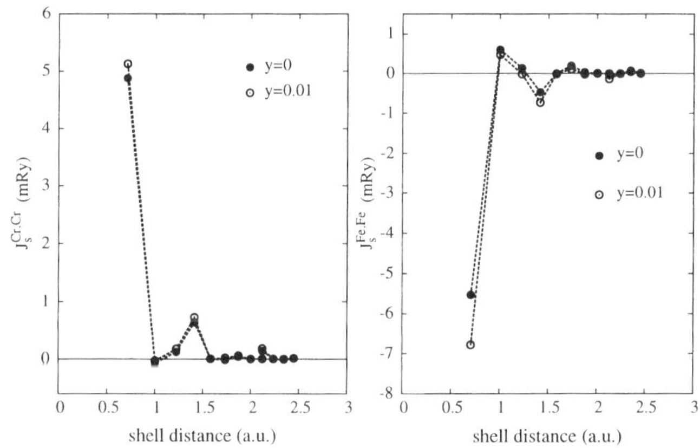

Figure 14: Exchange interactions $\bar{J}_s^{\rm Cr,Cr}$ and $\bar{J}_s^{\rm Fe,Fr}$ as a function of the shell distance for the disordered ferromagnetic semiconductor alloys: (a) $(Ga_{0.95-y}Cr_{0.05}As_y)As$ and (b) $(Ga_{0.95-y}Fe_{0.05}As_y)As$ evaluated for $y = 0$ (full circles) and for $y = 0.01$ (empty circles). The lines serve as a guide for eye.

creasing concentration of the As-antisites; (iii) the exchange interactions between Mn atoms are reduced with increasing concentrations of both Mn- and As-impurities. The first nearest-neighbor interactions $\bar{J}_1^{\rm Mn,Mn}$ are ferromagnetic but become antiferromagnetic in a highly compensated sample; (iv) a comparison of calculated and measured concentration dependences of the Curie temperature indicates a correlation between concentrations of Mn-impurities and As-antisites, namely an increase of the donor concentration with an increase of the Mn-content. This conclusion is supported by a recent evaluation of the formation energy of As-antisite defects in GaAs[30]; (v) doping of (Ga,Mn)As alloys with Zn-acceptors compensates the effect of As-antisites and increases the Curie temperature; and (vi) GaAs with Cr-impurities is semimetallic ferromagnet with the total spin moment of 3 μ_B per Cr atom. Cr-impurities seem to strengthen the tendency towards the ferromagnetism and the system is less sensitive to compensating effects of As-antisites. On the other hand, Fe doping gives rise to the metal with dominating antiferromagnetic couplings.

Acknowledgements

The financial support provided by the Grant Agency of the Czech Republic (No. 202/00/0122), the Grant Agency of the Academy of Sciences of the Czech Republic (No. A1010203, A1010214), the Czech Ministry of Education, Youth and Sports (COST P5.30, MSM 113200002), the CMS Vienna (GZ 45.504), and the RTN *Computational Magnetoelectronics* (HPRN-CT-2000-00143) is acknowledged.

REFERENCES

1. H. Ohno, Science **281**, 951 (1998).
2. Y.D. Park, A.T. Hanbicki, S.C. Erwin, C.S. Hellberg, J.M. Sullivan, J.E. Mattson, T.F. Ambrose, A. Wilson, G. Spanos, and B.T. Jonker, Science **295**, 651 (2002).
3. K. König, J. Schliemann, T. Jungwirth, and A.H.MacDonald, in *Electronic structure and magnetism in complex materials*, eds. D.J. Singh and D.A. Papaconstantopoulos (Springer-Verlag, Berlin, 2002) and cond-mat/0111314.
4. T. Dietl, H. Oho, F. Matsukara, J. Cibert, and D. Ferrand, Science **287**, 1019 (2000).
5. J. Schliemann and A.H. MacDonald, Phys. Rev. Lett. **88**, 137201 (2002).
6. For a recent review see: S. Sanvito, G. Theurich, and N. Hill, cond-mat/0111300.
7. J. Mašek and F. Máca, Phys. Rev. B **65**, 235209 (2002); F. Máca and J. Mašek, Acta Phys. Pol. A **100**, 319 (2001).
8. H. Akai, Phys. Rev. Lett. **81**, 3002 (1998).
9. M. Pajda, J. Kudrnovský, V. Drchal, I. Turek, and P. Bruno, Phys. Rev. B **64**, 174402 (2001).
10. M. Pajda, J. Kudrnovský, V. Drchal, I. Turek, and P. Bruno, Phys. Rev. Lett. **85**, 5424 (2000).
11. A.I. Liechtenstein, M.I. Katsnelson, V.P. Antropov, and V.A. Gubanov, J. Magn. Magn. Mater. **67**, 65 (1987).
12. V.P. Antropov, B.N. Harmon, and A.N. Smirnov, J. Magn. Magn. Mater. **200**, 148 (1999).
13. S.J. Potashnik, K.C. Ku, S.H. Chun, J.J. Berry, N. Samarth, and P. Schiffer, Appl. Phys. Lett. **79**, 1495 (2001).
14. P.A. Korzhavyi, I.A. Abrikosov, E.A. Smirnova, L. Bergqvist, P. Mohn, R. Mathieu, P. Svedlindh, J. Sadowski, E.I. Isaev. Yu.Kh. Vekhilov, and O. Eriksson, Phys. Rev. Lett. **88**, 187202 (2002).
15. M. Shirai, Physica **E10**, 143 (2001).
16. M. van Schilfgaarde and O.N. Mryasov, Phys. Rev. B **63**, 233205 (2001).
17. I. Turek, V. Drchal, J. Kudrnovský, M. Šob, and P. Weinberger, *Electronic Structure of Disordered Alloys, Surfaces and Interfaces*, (Kluwer, Boston, 1997); I. Turek, J. Kudrnovský, and V. Drchal, in *Electronic Structure and Physical Properties of Solids*, ed. H. Dreyssé, Lecture Notes in Physics 535, (Springer, Berlin, 2000), p. 349.
18. B.L. Gyorffy, J. Pindor, J.B. Staunton, G.M. Stocks, and H. Winter, J. Phys. F **15**, 1337 (1985).
19. H. Akai and P.H. Dederichs, Phys. Rev. B **47**, 8739 (1993).
20. M. Linnarsson, E. Janzén, B. Monemar, M. Kleverman, and A. Thilderkvist, Phys. Rev. B **55**, 6938 (1997).

21. A. Oswald, R. Zeller, P.J. Braspenning, and P.H. Dederichs, J. Phys. F: Met. Phys. **15**, 193 (1985).
22. P. Bruno, J. Kudrnovský, V. Drchal, and I. Turek, Phys. Rev. Lett. **76**, 4254 (1996).
23. P.M. Levy, S. Maekawa, and P. Bruno, Phys. Rev. B **58**, 5588 (1998).
24. J.A. Blackman and R.J. Elliott, J. Phys C **2**, 1670 (1969).
25. G. Bouzerar and T.P. Pareek, Phys. Rev. B **65**, 153203 (2002).
26. Strictly speaking, for $x > 0.07$ the ground state has a small amount of disorder in spin-polarization directions and a simple expression (6) is still a reasonable approximation. For example, for $x = 0.08$ and $y = 0.015$ there is $x^+ = 0.078$ and $x^- = 0.002$.
27. Q.X.Zhao, M. Willander, P.O. Holtz, W. Lu, H.F. Don, S.C.Shen, G. Li, and C. Jagadish, Phys. Rev. B **60**, R2193 (1999).
28. H. Akinaga, T. Manago, and M. Shirai, Jpn. J. Appl. Phys. **39**, L1118 (2000).
29. I. Galanakis (submitted to Phys. Rev. B, cond-mat/0203535).
30. J. Mašek, private communication.

VARIATION OF ELASTIC SHEAR CONSTANTS IN TRANSITION METAL ALLOYS

Göran Grimvall

Theory of Materials, Stockholm Centre for Physics, Astronomy and Biotechnology, Royal Institute of Technology, SE-106 89 Stockholm Sweden

INTRODUCTION

The body centred cubic (bcc) and the face centred cubic (fcc) lattice structures are the most common structures among the elements. With few exceptions only one, or none, of these structures is available for experimental studies for a given element at ambient pressure. Ab initio electron structure calculations can be performed for any lattice configuration. Thus they offer a possibility to investigate systematically how physical properties depend on the crystal structure. The present paper focuses on the elastic shear constants $C' = (C_{11} - C_{12})/2$ and C_{44} for elements in the 3rd, 4th and 5th row in the Periodic Table. This information is of particular interest because it is common among the transition metals that only one of the bcc and fcc structures exists as a stable, or possible metastable, phase. The other structure then is dynamically (mechanically) unstable, i.e., the stability requirement

$$C' = (C_{11} - C_{12})/2 > 0, \tag{1a}$$

$$C_{44} > 0, \tag{1b}$$

is violated. The inequalities (1a) and (1b) are the long-wavelength formulations of the condition that the phonon frequencies $\omega(\boldsymbol{q},s)$ of a dynamically stable solid structure must all be real;

$$\omega^2(\boldsymbol{q},s) > 0. \tag{2}$$

Here we are interested in cubic structures. In the long-wavelength limit $\boldsymbol{q} \to 0$ and for the longitudinal (s=L) and the two transverse ($s=T_1, T_2$) phonon branches, we can write $\omega = vq$, where v is the sound velocity. If ρ is the mass density, ρv^2 in the three high-symmetry crystallographic directions [*hkl*] is expressed in the elastic constants C_{11}, C_{12} and C_{44} as in Table 1.

Dynamical instabilities have profound consequences for, e.g., the modelling of alloy phase diagrams as a function of composition because then the vibrational entropy S, and hence the Gibbs energy $G = U - TS$, is not a meaningful thermodynamic quantity (Grimvall 1998, 2002).

Table 1: ρv^2 in the crystallographic directions [*hkl*]

Mode	[100]	[110]	[111]
L	C_{11}	$(C_{11} + C_{12} + 2C_{44})/2$	$(C_{11} + 2C_{12} + 4C_{44})/3$
T_1	C_{44}	C_{44}	$(C_{11} - C_{12} + C_{44})/3$
T_2	C_{44}	$(C_{11} - C_{12})/2$	$(C_{11} - C_{12} + C_{44})/3$

LATTICE ENERGIES AND BAIN PATHS

Consider a bcc lattice structure, with lattice parameter a (see the central part with short-dashed sides in Fig. 1). Through a tetragonal strain, one of the lattice parameters is changed by a factor $c/a = 2^{1/2}$, resulting in an fcc structure. An electron-structure calculation giving the total energy $U_{\mathrm{coh}}(c/a)$ can be performed for a lattice, with c/a varying from 1 (i.e., a bcc structure) to $2^{1/2}$ (fcc). The deformation path is referred to as a Bain path (cf. Milstein et al. (1994) and references there). Figure 2 shows the relative cohesive energy for such a Bain path connecting bcc and fcc tungsten, with data taken from Einarsdotter et al. (1997). The elastic constants C' for the bcc and fcc structures are directly related to the volume V per atom and the curvature of U_{coh} along the Bain path, taken at the corresponding c/a. Considering small variations $\delta(c/a)$ and δV in the cubic structures with bulk modulus B, the corresponding variation δU_{coh} is (Kraft et al. 1993)

$$\frac{\delta U_{\mathrm{coh}}}{V} = \frac{B}{2}\left(\frac{\delta V}{V}\right)^2 + \frac{2C'}{3}\left(\frac{\delta(c/a)}{c/a}\right)^2. \tag{3}$$

For a volume-conserving Bain path, $\delta V = 0$, the elastic constant C' is obtained from the curvature of U_{coh} at the bcc and fcc structures, respectively.

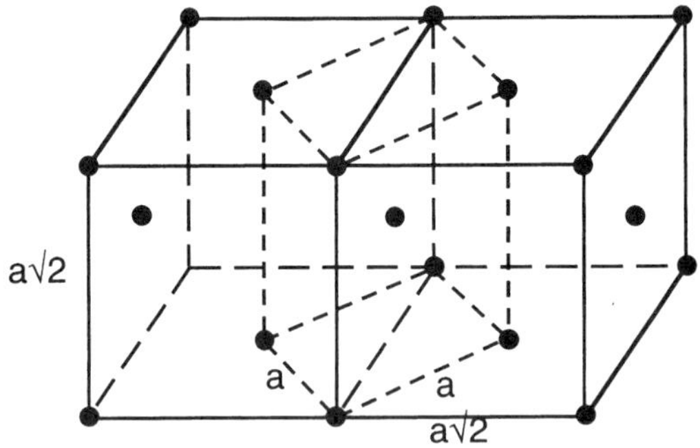

Figure 1. The fcc lattice structure can be obtained by a strain of a bcc lattice structure.

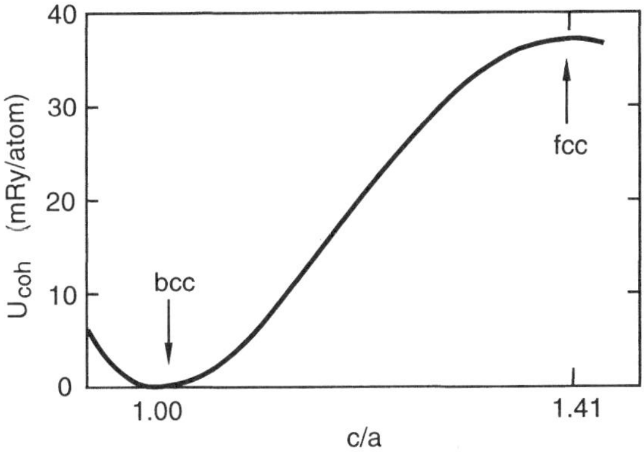

Figure 2. The energy along a Bain path connecting the bcc and fcc structures of tungsten.

Similarly, other deformations of a lattice can give other elastic constants, or combinations of them. The deformation in Fig. 1, leading to Eq. (3), is the most common version of a Bain path. In the volume-conserving approximation, the volume of the system is kept constant, while *c/a* is varied. A more accurate calculation allows also the volume to vary, so that the energy is minimised for each *c/a*.

COHESIVE ENERGIES AND LATTICE INSTABILITIES

During the 1980's a major controversy arose when ab initio cohesive energy calculations for transition metals were confronted with semiempirical data obtained from a so called CALPHAD approach, in which Gibbs energies are constructed to reproduce and predict alloy phase diagrams. The issue is often illustrated with reference to a plot such as in Fig. 3, where the results for the enthalpy difference at 0 K from ab initio calculations (solid line) are shown together with semiempirical data (points).

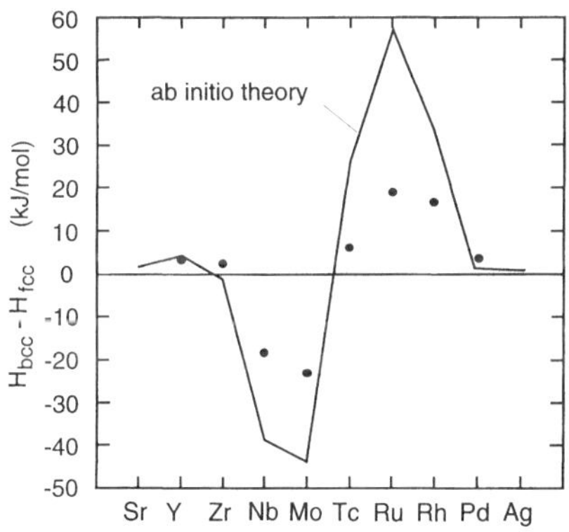

Figure 3. The enthalpy difference between the bcc and fcc lattice structures, as it appeared in the significant discrepancy between 'semiempirical' data (points) and the results of ab initio electron structure calculations in the 1980's. After Grimvall (1999).

The difference between the ab initio and semiempirical results is much too large to be ignored. The problem was found to have an unexpected but simple solution, as has been described in historical accounts by Grimvall (1998, 2002). Take, as an example, the Pt-W alloy system. Pure Pt has the fcc lattice structure and pure W the bcc structure. The CALPHAD method assumed that a Gibbs energy function can be assigned to Pt-W alloys of all compositions, having a bcc as well as an fcc structure. Pure Pt has the fcc structure. However, ab initio calculations show that pure W is dynamically unstable in the fcc lattice (cf. Fig. 2). Therefore, the Gibbs energy is not a thermodynamically well-defined quantity for W-rich fcc Pt-W alloys. It has been argued (Grimvall 1998, 2002) that the CALPHAD method is still a very valuable approach to the prediction of alloy phase diagrams, but in its common form it may not give correct cohesive energies of dynamically unstable structures. (Theoretically one may define these energies by requiring that the atoms form a *static* structure.)

A problem in the CALPHAD prediction of phase diagrams is that one cannot tell if an assumed non-observed phase is *metastable* (i.e., thermodynamically unstable but with a well-defined Gibbs energy), or if it is dynamically *unstable*. Intuitively, and with reference to Fig. 2 and Eq. 3, one may assume that if the energy difference between the static fcc and bcc structures is large, the phase with the lowest energy will have a large C'. This has been confirmed in ab initio calculations by Wills et al. (1992) for some transition metal elements and by Craievich et al. (1997) for disordered Ni-Cr alloys. Furthermore, the phase having the higher energy may tend to be dynamically unstable (i.e., have negative C'). As a consequence, large and *positive* values of C' for either the bcc or the fcc structure will tend to correlate with large and *negative* values for C' of the other structure. That trend will now be studied.

SELECTION OF DATA

The survey in this section refers to solid phases at ambient pressure. Information about the structure of the thermodynamically stable phases is taken from Young (1991). For those phases, the values of the elastic constants used in Table 2 and Fig. 4 are taken from the Landolt-Börnstein compilation of experimental data, by Every and McCurdy (1992). The survey below also gives references to works where C' and C_{44} are obtained in electronic structure calculations of the ab initio type. It is only recently that elastic constants have been calculated for bcc or fcc structures, when such phases are not observed experimentally. Therefore data are missing for many elements. Much work has been focussed on the tetragonal Bain path, from which C', but not C_{44}, is obtained. Often those C' refer to the approximate volume-conserving path. Furthermore, papers dealing with Bain paths usually only give the result as a graph of the energy versus the deformation, without any explicit values for the elastic constants, e.g., in Wills (1992) and in the papers by Šob et al. In the latter case, numbers derived from a fit to the Bain paths have been provided by Šob in private correspondence, and are quoted in Table 1. In some cases the shear elastic constants are small and just barely positive or negative. When it has not been possible to obtain a more precise numerical value, those elastic constants are given in Table 2 as ≥ 0 and ≤ 0, respectively.

Potassium, rubidium, cesium. All three elements have the bcc structure. Sliwko et al. (1996) obtained C' for K and Rb from a calculation of the tetragonal Bain path. Jianjun Xie et al. (2000) calculated C' and C_{44} for Cs. Grimvall and Ebbsjö (1975) found fcc K to be dynamically stable.

Calcium, strontium, barium. Calcium and strontium have fcc structures that transform to the bcc structure at 721 K and 830 K, respectively. Barium has the bcc structure at all temperatures. Sliwko et al. (1996) obtained C' for Ca and Sr from a calculation of the tetragonal Bain path. Moriarty (1973) obtained $C_{44} > 0$ for Ca. The fcc data for Ba in Table 1 are subjective estimations based on a comparison with the experimental data for fcc Ca and Sr.

Scandium, yttrium, lanthanum. Scandium and yttrium have the hcp structure, with a transition to bcc structure at 1608 K and 1752 K, respectively. Lanthanum is hcp at low temperatures, transforms to fcc at 550 K and to bcc at 1134 K. From measurements of the phonon spectrum by neutron scattering, Petry at al. (1993) obtained C' and C_{44} for bcc Sc at 1673 K, and similarly Güthoff et al. (1993) obtained C' and C_{44} for bcc La at 1673 K. Bain path calculations by Wills et al. (1992) show that C' for bcc La is slightly negative. Calculations by Craievich et al. (1997) for bcc Sc give $C' \approx 0$. Calculations by Persson et al. (2000) give $C' = -10$ GPa, $C_{44} = 30$ GPa for bcc Sc, and $C' = -5$ GPa, $C_{44} = 10$ GPa for bcc La.

Titanium, zirconium, hafnium. All three metals have an hcp structure, which is followed by a bcc structure above 1155 K, 1136 K and 2030 K, respectively. At low temperatures, the bcc structure is thought to be dynamical unstable ($C' < 0$); cf. e.g., calculations by Craievich et al. (1994) and Sliwko et al. (1996) for Ti and by Wills et al. (1992) for Hf. The nature of the bcc stabilisation at high temperatures is controversial. Aguayo et al. (2002) calculated C' and C_{44} for fcc Ti, Zr and Hf. Sliwko et al. (1996) obtained C' for fcc Ti from a Bain path calculation. Calculations by Persson et al. (2000) give $C' = -10$ GPa, $C_{44} = 20$ GPa for bcc Ti, and $C' = -20$ GPa, $C_{44} = 40$ GPa for bcc Zr.

Vanadium, niobium, tantalum. All three metals have the bcc structure. A Bain path calculation by Sliwko et al. (1996) gave $C' < 0$ for fcc V. Analogous calculations by Wills et al. (1992) and by Söderlind and Moriarty (1998) gave $C' < 0$ for fcc Ta. Mrovec et al. (1999) calculated $C' = -152$ GPa and $C_{44} = -57$ GPa for fcc Nb, and Wang and Šob (private communication 2002) obtained $C' = -102$ GPa for fcc Ta. Bain path calculations by Craievich et al. (1994) gave $C' < 0$ for fcc V, Nb and Ta.

Chromium, molybdenum, tungsten. All three metals have the bcc structure. Their fcc phases are dynamically unstable, as shown by the negative values of C' obtained in calculations by Craievich et al. (1994) for Cr, Mo and W, by Einarsdotter et al. (1977) and Šob et al. (1997) for W, and by Mrovec et al. (1999) for Mo. The two latter works also give C_{44} for fcc Mo and W. While Einarsdotter et al. (1997) and Šob et al. (1997) essentially agree on C' for fcc W (–159 GPa and –142 GPa), they differ significantly for C_{44} (–128 GPa and –60 GPa). In Table 2 the value $C_{44} = -100$ GPa is chosen.

Manganese, technetium, rhenium. Manganese has the bcc structure up to about 1000 K, followed by simple cubic, fcc and again bcc structures. Technetium and rhenium have the hcp structure at all temperatures. Bain path calculations by Wills et al. (1992) for Re and by Craievich et al. (1994) for Mn, Tc and Re gave $C' > 0$ for the fcc structure and $C' < 0$ for the bcc structure. From the calculated phonon spectrum of fcc and bcc Re by Persson et al. (1999) and Bain path by Craievich et al. (1994), C' and C_{44} are crudely estimated here.

Iron, ruthenium, osmium. Iron has the bcc structure up to 1173 K, followed by fcc structure and then a return to bcc structure at 1660 K. C_{11}, C_{12} and C_{44} do not change much on the transition from bcc to fcc structure (Every and McCurdy, 1992). Ruthenium and osmium have the hcp structure at all temperatures. Bain path calculations by Craievich et al. (1994) for Fe (non-magnetic), Ru and Os gave $C' > 0$ for the fcc structure and $C' < 0$ for the bcc structure. C' for fcc and bcc Os in Table 2 are estimated here from the Bain path calculation by Wills et al. (1992) and Craievich et al. (1994), after comparison with Ir (see below), and are uncertain.

Cobalt, rhodium, iridium. Cobalt has the hcp structure up to 695 K, followed by the fcc structure. Rhodium and iridium have the hcp structure at all temperatures. Liu and Singh (1993) calculated C' for bcc Co. Šob et al. (1997) calculated $C' = -383$ GPa and $C_{44} = 217$ GPa for bcc Ir. The result that $C' < 0$ for Ir is in agreement with Bain path calculations by Wills (1992) and Craievich et al. (1994). The entries for fcc Ir in Table 2 are from Every and McCurdy (1992). Bain path calculations by Craievich et al. (1994) for Co (non-magnetic), Rh and Ir gave $C' > 0$ for the fcc structure and $C' < 0$ for the bcc structure.

Nickel, palladium, platinum. All three metals have the fcc structure. Craievich et al. (1994) calculated C' for bcc Ni and obtained a negative value. The Bain path calculation by Wills et al. (1992) for bcc Pt shows that $C' \approx 0$.

Copper, silver, gold. All three metals have the fcc structure. Calculations show very small values for C' in the bcc lattice, with C' being negative for Cu (Müller et al., 1999; Šob 2002, private communication), barely positive for Ag (Magyari-Köpe et al. 2002), and approximately zero for Au (Wills et al. 1992). Magyari-Köpe et al. (2002) also calculated C_{44} for Ag. That result is used in the present subjective estimate of C_{44} for bcc Au in Table 2.

Zinc, cadmium, mercury. Zinc and cadmium have the hcp structure with a *c/a* axis ratio that is significantly higher than the ideal value. Mercury has a low-temperature tetragonal structure followed by a rhombohedral structure. Müller et al. (1999) found C' to be slightly negative in fcc Zn, while Magyari-Köpe et al. (2002) obtained a slightly positive value. Magyari-Köpe et al. (2002) calculated both C' and C_{44} for bcc and fcc Zn.

Gallium, indium, thallium. Gallium has an orthorhombic structure and indium a tetragonal structure. Thallium has the hcp structure, followed by a bcc structure at high temperatures. Nothing seems to be known about elastic constants in metastable or dynamically unstable bcc and fcc phases for Ga and In.

Germanium, tin, lead. Germanium has the diamond-type structure. Tin has a low-temperature diamond-type structure followed by a tetragonal structure. Lead has the fcc structure. Nothing seems to be known about elastic constants in metastable or dynamically unstable bcc and fcc phases for these metals. It can be remarked, for a comparison with Ge, that Ekman et al. (2000) found bcc and fcc Si to be metallic and with $C' < 0$.

Table 2: Elastic constants $C' = (C_{11} - C_{12})/2$ and C_{44} in the 5th row in the Periodic Table. Upright numbers refer to experiments. Numbers in italics are mostly from ab initio electron structure calculations. Parentheses denote that a subjective evaluation of available information is made in the present paper. For C_{44} in bcc Hf, Os and Pt, and in fcc Os, not enough seems to be known to motivate an estimate. See the text for references and discussion on entries in the table.

Element	structure at 0 K	bcc C' (GPa)	bcc C_{44} (GPa)	fcc C' (GPa)	fcc C_{44} (GPa)
Cs	bcc	0.2	1.5	*0.2*	*1.9*
Ba	bcc	3	10	(*3*)	(*10*)
La	fcc	1	10	7	18
Hf	hcp	(*≤0*)		*28*	*67*
Ta	bcc	53	83	*-102*	*19*
W	bcc	160	160	(*-150*)	(*-100*)
Re	hcp	(-75)	(*100*)	(*100*)	(*115*)
Os	hcp	(*-400*)		(*200*)	
Ir	fcc	*-383*	(*217*)	170	263
Pt	fcc	(*≥0*)		48	77
Au	fcc	(*≥0*)	(*50*)	15	42

RESULTS AND DISCUSSION

Looking at Table 2, we see a clear difference between on one hand the elements in the beginning (Cs, Ba, La, Hf) and the end (Pt, Au) of the 5th row of the Periodic Table, and on the other hand the intermediate transition metal series (Ta, W, Re, Os, Ir). Disregarding the results from ab initio calculations that C' is slightly negative at 0 K for La and Hf (with a stabilisation due to temperature effects), and very small (or perhaps even negative) for bcc Pt and Au, we note that in the first group both the bcc and the fcc structures are dynamically stable. In contrast, the stability condition on C' for Ta, W, Re, Os and Ir is strongly violated in either the bcc or the fcc structure. This is also shown in Fig. 4. That figure includes C' for the 4th-row alloys Zr-Nb and Nb-Mo bcc alloys (Every and McCurdy, 1992), plotted as a function of the number of electrons per atom. (There seem to be no experimental data for elastic constants of the corresponding 5th-row alloys.)

We note in Table 2 that while there are many examples where $C' < 0$, there is only one entry with $C_{44} < 0$. This can be related to the fact that $C_{44} < 0$ implies a more severe lattice instability than $C' < 0$. In the former case, the lattice is unstable under a shear for both the [100] and the [110] modes; see Table 1. The only case with $C_{44} < 0$ in Table 1 is fcc tungsten, for which also $C' < 0$. Then the fcc lattice is unstable under *all* shear modes; cf. Table 1.

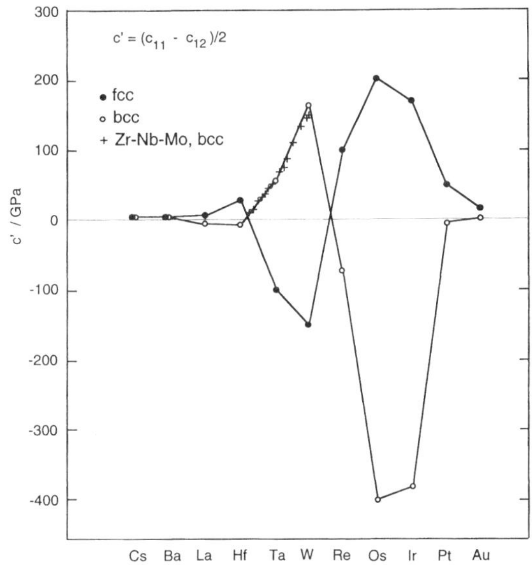

Figure 4. The elastic shear constant C' for some 5th-row elements in bcc and fcc lattice structures. See the text for details. Also shown (crosses) are experimental C' for Zr-Nb and Nb-Mo bcc alloys.

The elastic constants for the 3rd and 4th rows in the Periodic Table show a pattern very similar to that of the 5th row elements, but magnetism plays an important role for the 3rd row elements. For instance, bcc Fe is stable in both the bcc and fcc structures, while Os (which is in the same column as Fe in the Periodic Table) has a strongly negative C' in the bcc structure. For the 4th row elements, which are all non-magnetic, the available data are not as complete as for the 5th row.

A comment should be made regarding elastic anisotropy. It is usually expressed by the Zener parameter A_Z;

$$A_Z = 2\, C_{44}/(C_{11} - C_{12}) = C_{44}/C'. \tag{4}$$

A physically more relevant anisotropy parameter, A_E, was introduced by Every (Every 1980, Grimvall 1999);

$$A_E = (C_{11} - C_{12} - 2\, C_{44})/(C_{11} - C_{12}). \tag{5}$$

For an elastically isotropic system, $A_Z = 1$ and $A_E = 0$. When C' tends to zero, A_Z diverges while A_E is well defined. The bcc phase of pure Zr is dynamically stable only at high temperatures. It is of interest to consider the Zr-Nb-Mo binary random solid solution series, for which there are measured elastic constants in the bcc structure, from $Zr_{0,8}Nb_{0,2}$ to pure Mo (Every and McCurdy 1992). Figure 5 shows a smooth variation in the two anisotropy parameters, with no precursor effect in A_E as one approaches pure Zr for which $C' \approx 0$ in the bcc phase and hence A_Z diverges. Similar plots are given by Magyari Köpe et al. (2002) for fcc and bcc Ag-Zn alloys. As we noted above, there are elastic shear instabilities in those systems.

Returning to the relation between lattice instabilities and CALPHAD calculations, we note the similarity between Figs. 3 and 4. The discrepancies between the semi-empirical CALPHAD enthalpy differences $H_{bcc} - H_{fcc}$, and the corresponding values calculated ab initio are large only in the cases where either the bcc or the fcc structure is dynamically unstable.

The main part of Figure 4 refers to data for pure elements. It is so regular that one expects it to be relevant also for, e.g., binary (thermodynamically stable or hypothetical) alloys connecting two adjacent elements. There are few extensive sets of data for transition-metal alloys with components from the same row in the Periodic Table, apart from systems where magnetism may strongly affect the composition-dependence of the elastic constants. However, for the 4th-row Zr-Nb and Nb-Mo bcc alloys there are detailed experimental data for all three elastic constants. The results for C', marked with plus-signs in Fig. 4, fit well the general trend for the 5d elements.

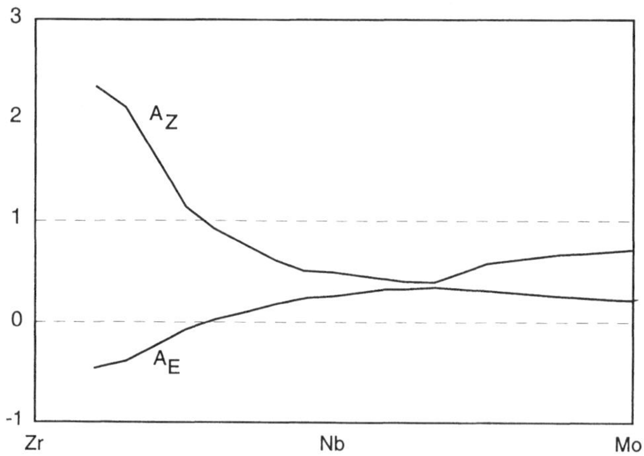

Figure 5. The Zener and Every anisotropy parameters, A_Z and A_E, for random binary Zr-Nb-Mo alloys in a sequence of increasing number of electrons per atom, and in the bcc lattice structure.

The temperature dependence of phonon frequencies usually has the form of a rather small and gradual decrease related to thermal expansion. However, in some cases of interest here the temperature dependence is anomalous and has drastic consequences. A good example is provided by Ti, Zr and Hf. They are thought to be dynamically unstable in the bcc lattice at low temperatures, but the bcc phase is not only dynamically stabilised, but also becomes the thermodynamically most stable phase with increasing temperature or dilute alloying with certain elements. The question of the precise origin of this temperature dependence has not yet been settled. A plausible mechanism is a change in the electron structure near the Fermi level, induced by increasing vibrational amplitude of the atoms, or by alloying. Features in the electronic spectrum near the Fermi level seem to be decisive also for the elastic constants of Nb-Mo alloys, possibly related to electronic topological transitions, cf. Bruno et al. (1994). A closer inspection of the Nb-Mo curve in Fig. 4 reveals significant changes in slope around 20 % Nb. Similar anomalous features are seen in C_{44} (Every and McCurdy, 1992). At these compositions there is also an anomalous temperature dependence, which is particularly pronounced for C_{44} (Every and McCurdy, 1992).

In conclusion, when data for elastic shear constants of metals obtained from experiments are combined with data from ab initio electron structure calculations, a striking and regular pattern emerges. For the elements in the middle of the transition metal series in the Periodic Table, the shear constants $C' = (C_{11} - C_{12})/2$ in assumed bcc and fcc lattice structures co-vary in such a way that a large and positive value of C' in the bcc structure implies a large and negative value in the fcc structure, and vice versa. Of the two stability criteria on the elastic shear constants in cubic structures, i.e., $C' > 0$ and $C_{44} > 0$, violating the latter means a more severe instability. Among the transition metals, it has so far been noted only for the Cr, Mo, W column in the Periodic Table.

ACKNOWLEDGMENT

This work has been supported in part by the Swedish research organisation SSF, through the project ATOMICS. I particularly thank Mojmir Šob for correspondence and for providing numerical values extracted from his works. I also thank Blanka Magyari-Köpe, Kristin Persson (née Einarsdotter) and Armando Fernández Guillermet for discussions and help.

REFERENCES

Aguayo, A., Murrieta, G., and de Coss, R., 2002, Elastic stability and electronic structure of fcc Ti, Zr, and Hf: A first-principles study, *Phys. Rev. B* **65**, 092106-1-092106-4.

Alippi, P., Marcus, P. M., and Scheffler, M., 1997, Strained tetragonal states and Bain paths in metals, *Phys. Rev. Lett.* **78**, 3892-3895.

Bruno, E., Ginatempo, B., Guiliano, E. S., Ruban, A. V., and Vekilov, Yu. Kh., 1994, Fermi surfaces and electronic topological transitions in metallic solid solutions, *Phys. Reports* **249**, 353-419.

Chen, Y., Ho, K. M., and Harmon, B. N., 1988, First-principles study of the pressure-induced bcc-hcp transition in Ba, *Phys. Rev. B* **37**, 283-288.

Craievich, P. J. and Sanchez, J. M., 1997, Vibrational free energy of the Ni-Cr system, *Comp. Mater. Science* **8**, 92-99.

Craievich, P. J., Sanchez, J. M., Watson, R. E., and Weinert, M., 1997, Structural instabilities of excited phases, *Phys. Rev. B* **55**, 787-797.

Craievich, P. J., Weinert, J., Sanchez, J. M., and Watson, R. E., 1994, Local stability of nonequilibrium phases, *Phys. Rev. Lett.* **72**, 3076-3079.

Einarsdotter, K., Sadigh, B., Grimvall, G., and Ozoliņš, V., 1997, Phonon instabilities in fcc and bcc tungsten, *Phys. Rev. Lett.* **79**, 2073-2076.

Ekman, K., Persson, K., and Grimvall, G., 2000, Lattice dynamics and thermodynamic properties of β-Sn phase in Si, *Phys. Rev. B* **62**, 14784-14789.

Every, A. G., 1980, General closed-form expressions for acoustic waves in elastically anisotropic solids, *Phys. Rev. B* 22, 1746-1760.

Every, A. G. and McCurdy, A. K., 1992, Second and higher order elastic constants, in: *Landolt-Börnstein Tables*, New Series III/29, D. F. Nelson, ed., Springer-Verlag, Berlin

Grimvall, G. and Ebbsjö, I., 1975, Polymorphism in metals, *Phys. Scr.* **12**, 168-172.

Grimvall, G., 1998, Reconciling ab initio and semiempirical approaches to lattice stabilities, *Ber. Bunsenges. Phys. Chem.* **102**, 1083-1087.

Grimvall, G., 1999, *Thermophysical Properties of Materials. Enlarged and Revised Edition*, North-Holland, Amsterdam.

Grimvall, G., 2002, Calphad and ab initio approaches to lattice stabilities, in: *CALPHAD and Alloy Thermodynamics*, Eds. P. E. A. Turchi, A. Gonis and R. D. Shull, The minerals, metals & materials society, pp. 81-91.

Güthoff, F., Petry, W., Stassis, C., Heiming, A., Hennion, B., Herzig, C., and Trampenau, J., 1993, Phonon dispersion of bcc La, *Phys. Rev. B* 47, 2563-2572.

Jianjun Xie, Chen, S. P., Tse, J. S., Klug, D. D., Zhiqiang Li, Uehara, K., and Wang, L. G., 2000, Phonon instabilities in high-pressure bcc-fcc and the isostructural fcc-fcc phase transitions of Cs, *Phys. Rev. B* **62**, 3624-3629.

Kraft, T., Marcus, P. M., Methfessel, M., and Scheffler, M., 1993, Elastic constants of Cu and the instability of its bcc structure, *Phys. Rev. B* **48**, 5886-5890.

Liu, A. Y. and Singh, D. J., 1993, Elastic instability in bcc cobalt, *Phys. Rev. B* 47, 8515-8519.

Magyari-Köpe, B., Grimvall, G., and Vitos, L., 2002, Elastic anomalies in Ag-Zn allolys, *Phys. Rev. B* **66**, 064210-7.

Milstein, F., Fang, H. E., and Marschall, J., 1994, Mechanics and energetics of the Bain transformation, *Phil. Mag. A* **70**, 621-639.

Moriarty, J. A., 1973, Hybridization and the fcc-bcc phase transitions in calcium and strontium, *Phys. Rev. B* **8**, 1338-1345.

Mrovec, M., Vitek, V., Nguyen-Manh, D., Pettifor, D., Wang, L. G., and Šob, M., 1999, Bond-order potentials for molybdenum and niobium: An assessment of their quality, in: *Multiscale Modelling of Materials*, eds. V. V. Bulatov, T. Diaz de la Rubia, R. Phillips, E. Kaxiras, and N. Ghoniem, MRS Symposium Proc. Vol. 538, Materials Research Society, Pittsburgh, pp. 529-534.

Müller, S., Wang, L.-W., Zunger, A., and Wolverton, C., 1999, Coherent phase stability in Al-Zn and Al-Cu fcc alloys: The role of the instability of fcc Zn, *Phys. Rev. B* **60**, 16 448-16462.

Persson, K., Ekman, M., and Grimvall, G., 1999, Dynamical and thermodynamical instabilities in the disordered Re_xW_{1-x} system, *Phys. Rev. B* **60**, 9999-10007.

Persson, K., Ekman, M., and Ozoliņš, V., 2000, Phonon instabilities in bcc Sc, Ti, La, and Hf, *Phys. Rev. B* **61**, 11221-11224.

Petry, W., Trampenau, J., and Herzig, C., 1993, Phonon dispersion of β-Sc, *Phys. Rev. B* **48**, 881-886.

Sliwko, V. L., Mohn, P., Schwarz, K., and Blaha, P., 1996, The fcc-bcc structural transition: I. A band theoretical study for Li, K, Rb, Ca, Sr, and the transition metals Ti and V, *J. Phys. Condens. Matter* **8**, 799-815.

Šob, M., Wang, L. G., and Vitek, V., 1997, Local stability of higher-energy phases in metallic materials and its relation to the structure of extended defects, *Comp. Mater. Sci.* 8, 100-106.

Šob, M., Friak, F., Wang, L. G., and Vitek, V., 2002, The role of ab initio electronic structure calculations in contemporary materials science, in: *Symp. L of the Int. Conf. on Materials for Advanced Technologies* (ICMAT 2002, Singapore, 2001), eds. C.-H. Chiu, Z. Chen, H. Gao, K. Y. Lam, and A. A. O. Tay, Key Engineering materials 227, p. 261-272.

Söderlind, P. and Moriarty, J. A., 1998, First-principles theory of Ta up to 10 Mbar pressure: Structural and mechanical properties, *Phys. Rev. B* **57**, 10340-10350.

Wang, L. G. and Šob, M., 1999, Structural stability of higher-energy phases and its relation to the atomic configurations of extended defects: The example of Cu, *Phys. Rev. B* **60**, 844-850.

Wills, J. M., Eriksson, O., Söderlind, P., and Boring, A. M., 1992, Trends of the elastic constants of cubic transition metals, *Phys. Rev. Lett.* **68**, 2802-2005.

Young, D.A., 1991, *Phase Diagrams of the Elements*, Univ. Calif. Press, Berkeley.

THEORETICAL STRENGTH, MAGNETISM AND STABILITY OF METALS AND INTERMETALLICS

Mojmír Šob,[1] Martin Friák,[1,2] Dominik Legut,[1,3] and Václav Vitek[4]

[1]Institute of Physics of Materials, Academy of Sciences of the Czech Republic, Žižkova 22, CZ-616 62 Brno, Czech Republic
[2]Institute of Condensed Matter Physics, Faculty of Science, Masaryk University, Kotlářská 2, CZ-611 37 Brno, Czech Republic
[3]Department of Chemistry of Materials, Faculty of Chemistry, Brno University of Technology, Purkyňova 118, CZ-612 00 Brno, Czech Republic
[4]Department of Materials Science and Engineering, University of Pennsylvania, 3231 Walnut St., Philadelphia, PA 19104-6272, U.S.A.

1. INTRODUCTION

The electronic structure (ES) of materials, which in the general sense determines all their physical properties, can be obtained accurately from ab initio (first-principles) ES calculations, i.e. from fundamental quantum theory. Here the atomic numbers of constituent atoms and, usually, some structural information are employed as the only input data. Such calculations are routinely performed within the framework of the density functional theory in which the complicated many-body interaction of all electrons is replaced by an equivalent but simpler problem of a single electron moving in an effective potential. For a given material, the calculated total energies are used to obtain equilibrium lattice parameters, elastic moduli, relative stabilities of competing crystal structures, energies associated with point and planar defects, etc. In addition, we also obtain information about electronic densities of states and charge densities that enables us to attain a deeper insight and learn which aspects of the problem are important.

Recently, theoretical calculations of strength of materials became possible using ab initio ES calculations. In most engineering applications, the strength of materials is limited by the presence of internal defects. If there were no defects, then, for example, the tensile strength could be several orders of magnitude higher and its value would be comparable with that of the Young modulus. Therefore, the strength of ideal (defect-free) crystals sets up an upper limit of attainable stresses, and may be called the ideal strength. While it may not be possible to achieve the ideal strength in practice, it is not possible to exceed it. The ideal strength has already been approached in situations that are technologically relevant. These include the low-temperature deformation of inherently strong materials, such as diamond, Si, Ge and some of the transition-metal carbonitrides, deformation of whiskers, nanoindentation of materials with low defect densities, hardened thin films and coatings etc. Its knowledge is useful for estimation of the ideal work of fracture, stresses needed for homogeneous nucleation of dislocations and for modeling of crack propagation. Therefore, the determination of the ideal (theoretical) strength is of principal interest in many applications. Understanding its source and characteristics can help to identify those aspects of mechanical behavior that are fundamental consequences of crystal structure and bonding.

Until now, most calculations of the theoretical strength of materials have been based on empirical or semiempirical interatomic potentials (for a review see e.g. Ref. 1 and the references therein; ideal shear strengths for all basic cubic structures calculated by means of semiempirical potentials may be found in Ref. 2). However, these interatomic potentials are fitted to the properties of the equilibrium ground state and, therefore, it is not guaranteed that they are applicable when the material is loaded close to its theoretical strength limit, very far from the equilibrium state.

In contrast, this is not a problem for ab initio ES calculations. Nevertheless, most of the ES calculations were directed towards finding the equilibrium state of a given material that corresponds to the minimum of the total energy or towards analysis of relatively small deviations from that state. On the other hand, theoretical strength is related to the maximum force that may be applied to the material without perturbing its stability. It is usually connected with an inflexion point on the dependence of the total energy on deformation parameters.

The first paper dealing with the ideal tensile strength from the first principles was that of Esposito et al.[3]. However, those authors have not performed relaxations of dimensions of the loaded crystal in the directions perpendicular to the loading axis. Paxton et al.[4] and Xu and Moriarty[5] calculated shear strength for unrelaxed shear deformation. Other ab initio calculations of properties of the systems far from equilibrium have also been made, such as exploration of the structural stability, but the results were not employed to evaluate the strength[6-10].

Our group at the Institute of Physics of Materials in Brno performed the first ab initio simulation of a tensile test (including the relaxation in perpendicular directions to the loading axis) and obtained the theoretical tensile strength in tungsten[11]. The results compared very well with experiments performed on tungsten whiskers by Mikhailovskii et al.[12] Further, we calculated ideal tensile strength in NiAl[13] and Cu[14]. These results found a very good response in the international solid state physics and materials science community and established a basis for further calculations of ideal tensile strength. Li and Wang computed the ideal tensile strength in Al[15] and in SiC[16]. The group at the University of California at Berkeley calculated ideal shear strength in Al and Cu[17], performed a thorough theoretical analysis of the problem of strength and elastic stability[18] and, among others, verified our values of ideal tensile strength for tungsten[19].

From 1997, ab initio calculations of theoretical strength under isotropic triaxial (hydrostatic) tension (i.e., negative hydrostatic pressure) also appear[20-25]. As the symmetry of the structure does not change during this deformation, simpler ab initio approaches may be applied.

Very recently, we have simulated a tensile test in prospective high-temperature materials, namely in transition metal disilicides $MoSi_2$ and WSi_2 with the $C11_b$ structure. This study included calculation of the tensile strength for [001] loading and analysis of bonds and their changes during the test[26]. Theoretical tensile strength of iron in the loading direction [001] was determined in Refs. 27 and 28; in Ref. 29, we compared those results to each other and calculated the tensile strength of iron for uniaxial loading in the [111] direction. Theoretical tensile strength of C, Si and Ge for uniaxial loading in the [001] direction was calculated in Ref. 30. Tables summarizing ab initio values of theoretical tensile strengths for various materials are given in Refs. 29, 31 and 32; Ref. 32 includes also ab initio values of shear strengths and some semiempirical results. An extensive review of the semiempirical and ab initio calculated values of uniaxial and isotropic triaxial tensile strengths as well as of shear strengths calculated up to 1999 can be found in Ref. 33.

There are more general problems of stability of materials and of phase transformations that are closely related to the tensile tests described above. Namely, the tensile test may be considered as a special case of so-called displacive phase transformation path[6,8]. These paths are well known in studies of martensitic transformations. Such transformations play a major role in the theory of phase transitions. They proceed by means of cooperative displacements of atoms away from their lattice sites that alter crystal symmetry without changing the atomic order or composition. A microscopic understanding of the mechanisms of these transformations is vital since they occur prominently in many materials.

Displacive phase transformations are also of interest in studies of epitaxial thin films. Pseudomorphic epitaxy of a cubic or tetragonal (001) film typically results in a strained tetragonal structure. In this case, there is a stress in the (001) plane keeping the structure of the film and of the substrate coherent, and the stress perpendicular to this plane vanishes. A tetragonal phase arises that may be stable or metastable[10]. Similarly, an epitaxial film grown on the (111) plane of a cubic substrate exhibits a trigonal deformation of its lattice, which may be considered as a uniaxial deformation along the [111] axis. In this context it is interesting to consider the trigonal deformation path, which comprises the bcc, fcc and simple cubic structures as special cases[34-36,8]. We have also proposed the bcc to hcp transformation path, which we investigated in iron[37] and in intermetallic compounds TiAl and NiAl[38]. Very recently, we have studied the tetragonal bcc-fcc transformation path in iron and have shown how these results may be used to predict the lattice constants and magnetic order of iron overlayers on various metallic substrates[39,40]. This research is closely connected with the studies of iron stability performed in Refs. 41-46.

Another area of importance of the local stability of non-equilibrium phases and phase transformation paths is the structure of extended defects in solids. It was found in recent studies that atomic configurations in grain boundary (GB) regions, or at other interfaces, may contain certain metastable structures, different from the ground-state structures. For example, the 9R (α-Sm) structure was theoretically predicted and verified by high-resolution electron microscopy (HREM) at GBs in silver and copper[47,48]. Similarly, the bcc structure was found at certain grain boundaries in copper[49]. We have shown[50] that the bcc Cu at grain boundaries is stabilized by external constraints exerted by the surrounding fcc grains.

Occurrence of such phases at interfaces is even more likely in more complex non-cubic alloys. For example, new structural features of TiAl, which crystallizes in tetragonal $L1_0$ structure, have been discovered recently. Abe et al.[51] found a B19-type hcp-based structure in a Ti-48at.%Al alloy quenched from the disordered phase, and Banerjee et al.[52] observed a series of structural transitions in the form of changes in the stacking sequence of the close-packed atomic planes in the Ti and Al layers in Ti/Al multilayered thin films.

Consequently, in order to explore adequately extended defects both in pure metals and alloys, in particular in intermetallics, detailed information about possible metastable structures, as well as lattice transformations connecting them, is needed. Armed with this knowledge one can predict whether an interface may be associated with a metastable structure and assess thus its stability and ability to transform to other structures (for example during deformation or due to changes in stoichiometry).

The purpose of the present paper is to study lattice configurations found in the course of tetragonal and trigonal displacive transformation (deformation) paths. These configurations are produced by large homogeneous distortions that transform the initial (ground-state) structure into new (higher-energy) structures with different symmetries. Such investigations are closely linked with theoretical strength and phase transformations and constitute a basis for future analyses of various configurations of extended defects in metallic materials. As specific examples, we study iron and the intermetallic compound Ni_3Al. Iron exists in both bcc and fcc modifications and has many magnetic phases, especially in thin films. Notably, fcc iron films exhibit a large variety of structural and magnetic properties that depend delicately on the thickness of the iron layer and preparation conditions[53]. Ni_3Al is the most important strengthening constituent of commercial nickel-based superalloys used extensively as structural materials for elevated temperatures applications. This phase is responsible for the high-temperature strength and creep resistance of the superalloys. Ni_3Al and a number of other intermetallic compounds with the $L1_2$ structure exhibit so called anomalous yield behavior, when their yield strength increases rather than decreases with increasing temperature. This behavior is not the result of a change in long-range order with temperature since the (Bragg) long-range parameter, S, is almost constant in Ni_3Al up to 1000 °C. There is now near-universal agreement that the anomaly results from the special properties of screw dislocation cores and the anisotropy in antiphase boundary energies; a review and comparison of various models are presented in Ref. 54. Single crystals of Ni_3Al are ductile, but pure polycrystalline Ni_3Al is very brittle at room temperature because of intergranular fracture. Both iron and Ni_3Al exhibit magnetic ordering; therefore, we also study the changes of the magnetic state of these materials during deformation.

2. DISPLACIVE PHASE TRANSFORMATION PATHS

We consider two simple transformation paths connecting cubic structures. They are the bcc-fcc transformation path via tetragonal deformation corresponding to extension along the [001] axis (the usual Bain's path) and the trigonal deformation path that corresponds to uniaxial deformation along the [111] axis (Figs. 1 and 2).

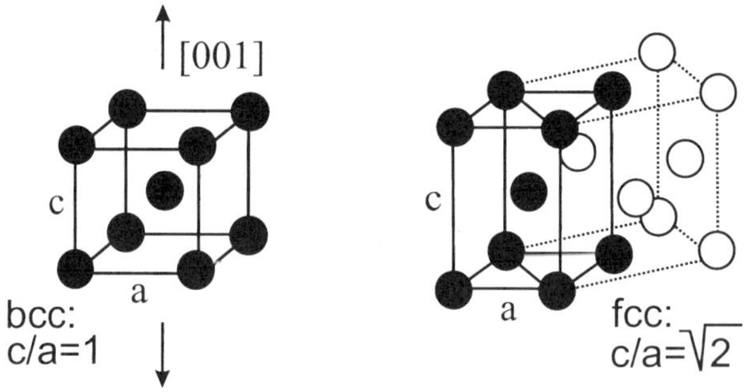

Figure 1: High-symmetry structures obtained along the tetragonal deformation path. The c and a are the length scales along the [001] and [100] directions, respectively. The original bcc cell is indicated by filled circles and heavy solid lines.

In the case of tetragonal deformation path, we start with the bcc structure considered as a tetragonal one with the ratio $c/a = 1$, where c is measured along the [001] direction and a along a [100] direction. When c/a is varied, we arrive at body-centered tetragonal structures. There is one exception: for $c/a = \sqrt{2}$ the structure becomes fcc (Fig. 1).

Similarly, we may consider the bcc structure as trigonal with the ratio of $c/a = 1$, where c is measured along the [111] direction and a along a direction perpendicular to [111]. If $c/a \neq 1$, the structure becomes trigonal except for $c/a = 2$, when we attain the simple cubic (sc) structure, and $c/a = 4$, which again corresponds to the fcc structure (see Fig. 2).

When studying the behavior of the total energy along the deformation paths, one usually assumes that the atomic volume is constant. Then both deformation paths discussed above may be fully parameterized by the ratio c/a.

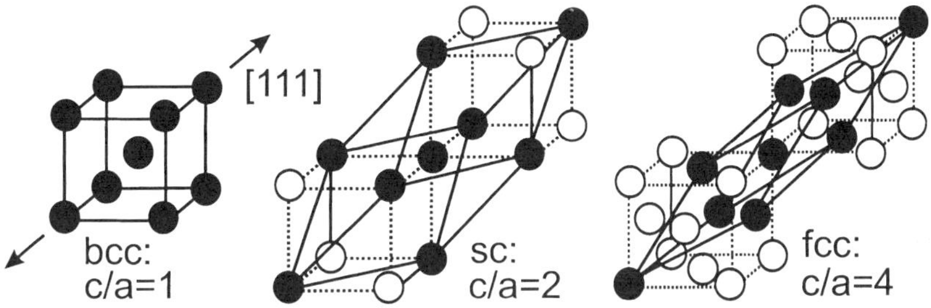

Figure 2: High-symmetry structures obtained along the trigonal deformation path. The c and a are the length scales along the [111] direction and along a direction perpendicular to [111], respectively.

Analogous deformation paths may be devised for intermetallic compounds with B2, $L1_2$ or $D0_3$ structures[8,55]. In the $L1_2$ (Cu_3Au) structure, the atoms are at the fcc positions with the (002) planes occupied alternatively by Cu atoms and by Cu and Au atoms in the same ratio. We may consider this structure as tetragonal or trigonal with the ratio $c/a = 1$. Now, performing a tetragonal deformation, the cubic symmetry of the $L1_2$ configuration is lost and becomes tetragonal, even for $c/a = \sqrt{2}/2$, when atoms are at the bcc-like positions, but because we have two kinds of atoms, the structure does not attain the cubic symmetry. When we perform the trigonal deformation, the structure becomes trigonal except for the case of $c/a = 0.5$, when we encounter a simple-cubic-based structure that, indeed, has a cubic symmetry. For $c/a = 0.25$, the atoms adopt the bcc-like positions, but the symmetry of the structure remains trigonal. All these structures will be characterized in more details in a subsequent publication[55]. (Note different "normalization" of the ratio c/a for the $L1_2$ structure. Here we ascribed the value of $c/a = 1$ to the fcc-based configuration. Consequently, as it may be seen from Figs. 1 and 2, the bcc-based configuration, obtained by the tetragonal deformation, corresponds to $c/a = \sqrt{2}/2$, and the sc- and bcc-based structures, obtained by the trigonal deformation, correspond to $c/a = 0.5$ and $c/a = 0.25$, respectively.)

Craievich et al.[6] have shown that some energy extrema on constant-volume transformation paths are dictated by the symmetry. Namely, most of the structures encountered along the transformation paths between some higher-symmetry structures, say between bcc and fcc at the Bain's path, have a symmetry that is lower than cubic. At those points of the transformation path where the symmetry of the structure is higher, the derivative of the total energy with respect to the parameter describing the path must be zero. These are the so-called symmetry-dictated extrema. However, other extrema may occur that are not dictated by the symmetry and reflect properties of the specific material. The same is true for the transformation paths corresponding to uniaxial loading[10,56]. Configurations corresponding to energy minima at the transformation paths represent stable or metastable structures and may mimic atomic arrangements that could be encountered when investigating thin films[10] and extended defects such as interfaces or dislocations[8,38]. For iron and Ni_3Al, we will discuss these configurations below.

3. TENSILE TEST SIMULATION

To simulate a uniaxial tensile test, we start by determining the structure and total energy of the material in the ground state. Then, in the second step, we apply an elongation along the loading axis by a fixed amount ε that is equivalent to application of a certain tensile stress σ. For each value of ε, we minimize the total energy by relaxing the stresses σ_1 and σ_2 in the directions perpendicular to the loading axis. The stress σ is given by

$$\sigma = \frac{c}{V}\frac{\partial E}{\partial c} = \frac{1}{Ac_0}\frac{\partial E}{\partial \varepsilon}, \qquad (1)$$

where E is the total energy per repeat cell, V is the volume of the repeat cell, c is the dimension of the repeat cell in the direction of loading, A (equal to V/c ratio) is the area of the basis of the repeat cell in the plane perpendicular to the loading axis, and c_0 is the value of c in the undeformed state.

We are also interested in tensile strength at isotropic triaxial (hydrostatic) tension. In this case, we start again with the material in its ground-state structure, but the dimension of the crystal is gradually increased homogeneously in all directions. The hydrostatic stress σ is then calculated using the formula $\sigma = dE/dV$.

The inflexion point in the dependence of the total energy on elongation yields the maximum of the tensile stress; if any other instability (violation of some stability condition, soft phonon modes, magnetic spin arrangement etc.) does not occur prior to reaching the inflexion point, it also corresponds to the theoretical tensile strength, σ_{th}.

4. METHODS OF CALCULATION

The atomic configurations corresponding to the deformed structures have usually a lower symmetry and, at the strength limit, they are very far from the ground state. Therefore, to get reliable structural energy differences, we must use a full-potential method for the calculations. Here we use the full-potential linearized augmented plane wave (FLAPW) code WIEN97 described in detail in Ref. 57. In the FLAPW method, no assumptions are made about the potential or charge density and the muffin-tin geometry is used only when constructing the basis functions. The total energy functional is evaluated for the full charge density and no spheroidization is introduced. The calculations are performed self-consistently, including all electrons present in the material. The exchange-correlation energy is evaluated within the generalized-gradient approximation (GGA)[58]. This is important especially for iron, since the local density approximation does not render the ground state of iron correctly. The muffin-tin radius of iron atoms of 1.90 au is kept constant for all calculations. The number of $\boldsymbol{k}$-points in the whole Brillouin zone is equal to 6000 and the product of the muffin-tin radius and the maximum reciprocal space vector,

$R_{MT}\, k_{max}$, is set to 10. The maximum l value for the waves inside the atomic spheres, l_{max}, and the largest reciprocal vector in the charge Fourier expansion, G_{max}, is equal to 12 and 15, respectively. In the case of Ni_3Al, the muffin-tin radii of both Ni and Al atoms are equal to 2.0 au, number of $\boldsymbol{k}$-points in the whole Brillouin zone is 4000, and the product $R_{MT}\, k_{max} = 8$. The values of l_{max} and G_{max} are 12 and 10, respectively.

5. RESULTS AND DISCUSSION

5.1. Iron

5.1.1. Total energies and magnetic states of tetragonally and trigonally deformed iron. We have calculated the total energy and magnetic moment of iron deformed along the tetragonal and trigonal paths at constant atomic volumes ranging from $V/V_{exp} = 0.84$ to $V/V_{exp} = 1.05$, where V_{exp} is the experimental equilibrium atomic volume of the ferromagnetic bcc iron corresponding to the lattice constant $a_{bcc} = 5.408$ au. As shown in Figs. 3 and 4, we include non-magnetic (NM), ferromagnetic (FM) and two antiferromagnetic states, namely the single-layer antiferromagnetic state (AFM1), in which the (001) or (111) planes have alternating magnetic moments (↑↓↑↓ ...), and the double-layer antiferromagnetic state (AFMD), where the pairs of (001) or (111) planes have alternating magnetic moments (↑↑↓↓ ...). The total energy of iron is plotted as a function of volume and the c/a ratio in Figs. 5 and 6. We show only those states the energies of which are the lowest for a given configuration. In Fig. 5, we can clearly see the "horseshoes" dividing the plane into the AFM1, AFMD and FM regions whereas the area of Fig. 6 is dominated by the FM states. The global minimum of energy is in the FM region at $c/a = 1$, $V/V_{exp} = 0.985$, which corresponds to the bcc structure. The calculated equilibrium volume is about 1.5 % lower than the experimental value, which may be considered as a very good agreement.

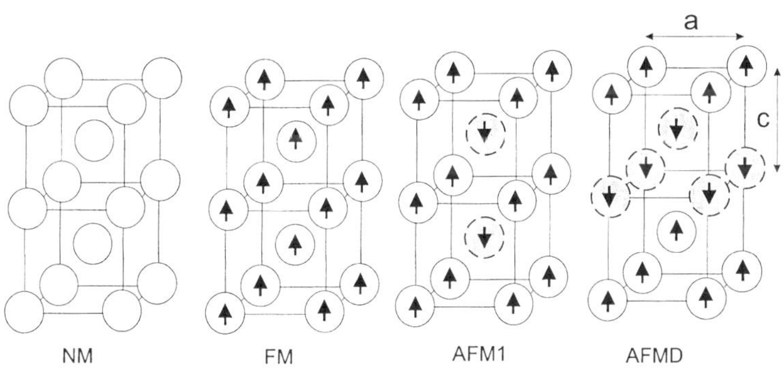

Figure 3. Non-magnetic (NM), ferromagnetic (FM), antiferromagnetic single-layer (AFM1) and antiferromagnetic double layer (AFMD) states of iron included in calculations of total energy profiles along tetragonal deformation paths.

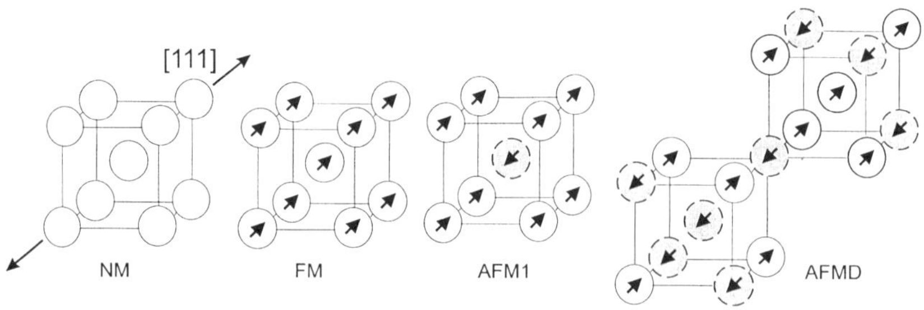

Figure 4. Non-magnetic (NM), ferromagnetic (FM), antiferromagnetic single-layer (AFM1) and antiferromagnetic double layer (AFMD) states of iron included in calculations of total energy profiles along trigonal deformation paths. (Note that these figures do not display all the (111) planes in the lattices shown).

The ground-state energy minimum is dictated by the symmetry. Any energy profile at the constant volume, obtained from Figs. 5 and 6, also exhibits the minimum at $c/a = 1$.

Let us discuss the tetragonal case first (Fig. 5). Apart of the large FM area, there are AFMD and AFM1 regions in the neighborhood of the fcc structure, which corresponds to the line $c/a = \sqrt{2}$. Note that the lattice symmetry of the fcc iron with the AFM1 and AFMD spin ordering is tetragonal and, therefore, we do not find any extremum of the total energy of these states (dictated by symmetry) at $c/a= \sqrt{2}$. In accordance with Ref. 42, we found that the fcc iron with the AFM1 or AFMD spin ordering is unstable with respect to the tetragonal deformation. A more detailed discussion of the tetragonal case is presented in Refs. 39 and 40. In those papers, we also showed how the contour plot presented in Fig. 5 may be used to predict the lattice parameters and magnetic states of iron overlayers at (001) substrates.

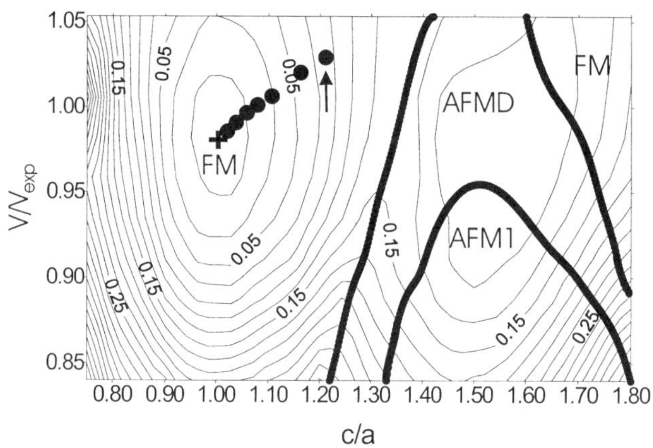

Figure 5. Total energy (per atom) of iron as a function of the tetragonal *c/a* ratio and volume relative to the energy of the FM bcc equilibrium state calculated within the GGA. Only states with the minimum energy are shown. The contour interval is equal to 20 meV. Thick lines show the FM/AFMD and AFMD/AFM1 phase boundaries. The cross corresponds to the global, symmetry-dictated minimum (ground state). The path representing the simulation of the tensile test for loading along the [001] direction is denoted by full circles; the highest circle marked by an arrow corresponds to the maximum stress obtained in the simulation of the tensile test.

The AFMD structure with the tetragonal symmetry may be considered as a close approximation of the spin-spiral state with $\mathbf{q} = (2\pi/a)\ (0, 0, 0.6)$, found as the ground state of the fcc iron[59]. It will be the topic of future studies to ascertain how the non-collinearity of magnetic moments changes the borders between various magnetic phases in the $(c/a, V/V_{exp})$ plane. We surmise that the region of non-collinear magnetism will not be too different from the AFMD region shown in Fig. 5.

The AFM1 and AFMD states with the trigonal symmetry (Fig. 4) have mostly higher energy than the FM states and, consequently, they are nearly invisible in Fig. 6, except for the lower right corner. However, two regions of the FM states may be found in Fig. 6: FM(HS), the high-spin states (with magnetic moment higher than about 2 μ_B) and FM(LS), the low-spin states (with magnetic moment lower than about 1.2 μ_B). There is a sharp discontinuity in the magnetic moment at the border FM(HS)/FM(LS). Nonetheless, the total energy remains surprisingly smooth. The triangles in Fig. 6 denote local energy minima of fcc FM states and the square marks the point where the volume dependencies of the total energies of the fcc FM(HS) and FM(LS) states, displayed in Fig. 7, intersect. From Fig. 7 we see that the square represents a "sharp" saddle point in Fig. 6.

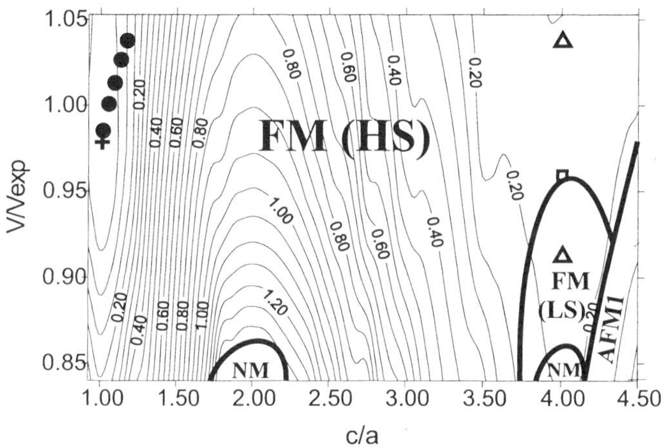

Figure 6. Total energy (per atom) of iron as a function of the trigonal c/a ratio and volume relative to the energy of the FM bcc equilibrium state, calculated within the GGA. Only states with the minimum energy are shown. The contour interval is equal to 50 meV. Thick lines show the FM(HS)/NM, FM(HS)/FM(LS), FM(LS)/NM and FM(LS)/AFM1 phase boundaries. The cross corresponds to the global symmetry-dictated minimum (ground state), the triangles show the local minima of the total energy of the fcc states in the FM(HS) and FM(LS) region at $V/V_{exp} = 1.037$ and 0.911, respectively. The square at $V/V_{exp} = 0.955$ denotes the crossing point of the dependencies of the total energy of the FM(HS) and FM(LS) fcc states on volume, presented in Fig. 7. As Fig. 7 shows, this square represents a "sharp" saddle point. The path representing simulation of the tensile test for loading along the [111] direction is denoted by full circles; the state corresponding to the maximum stress attained in the tensile test simulation ($V/V_{exp} = 1.114$, $c/a = 1.356$) lies outside the area of the figure.

All total energy profiles at a constant volume $V > 0.955\ V_{exp}$ exhibit three symmetry-dictated extrema: a minimum at $c/a = 1$ (bcc structure), a maximum at $c/a = 2$ (sc structure) and a minimum at $c/a = 4$ (fcc structure). The reason is that both FM and NM structures exhibit a higher (cubic) symmetry at those values of c/a. The profile of the total energy at the constant volume $V = V_{exp}$, obtained from Fig. 6, is shown in Fig. 8. It is qualitatively similar to the total energy profiles of trigonally deformed Ta[60,61] or Ir[8] and Cu[50] (the ground-state structure of Ir and Cu is fcc and, therefore the fcc minimum for those metals is lower than the bcc minimum).

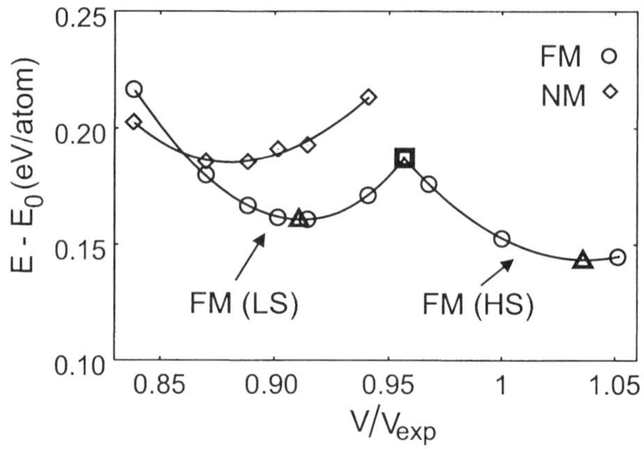

Figure 7. Total energies of the fcc FM and NM states of iron as functions of volume relative to the total energy of the FM bcc ground state. The triangles denote local energy minima (for their exact position see the description of Fig. 6), and the square corresponds to the intersection of the FM(HS) and FM(LS) curves.

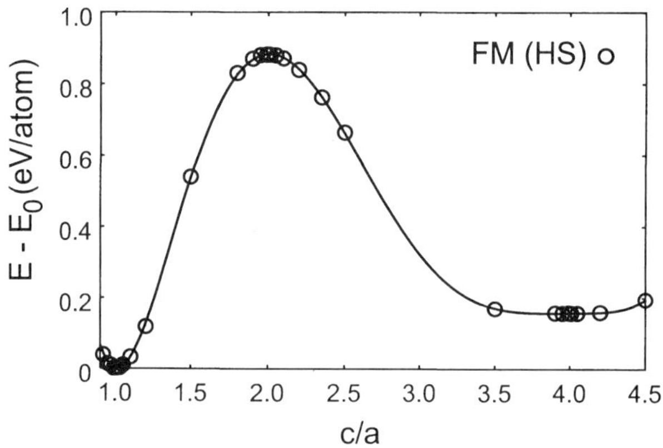

Figure 8. The profile of the total energy of trigonally deformed iron at $V/V_{exp} = 1$ (cf. with Fig. 6). All energy extrema (at c/a = 1, 2 and 4) are dictated by symmetry.

5.1.2. Uniaxial and isotropic triaxial tensile tests. In accordance with methodology described in Sec. 3, we performed the simulation of a tensile test in iron for uniaxial loading along the [001] and [111] directions, respectively, as well as for isotropic triaxial loading corresponding to the negative hydrostatic pressure. The corresponding total energies as functions of relative elongation ε are displayed in Fig. 9(a). In case of the isotropic triaxial loading, ε corresponds to a relative extension of the bcc lattice parameter

It is seen from Fig. 9(a) that the total energy profiles have a parabolic, convex character in the neighborhood of the ferromagnetic (FM), symmetry-dictated, minimum that corresponds to the bcc structure (ground state). With increasing value of ε the curves reach (due to non-linear effects) their inflexion points (marked by vertical lines in Fig. 9(a)) and become concave. The inflexion point for [001] uniaxial loading occurs (most likely incidentally) for nearly the same elongation of ε = 0.15 as for the isotropic triaxial loading. In the case of the [001] tensile test, this elongation corresponds to the lattice parameter in the direction of loading equal to 6.20 au (accompanied by relaxation in [100] and [010] directions in which the lattice constant decreases to 5.12 au) and, in the case of isotropic triaxial strain, to the bcc structure with the lattice constant of 6.20 au.

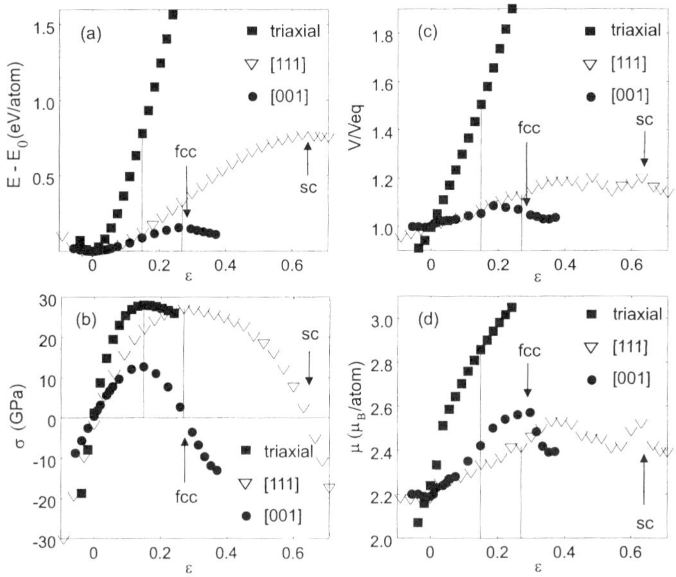

Figure 9. Total energy per atom measured with respect to the energy of the equilibrium state (a), tensile stress (b), relative atomic volume ratio measured with respect to the equilibrium volume V_{eq} (c), and magnetic moment per atom μ (d) of FM iron loaded triaxially (full squares) and uniaxially along the [001] (full circles) and [111] (empty triangles) directions vs. elongation ε. The relative elongation ε reflects the changes of the lattice parameter a_{bcc} for isotropic triaxial loading and, in the case of uniaxial tensile tests, the increase/decrease of the crystal dimension in the directions of loading. The thin vertical lines mark the states exhibiting maximum stress (i.e. theoretical tensile strength). Incidentally, the maximum stresses for [001] uniaxial and isotropic triaxial loading are reached at nearly the same strain ε.

The tensile stresses calculated according to formulas given in Sec. 3 are shown in Fig. 9(b). The inflexion points on the total energy profiles correspond to maximum stresses which the material may accommodate if its structure type does not change during the deformation. They are equal to $\sigma^{[001]}_{max}$ = 12.7 GPa (this value was reported in our previous work[27] and is not very different from 14.2 GPa found in Ref. 28), $\sigma^{[111]}_{max}$ = 27.3 GPa and $\sigma^{[triaxial]}_{max}$ = 27.9 GPa for uniaxial tensile test along the [001] and [111] direction and for isotropic triaxial loading, respectively. These values represent the theoretical tensile strengths provided other instabilities (soft phonon modes, etc.) do not come forth before reaching the inflexion point. In the case of iron with its large variety of magnetic phases, another instability may originate from transitions between those phases. However, as it is seen from Figs. 5 and 6, no such transition appears during tensile tests along [001] and [111] directions (all states involved up to the maximum stress lie in the FM region). A similar situation arises for isotropic triaxial deformation[25]. Other conditions of stability[62,63] will be analyzed in a subsequent publication, but our preliminary calculations indicate that they will not be violated. It should be noted that the theoretical strength for loading in the [111] direction, equal to 27.3 GPa, is nearly the same as that obtained for isotropic triaxial loading, 27.9 GPa. At present, we do not have any plausible explanation of this fact.

In Fig. 9(a), it is seen that there are also other extrema of the total energy dictated by symmetry – maxima corresponding to the fcc and sc structures when simulating tensile tests with loading along the [001] and [111] directions, respectively. These extrema are denoted by arrows in Fig. 9(a). Their presence dictates that the corresponding dependence of the energy on elongation must bend, which imposes certain limitations on the maximum stress[11]. In the cases when there is no symmetry-dictated maximum (e.g. in the uniaxial tensile test along the [001] direction of NiAl with the B2 structure in the ground state[13]), the maximum stress is usually higher.

Since the structural energy difference E_{sc}–E_{bcc} is about five times higher than the difference E_{fcc}–E_{bcc} (755 meV/atom compared to 155 meV/atom), the E vs. ε curve for the [111] loading must rise much higher, albeit for larger strains, than that for the [001] loading (see Fig. 9(a)). Consequently, for the tensile test in the [111] direction the inflexion point occurs at a higher strain and for a higher stress than in the test with loading in the [001] direction. Thus, similarly as for W[11], a marked anisotropy of ideal tensile strengths for the [001] and [111] loading directions may be understood in terms of structural energy differences of nearby higher-symmetry structures found at the deformation path.

Relative changes of atomic volume and the dependences of the magnetic moment of FM iron per atom are shown as functions of elongation in Figs. 9(c) and 9(d), respectively. In the neighborhood of the ground state structure the atomic volume increases with increasing elongation but it exhibits a more complex behavior at larger deformations. For isotropic triaxial loading, the magnetic moment shows monotonous increase with increasing volume (in agreement with Herper et al.[64]) while in tensile tests it exhibits local extrema at points corresponding to both higher-symmetry structures (maxima for fcc and simple cubic) as well as at some other points along the paths.

5.2. Intermetallic compound Ni_3Al

In contrast with iron, in the case of Ni_3Al we start with the fcc-based $L1_2$ structure and, therefore, as mentioned in Sec. 2, we renormalize the ratio c/a by ascribing the value of $c/a = 1$ to the $L1_2$ structure. As a result, the c/a for the tetragonal path is by a factor of $\sqrt{2}$ smaller and for the trigonal path by a factor of 4 smaller than in the case of iron.

Using the GGA, the minimum of the total energy is obtained for the ferromagnetic state with the lattice constant equal to 3.561 Å (6.729 au) and magnetic moment of 0.80 μ_B per formula unit. The lattice constant agrees very well with the experimental value[65] of 3.568 Å (6.743 au) whereas the experimental magnetic moment, 0.23 μ_B per formula unit, is much lower. When including the spin-orbital coupling, Xu et al.[66] obtained a value of 0.46 μ_B per formula unit, which is closer to the experimental value. At present, we are verifying this conclusion.

Figure 10 shows the total energy of Ni_3Al as a function of c/a for the trigonal deformation at the experimental lattice volume. This dependence displays a symmetry-dictated minimum at $c/a = 1$ (the ground-state, $L1_2$ structure) and a symmetry-dictated maximum at $c/a = 0.5$ (a sc-based structure exhibiting cubic symmetry). A subsidiary minimum occurs at $c/a \approx 0.27$, which is not dictated by the symmetry. In the structure obtained for $c/a = 0.25$, the atoms are at the bcc-like positions, but the symmetry of this structure remains trigonal.

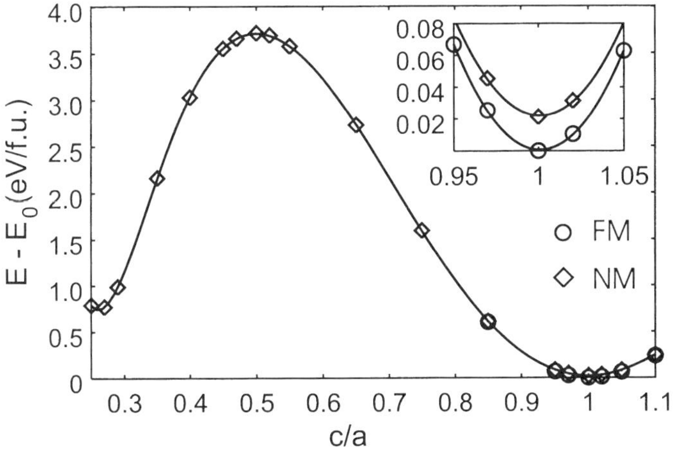

Figure 10. Total energy of Ni_3Al (per formula unit, f.u.) as a function of *c/a* for the trigonal deformation at the experimental lattice volume. The insert shows the details in the neighborhood of the ground state.

There is a very small energy difference between the FM and NM state of the $L1_2$ structure – only 21.3 meV/formula unit (see the insert in Fig. 10). This is consistent with the results of Xu et al.[66] (0.2-0.5 mRy/f.u.) and Min et al.[67] (~ 1 mRy/f.u.).

It is seen from Figs. 10 and 11 that the region of existence of FM state is limited. For $c/a \leq 0.75$, the magnetic moment is equal to zero (Fig. 11) and the compound is in a non-spin-polarized state.

Fig. 12 displays the total energy of Ni_3Al as a function of the volume and *c/a* for the trigonal deformation. Again, we show only those states the energies of which are the lowest for a given configuration. The total energy profile presented in Fig. 10 is contained in Fig. 12 as a profile for $V/V_{exp} = 1$. A nearly vertical border divides the area of Fig. 12 into FM and NM regions. All energy profiles corresponding to a constant volume exhibit the symmetry-dictated maximum at $c/a = 0.5$. In the contour plot (Fig. 12), there is a saddle point for $c/a = 0.5$ and $V/V_{exp} \sim 1.2$ (outside the area of the figure). The minimum at $c/a \approx 0.27$, $V/V_{exp} \approx 1.01$ is not dictated by symmetry.

Fig. 13 shows the total energy of Ni_3Al as a function of the volume and *c/a* for the tetragonal deformation. Here NM regions extend to both sides of the FM ground state. However, there are no energy extrema and saddle points in those NM regions. It is interesting that the transition from the FM to NM state during both the trigonal and tetragonal deformation is essentially continuous, without any discontinuities in magnetic moment (see e.g. Fig. 11). Xu et al.[66] have shown that the energy gain in Ni_3Al associated with magnetism is about an order of magnitude smaller than that due to the structural differences. Our calculations show that the NM Ni_3Al in the $L1_2$ structure is stable with respect to tetragonal and trigonal deformations (the shear moduli C' and C_{44} are nearly the same for the NM and FM states). Therefore, magnetism does not appear to play an important role in the control of phase stability. This is in sharp contrast with iron, where the onset of ferromagnetism stabilizes the bcc structure and NM bcc states are not stable with respect to tetragonal deformation [39,40].

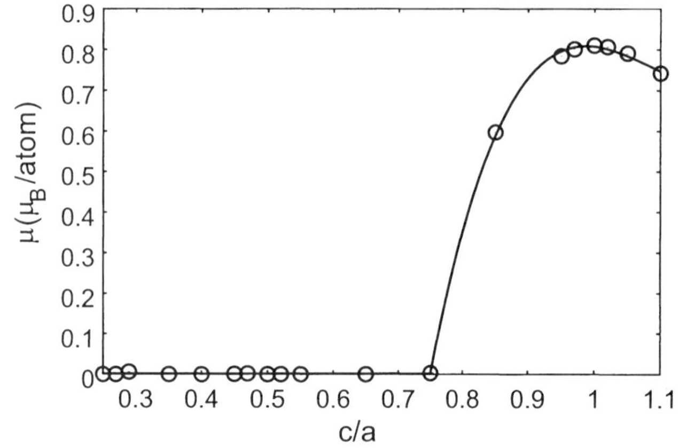

Figure 11. Magnetic moment of Ni_3Al as a function of *c/a* for the trigonal deformation at the experimental lattice volume.

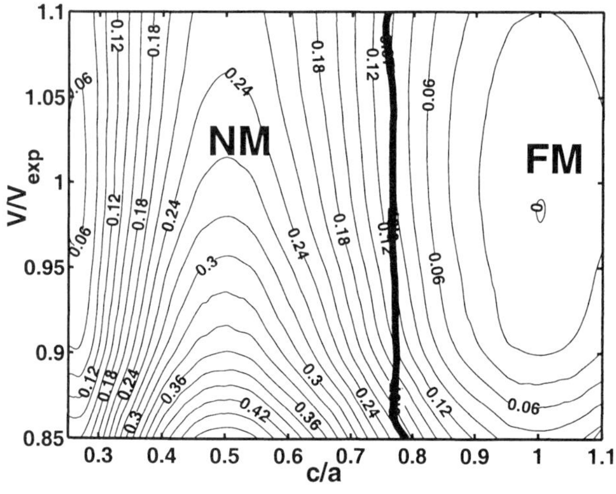

Figure 12. Total energy (per formula unit) of Ni_3Al as a function of volume and *c/a* ratio, characterizing the trigonal deformation, calculated within the GGA. The energy is measured relative to the energy of the equilibrium FM $L1_2$ state (the minimum at $c/a = 1$). Only states with the minimum energy are shown. The contour interval is 20 mRy. Thick line shows the NM/FM phase boundary. The ground-state minimum at $c/a = 1$ and the saddle point at $c/a = 0.5$ (outside the figure area) are dictated by symmetry.

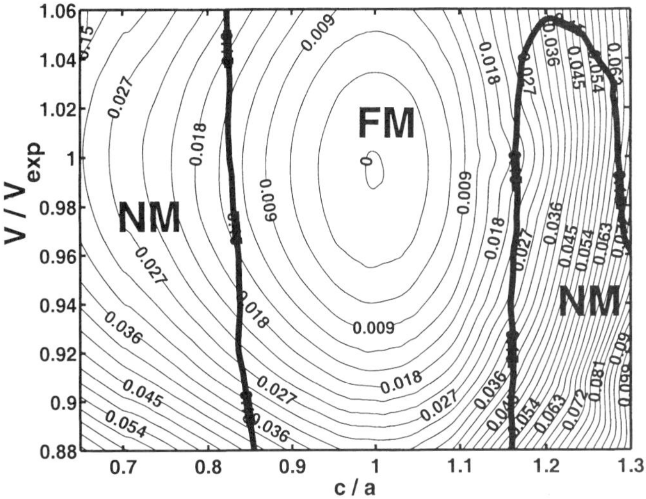

Figure 13. Total energy (per formula unit) of Ni_3Al as a function of volume and *c/a* ratio, characterizing the tetragonal deformation, calculated within the GGA. The energy is measured relative to the energy of the equilibrium FM $L1_2$ state (the minimum at *c/a* = 1). Only states with the minimum energy are shown. The contour interval is 3 mRy. Thick lines show the NM/FM phase boundaries. The only symmetry-dictated extremum is at *c/a* = 1.

Now, we can also simulate a tensile test in Ni_3Al to get theoretical tensile strengths for uniaxial loading along the [001] and [111] directions. These calculations are presently carried out.

6. AB INITIO CALCULATED VALUES OF THEORETICAL TENSILE STRENGTH

For the sake of completeness, we summarize in the Table 1 all ab initio calculated values of the theoretical tensile strength (including relaxation in directions perpendicular to the loading axis and, if applicable, of internal structure parameters) that have been calculated until now. Most of them correspond to the inflexion point on the strain dependence of the total energy. As for the strength of W for [110] loading, the material probably breaks down due to some other instability before reaching the inflexion point and, therefore, the true theoretical tensile strength will be lower than that given in the Table. The situation is most likely the same in the case of Cu where the experimental ideal strengths are about an order of magnitude lower than the calculated ones[14,68]. Semiempirical calculations[56] indeed suggest that, for the [001] direction, the tetragonal shear modulus becomes zero well before reaching the inflexion point. It may be expected that similar instabilities will occur for the [110] and [111] orientations. This will be the subject of further investigations.

Table 1. Theoretical tensile strengths σ_{th} calculated ab initio (results discussed in this paper are marked by asterisk).

material	structure	direction	σ_{th} (GPa)
Fe	A2	[111]	27.3[29,*]
Fe	A2	[001]	12.7[27,29,*]
Fe	A2	[001]	14.2[28]
W	A2	[001]	28.9[11]
W	A2	[001]	29.5[19]
W	A2	[111]	40.1[11]
W	A2	[110]	54.3[11]
Al	A1	[001]	12.1[15]
Al	A1	[111]	11.05[15]
Cu	A1	[001]	33[14]
Cu	A1	[110]	31[14]
Cu	A1	[111]	29[14]
NiAl	B2	[001]	46[13,14]
NiAl	B2	[111]	25[13,14]
β-SiC	B3 (3C)	[001]	101[16]
β-SiC	B3 (3C)	[111]	50.8[16]
$MoSi_2$	$C11_b$	[001]	37[26]
WSi_2	$C11_b$	[001]	38[26]
C	A4	[001]	95[30]
Si	A4	[001]	22[30]
Ge	A4	[001]	14[30]

7. CONCLUSIONS

We analyzed the energetics of iron and the intermetallic compound Ni_3Al subjected to tetragonal and trigonal deformation by means of full-potential ab initio electronic structure calculations and found borders between various phases with different spin polarizations. Whereas in iron the magnetic effects are vital for understanding the deformation behavior and a variety of magnetic orderings occurs, it transpires that in Ni_3Al magnetism is not very important in phase stability considerations. The $L1_2$ ground state is ferromagnetic, but the energy difference between the FM and NM state is quite small, about 21 meV/formula unit. It is interesting that during tetragonal deformation, iron transforms to AFMD and AFM1 states (Fig. 5), whereas during trigonal deformation, it is mostly ferromagnetic (Fig. 6).

For iron, we analyzed uniaxial tensile tests and discussed the anisotropy of the theoretical tensile strength, namely 12.7 GPa for [001] and 27.3 GPa for the [111] direction of loading. This marked anisotropy may be understood in terms of the symmetry-dictated extrema that are present along the deformation paths. Also the isotropic triaxial (hydrostatic) tension was analyzed and theoretical tensile strength of iron for this mode of loading was found to be 27.9 GPa, very close to the value for uniaxial [111] loading.

It should be noted that the calculated dependence of the total energy on parameters of the transformation paths provides useful information when constructing semi-empirical interatomic potentials that may be used for computer simulation of atomic configurations of various extended defects for which the first-principles calculations are intractable. An example is bond-order potentials (BOPs)[69] for which we have shown[60,70] recently how such first-principles results may be employed in their construction and testing.

Stability of higher-energy structures is also an important issue in the theoretical basis of the CALPHAD (CALculation of PHAse Diagrams) method[71]. Grimvall[72,73] concludes that when either the bcc or fcc structure of a metal is dynamically unstable, i.e. unstable, for example, with respect to the tetragonal or trigonal deformation, then there are large discrepancies between the semiempirical enthalpy differences H_{bcc}-H_{fcc} obtained from the CALPHAD method and ab initio results. However, as we can see from Refs. 6, 8 and 74, this is the case in most transition metals. In ab initio calculations, the dynamical instability is suppressed since we impose a rigid lattice (in reality, this might be stabilized by some external constraints), and the energy and enthalpy of such structure have well defined physical meaning. However, it appears that it is not certain how such values may be compared with those obtained from semiempirical CALPHAD method.

8. ACKNOWLEDGEMENTS

This research was supported by the Grant Agency of the Academy of Sciences of the Czech Republic (Project No. A1010817), by the Grant Agency of the Czech Republic (Project No. 106/02/0877), by the Research Project Z2041904 of the Academy of Sciences of the Czech Republic, and by the U.S. Department of Energy, Basic Energy Sciences (Grant No. DE-FG02-98ER45702). A part of this study has been performed in the framework of the COST Project No. OC 523.90. The use of the computer facilities at the MetaCenter of the Masaryk University, Brno, and at the Boston University Scientific Computing and Visualization Center is acknowledged.

9. REFERENCES

1. F. Milstein and S. Chantasiriwan, Phys. Rev. B **58,** 6006 (1998).
2. P. Šandera and J. Pokluda, Scripta Metall. Mater. **29**, 1445 (1993).
3. E. Esposito, A.E. Carlsson, D.D. Ling, H. Ehrenreich, and C.D. Gelatt, Jr., Phil. Mag. **41**, 251 (1980).
4. A.T. Paxton, P. Gumbsch, and M. Methfessel, Phil. Mag. Lett. **63**, 267 (1991).
5. W. Xu, and J.A. Moriarty, Phys. Rev. B **54**, 6941 (1996).
6. P.J. Craievich, M. Weinert, J.M. Sanchez, and R.E. Watson, Phys. Rev. Lett. **72**, 3076 (1994).
7. L. Vitos, J. Kollár, and H.L. Skriver, in Stability of Materials, eds. A. Gonis, P.E.A. Turchi, and J. Kudrnovský, Plenum Press, New York-London 1996, p. 393.
8. M. Šob, L.G. Wang, and V. Vitek, Comp. Mat. Sci. **8**, 100 (1997).
9. P.J. Craievich, J.M. Sanchez, R.E. Watson, and M. Weinert, Phys. Rev. B **55**, 787 (1997).
10. P. Alippi, P.M. Marcus, and M. Scheffler, Phys. Rev. Lett. **78**, 3892 (1997).
11. M. Šob, L.G. Wang, and V. Vitek, Mat.Sci. Eng. A **234-236**, 1075 (1997).
12. I.M. Mikhailovskii, I. Ya. Poltinin, and L.I. Fedorova, Fizika Tverdogo Tela **23** 1291 (1981) (English translation in Sov. Phys. Solid State **23** 757 (1981)).
13. M. Šob, L.G. Wang, and V. Vitek, Phil. Mag. B **78**, 653 (1998).
14. M. Šob, L.G. Wang, V. Vitek, Kovové materiály (Metallic Materials) **36**, 145 (1998).
15. W. Li and T. Wang, J. Phys.: Condens. Matter **10**, 9889 (1998).
16. W. Li and T. Wang, Phys. Rev. B **59**, 3993 (1999).
17. D. Roundy, C.R. Krenn, M.L. Cohen, and J.W. Morris, Jr., Phys. Rev. Lett. **82**, 2713 (1999).
18. J.W. Morris, Jr. and C.R. Krenn, Phil. Mag. A **80**, 2827 (2000).
19. D. Roundy, C.R. Krenn, M.L. Cohen, and J.W. Morris, Jr., Phil. Mag. A **81**, 1725 (2001).
20. P. Šandera, J. Pokluda, L.G. Wang, and M. Šob, Mat. Sci. Eng. A **234-236**, 370 (1997).
21. Y. Song, R. Yang, D. Li, W.T. Wu, and Z.X. Guo, Phys. Rev. B **59**, 14220 (1999).
22. M. Černý, P. Šandera, and J. Pokluda, Czech. J. Phys. **49**, 1495 (1999).
23. Y. Song, R.Yang, D. Li, and Z.X. Guo, Phil. Mag. A **81**, 321 (2001).
24. Y. Song, Z.X. Guo, and R. Yang, Phil. Mag. A **82**, 1345 (2002).

25. M. Černý, J. Pokluda, P. Šandera, M. Friák, and M. Šob, Phys. Rev. B, submitted.
26. M. Friák, M. Šob, and V. Vitek, in High-Temperature Ordered Intermetallic Alloys IX, eds. J.H. Schneibel, K.J. Hemker, R.D. Noebe, S. Hanada, and G. Sauthoff, Mater. Res. Soc. Symp. Proc. vol. **646**, Materials Reseach Society, Warrendale, PA, 2001, paper N4.8.
27. M. Friák, M. Šob, and V. Vitek, in Proceedings of the International Conference Juniormat'01, Institute of Materials Engineering, Brno University of Technology, Brno 2001, pp. 117-120.
28. D. M. Clatterbuck, D.C. Chrzan, and J. W. Morris, Jr., Phil. Mag. Lett. **82**, 141 (2002).
29. M. Friák, M. Šob, and V. Vitek, Phil. Mag. A, submitted.
30. D. Roundy and M. L. Cohen, Phys. Rev. B **64**, art. no. 212103 (2001).
31. M. Šob, L.G. Wang, M. Friák and V. Vitek, in Computational Modeling of Materials, Minerals, and Metals Processing, eds. M. Cross, J.W. Evans, and C. Bailey, The Minerals, Metals & Materials Society, Warrendale, PA, 2001, pp. 715-724.
32. J.W. Morris, Jr., C.R. Krenn, D. Roundy, and M.L. Cohen, in Phase Transformations and Evolution in Materials, ed. P.E.A. Turchi and A. Gonis, The Minerals, Metals & Materials Society, Warrendale, PA, 2000, pp. 187-207.
33. J. Pokluda and P. Šandera, in METAL 2000 (Proc. 9th Int. Metallurgical Conf., Ostrava, Czech Republic, May 16-18, 2000), ed. T. Prnka, Tanger, Ostrava 2000.
34. J.A. Moriarty, Phys. Lett. A **131**, 41 (1988).
35. S. Fox and H.J.F. Jansen, Phys. Rev. B **53**, 5119 (1996).
36. M.J. Mehl and D.A. Papaconstantopoulos, Phys. Rev. B **54**, 4519 (1996).
37. M. Šob, M. Friák, L.G. Wang, and V. Vitek, in Multiscale Modelling of Materials, eds. V.V. Bulatov, T. Diaz de la Rubia, R. Phillips, E. Kaxiras, N. Ghoniem, Mater. Res. Soc. Symp. Proc. vol. **538**, Materials Reseach Society, Warrendale, PA, 1999, pp. 523-527.
38. V. Paidar, L.G. Wang, Šob, V. Vitek, Modelling Simul. Mater. Sci. **7**, 369 (1999).
39. M. Friák, M. Šob, and V. Vitek, Phys. Rev. B **63**, 052405 (2001).
40. M. Friák, M. Šob, and V. Vitek, in Electron Correlations and Material Properties 2, ed. A. Gonis et al., Kluwer Academic Publishers, Boston-London-Dordrecht, 2002, in print.
41. D. Spišák and J. Hafner, Phys. Rev. B **61**, 16129 (2000).
42. S.L. Qiu, P.M. Marcus, and H.Ma, J. Appl. Phys. **87**, 5932 (2000).
43. S.L. Qiu, P.M. Marcus, and H.Ma, Phys. Rev. B **64**, 104431 (2001).
44. D. Spišák and J. Hafner, Phys. Rev. Lett. **88**, 056101 (2002).
45. D. Spišák and J. Hafner, Phys. Rev. B **65**, 235405 (2002).
46. H. Ma, S.L. Qiu, and P.M. Marcus, Phys. Rev. B **66**, 024113 (2002).
47. F. Ernst, M.W. Finnis, D. Hofmann, T. Muschik, U. Schönberger, and U. Wolf, Phys. Rev. Lett. **69**, 620 (1992).
48. D. Hofmann and M.W. Finnis, Acta Metall. Mater. **42**, 3555 (1994).
49. C. Schmidt, F. Ernst, M.W. Finnis, and V. Vitek, Phys. Rev. Lett. **75**, 2160 (1995).
50. L.G. Wang and M. Šob, Phys. Rev. B **60**, 844 (1999).
51. E. Abe, T. Kumagai, and M. Nakamura, Intermetallics **4**, 327 (1996).
52. R. Banerjee, R. Ahuja, and H.L. Fraser, Phys. Rev. Lett. **76**, 3778 (1996).
53. H. Jenniches, J. Shen,, Ch.V. Mohan, S.S. Manoran, J. Barthel, P. Ohresser, M. Klaua, and J. Kirschner, Phys. Rev. B **59**, 1196 (1999).
54. V. Vitek, D.P. Pope, J, Bassani, in Dislocations in Solids **10** (Plasticity of Gamma Alloys), ed. F.R.N. Nabarro and M.S. Duesberry, Elsevier Science Publishers, Amsterdam 1996, p. 135.
55. D. Legut, M. Friák, L.G. Wang, and M. Šob, to be published.
56. F. Milstein and B. Farber, Phys. Rev. Lett. **44**, 277 (1980).
57. P. Blaha, K. Schwarz, J. Luitz, WIEN97, Technical University of Vienna 1997 -improved and updated Unix version of the original copyrighted WIEN-code, which was published by P. Blaha, K. Schwarz, P. Sorantin, S. B. Trickey, in: Comput. Phys. Commun. **59**, 399 (1990).
58. J. P. Perdew, S. Burke, M. Ernzerhof, Phys. Rev. Lett. **77**, 3865 (1996).
59. L.M. Sandratskii, Adv. Phys. **47**, 91 (1998).
60. M. Mrovĕc, V. Vitek, D. Nguyen-Manh, D.G. Pettifor, L.G. Wang, and M. Šob, in Multiscale Modeling of Materials, eds. V.V. Bulatov, T. Diaz de Rubia, R. Phillips, E. Kaxiras, and N. Ghoniem, MRS Symposium Proc. vol. **538**, Materials Research Society, Warrendale, PA, 1999, pp. 529-534.
61. M. Šob, M. Friák, L.G. Wang, and V. Vitek, Key Eng. Materials **227**, 261 (2002).
62. D.J. Rasky and F. Milstein, Phys. Rev. B **33**, 2765 (1986).
63. J. Wang, J. Li, S. Yip, S. Phillpot, and D. Wolf, Phys. Rev. B **52**, 12627 (1995).
64. H. C. Herper, E. Hoffmann, P. Entel, Phys. Rev. B **60**, 3839 (1999).
65. F.R. de Boer, C.J. Schinkel, J. Biesterbos, and S. Proost, J. Appl. Phys. **40**, 1049 (1969).
66. J. Xu, B.I. Min, A.J. Freeman, and T. Oguchi, Phys. Rev. B **41**, 5010 (1990).
67. B.I. Min, A.J. Freeman, and H.J.F. Jansen, Phys. Rev. B **37**, 6757 (1988).
68. H. Kobayashi and I. Hiki, Phys. Rev. B **7**, 594 (1973).
69. D.G. Pettifor, Bonding and Structure of Molecules and Solids, Oxford University Press, Oxford 1995.

70. M. Mrovĕc, V. Vitek, D. Nguyen-Manh, D.G. Pettifor, L.G. Wang, and M. Šob, in Multiscale Phenomena in Materials - Experiment and Modeling, eds. D.H. Lassila, I.M. Robertson, R. Phillips, and B. Devincre, MRS Symp. Proc. vol. **578**, Materials Research Society, Warrendale, PA, 2000, pp. 199-204.
71. N. Saunders and P. Miodownik, CALPHAD (Calculation of Phase Diagrams): A Comprehensive Guide, Pergamon Materials Series vol. **1**, ed. R.W. Cahn, Elsevier Science Publishers, Oxford-New York-Tokyo, 1998.
72. G. Grimvall, Ber. Bunsenges. Phys. Chem. **102**, 1083 (1998).
73. G. Grimvall, this issue.
74. L.G. Wang and M. Šob, J. Phys. Chem. Solids, accepted for publication.

REJUVENATION OF DEFORMATION-DAMAGED MATERIAL BY MAGNETIC ANNEALING
- A NEW APPROACH TO GRAIN BOUNDARY ENGINEERING -

Tadao Watanabe, Shuichi Nishizawa and Sadahiro Tsurekawa

Laboratory of Materials Design and Interface Engineering
Department of Machine Intelligence and Systems Engineering
Graduate School of Engineering, Tohoku University
Aramaki-Aza-Aoba 01, Sendai 980-8579, Japan

1. INTRODUCTION

The nucleation and growth of cavities and microcracks at grain boundaries cause different types of brittleness, such as creep embrittlement at high temperature, stress corrosion cracking in reactive environment and under stress, oxidation embrittlement in air, radiation embrittlement by irradiation, occurring in engineering polycrystalline materials in service. For example, high temperature creep fracture often becomes an important source of materials embrittlement, leading to serious accidents occurring in power station, aircraft engine, chemical plants operating at high temperatures[1-3]. In addition the segregation of detrimental elements to grain boundary promotes intergranular fracture, leading to segregation-induced brittleness[4, 5]. High temperature intergranular fracture is of particular importance and has been extensively studied so far in order to control the occurrence of intergranular fracture, by using existing disciplines, and more recently by grain boundary engineering proposed by one of the present authors at the beginning of 1980's[6-8]. However even now there exist many unsolved problems of materials design and development: one of long standing problems and dilemmas is that the material becomes more brittle when the strength is increased by some means. So strengthening of material does not always bring advantage and benefit of materials design and engineering. We need to develop structural materials with higher strength and higher fracture resistance, *ie.* higher fracture toughness. We have already proved that grain boundary engineering based on grain boundary design and control can provide us a solution of the dilemma[9], on the basis of basic studies of the effects of grain boundary character and structure on intergranular sliding and fracture at high temperature[10].

In this paper we want to propose another possibility of grain boundary engineering which is a new approach to rejuvenation of damaged materials, in other words, how we material scientists and engineers can save 'dying material' by healing mortal damage existing in the material already served for many years. In particular this paper first introduce our most recent finding of the rejuvenation effect of magnetic annealing on damaged ferromagnetic polycrystalline material deformed at high temperature. This work associated with a new approach to grain boundary engineering may give the field of materials science and engineering a new potential, particularly to high temperature materials used for nuclear reactor and jet craft engine used in much more severe operating conditions than ordinary one, in order to improve the efficiency of an operating machine system. The application of a magnetic field has been already attempted by the present authors and coworkers in annealing of deformed materials[11] and sintering of powder compacts[12, 13].

2. EXPERIMENTAL PROCEDURE

2.1. Specimen Preparation and Grain Boundary Microstructure Analysis

The material used in this work was an iron-2.9at.%cobalt alloy which was prepared by vacuum melting (6.5×10^{-4} Torr) electrolytic iron (99.9% pure) and cobalt pellet (99.5% pure) in alumina crucible. The ingot was hot-forged to plate 10mm thick and then hot rolled at about 973K into 1mm thick sheet. Tensile specimens which had 5mm wide and 13 mm long gage part were machined from the sheet and were annealed at 1123K for 3 in vacuum (2×10^{-5} Torr). The average grain size was 59.3μm.

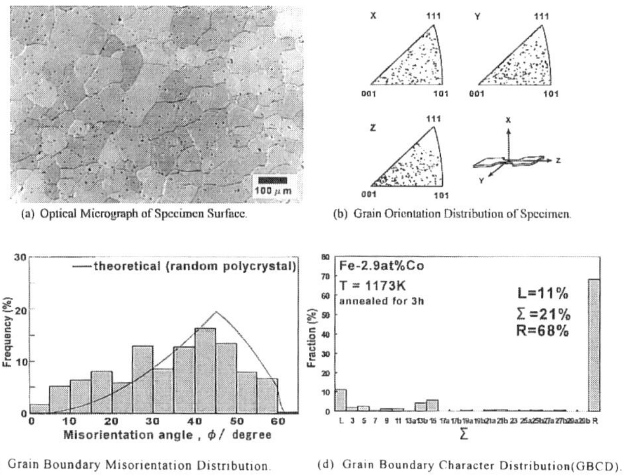

(a) Optical Micrograph of Specimen Surface. (b) Grain Orientation Distribution of Specimen.

(c) Grain Boundary Misorientation Distribution. (d) Grain Boundary Character Distribution(GBCD).

Figure 1 Grain boundary microstructure in Fe-2.9at%Co polycrystalline specimen annealed at 1173K for 3h.

The statistical analyses of grain orientation distribution, grain boundary misorientation distribution and grain boundary character distribution (GBCD) in the annealed specimens were carried out by the orientation imaging microscopy (OIM) at accelerating voltage 20kV. The optical micrograph of annealed specimen surface and the result of OIM analyses are given in Fig. 1. The distribution of grain orientations in the specimens was not ideally random but localized between <100> and <101> to some extent, particularly for Z direction parallel to the rolling direction of sheet, as indicated in Fig. 1(b). Accordingly the grain boundary misorientation distribution showed some deviation from the distribution for a random polycrystal indicated by the theoretical curve. It was found that the annealed specimen had the frequency of high energy random boundaries (68%) lower than the theoretical value (86%) for random polycrystal, and the total frequency of low angle boundaries and low Σ(3~29) coincidence boundaries was 32%, both types of which are known to be low energy type and more fracture-resistant.

2.2. High Temperature Deformation Test

In order to introduce intergranular cavities or microcracks, the annealed tensile specimens were deformed up to a given amount of plastic strain of ε=0.03, 0.06 and 0.1 at a strain rate of $3x10^{-4}$/s at 1023K below the Curie temperature Tc (1083K) of the alloy in argon. Figure 2 shows a typical stress-strain curve obtained from high temperature tensile test carried out for the alloy in the above mentioned condition. The total elongation without interruption test was about 50%. The high temperature deformation test was interrupted at different amounts of strain (0.03, 0.06, 0.1) for each specimen to be used for magnetic annealing and then the specimen was immediately cooled down. There was no significant difference among the stress-strain curves for the specimens deformed under the same condition.

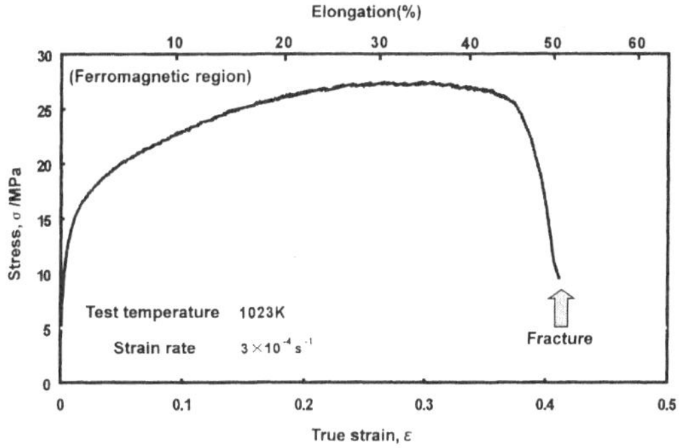

Figure 2 Stress-strain curve obtained at 1023K at a strain rate of $3x10^{-4}s^{-1}$.

2.3. Magnetic Annealing

After received high temperature deformation, the gauge part of the specimen (~ 5x13x1mm) was cut out with a spark machine and prepared as the deformed sample for magnetic annealing. A helium free superconducting magnet-installed heat treatment system (made by Sumitomo Heavy Machine Corporation, the maximum magnetic field strength=6T) was used for magnetic annealing in this work. A schematic illustration of the system is shown in Fig. 3. Molybdenum sheet heating element was used for the furnace. A specially designed locking port was attached in order to insert the sample into the center of the magnetic field operating at high temperature within a few minutes without breaking high vacuum. The maximum annealing temperature of this system was 1773K in vacuum. A specially designed carbon holder was used in which the deformed sheet sample was put for magnetic annealing. Mica sheets were inserted between the sample and the carbon holder to avoid carburization at sample surface due to a contact with the carbon holder. Magnetic annealing was carried out at 1023K below the Curie temperature *Tc* (1083K) in a direct current magnetic field whose strength was chosen as 0, 3T, 6T, in vacuum of 2x10-5Torr for maximum annealing time of 50 hr.

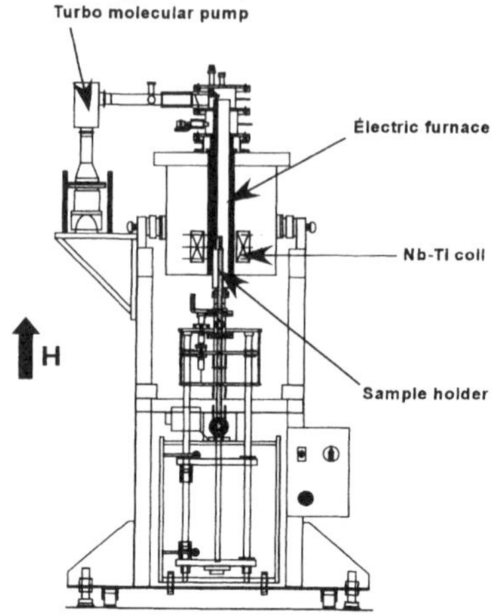

Figure 3. Schematic illustration of superconducting magnetic field heat treatment system.

2.4. Quantitative Evaluation of Cavitation and Rejuvenation

To make quantitative evaluation of cavitation by high temperature deformation and rejuvenation by magnetic annealing, two types of measurements were made; one was micrographic analysis of the size and the area of intergranular cavities by using SEM micrographs. The mean cavity area was determined for more than 200 intergranular cavities. The area density of cavities was studied as a function of the cavity area, and the aspect ratio between short and long diameter. The other method of evaluation was by measurement of the density by the Archimedes method. Careful measurements of the density were made for annealed and deformed specimens, and also for the specimens deformed and subsequently annealed in a magnetic field, taking particular care to avoid penetration of water into cavitated boundaries from the surface by putting paraffin film on the surface.

3. CAVITATION BY HIGH TEMPERATURE DEFORMATION

3.1. Effect of Grain Boundary Microstructure on Intergranular Cavitation

From SEM microscopic observations on cavitation in the deformed specimens whose grain boundary microstructure had been analyzed by the OIM technique before high temperature deformation test, it was found that cavitation occurred preferentially at random boundaries, particularly at triple junctions at which two or three random boundaries met, as shown in Fig. 4. Most cavities (88%~94%) are associated with more than one random boundaries. Moreover there was the tendency that cavi-

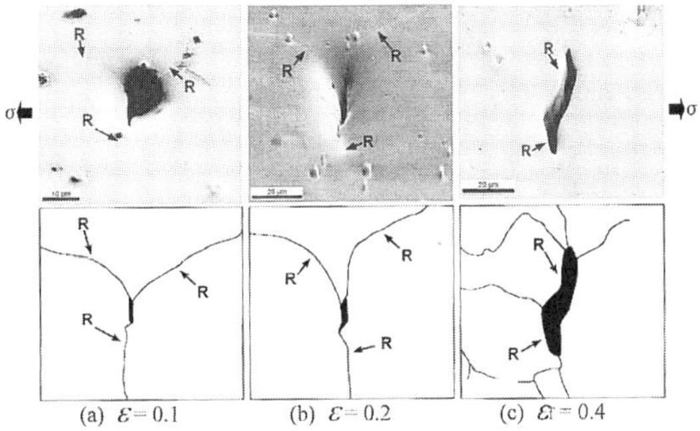

Figure 4. SEM micrographs showing cavities at grain boundaries.

ties formed more preferentially at grain boundaries aligned perpendicularly to the tensile axis. The propensity to preferential cavitation at random boundaries in high temperature creep deformation in alpha iron-alloys was reported very early by one of the present authors[10] and it was also confirmed that low angle boundary and low Σ coincidence boundaries are very resistant to grain boundary sliding and fracture. In fact, this is the effect of grain boundary character/structure on high temperature intergranular fracture. So the importance of the grain boundary character distribution (GBCD) and the grain boundary connectivity was pointed out[8]. A schematic illustration of structure-dependent intergranular fracture controlled by sliding or vacancy condensation mechanism is shown in Fig. 5.

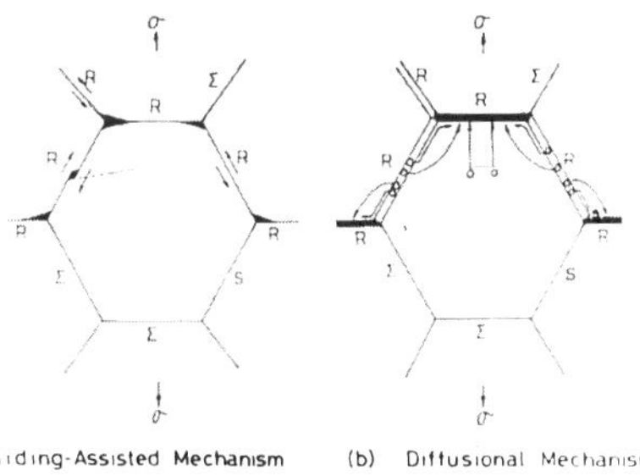

Figure 5. Schematic representation of mechanism of structure-dependent intergranular fracture in polycrystals at high temperature[8].

3.2. Effects of the Plastic Strain on Cavitation and the Density

Figures 6 and 7 show the effect of the amount of plastic strain on the mean cavity area and the relative bulk density of deformed specimens, respectively. It appears that cavitation proceeds rather slowly at early stage of deformation up to a certain level of plastic strain, then the rate of cavitation goes up, leading to the final rupture. A similar change of the density during creep deformation in oxygen-free copper polycrystals was reported very early by Bottner and Robertson[14]. They found that the density decreased almost continuously at the primary and secondary creep stage then showed a rapidly increasing rate of the decrease of the density after the on-set of the tertiary creep stage. As shown in Fig. 7, a rapid decrease of the density occurred beyond the strain 0.2~0.3, which almost corresponds to the value of creep strain at which the on-set of tertiary creep occurred in alpha iron and iron 3at%cobalt alloy at 1023K at stress 14MPa, on the effect of ferromagnetism on creep deformation[15]. It is reasonable to think that a rapid decrease of the density may occur due to the interlinkage of isolated cavities on random boundaries, resulting in the formation of a long intergarnular crack over the length of several grain boundaries. Again this may be controlled by the grain boundary character distribution and the grain boundary connectivity, as schematically shown in Fig. 5.

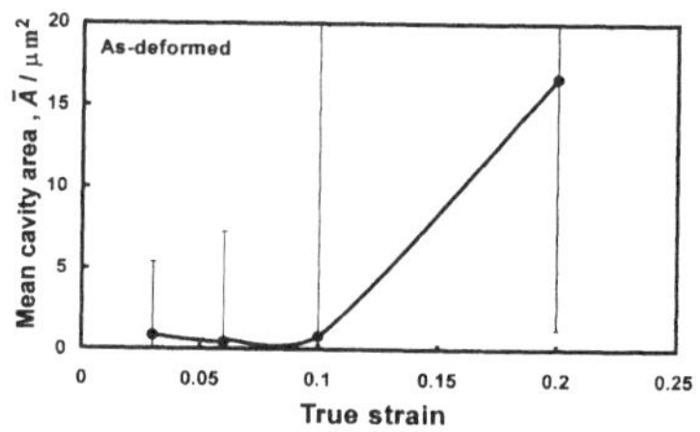

Figure 6. Strain dependence of mean cavity area obtained from tensile test at 1023K.

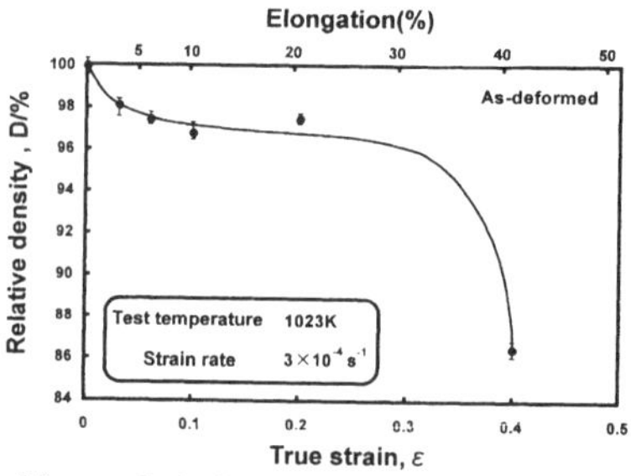

Figure 7. Relationship between relative density and true strain for Fe-3%Co alloy deformed at 1023K.

4. REJUVENATION BY MAGNETIC ANNEALING

4.1. Observations of Cavities after Magnetic Annealing

Direct observations of cavities in the specimen deformed and subsequently annealed in a magnetic field were made by scanning electron microscopy (SEM). The changes of the area density and of the morphology of intergranular cavities after magnetic annealing were carefully studied. Figures 8(a) and 8(b) show SEM micrographs of cavitated boundary in as deformed specimen and in the specimen deformed and annealed in a magnetic field. It is evident that many isolated cavities exist along grain boundary in as-deformed specimen, while after magnetic annealing, a lower density of cavities were observed mostly at triple junctions.

From, SEM micrograph analyses, the mean cavity area was determined for the specimens deformed up to different strain levels and subsequently annealed in a magnetic field with different field strengths. It was found that the mean cavitated area decreased rapidly then reached certain level with increasing magnetic field strength, almost irrespective of the amount of plastic strain before magnetic annealing, as shown in Fig. 9.

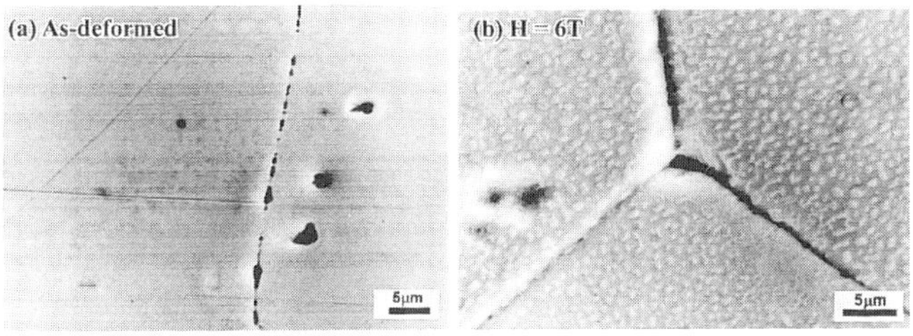

Figure 8. SEM micrographs showing cavities before annealing and after 50h annealing at 1023K in a magnetic field (H=6T).

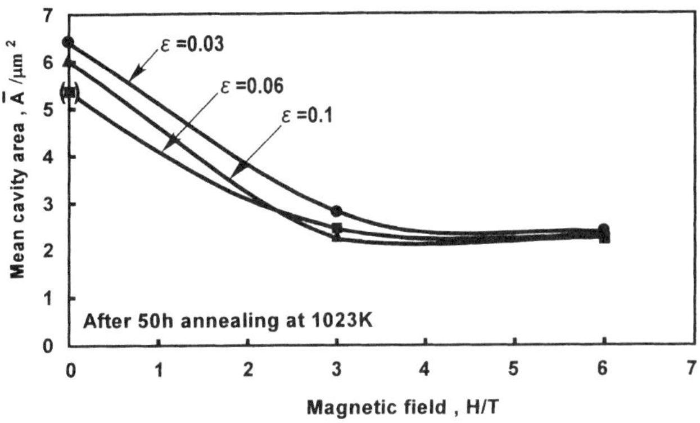

Figure 9. Magnetic field dependence of mean cavity area in the specimens deformed to different plastic strains and annealed at 1023K for 50h.

4. 2. The Degree of Rejuvenation

In order to evaluate the effect of magnetic field strength on rejuvenation, we define a new parameter which can simply describe the degree of rejuvenation, *D.R.*, as follows.

$$D.R. = (D_{mag} - D_p) / (D_0 - D_p) \times 100 \qquad (1)$$

where D_{mag} is the density of the specimen deformed and magnetically annealed, D_p the density of as-deformed specimen, D_0 the density of annealed specimen before deformation.

Figure 10 shows clearly that the degree of rejuvenation, (*D.R.*) increases with increasing annealing time and tends to saturate for the specimens deformed up to the true strain 0.1 and subsequently annealed in a magnetic field. However it is interesting to see that the degree of rejuvenation tends to go up first and then go down with annealing time for the specimens ordinarily annealed without a magnetic field. The reason why longer ordinary annealing is less effective than shorter annealing, is not clear. On the other hand by magnetic annealing the degree of rejuvenation always increases with increasing annealing time although it tends to saturate.

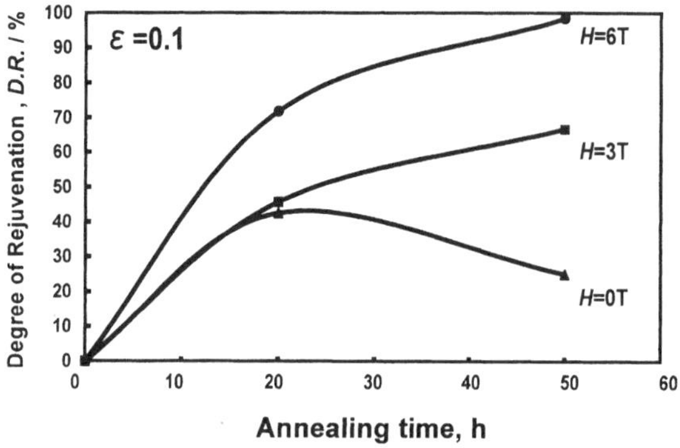

Figure 10. Degree of rejuvenation as a function of annealing time for the specimens strained to 0.1 and annealed at 1023K.

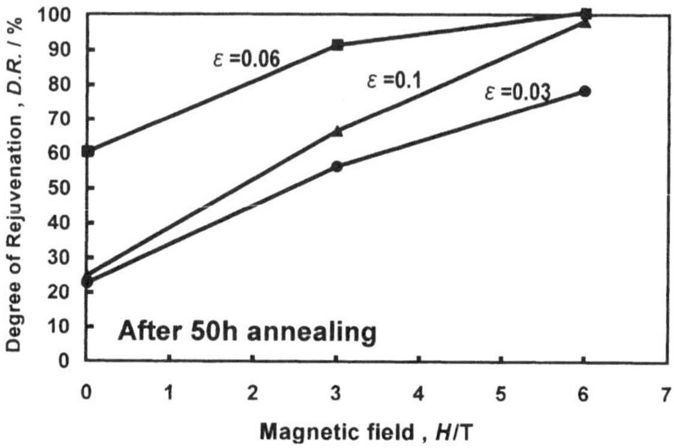

Figure 11. Magnetic field dependence of the degree of rejuvenation, *D.R.*, for the specimens deformed to different strains and annealed at 1023K.

Figure 11 shows the magnetic field dependence of the degree of rejuvenation *D.R.* observed for the specimens magnetically annealed for 50 hr. We see that the degree of rejuvenation is higher at higher magnetic field, in other words, the effect of magnetic annealing on rejuvenation of damaged material is more effective at higher magnetic field, almost irrespective of the amount of plastic strain which may determine the density of cavities in deformed material.

Figure 12 shows schematically the effects of magnetic field strength and an-

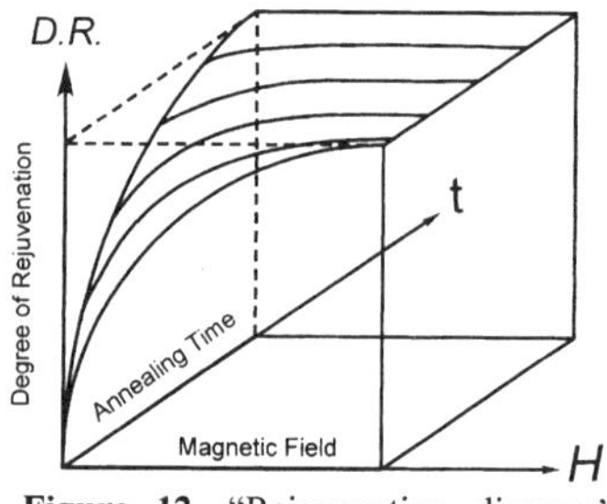

Figure 12. "Rejuvenation diagram" for high-temperature deformed and magnetically annealed ferromagnetic iron alloy.

nealing time on the degree of rejuvenation in 3D diagram, termed "Rejuvenation Diagram". This diagram clearly shows that a higher magnetic field and longer annealing time are desirable condition of magnetic annealing which can produce more significant rejuvenation, at least in damaged iron-cobalt alloy used in this work.

5. MECHANISM OF REJUVENATION BY MAGNETIC ANNEALING

This work has revealed that magnetic annealing is very effective in rejuvenation of damaged material particularly containing intergranular cavities introduced by high temperature deformation. Now we consider possible mechanism of rejuvenation effect by magnetic annealing observed in the iron-cobalt alloy in connection with our recent work on the effect of the application of magnetic field on sintering process of iron powder compact.

Quite recently we found that the application of a direct current (dc) magnetic field (maximum field strength 15KOe) could drastically enhance sintering process, particularly at early stage of sintering, in iron powder compact[12, 13]. Sintering process involves the densification controlled by diffusional mechanism to eliminate pores existing in the space between contacted particles. So it is not difficult to consider that there is some similarity between sintering and densification by the elimination of pores and rejuvenation by elimination of cavities by magnetic annealing. In sintering the migration of grain boundary is important to the densification by elimination of remaining pores or voids. In fact it was recently found that the application of a magnetic field could enhance grain boundary migration which is known to play an important role in elimination of pores during sintering , in iron-silicon alloy[13] and bismuth[16] bicrystals.

On the other hand when a ferromagnetic polycrystal is kept in a magnetic field, the presence of grain boundaries is considered to generate free poles distributing along grain boundary because of anisotropic magnetizationin in neighbouring grains[17]. The occurrence of such local magnetic free poles along the grain boundary may affect grain boundary properties such as grain boundary energy and grain boundary diffusion associated with the elimination of cavities by diffusional process in our case. However the detail of possible mechanism of rejuvenation by magnetic annealing studied in this work has not been fully understood and we need more basic knowledge about effects of magnetic field on metallurgical phenomena, particularly there is very few literature on the effect of such high magnetic field used in this work on grain boundary related phenomena.

6. CONCLUSION

A new approach to grain boundary engineering has been extended to unsolved engineering materials problem, that is "rejuvenation of damaged materials". The application of a high magnetic field during annealing of deformed ferromagnetic iron cobalt alloy was found to enhance the elimination of intergranular cavities introduced by high temperature deformation and the restoration of the density. Relative density decreased slowly with increasing plastic strain up to certain level then rapidly decreased leading to the final fracture. The formation and interlinkage of intergranular cavities are very likely controlled by the grain boundary character distribution (GBCD) and the grain boundary connectivity. Intergranular cavities are mostly associated with random boundaries. It was also found that the degree of rejuvenation increased with increasing the magnetic field strength and annealing time.

ACKNOWLEDGEMENTS

This work was supported by a Grant-in-Aid for Fundamental Research (B)(2) 13555180 and by a Grant-in-Aid for COE Research (11CE2003) from the Ministry of Education Science, Sports and Culture of Japan. The authors thank Dr.Kawahara for his help in preparing the manuscript.

REFERENCES

1. A. J. Perry, Review: Cavitation in creep, *J. Mater. Sci.*, **9**, 1016-1039 (1974).
2. L. E. Svensson and G. L. Dunlop, Growth of Intergranular Creep Cavities, *Intern.Metals Reviews*, No.2, 109-131 (1981).
3. D. M. R. Taplin and A. L. W. Collins, Fracture at High Temperature under Cyclic Loading, *Ann. Rev. Mater. Sci.*, **8**, 235-268 (1978).
4. C.L.Briant and S.K.Banerji, Intergranular Failurein steel: the role of grain boundary composition, *Intern.Metals Review,* No.4, 164-199 (1978).
5. M. P. Seah and E. D. Hondros, in: *Atomistics of Fracture*, ed by R.M.Latanision and J. R. Pickens, (Plenum Pub., 1983), pp. 855-887.
6. T. Watanabe, An Approach to Grain Boundary Design for Strong and Ductile Polycrystals, *Res Mechanica*, **11**, 47-86 (1984).
7. T. Watanabe, Grain Boundary Design for the Control of Intergranular Fracture, *Mater. Sci. Forum*, **46**, 25-48 (1989).
8. T. Watanabe, Grain Boundary Design for High Temperature Materials, *Mater. Sci. Eng.*, **A166**, 11-28 (1993).
9. T. Watanabe and S. Tsurekawa, The Control of Brittleness and Development of Desirable Mechanical Properties in Polycrystalline Systems by Grain Boundary Engineering, *Acta Mater.*, **47**(15), 4171-4189 (1999).
10. T. Watanabe, Grain Boundary Sliding and Stress Concentration during Creep, *Met. Trans.*, **14A**, 531-545, (1983).
11. T. Watanabe, Y. Suzuki, S. Tanii and H. Oikawa, The Effect of Magnetic Annealing on Recrystallization and Grain Boundary Character Distribution (GBCD) in Iron-Cobalt Alloy Polycrystals, Phil. Mag. Letters, **62**, 9-17 (1990).
12. T. Matsuzaki, T. Sasaki, S. Tsurekawa and T. Watanabe, Grain Boundary Migration and Grain Growth during Sintering and Annealing in Magnetic Field in Iron and Its Alloy, *JIM. Proc. Vol.*,**13**, 529-534 (1999).
13. S. Tsurekawa, K. Harada, T. Sasaki, T. Matsuzaki and T. Watanabe, Magnetic Sintering of Ferromagnetic Metal Powder Compacts, *Mater.Trans. JIM.*, **41**, 991-999 (2000).
14. R. C. Boettner and W. D. Robertson, A Study of the Growth of Voids in Copper during the Creep Process by Measurement of the Accompanying Change in Density, *Trans. Met. Soc. AIME.*, **221**, 613-622 (1961).
15. S. Karashima, H. Oikawa and T. Watanabe, The Effect of Ferromagnetism upon Creep Deformation of Alpha Iron and Its Solid Solution Alloys, *Trans. Met. Soc. AIME.*,**242**,1703-1708 (1968).
16. D. A. Molodov, G. Gottstein, F. Herighaus and L. S. Shvindlerman, Motion of Planar Grain Boundaries in Bismuth Bicrystals Driven by a Magnetic Field, *Scripta Materialia*, **37**, 1207-1213 (1997).
17. L. J. Dijkstra, Relation of Magnetic Properties to Microstructure, *Relation of Properties to Microstructure,ASM*, 209-232 (1954).

THEORY

COHERENT POTENTIAL APPROXIMATION WITHIN THE EXACT MUFFIN-TIN ORBITALS THEORY

L. Vitos[1,2], I. A. Abrikosov[3], and B. Johansson[1,3]

[1]Applied Materials Physics, Department of Materials Science and Engineering, Royal Institute of Technology, SE-10044 Stockholm Sweden

[2]Research Institute for Solid State Physics, H-1525 Budapest P.O.Box 49, Hungary

[3]Condensed Matter Theory Group, Physics Department Uppsala University, S-75121 Uppsala, Sweden

ABSTRACT

An implementation of the *coherent potential approximation* (CPA) is carried out within the frameworks of the *exact muffin-tin orbitals* (EMTO) theory. During the self-consistent iterations the Poisson equation is solved using the spherical cell approximation, and the charge transfer between alloy components is treated within the screened impurity model. The total energy is calculated using the *full charge density* (FCD) technique. The FCD-EMTO-CPA method is suitable for accurate determination of the electronic structure and total energy of completely random alloys with a substitutional disorder on any kind of underlying crystal lattice. The accuracy of the method is demonstrated through test calculations performed on face centered cubic (*fcc*), body centered cubic (*bcc*), and hexagonal close packed (*hcp*) Cu-Zn binary alloys.

INTRODUCTION

One of the most successful approximations for calculations of the electronic structure and total energy of random substitutional alloys is the *coherent potential approximation* (CPA)[1]. It is based on the assumption that the alloy may be replaced by an ordered effective medium, the parameters of which must be determined self-consistently. The CPA was originally introduced by Soven[2] for the electronic structure problem and by Taylor[3] for phonons in random alloys. Combined with the multiple scattering theory[4], the CPA became particularly popular and suitable for many practical applications. Based on the local *density functional theory* (DFT)[5], Winter and Stocks[6] have shown that the Kohn-Sham potential for alloy components can be calculated self-consistently, and Johnson *et al.*[7,8] have derived the CPA-DFT expression for the total energy of a random alloys. Recently, this expression was modified to include correctly the electrostatic contribution to the total energy in random alloys due to the charge transfer effects[9–13]. The CPA has been used for calculations of bulk electronic structure, ground state thermodynamic properties, phase stabilities, magnetic properties, surface electronic structure, segregation, and many other characteristics of alloys[14]. In these applications, as well as from a comparison with other theoretical methods for the calculations of the electronic properties of random alloys[13–16], the reliability of the CPA was well established.

At the same time, the CPA, being a single site approximation to the impurity problem, has limited applicability. For example, one cannot treat directly within the CPA alloys with short-range order. Also, systems with a large size mismatch between the alloy components are difficult to describe because of the local lattice relaxations. However, certain limitations of the CPA are not directly related to the approximation itself. Rather, they originate from additional approximations introduced by particular implementations. The most common electronic structure calculations methods used for the implementations of the CPA are the Korringa-Kohn-Rostoker (KKR)[1,4] and the linear muffin-tin orbital (LMTO)[17] methods. The shape approximation for the one-electron density and potential, used in these methods, are insufficient for the accurate description of the behavior of the total energy upon, e.g., anisotropic lattice distortions. Thus, one cannot calculate, for example, elastic constants in random alloys or relax c/a ratio in alloys with a tetragonal or hexagonal symmetry. In addition, the LMTO method (without correction terms) may not give a proper description of the open structures or structural energy differences between structures with different packing fraction, to the extent that the energy difference between the *bcc* and *fcc* structures of Cu has a wrong sign[18].

Recently, we have shown[19] that the accuracy of the CPA is greatly improved via an implementation within the basis set of the so-called *exact muffin-tin orbitals* (EMTO). The EMTO theory has been developed by Andersen[20–22]. We have shown that the EMTO-CPA method, combined with the *full charge density* (FCD) formalism[23], allows one to calculate the energy of a random alloy with the same accuracy, as that of modern full-potential methods in the case of pure elements or ordered compounds. In this paper we present a detailed description of the FCD-EMTO-CPA method.

OVERVIEW OF THE EMTO THEORY

The basic idea of the EMTO method is illustrated in Fig. 1. In the EMTO theory[20–22] the one-electron Kohn-Sham equation is solved within the muffin-tin approximation for the effective potential

$$v(\mathbf{r}) \approx v_{mt}(\mathbf{r}) \equiv v_0 + \sum_R [v_R(r_R) - v_0]. \tag{1}$$

R runs over the lattice sites and $\mathbf{r}_R \equiv \mathbf{r} - \mathbf{R}$. $v_R(r_R)$ are spherical potentials and they become equal to v_0 outside *potential spheres* of radii s_R, shown by large unfilled circles in Fig. 1. It has been shown that for accurate representation of the full potential $v(\mathbf{r})$ the potential spheres should overlap[21,24]. $v_R(r_R), v_0$ in Eq. (1), as well as s_R are determined by minimizing (a) the deviation between $v(\mathbf{r})$ and $v_{mt}(\mathbf{r})$ and (b) the errors coming from the overlap region.

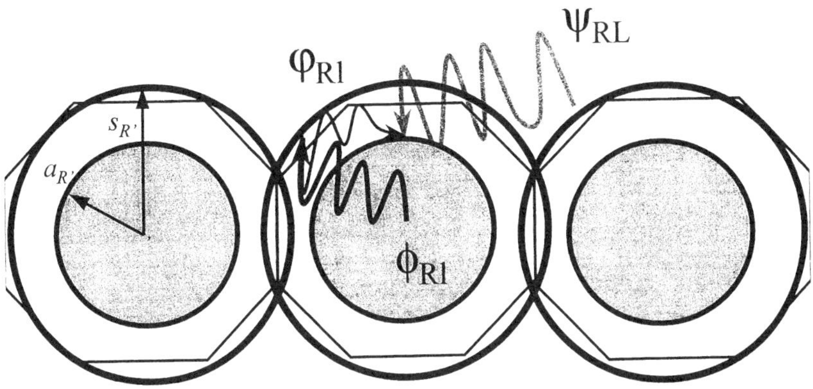

Figure 1. Basic idea of the exact muffin-tin orbitals method. The effective one-electron potential is constructed in the muffin-tin approximation, Eq. (1), i.e. it is spherical inside *potential spheres* of radii $s_{R'}$, shown by large unfilled circles, and it is flat and equal to the muffin-tin zero v_0 outside the sphere. The energy dependent EMTO's are constructed from the *screened spherical waves*, $\psi_{RL}(\kappa_j, \mathbf{r})$, Eq. (3), shown with gray line, with boundary conditions given in conjunction with non overlapping *hard spheres* with radii a_R. The hard spheres are shown with filled circles. Inside the potential spheres the low l ($l \leq l_{max}$) projections of the EMTO's onto the spherical harmonics $Y_L(\hat{r})$ are substituted by the *partial waves*, $\phi_{Rl}(\epsilon_j, r_R)$, Eq. (4), shown with a thick black line. The matching between them is realized by backwards extrapolated *free-electron solutions* $\varphi_{Rl}(\epsilon, r_R)$, thin black line, which joins continuously and differentiable onto the partial waves at the boundary of the potential sphere, and continuously but not differentiable onto the screened spherical waves at the boundary of the hard sphere. The full nonspherical charge density, Eqs. (7) and (14), and the full potential are constructed inside the *Wigner-Seitz cells*, shown by hexagons. At the last iteration they are used for the total energy calculations in the framework of the *full charge density* method.

The one-electron wave functions, Ψ_j, are expanded in terms of the *exact muffin-tin orbitals* (EMTO's)[20,24,25], $\bar{\psi}_{RL}$, i.e.

$$\Psi_j(\mathbf{r}) = \sum_{RL} \bar{\psi}_{RL}(\epsilon_j, \mathbf{r}_R)\, v_{RL,j}. \tag{2}$$

The energy dependent EMTO's are defined for each site R and for each $L \equiv (l, m)$ with $l \leq l_{max}$ (usually $l_{max} = 3$). They are constructed from the *screened spherical waves*, $\psi_{RL}(\kappa_j, \mathbf{r})$ (the gray line in Fig. 1), which are solutions of the wave equation

$$[\nabla^2 + \kappa_j^2]\psi_{RL}(\kappa_j, \mathbf{r}) = 0 \tag{3}$$

for $\kappa_j^2 = \epsilon_j - v_0$, with boundary conditions given in conjunction with non overlapping hard spheres with radii a_R[20] (shown with filled circles in Fig. 1). These boundary conditions require that the screened spherical waves behave like a pure real spherical harmonic $Y_L(\hat{r}_R)$ on their own a-spheres, while the $Y_{L'}(\hat{r}_{R'})$ projections on all the other hard spheres vanish. Inside the potential spheres the low l ($l \leq l_{max}$) projections of the

EMTO's onto the spherical harmonics are substituted by the *partial waves*, $\phi_{Rl}(\epsilon_j, r_R)$ (thick black line in Fig. 1), defined as the regular solutions of the radial Schrödinger equation for potential $v_R(r_R)$ and energy ϵ_j

$$\frac{\partial^2[r_R\phi_{Rl}(\epsilon_j, r_R)]}{\partial r_R^2} = \left[\frac{l(l+1)}{r_R^2} + v_R(r_R) - \epsilon_j\right] r_R\phi_{Rl}(\epsilon_j, r_R). \tag{4}$$

Because the screened spherical waves $\psi_{RL}(\kappa_j, \mathbf{r})$ have pure lm-character at the hard spheres, the matching between them and the partial waves $\phi_{Rl}(\epsilon_j, r_R)Y_L(\hat{r}_R)$ is realized by the backwards extrapolated free-electron solutions[20], $\varphi_{Rl}(\epsilon, r_R)Y_L(\hat{r}_R)$ (thin black line in Fig. 1), which joins continuously and differentiable onto the partial waves at the boundary of the potential sphere, and continuously but not differentiable onto the screened spherical waves at the boundary of the hard sphere, introducing a kink.

The expansion coefficients $v_{RL,j}$ and the one-electron energies ϵ_j in Eq. (2) are determined from the condition that $\Psi_j(\mathbf{r})$ should be a solution of the Kohn-Sham equation in the entire space. This condition leads to the *kink cancellation* equation

$$\sum_{RL} a_{R'} \left[S_{R'L'RL}(\kappa_j, \mathbf{k}) - \delta_{R'R}\delta_{L'L}\, D_{Rl}(\epsilon_j)\right] v_{RL,j}(\mathbf{k}) = 0, \tag{5}$$

where $l', l \leq l_{max}$. $S_{R'L'RL}$ are the elements of the *slope matrix*[20], and they represent the expansion coefficients of $\psi_{RL}(\kappa_j, \mathbf{r})$ around site R'. With properly chosen energy independent boundary conditions and for κ_j^2 below the bottom of the a−spheres continuum, $S_{R'L'RL}$ have short range and week energy dependence[20]. In Eq. (5) $D_{Rl}(\epsilon_j)$ is the logarithmic derivative of $\varphi_{Rl}(\epsilon, r_R)$ at $r_R = a_R$[25,24]. For periodic systems the slope matrix and the expansion coefficients depend on the Bloch vector $\mathbf{k}$ from the first Brillouin zone, and the summation in Eq. (5) runs therefore only over lattice vectors R that belong to a unit cell. In practice, Eq. (5) is solved using the Green function formalism.

The total electron density is given in terms of the wave functions

$$n(\mathbf{r}) = \sum_{j}^{\epsilon_j \leq \epsilon_F} |\Psi_j(\mathbf{r})|^2, \tag{6}$$

where the summation includes the states below the Fermi level ϵ_F. Using the two center expansions of the EMTO's[20,24] the multicenter expansion (6) can be transformed to one-center form

$$n(\mathbf{r}) = \sum_R n_R(\mathbf{r}_R) = \sum_{RL} n_{RL}(r_R)Y_L(\hat{r}_R). \tag{7}$$

Here the densities $n_R(\mathbf{r}_R)$ are defined inside the Wigner-Seitz cell at R, shown by hexagons in Fig. 1. The partial components $n_{RL}(r_R)$ are expressed in terms of the EMTO's and the EMTO path operator[24].

For self-consistent calculations one constructs the overlapping potential (1) from the total charge density (7). This procedure, within the EMTO, assumes[26] the calculation of the full-potential and the construction of the optimized overlapping muffin-tin wells. The former step, using the one-center formalism for the charge density and potential, is very demanding and converges very slowly in the corners of the unit cell. However, within the spherical cell approximation (SCA)[24,25] the overlapping potential depends only on the spherical part of the full-potential and this can be computed efficiently and with high accuracy. The SCA represents two approximations: (i) for the calculation of the Madelung contribution to the electrostatic potential around site R the Wigner-Seitz cell centered at R' is substituted by a spherical cell of radius $w_{R'}$ and volume equal to the volume of the cell (ii) the potential inside the sphere is fixed to the spherical part of the full-potential, *viz.* $v_R(r_R) = \frac{1}{4\pi}\int v(\mathbf{r})d\hat{\mathbf{r}}_R$, and, thus, the expression for v_0 reduces to[24]

$$v_0 = \sum_R \int_{s_R}^{w_R} r_R^2\, v_R(r_R)\, dr_R / \sum_R \left[(w_R^3 - s_R^3)/3\right]. \tag{8}$$

The solution of the Poisson equation within the SCA for the spherical part of the full-potential is described in Refs.[24,25]. Finally, the self-consistent total charge density is used to compute the total energy. The EMTO kinetic energy is obtained from the one-electron equations[24], and it is exact for the optimized overlapping potential. The Coulomb and exchange-correlation part of the energy functional are evaluated without shape approximation to the potential or density inside the Wigner-Seitz cells. Details about the total energy calculation technique for ordered systems are discussed in Refs.[23,24].

EMTO-CPA METHOD: THE AVERAGE ONE-ELECTRON GREEN FUNCTION

We consider a substitutional alloy with a fixed underlying lattice, like the *fcc*, *bcc*, or any other more complicated crystal lattice. We denote unit cell sites of the underlying lattice like U, U', etc. On each site U we have N_U alloy components i with concentrations c_U^i and spherical potentials $v_U^i(r_U)$ with $i = 1, 2, ..., N_U$. In real random alloys the potentials $v_R^i(r_R)$ at each lattice site R that belong to sublattice U are somewhat different due to different local environment. Moreover, $v_U^i(r_U)$ is not an average of the potentials $v_R^i(r_R)$, because the one-electron potential is not a self-averaging quantity, but it is calculated from the average Green function, which has a property of the self-averaging. Thus, we make an approximation introducing the potential $v_U^i(r_U)$. However, this approximation is quite accurate[14].

The radial Schrödinger equations are solved for each sort of atoms and from the matching conditions at the $s-$spheres of the $i-$th component, of radii s_U^i, and $a-$spheres we set up the backwards extrapolated free-electron solutions and the corresponding logarithmic derivatives D_{Ul}^i.

In order to solve the multiple-scattering problem for the alloy, one replaces the original random alloy by an ordered lattice of effective scatterers, or the so-called effective medium. The effective medium is described by a site U (but not a sort i) dependent (complex) coherent potential. Therefore, the effective medium has the symmetry of the underlying crystal lattice. In the single site approximation the properties of these effective atoms have to be determined self-consistently by the condition that the scattering of electrons of real atoms embedded in the effective medium vanish on the average. For a disordered binary alloy A_cB_{1-c} this idea is schematically illustrated in Fig. 2.

In the EMTO formalism the coherent potential is introduced via the site-diagonal logarithmic derivative $\tilde{D}_{UL'UL}(z)$ of the effective scatterers, and, therefore, the coherent Green function or the path operator is given by[19]

$$\sum_{U''L''} a_{U'} \left[S_{U'L'U''L''}(\kappa, \mathbf{k}) - \delta_{U'U''}\tilde{D}_{U'L'U'L''}(z)\right] \tilde{g}_{U''L''UL}(z, \mathbf{k}) = \delta_{U'U}\delta_{L'L}, \tag{9}$$

where $l, l', l'' \leq l_{max}$. The (restricted) average of the on-site (UU) elements of the Green function for alloy component i, $g_{ULUL'}^i$, is calculated as an impurity Green function of the $i-$th alloy component embedded in the effective medium (see Fig. 2). In the single-site approximation, it is obtained from the real space Dyson equation as a single site perturbation on the coherent potential

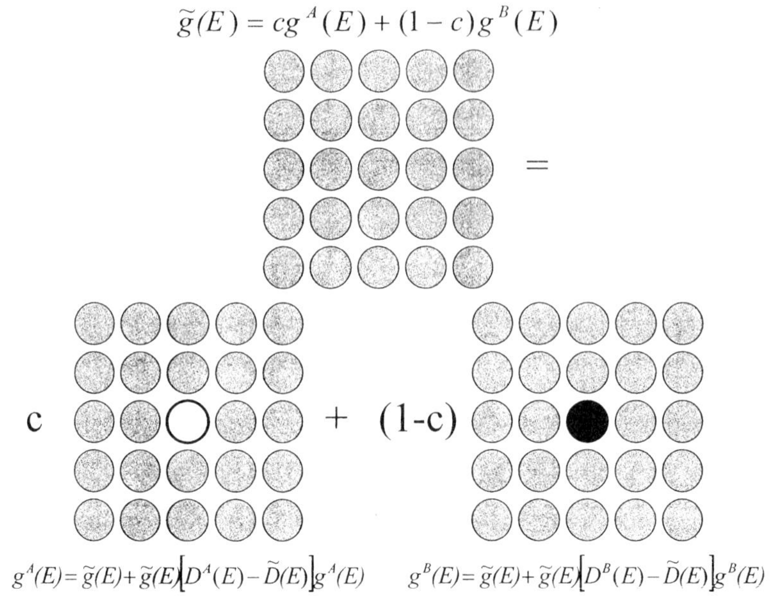

Figure 2. Schematic illustration of the basic idea of the coherent potential approximation for a disordered binary alloy A_cB_{1-c}. The original random alloy is replaced by an ordered lattice of effective scatterers (top panel), or the so-called effective medium. The properties of the effective atoms (gray circles) are determined self-consistently. The restricted averages of the on- site elements of the Green function for alloy components A (white circle) and B (black circle), g^A and g^B, respectively, are calculated considering them as impurities embedded in the effective medium (bottom panel). This is done by solving the corresponding Dyson equations, given at the bottom of the figure. The condition of vanishing on the average of scattering of electrons off the alloy components leads to the relation between the Green function of the effective medium $\tilde{g}$ and the Green functions of alloy components, given at the top of the figure.

$$g^i_{ULUL'}(z) = \tilde{g}_{ULUL'}(z) + + \sum_{L''L'''} \tilde{g}_{ULUL''}(z) \left[D^i_{Ul''}(z)\delta_{L''L'''} - \tilde{D}_{UL''UL'''}(z)\right] g^i_{UL'''UL'}(z). \tag{10}$$

The condition of vanishing on the average scattering leads to the following relation between the site diagonal part of the **k**−integrated coherent Green function $\tilde{g}_{ULUL'}(z)$ and the Green functions of alloy components within the CPA

$$\tilde{g}_{ULUL'}(z) = \sum_i c^i_U \, g^i_{ULUL'}(z), \tag{11}$$

Equations (9),(10) and (11) are solved self-consistently for $\tilde{D}(z), \tilde{g}(z, \mathbf{k})$ and $g^i(z)$. The total number of states below the Fermi level,

$$N(\epsilon_F) = \frac{1}{2\pi i} \oint_{\epsilon_F} < G(z) > dz, \tag{12}$$

is obtained from the average Green function

$$< G(z) > \;\equiv\; \int_{BZ} \sum_{U'L'UL} \tilde{g}_{U'L'UL}(z,\mathbf{k})\, \dot{S}_{ULU'L'}(z,\mathbf{k}) d\mathbf{k} \; - \\ - \sum_{Ui} c_U^i \sum_L \left[g^i_{ULUL}(z)\, \dot{D}^i_{Ul}(z) + \left(\frac{\dot{D}^i_{Ul}(z)}{D^i_{Ul}(z)} - \frac{1}{z - e^i_{Ul}} \right) \right], \tag{13}$$

where the *over dot* stands for the energy derivative, and $l, l' \leq l_{max}$. The site off-diagonal elements of the coherent Green function $\tilde{g}_{U'L'UL}(z,\mathbf{k})$ are calculated from Eq. (9) with the self-consistent logarithmic derivative $\tilde{D}_{UL'UL}(z)$ of the effective scatterers. The energy integral from Eq. (12) is performed on a complex contour that cuts the real axis below the bottom of the valence band and at ϵ_F. The $g^i(z)\dot{D}^i(z)$ term from the right hand side of Eq. (13) may include the unphysical poles of $\dot{D}^i(z)$, which, however, are canceled by the last term, where e^i denote the real zeros of the logarithmic derivative function.

The first term from the right hand side of Eq. (13) assures the proper normalization of the one-electron states for the optimized overlapping potential. In fact, within the single site approximation for the impurity Green function, Eq. (12) gives the exact number of states at the Fermi level[19,24].

EMTO-CPA METHOD: THE FULL CHARGE DENSITY

The full charge density of each alloy component is represented separately in one-center form[19], *viz.* Eq. (7). The partial components $n^i_{RL}(r_R)$ of the average density $n^i_R(\mathbf{r}_R)$ on a site R belonging to the sublattice U are determined from the restricted average of the on-site element of the Green function for the i−th alloy component $g^i_{ULUL'}$ given by Eq. (10) via the density matrix $\mathcal{D}$ of the alloy component[24]

$$\mathcal{D}^i_{RL'L}(z) \equiv \begin{cases} g^i_{UL'UL}(z) + \frac{\delta_{L'L}}{a_U\, D^i_{Ul}(z)} \left(\frac{\dot{D}^i_{Ul}(z)}{D^i_{Ul}(z)} - \frac{1}{z - e^i_{Ul}} \right) & \text{if } l, l' \leq l_{max} \\ \sum_{U''L''} \int_{BZ} \tilde{g}_{UL'U''L''}(z,\mathbf{k}) S_{U''L''UL}(\kappa,\mathbf{k}) d\mathbf{k} & \text{if } l' \leq l_{max} \text{ and } l > l_{max} \\ \sum_{U''L''U'''L'''} \int_{BZ} S_{UL'U''L''}(\kappa,\mathbf{k}) \times & \\ \times \tilde{g}_{U''L''U'''L'''}(z,\mathbf{k}) S_{U'''L'''UL}(\kappa,\mathbf{k}) d\mathbf{k} & \text{if } l', l > l_{max} \end{cases}$$

as

$$n^i_{RL}(r_R) \;=\; \frac{1}{2\pi i} \oint_{\epsilon_F} \sum_{L''L'} C^L_{L''L'} Z^i_{Rl''}(z, r_R)\, \mathcal{D}^i_{RL''L'}(z)\, Z^i_{Rl'}(z, r_R) dz, \tag{14}$$

where $C^L_{L''L'}$ are the real harmonic Gaunt coefficients. The $Z^i_{Rl}(z, r_R)$ functions denote the $Y_L(\hat{r}_R)$ projections of the exact muffin-tin orbitals[20,24], i.e.

$$Z^i_{Rl}(z, r_R) \;=\; \begin{cases} N^i_{Rl}(z)\phi^i_{Rl}(z, r_R) & \text{if } l \leq l_{max} \text{ and } r_R \leq s_R \\ \varphi^i_{Rl}(z, r_R) & \text{if } l \leq l_{max} \text{ and } r_R > s_R \\ -j_l(\kappa\, r_R) & \text{if } l > l_{max} \text{ for all } r_R \end{cases}, \tag{15}$$

where $j_l(\kappa\, r_R)$ are the spherical Bessel functions[27]. In Eq. (15) $N^i_{Rl}(z)$ represents the normalization function of the partial waves. This is determined from the matching conditions at the a and s−spheres[25,24]. Note that in Eq. (14), due to the one-center form, the l'' and l' summations include the *higher* terms as well. In applications the higher $Z^i_{Rl}(z, r_R)$ functions are truncated at $l^h_{max} = 8 - 12$[24].

EMTO-CPA METHOD: THE POISSON EQUATION

In the case of an alloy we define an overlapping muffin-tin potential $v^i_{mt}(\mathbf{r})$, Eq. (1), for each alloy component i at sublattice U of the underlying crystal lattice, and assume that the potential for this element is the same at any site R that belongs to the sublattice U. Remember that we denote the spherically symmetric part of the total potential $v^i(\mathbf{r})$ as $v^i_R(r_R)$. It is calculated from the restricted average Green function for the $i-$th alloy component, Eq. (10), considered according to the main idea of the CPA, Fig. 2, as an impurity embedded in the effective medium. We employ the SCA to solve the Poisson equation for $v^i_R(r_R)$. The spherical potential due to the charges inside of the potential sphere (the electronic, n^i, and protonic, Z^i, charges) is given by[24]

$$v_R^{I,i}(r_R) = 8\pi\frac{1}{r_R}\int_0^{r_R} {r'_R}^2 n^i_{RL_0}(r'_R)dr'_R + 8\pi\int_{r_R}^{s^i_R} r'_R n^i_{RL_0}(r'_R)dr'_R - \frac{2Z^i_R}{r_R}, \tag{16}$$

where $L_0 \equiv (0,0)$. The net charges from the outside of the potential sphere are taken into account by the average Madelung potential, whose spherically symmetric part is

$$v_R^M = \frac{1}{w}\sum_{R'L'} M_{RL_0R'L'}\, Q^{SCA}_{R'L'}, \tag{17}$$

where $M_{RLR'L'}$ are the elements of the Madelung matrix, the sum runs over the unit cell vectors of the underlying lattice, w is the average atomic radius, and Q^{SCA}_{RL} are the average multipole moments calculated within the spherical cells,

$$Q^{SCA}_{RL} = \sum_i c^i_R\Big[\frac{\sqrt{4\pi}}{2l+1}\int_0^{w^i_R}\left(\frac{r_R}{w}\right)^l n^i_{RL}(r_R)\, r_R^2\, dr_R - Z^i_R\,\delta_{L,L_0}\Big] + \delta^{SCA}\,\delta_{L,L_0}. \tag{18}$$

The site independent normalization constant δ^{SCA} appears because the integral from (18) is performed over the spherical cell rather than over the unit cell and it is determined from the condition of charge neutrality $\sum_R Q^{SCA}_{RL_0} = 0$.

Within the SCA a correction to the Madelung potential (17) has to be considered[24]. The average number of electrons inside the $s-$spheres at R,

$$Q^s_R = \sum_i c^i_R\, Q^{i,s}_R = \sum_i c^i_R\sqrt{4\pi}\int_0^{s^i_R} r_R^2 n^i_{RL_0}(r_R)dr_R, \tag{19}$$

is usually different from the average number of electrons inside the cell, $Q^{SCA}_{RL_0}+\sum_i c^i_R Z^i_R$. In other words, the potential spheres are not charge neutral. The above difference contributes with a constant shift, Δv^{SCA}_R, to the spherical potential. In the EMTO-CPA method this extra or missing charge is redistributed equally on the N_{NN} nearest neighbor cells, i.e. we make the approximation

$$\Delta v^{SCA}_R \approx \frac{1}{w}\sum_{R_{NN}} M_{RL_0R_{NN}L_0}\,\Delta Q_{R_{NN}}, \tag{20}$$

where $\Delta Q_{R_{NN}} \equiv \frac{1}{N_{NN}}\left(Q^{SCA}_{RL_0} + \sum_i\, c^i_R Z^i_R - Q^s_R\right)$.

Since the impurity problem in both the CPA and the Poisson equations is treated within the single site approximation, the Coulomb system of a particular alloy component may contain a non-zero net charge. The effect of the charge misfit on the spherical potential is taken into account using the *screened impurity model*[11,13], i.e. an additional shift of

$$\Delta v_R^{CPA,i} = -\frac{2\alpha_c}{w}\left(Q_R^{i,s} - Q_R^s\right), \tag{21}$$

is added to the spherical part of the full-potential around site R. Here $\alpha_c \approx 0.6$ and $Q_R^{i,s}$ and Q_R^s are defined in (19). Note in the case of ordered systems we have $\Delta v_R^{CPA,i} = 0$.

The total potential within the potential sphere of the i-th alloy component is obtained as the sum of Eqs. (16), (17), (20), (21) and the spherical symmetric exchange-correlation potential, namely

$$v_R^i(r_R) = v_R^{I,i}(r_R) + v_R^M + \Delta v_R^{SCA} + \Delta v_R^{CPA,i} + \mu_{xc}[n_{RL_0}^i(r_R)]. \tag{22}$$

Finally, the interstitial potential in the case of EMTO-CPA is obtained as the average interstitial potential calculated from $v_R^i(r_R)$

$$v_0 = \sum_{Ri} c_R^i \int_{s_R^i}^{w_R^i} r_R^2\, v_R^i(r_R)\, dr_R / \sum_{Ri} c_R^i \left[({w^i}_R^3 - {s^i}_R^3)/3\right]. \tag{23}$$

EMTO-CPA METHOD: THE TOTAL ENERGY

The total energy of the random alloy is calculated as

$$\begin{aligned} E_{tot} &= T[n] + \sum_R \sum_i c^i \left(F_{intraR}^i[n_R^i] + E_{xcR}^i[n_R^i]\right) + F_{inter}[Q] - \\ &- \sum_i c^i \frac{\alpha_c}{w}\left(Q_R^{i,s} - Q_R^s\right)^2, \end{aligned} \tag{24}$$

where T is the total kinetic energy, F_{intraR}^i and E_{xcR}^i are the electrostatic and exchange-correlation energies due to the charges from the Wigner-Seitz cell at R, and F_{inter} is the average Madelung energy. The last term from (24) is the SIM correction to the electrostatic energy[11,13]. The individual energy functionals are evaluated using the *Full Charge Density* method in combination with the shape function technique[23,24]. The total kinetic energy is determined from the one-electron equations,

$$T[n] = \frac{1}{2\pi i}\oint_{\epsilon_F} z < G(z) > dz - \sum_R \sum_i c^i \int_{\Omega_R} v_{mt}^i(\mathbf{r}_R) n_R^i(\mathbf{r}_R) d\mathbf{r}_R, \tag{25}$$

where the first term from the right hand side is the sum of the average one-electron energies and $< G(z) >$ is given by Eq. (13). The Madelung energy is the average interaction energy between the Wigner-Seitz cells and it involves the average multipole moments

$$Q_{RL} \equiv \frac{\sqrt{4\pi}}{2l+1}\int_{cell} \sum_i c^i\, n_R^i(\mathbf{r}_R) \left(\frac{r_R}{w}\right)^l Y_L(\hat{r}_R) d\mathbf{r}_R - \delta_{L,(0,0)} \sum_i c^i Z_R^i. \tag{26}$$

Explicit expressions for $F_{intraR}^i[n_R^i]$ and $F_{inter}[Q]$ can be found in Refs.[23,24]. $E_{xcR}^i[n_R^i]$ is calculated within the *local density* (LDA) or a *gradient level* approximation[28–30].

APPLICATION: $Cu_{1-x}Zn_x$ ALLOY

In this section we demonstrate the application of the FCD-EMTO-CPA method in the description of thermodynamic properties of $Cu_{1-x}Zn_x$ random alloy. We have applied the method for the calculation of the mixing enthalpies of α, β, β', ϵ, and

$\eta-$phases of the Cu-Zn system, covering almost the whole phase diagram[31]. The random alloys were considered as completely random, i.e. we neglected any short-range order effects in the system, while the $\beta'-$ phase was considered as fully ordered phase with B2 (CsCl) structure and exact stoichiometry. The α phase has fcc structure and at low temperatures it is stable for $x \lesssim 0.38$. The $\beta-$CuZn with the *bcc* (A2) structure is stable at high temperatures for $0.4 \lesssim x \lesssim 0.55$. At low temperatures it undergoes disorder to order phase transition into the $\beta'-$phase at a temperature of about 730 K. The ϵ phase is stable from $x \sim 0.78$ to 0.86 and it has *hcp* (A3) structure with $c/a \approx 1.56$. The Zn solid solution, the so-called $\eta-$Brass, extends to 3 at. % Cu, and it has *hcp* structure with anomalously large c/a ratio, $1.80 < c/a \lesssim 1.856$[32,33].

Due to the richness of the phase diagram, the Cu-Zn system has become a classical subject for the theory of random alloys. It is a Hume-Rothery system, where the phase stability is determined by the so-called electron concentration[32]. The mixing enthalpies of this system were mainly studied within the CPA for the α-CuZn alloys. The first calculations in the framework of the local density approximation and the CPA were carried out by Johnson *et al.*[7,8]. Then, Johnson and Pinski have shown that the agreement with experiment is improved by including the electrostatic contribution to the total energy of a random alloy[10]. Abrikosov *et al.*[34] have calculated the mixing energies in fcc Cu-Zn alloys including the short-range order effects, and have shown that the latter gives a substantial contribution to the alloy total energy. This was confirmed by Faulkner *et al.*[35] and by Müller and Zunger[36]. The latter authors have also shown that the local lattice relaxations account for some lowering of the mixing enthalpy of a random alloy. On the other hand, much less calculations were done for the *bcc*, and especially for the *hcp* phases.

The latter may be explained by the fact that the c/a ratio in the *hcp* Cu-Zn alloys shows an anomalous behavior, with very strong concentration dependence, as well as with a discontinuity at about 8 at. % Cu dissolved into Zn[32,37]. The implementations of the CPA within the KKR or LMTO basis sets are not suitable for the treatment of this problem. Using the FCD-EMTO-CPA method we are able to calculate theoretical c/a axial ratios of the *hcp* structures and the corresponding equilibrium atomic radii[19]. Therefore, we are able to calculate the mixing energy for all the phases mentioned above in the complete interval of concentrations.

In all our calculations the exchange-correlation term was treated within the local density approximation[28,38]. The EMTO basis set included s, p, d and f orbitals and the one-electron equations were solved within the scalar relativistic and frozen core approximations. The Green function was calculated for 16 complex energy points distributed exponentially on a semi-circular contour. The equilibrium total energy of alloys were determined at the theoretical equilibrium volumes, which were calculated for each phase considered. The k-space integral were well converged for all underlying lattices. The c/a axial ratios of *hcp* structures were calculated by 2$-$dimensional total energy minimizations. The atomic volumes for the Cu-rich fcc alloys and for the Zn-rich *hcp* alloys, as well as the concentration dependent c/a ratios, used in the present work may be found in Ref.[19].

Calculated mixing enthalpies of α, β, β', ϵ, and $\eta-$phases of the Cu-Zn system are shown in Fig. 3. As standard states we used the pure fcc Cu and *hcp* Zn with theoretically determined c/a ratio 1.82, which compares well with experimental $c/a = 1.856$[33]. One can see that calculations correctly reproduce the principal features of the Cu-Zn phase diagram[31]. Considering the random alloys alone we observe that for the Cu-rich alloys the fcc structure is stable. When Cu concentration increases, the fcc alloys are almost degenerate with the *hcp* alloys, followed by the region where *bcc* alloys are more

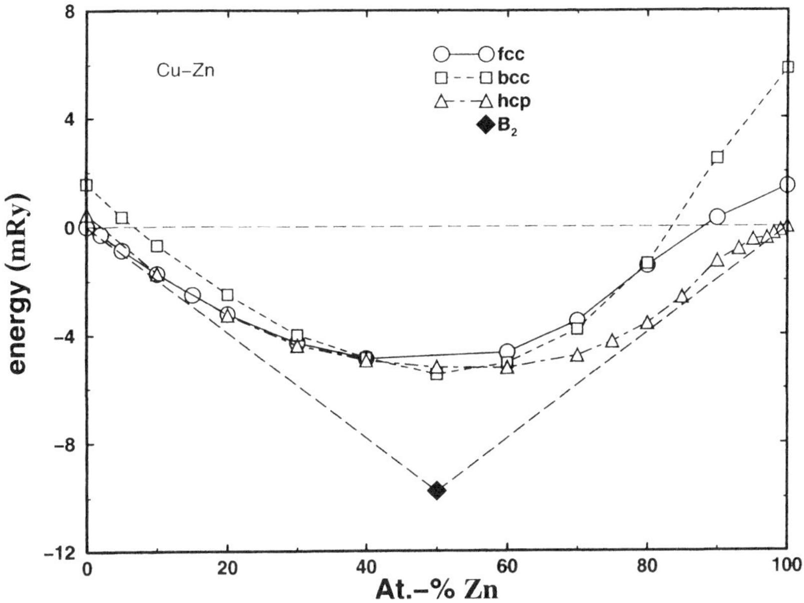

Figure 3. Calculated mixing enthalpies of completely random *fcc* (circles, full line), *bcc* (squares, dashed line), and *hcp* (triangles, dot-dashed line) Cu-Zn alloys, as well as the ordered B2 CuZn compound (diamond). The pure *fcc* Cu and *hcp* Zn with theoretically determined c/a ratio 1.82 are used as standard states, and the so-called ground state lines (long-dashed lines) connect them to the B2 CuZn compound.

stable. For the Zn-rich alloys the *hcp* structure is more stable. The $\beta'-$phase (B2) is the most stable phase, and as a matter of fact the energies of all the completely random alloys considered in this study are above the ground-state line (the long-dashed line in Fig. 3). Of course, in reality the energy of a random alloy will be lowered by the short-range order that is known to be present in the system[39].

The most interesting behavior of the mixing enthalpy is observed in Zn-rich *hcp* alloys. One can clearly see a pronounced bump on the curve around 90 at.% of Zn, indicating the existence of two *hcp* based alloys, which are stable from $\sim$ 75 to 85 at. % Zn and from $\sim$ 95 to 100 at. % Zn, respectively. These concentration intervals essentially coincides with the stability fields of the ϵ and $\eta-$phases[31]. Our calculations show directly that at low Cu concentrations the $\eta-$phase, with large c/a ratio[10], is the ground state structure of the copper-zinc alloy. With increasing Cu concentration a second total energy minimum in the volume versus axial ratio plane starts to develop. For the case of $Cu_{0.07}Zn_{0.93}$ alloy this was demonstrated explicitly in Fig. 3 of Ref.[19]. For the Zn concentration $\lesssim$ 90 at. % the second energy minimum becomes stable relative to the first one, and the system stabilizes in *hcp* structure with $c/a \lesssim 1.6$, corresponding to the $\epsilon-$phase. The transition between two phases occurs discontinuously at around 90 at. % of Zn. The existence of the discontinuity was predicted from the analysis of experimental data by Massalski *et al.*[37].

As compared to the experimentally derived mixing enthalpies[31], the calculated values show the same trends, but have somewhat higher values. This is a direct consequence of the limitations of the CPA, which does not allow one to calculate directly the total energies of alloys with a short-range order, and also neglect the effects due to local lattice relaxations. For example, our calculated value of the mixing enthalpy for the completely random *fcc* $Cu_{60}Zn_{40}$ alloy is -4.8 mRy/atom, or about -1530 cal/mole. The experimental mixing enthalpy for the alloy with 38 at. % Zn is -1969 cal/mole[31], or -6.26 mRy/atom. The experimental values are given for temperature 773 K. At this temperature the short-range order decreases the mixing enthalpy of the alloy by 1 mRy/atom[34], and the local lattice relaxations add about -0.5 mRy/atom[36]. Therefore, including these effect we would receive the theoretical value -6.3 mRy/atom, or -2000 cal/mole, in perfect agreement with experiment.

Note, that though the short-range order effects, as well as the local lattice relaxations, cannot be directly treated by the CPA, they may be considered by means of complimentary CPA-based methods. For example, the screened generalized perturbation method[13] allows one to calculate quite accurately the short-range order parameters in random alloys, while the effect of lattice relaxation can be included by means of simple models[40]. Moreover, the FCD-EMTO-CPA method is a suitable starting point for the development of more powerful methods within the alloy theory, which go beyond the single site approximation, like the locally self-consistent Green function (LSGF) method[34]. In this case one will be able to treat directly alloys with short-range order[34] and with local lattice relaxations[41].

CONCLUSIONS

We have presented and tested an *ab initio* total energy method for random alloys. The method is based on the *exact muffin-tin orbitals theory* in conjunction with the *full charge density* technique and the *coherent potential approximation.* The EMTO kinetic energy, determined within the single site and the spherical cell approximations, is combined with the Coulomb and exchange-correlation energies calculated from the total charge density of the alloy components using the shape function technique. The FCD-EMTO-CPA method has been applied for the study of the thermodynamic properties of the copper-zinc system in complete interval of concentrations. The overall good agreement between the calculated and measured mixing enthalpies is obtained in the calculations, and the main features of the Cu-Zn phase diagram are well reproduced.

ACKNOWLEDGMENTS

The Swedish Research Council (VR), the Swedish Foundation for Strategic Research (SSF) and The Royal Swedish Academy of Sciences (KVA) are acknowledged for financial support. Part of this work was supported by the research project OTKA T035043 of the Hungarian Scientific Research Fund and by the Hungarian Academy of Science.

REFERENCES

1. For a review see J. S. Faulkner, Prog. Mater. Sci. **27**, 1 (1982).
2. P. Soven, Phys. Rev. **156**, 809 (1967).
3. D. W. Taylor, Phys. Rev. **156**, 1017 (1967).

4. B. L. Györffy Phys. Rev. B **5**, 2382 (1972).
5. P. Hohenberg and W. Kohn, Phys. Rev. **136B** 864 (1964); W. Kohn and L.J. Sham, Phys. Rev. **140A**, 1133 (1965).
6. H. Winter and G. M. Stocks, Phys. Rev. B **27**, 882 (1983).
7. D. D. Johnson, D. M. Nicholson, F. J. Pinski, B. L. Györffy, and G. M. Stocks, Phys. Rev. Lett. **56**, 2088 (1986).
8. D. D. Johnson, D. M. Nicholson, F. J. Pinski, B. L. Györffy, and G. M. Stocks, Phys. Rev. B **41**, 9701 (1990).
9. I. A. Abrikosov, Yu. H. Vekilov, P. A. Korzhavyi, A. V. Ruban, and L. E. Shilkrot, Solid State Commun. **83**, 867 (1992).
10. D. D. Johnson and F. J. Pinski, Phys. Rev. B **48**, 11553 (1993).
11. P. A. Korzhavyi, A. V. Ruban, I. A. Abrikosov, and H. L. Skriver, Phys. Rev. B **51**, 5773 (1995).
12. A. V. Ruban, A. I. Abrikosov, and H. L. Skriver, Phys. Rev. B **51**, 12958 (1995).
13. A. V. Ruban and H. L. Skriver, Phys. Rev. B **66**, 024201 (2002); A. V. Ruban, S. I. Simak, P. A. Korzhavyi, and H. L. Skriver, Phys. Rev. B **66**, 024202 (2002).
14. I. A. Abrikosov and B. Johansson, Phys. Rev. B **57**, 14164 (1998) and references therein.
15. I. A. Abrikosov, A. V. Ruban, B. Johansson, and H. L. Skriver, Comput. Mater. Sci. **10**, 302 (1998).
16. D. D. Johnson and M. Asta, Comput. Mater. Sci. **8**, 54 (1997).
17. O. K. Andersen, O. Jepsen, and D. Glötzel, in *Highlights of Condensed-Matter Theory*, edited by F. Bassani, F. Fumi, and M. P. Tosi, North-Holland, New York, (1985).
18. P. James, I. A. Abrikosov, O. Eriksson, and B. Johansson, in *Properties of Complex Inorganic Solids*, edited by A. Gonis, A. Meike, and P. E. A. Turchi (Plenum, NY, 1997), P. 57.
19. L. Vitos, I. A. Abrikosov, and B. Johansson, Phys. Rev. Lett. **87**, 156401 (2001).
20. O. K. Andersen, O. Jepsen, and G. Krier,in *Lectures on Methods of Electronic Structure Calculations*, edited by V. Kumar, O. K. Andersen, and A. Mookerjee, World Scientific Publishing Co., Singapore, pp. 63-124 (1994).
21. O. K. Andersen, C. Arcangeli, R. W. Tank, T. Saha-Dasgupta, G. Krier, O. Jepsen, and I. Dasgupta, in *Mat. Res. Soc. Symp. Proc.* **491**, pp. 3-34 (1998).
22. O. K. Andersen, T. Saha-Dasgupta, R. W. Tank, C. Arcangeli, O. Jepsen, G. Krier,in *Electronic Structure and Physical Properties of Solids: The uses of the LMTO method*, ed. H. Dreyssé, Lectures Notes in Physics, Springer-Verlag Berlin, pp. 3-84, (2000).
23. J. Kollár, L. Vitos, and H. L. Skriver, in *Electronic Structure and Physical Properties of Solids: the uses of the LMTO method*, ed. H. Dreyssé, Lectures Notes in Physics, Springer-Verlag Berlin, pp. 85-113, (2000).
24. L. Vitos, Phys. Rev. B **64** 014107 (2001).
25. L. Vitos, H.L. Skriver, B. Johansson, and J. Kollár, Comp. Mat. Sci., **18**, 24, (2000).
26. O. K. Andersen and C. Arcangeli, (unpublished).
27. *Handbook of Mathematical Functions*, edited by M. Abramowitz and I. A. Stegun, Dover Publications, Inc., New York (1970).
28. J. Perdew and Y. Wang, Phys. Rev. B **45**, 13244, (1992).
29. L. Vitos, B. Johansson, J. Kollár, and H. L. Skriver, Phys. Rev. B **62**, 10046 (2000).
30. J.P. Perdew, K. Burke, and M. Ernzerhof, Phys. Rev. Lett. **77**, 3865 (1996).
31. R. Hultgren, P. Desai, D. T. Hawkins, M. Gleiser, and K. K. Kelley, *Selected Values of Thermodynamic Properties of Binary Alloys*, (American Society for Metals, Ohio, 1973), p. 228.
32. T. B. Massalski and U. Mizutani, Progress in Materials Science **22**, 151 (1978).
33. W.B. Pearson, *A handbook of lattice spacings and structures of metals and alloys*, Perg-

amon Press, London (1958).
34. I. A. Abrikosov, A. M. N. Niklasson, S. I. Simak, B. Johansson, A. V. Ruban, and H. L. Skriver, Phys. Rev. Lett. **76**, 4203 (1996).
35. J. S. Faulkner, N. Y. Moghadam, Y. Wang, and G. M. Stocks, Phys. Rev. B **57**, 7653 (1998).
36. S. Müller and A. Zunger, Phys. Rev. B **63**, 094204 (2001).
37. T. B. Massalski, L. F. Vassamillet, and Y. Bienvenu, Acta Metall. **21**, 649 (1973).
38. D.M. Ceperley and B.J. Alder, Phys. Rev. Lett. **45**, 566 (1980).
39. L. Reinhard, B. Schönfeld, G. Kostorz, and W. Bührer, Phys. Rev. B 41, 1727 (1990).
40. L. V. Pourovskii, A. V. Ruban, I. A. Abrikosov, Y. Kh. Vekilov and B. Johansson, Phys. Rev. B **64**, 035421 (2001).
41. I. A. Abrikosov, P. A. Korzhavyi, and B. Johansson, in *Electronic structure and physical properties of solids: The uses of the LMTO method*, ed. H. Dreyssé, Lectures Notes in Physics, Springer-Verlag Berlin, pp. 379, (2000).

CHARGE DISTRIBUTIONS IN METALLIC ALLOYS: A CHARGE EXCESS FUNCTIONAL THEORY APPROACH

Ezio Bruno

Dipartimento di Fisica and Unità INFM, Università di Messina, Salita Sperone 31, 98166 Messina, Italy. E-mail: bruno@dsme01.unime.it.

1. INTRODUCTION

In the last decade, the availability of large parallel computing resources and the development of order N algorithms [1, 2] made feasible ab initio electronic structure calculations in extended metallic systems. A new, surprising, result in the theory of metallic alloys has been obtained by Faulkner, Wang and Stocks, who have analysed density functional theory calculations for unit cells containing hundred to thousand atoms and designed to simulate binary alloys with substitutional disorder. They discovered [3, 4] that the net charge at each crystal site, q_i, is related to V_i, that part of electrostatic potential at the same site that is due to the interactions with all the other charges in the system, through a simple linear law

$$a_i q_i + V_i = k_i \tag{1}$$

For a specified configuration of the binary alloy $A_{c_A}B_{c_B}$, the coefficients a_i and k_i in Eq. (1) take the values a_A and k_A if the i-th site is occupied by a A atom or a_B and k_B otherwise. Moreover, the sets of coefficients extracted from different samples corresponding to the same mean concentration show up little differences. In the following, these linear relations shall be referred to as the qV laws. The above new findings can be considered empirical in the sense that, although obtained from ab initio calculations, they have not yet been formally derived.

In spite of the simplicity of Eq. (1), for each of the alloying species, the local charge excesses take *any* value within a certain interval. The corresponding distribution, even for the random alloy model, appears complex and cannot be reproduced, without a proliferation of adjustable parameters, in terms of the number of unlike neighbours of each site [5, 6]. On the other hand, accurate calculations of the alloy total energies and phase equilibria must necessarily keep into account such a distribution. Recently, it has been shown that three coefficients of the qV laws for a binary alloy can be calculated within a single site theory, namely a Coherent Potential Approximation including local fields (CPA+LF) [7]. More precisely, the coefficients a_A and a_B can be viewed as the responses of impurity sites occupied by A or B atoms to local external fields, while the third parameter can be viewed as the difference between the electronegativities of the A and B impurities embedded in the 'mean field alloy' defined by the alloy CPA Green's function.

the local In the present paper, I shall demonstrate that the distribution of charges can be obtained from a variational principle, without any need of sophisticated electronic structure calculations for supercells. For this purpose I shall formulate a Ginzburg-Landau theory in which charge excesses, q_i, play the role of the order parameter field. Hereafter this phenomenological approach is referred to as the 'charge excess functional' (CEF) theory. As it will be seen below, the CEF is completely determined by only three concentration dependent, material specific, parameters. These parameters can be calculated by the CPA+LF theory or extracted from order N calculations. Given the atomic positions within a supercell, the CEF scheme determines the charge excesses at each site, and, hence, the electrostatic energy, with an excellent accuracy. Moreover, the above procedure, using a *single* set of parameters, can be applied to any ordered, partially ordered or disordered configuration corresponding to the same mean alloy concentration. Furthermore, CEF calculations require really modest computational efforts: 20 seconds CPU time on a 1 GHz Pentium III processor for a 1000 atoms sample. This is a particularly interesting feature as it opens new perspectives. In principle, one could take advantage from these performances and determine, in a parameter free theory, the equilibrium values of the short range order parameter for metallic alloys with an accuracy unprecedented for this kind of calculation. My group is currently developing a new computer simulation technique, based on the joint use of CEF and Metropolis' Monte Carlo that should allow the study of phase equilibria in metallic alloys.

The following of this paper is arranged as follows. In Section II, I shall present the CEF theory and its general solution for the site charge excesses. In Section III, the method will be applied to bcc $Cu_{0.50}Zn_{0.50}$ alloys and its accuracy will be tested through a comparison with order N Locally Self-consistent Multiple Scattering (LSMS) theory calculations [8]. The comparison will also show that the CEF describes the distribution of charges in metallic alloys with a surprisingly good accuracy, when the material specific parameters are obtained from order N calculations, while fairly good results are obtained using the parameters obtained from the above generalisation of the CPA theory. The final Section IV is devoted to a thorough analysis of the CEF method and of its possible extensions and applications to the study of phase transitions and ordering phenomena in the metallic state.

2. A CHARGE EXCESS FUNCTIONAL FORMALISM FOR CHARGE TRANSFERS IN METALLIC ALLOYS

2.1. The model

The binary alloy $A_{c_A}B_{c_B}$, $c_A + c_B = 1$, shall be studied by the means of supercells containing N 'atoms' with periodic boundary conditions. Each site of the cell can be occupied by a A or a B atom. If the chemical occupations are not considered, the lattice described by the sites within the supercell and their periodic replicas is a simple lattice, with one atom per unit cell. Below, it shall referred to as the 'geometrical lattice'. In order to have a theory flexible enough to deal on equal footing both with ordered and disordered alloys, in principle, one should consider all the $\frac{N!}{(c_A N)!(c_B N)!}$ different 'alloy configurations' that belong to the statistical ensemble specified by a given molar fraction, c_A. Each configuration is described by the set of 'occupation numbers',

$$X_i^\alpha = \begin{cases} 1 & \text{if the } i\text{-th site is occupied by a } \alpha \text{ atom} \\ 0 & \text{otherwise} \end{cases} \qquad (2)$$

The arrays $\boldsymbol{X}^A$ and $\boldsymbol{X}^B$ describe completely an alloy configuration with a certain redundancy, since, for any i, it holds $X_i^A + X_i^B = 1$. Below, the convention is used that Latin indices, i, j, ..., identify the sites in the supercell, and Greek indices, α, β, ..., the chemical species, A or B. Moreover, whenever their ranges are not indicated, the sums over the Latin indices run from 1 to N, while the Greek indices take only the values A and B.

A volume ω_i is associated with each crystal site. In the following it will be assumed that all the atomic volumes sum up to the supercell volume. There is some arbitrariety in the way in which these volumes can be chosen: they could be built using the Wigner-Seitz construction (and possibly approximated by spheres, as in the case of the Atomic Sphere Approximation), or they could be non-overlapping muffin-tin spheres to which an appropriate fraction of the interstitial volume is added.

Each site in the supercell is occupied by a nucleus of charge Z_i, and, for each site, a charge excess can be defined as follows:

$$q_i = \int_{\omega_i} d\vec{r}\rho(\vec{r}) - Z_i \tag{3}$$

where $\rho(\vec{r})$ is the electronic density. The above charge excesses satisfy a global electroneutrality condition

$$\sum_i q_i = 0 \tag{4}$$

Different models for ordered and disordered alloys will be discussed below. The *random alloy* model can be defined by saying the the occupations of different sites are not statistically correlated, i.e.

$$\langle X_i^\alpha X_j^\beta \rangle = \langle X_i^\alpha \rangle \langle X_j^\beta \rangle \tag{5}$$

or, equivalently, by assuming equal statistical weights for all the alloy configurations in a fixed concentration ensemble. This, evidently, corresponds to the $T \to \infty$ limit for the alloy site occupations. Real alloys, of course, should be studied at finite temperatures and, in order to describe *how much* they are ordered, it is customary to introduce the short range order parameters [9],

$$p(\vec{r}_{ij}) = \langle X_i^A X_j^B \rangle - \langle X_i^A \rangle \langle X_j^B \rangle \tag{6}$$

Also a charge correlation function can be defined as

$$g(\vec{r}_{ij}) = \langle q_i q_j \rangle - \langle q_i \rangle \langle q_j \rangle = \langle q_i q_j \rangle \tag{7}$$

Of course, even for random alloys, the excess charges are correlated, i.e. $g(\vec{r}_{ij}) \neq 0$.

2.2. The charge excess functional

Within the muffin-tin or the atomic sphere approximation, the electrostatic energy of the system can be written as the sum of site-diagonal terms plus a Madelung term [10, 4]. I shall concentrate on the latter,

$$E_M = \sum_{ij} M_{ij} q_i q_j = \frac{1}{2} \sum_i q_i V_i \tag{8}$$

The Madelung matrix elements M_{ij} in Eq. (8) are defined [11] as

$$M_{ij} = \sum_{\vec{R}} \frac{1}{|\vec{r}_{ij} + \vec{R}|} \tag{9}$$

where $\vec{r}_{ij}$ are the translations from the i-th to the j-th site within the supercell and $\vec{R}$ are the superlattice translation vectors.

Equation (8) defines also the Madelung potential at the i-th site,

$$V_i = 2 \sum_j M_{ij} q_j \tag{10}$$

Everywhere in this paper atomic units are used in which $e^2 = 2$.

The starting point of the model is the *assumption* that linear laws hold and relate the charge excess at the i-th site q_i, and the Madelung potential at the same site V_i. As discussed above, this is an evidence from basically exact order N calculations, although also the single site CPA+LF model [7] is able to provide a realistic estimate of the coefficients entering in the linear laws. Therefore, I shall assume that, for some specified configuration, the following equations are satisfied,

$$a_i q_i + 2 \sum_j M_{ij} q_j = a_i b_i = k_i \tag{11}$$

where a_i and b_i, the coefficients of the linear laws, are assumed to depend only on the occupation of the i-th site in the configuration given and then to take the values a_A and b_A or a_B and b_B depending on the chemical occupation of the i-th site. Moreover, it is required that the global electroneutrality condition, Eq. (4), must be satisfied.

If all the material specific coefficients, a_α and b_α, were specified by the mean alloy concentrations, one would have a set of N+1 equations, Eqs. (11) and (4), and N unknown quantities to be determined, the q_i. In general, the determinant of this set of equations is not singular and, hence, the problem would be overdetermined. This is not in contrast with the results of order N calculations: in Refs. [3, 4] it is found that all the four constants are determined *for a given configuration*, while different configurations corresponding to the same mean alloy concentration are characterised by slightly different sets of constants. Actually, in Ref. [3], the comparison between results for disordered and ordered alloy configurations shows that, for ordered alloys, the constants a_α have values very close to those found for random alloys at the same mean concentration, while larger discrepancies are found for the constants b_α.

As discussed in Ref. [7], an useful hint for solving the problem comes from the the CPA+LF model. This theory views the quantities a_α as the responses of the impurity sites, embedded in the CPA 'mean' alloy, to a local field. Hence, in the CPA+LF model, the same quantities depend *only* on the mean alloy concentrations. On the other hand, in the same theory, the zero-field charges, b_A and b_B, are related one to the other through the CPA 'electronegativity' condition. These facts suggest that, in different configurations corresponding to the same alloy concentration, the constants b_α are probably *renormalised* by the global constraint, Eq. (4), while much smaller effects, if any, are expected for a_α.

To make further progresses, consider the following functional of the site charge excesses,

$$\Omega([q],\mu) = \frac{1}{2}\sum_i a_i(q_i - b_i)^2 + \sum_{ij} M_{ij}q_iq_j - \mu\sum_i q_i \tag{12}$$

where the Lagrange multiplier μ has been introduced to impose the global electroneutrality constraint. By functional mimimization with respect to the order parameter field $\{q_i\}$, and to the multiplier μ, the following set of Euler-Lagrange equations is obtained,

$$a_i(q_i - b_i) + 2\sum_j M_{ij}q_j = \mu \tag{13}$$

$$\sum_i q_i = 0 \tag{14}$$

Equation (14) evidently coincides with the electroneutrality condition, Eq. (4). On the other hand, Eq. (13) reduces to Eq. (11) only when $\mu = 0$. When $\mu \neq 0$, one can think that the renormalization of constants, $a_ib_i \to a_ib_i + \mu$ occurs in Eq. (11) in order to ensure the global electroneutrality constraint to be satisfied. If the problem of determining the site charge excesses is redefined as a minimum principle for the Ginzburg-Landau functional Ω, the four constants a_α, b_α, obtained for *a given alloy configuration* can be used also for other configurations, since they will be properly renormalised: in other words, the information obtained from a specific configuration is *transferable* to other configurations belonging to the same fixed concentration ensemble.

The minimum principle for Ω leads to a scheme in which the constants a_α, related with the response to the external potential, are not affected by the electroneutrality constraint. Now, since Ω has the dimension of an energy and contains the electrostatic energy, $\sum_{ij} M_{ij}q_iq_j$, one can think that the minimum of the functional

$$E([q],\mu) = \frac{1}{2}\sum_i a_i(q_i - b_i)^2 + \sum_{ij} M_{ij}q_iq_j \tag{15}$$

corresponds the total electronic energy of the alloy configuration, except but an addictive constant. The quadratic terms in Eq. (15) can be considered as energy contributions associated with local charge rearrangements. Moreover, the quantity μ, introduced simply as a Lagrange multiplier, can be interpreted as a chemical potential ruling the charge transfers in metallic alloys. It is evidently related with the usual electronic chemical potential, μ_{el}, that in the ensemble representable version of the density functional theory [12], appropriate for random alloys, enters in the functional through the constraint,

$$\int d\vec{r}\,\rho(\vec{r}) = \sum_i Z_i \tag{16}$$

2.3. Explicit determination of the charge transfers

In the previous subsection, the functional $\Omega([q],\mu)$, hereafter referred to as the grand potential, has been introduced. Once minimised with respect to its variables, it provides the solution for the charge distribution and the chemical potential in the configuration considered, while the four constants entering in its definition, a_α, b_α, can be considered as the characteristic parameters for a specific alloy system at some specified concentration. In facts, the constants b_α can be evaluated for *any* alloy configuration corresponding to the given mean alloy concentration, while the arbitrariety introduced in this way is removed since they will enter in determining the charge distribution only through the combinations $a_A b_A + \mu$ and $a_B b_B + \mu$, as discussed in the previous subsection. Below I shall elaborate the explicit solution of the problem. For this purpose, however, an appropriate formalism is necessary.

Below I shall use a tensor notation and denote the set of all the site charges, q_i, simply as $\boldsymbol{q}$. Analogously the sets of the Madelung potentials, V_i, and the site occupations, X_i^α, shall be denoted as $\boldsymbol{V}$ and $\boldsymbol{X}^\alpha$. Thus, e.g., $\boldsymbol{V}=2\,\boldsymbol{M}\cdot\boldsymbol{q}$ should be read as $V_i = 2\sum_j M_{ij}q_j$. Moreover, I introduce the vector

$$\boldsymbol{f} = \sum_\alpha k_\alpha \boldsymbol{X}^\alpha \tag{17}$$

and the tensor $\boldsymbol{\Gamma}$ with matrix elements

$$\Gamma_{ij} = \sum_\alpha a_\alpha X_i^\alpha\, \delta_{ij} \tag{18}$$

where δ_{ij} is the Kroenecker delta. With this notation, Eqs. (13) and (14) can be rewritten as

$$(\boldsymbol{\Gamma} + 2\mathbf{M})\cdot\mathbf{q} = \mathbf{f} + \mu(\boldsymbol{X}^{\mathbf{A}} + \boldsymbol{X}^{\mathbf{B}}) \tag{19}$$

$$(\boldsymbol{X}^A + \boldsymbol{X}^B)\cdot\mathbf{q} = \mathbf{0} \tag{20}$$

Then, the solution for the charge distribution can be written in terms of $\boldsymbol{\Lambda} = (\boldsymbol{\Gamma}+2\boldsymbol{M})^{-1}$, as follows

$$\boldsymbol{q} = \boldsymbol{\Lambda}\cdot[(\boldsymbol{f} + \mu(\boldsymbol{X}^A + \boldsymbol{X}^B))] = (k_A+\mu)\boldsymbol{\Lambda}\cdot\boldsymbol{X}^A + (k_B+\mu)\boldsymbol{\Lambda}\cdot\boldsymbol{X}^B \tag{21}$$

The chemical potential can be determined by multiplying Eq. (21) on the left by $(\boldsymbol{X}^A + \boldsymbol{X}^B)$ and using Eq. (20), e.g.,

$$(k_A+\mu)(\Lambda_{AA}+\Lambda_{BA}) + (k_B+\mu)(\Lambda_{AB}+\Lambda_{BB}) = 0 \tag{22}$$

where, the quantities

$$\Lambda_{\alpha\beta} = \boldsymbol{X}^\alpha\cdot\boldsymbol{\Lambda}\cdot\boldsymbol{X}^\beta \tag{23}$$

have been introduced. Since the matrices Γ_{ij} and M_{ij} are real and symmetric, it follows that also Λ_{ij} is real and symmetric and, hence,

$$\Lambda_{\alpha\beta} = \Lambda_{\beta\alpha} \tag{24}$$

By substitution of Eqs. (22) and (24) in Eq. (21), I find the final result of this section,

$$\boldsymbol{q} = (k_A - k_B)\left[(1-y)\boldsymbol{\Lambda}\cdot\boldsymbol{X}^A - y\boldsymbol{\Lambda}\cdot\boldsymbol{X}^B\right] \tag{25}$$

where

$$y = \frac{\Lambda_{AA}+\Lambda_{AB}}{\Lambda_{AA}+2\Lambda_{AB}+\Lambda_{BB}} \tag{26}$$

TABLE 1. Material specific parameters of the CEF, a_{Cu}, a_{Zn} and $k_{Cu} - a_{Zn}$, for a bcc $Cu_{0.50}Zn_{0.50}$ alloy. The first set of parameters, indicated as LSMS, has been extracted from the qV data obtained by the 'exact' LSMS calculations in Ref.[8] for a 1024 atoms supercell that simulate a random alloy. The parameters in the second set have been calculated using the CPA+LF model of Ref.[7]. All the quantities are in atomic units.

	LSMS	CPA+LF
a_{Cu}	1.84284	1.22787
a_{Zn}	1.82459	1.21890
$k_{Cu} - k_{Zn}$	0.28957	0.14035

TABLE 2. CEF and CEF-CPA calculations (see the text for explanations) for the same bcc $Cu_{0.50}Zn_{0.50}$ sample as in Table I are compared with the 'exact' LSMS results of Ref.[8]. $\langle q\rangle_{Cu}$ and $\langle V\rangle_{Cu}$ are, respectively, the mean values of the charges and the Madelung potentials at the Cu sites, σ_{Cu} and σ_{Zn} the standard deviations of the charge distributions for Cu and Zn, E_{MAD}/N is the Madelung energy per atom and 'errors' stand for the mean square deviations between CEF (or CEF-CPA) charges and LSMS charges. All the quantities, unless otherwise stated, are in atomic units.

	CEF	CEF-CPA	LSMS
$\langle q\rangle_{Cu}$	0.099787	0.090649	0.099783
$\langle V\rangle_{Cu}$	-0.038197	0.039881	0.038188
σ_{Cu}	0.02507	0.03082	0.02523
σ_{Zn}	0.02801	0.03412	0.02814
E_{MAD}/N (mRy)	-2.552	-2.453	-2.557
'errors'	$2.7\ 10^{-6}$	$1.5\ 10^{-4}$	

Eq. (25) clarifies that, in the general solution of the CEF for the charge distribution, b_A and b_B enter only via the difference $k_A - k_B = a_A b_A - a_B b_B$, while the dependence on the alloy configuration is conveyed by the quantity y.

The formulation of the charge excess functional, Eq. (12), and the solution for the local charges, Eq. (25) are well defined both for ordered and disordered alloys. This has been possible because of the introduction of the chemical potential μ: due to this, only *three* of the four constants that characterise the qV linear laws enter in the final solution, Eq. (25). These three quantities together with y, determined by the actual alloy configuration, are equivalent to the original set of four constants.

In the next Section, the charge excess functional formalism will be applied to $Cu_{0.50}Zn_{0.50}$ alloys on a geometrical bcc lattice.

3. CHARGE DISTRIBUTIONS IN BCC CuZn ALLOYS

3.1. Testing the CEF model

As discussed above, Eqs. (25-26) completely determine the distribution of charge excesses for a given alloy configuration. The three material specific parameters contained in the CEF can be extracted from order N as well as from CPA+LF calculations. In this section I shall compare the charge excesses obtained by the CEF with order N Locally Self-consistent Multiple Scattering (LSMS) theory calculations [8].

I have selected a specific configuration of a supercell containing 1024 atoms and designed to simulate a $Cu_{0.50}Zn_{0.50}$ random alloy on a bcc geometrical lattice for which LSMS calculations were available [8]. Unless otherwise stated, the CEF calculations reported in the present paper have been performed on the same configuration. As reported in Table I, two distinct sets for the three CEF parameters (in this case a_{Cu}, a_{Zn} and $k_{Cu} - k_{Zn}$) have been used. The parameters in the first set have been extracted from linear fits of the qV data in Ref. [8], those in the second have been calculated by the CPA+LF model, as described in Ref. [7]. Accordingly, two different sets of CEF calculations have been executed that shall be referred below to as: CEF, for the first set, or CEF-CPA, for the second.

The differences between LSMS and CEF calculations are really small, as it is apparent from Table II: 5 parts over 10^5 for the mean values of the charges and of the Madelung po-

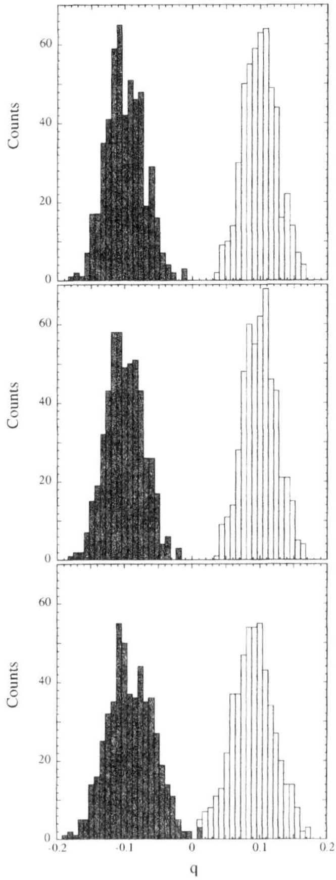

Figure 1. Cu (light histogram) and Zn (dark histogram) charge excesses distributions for the bcc $Cu_{0.50}Zn_{0.50}$ 1024 atoms sample of Table I. From top to bottom: LSMS results (Ref.[8]), CEF and CEF-CPA calculations. Atomic units are used.

tentials, 2 parts over 10^4 for the Madelung energies, less than 1 per cent for the widths of the charge distributions. The distributions, reported in Fig. (1), appear very similar. In order to have a more precise assessment of the accuracy of the CEF results, I have compared directly the charge excesses *at each lattice site*. In Fig. (2), the differences $\Delta q_i = q_i^{CEF} - q_i^{LSMS}$ are plotted. The absolute values of the Δq_i are always smaller than 0.005 electrons and no correlation is visible between the size of these 'errors' and the chemical occupation of the site. Interestingly, the mean square deviation between the two set of charges, reported in Table I, is of the order of 10^{-6}, i.e. its size is comparable with the numerical errors in LSMS calculations. The main source of the tiny differences found is that all the CEF charges, *by construction*, lie on the two straight lines corresponding to the qV laws for each of the alloying species, while the same laws hold only approximately for LSMS calculations. Tests about the *transferability* of the CEF parameters extracted from one sample to the other samples are currently being performed. Preliminary results already available [13] suggest that, when using

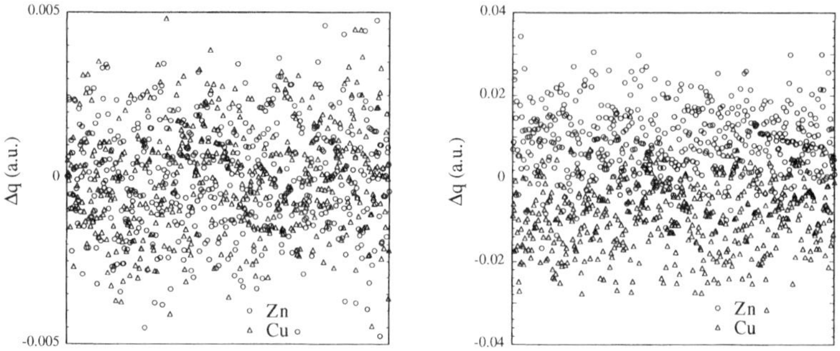

Figure 2. Left frame: $\Delta q_i = q_i^{CEF} - q_i^{LSMS}$; right frame: $\Delta q_i = q_i^{CEF-CPA} - q_i^{LSMS}$, where q_i^{CEF}, $q_i^{CEF-CPA}$ and q_i^{LSMS} are, respectively, the site charge excesses at the i-th site as obtained by CEF, CEF-CPA or LSMS (Ref.[8]) calculations for the bcc $Cu_{0.50}Zn_{0.50}$ 1024 atoms sample of Table I. In abscissa: site identifiers. Open circles: Zn sites; triangles: Cu sites. Atomic units are used.

parameters extracted from one sample, the CEF is able to reproduce the charge distribution in other samples maintaining the above mentioned accuracy, even when CEF parameters extracted from random samples are used in ordered, partially ordered or segregated samples and vice versa. This *transferability* of the CEF parameters is a very remarkable result: evidently the renormalization of the constants mentioned in the previous section is able to deal with very different samples and, more important, this implies that the CEF theory is generally applicable to metallic alloys, no matters whether they are ordered, disordered or segregated. Furthermore, the accuracy obtained for the Madelung energy demonstrates that the theory is able to describe very carefully the electrostatic contributions to the energetics of ordering phenomena.

The application of the CEF theory using parameters from CPA+LF calculations has a particular interest since the CPA+LF model is not based on a specific alloy configuration and, therefore, does not require expensive calculations on supercells. In this case, as it can be seen in Table I and Figs. (1-2), the agreement with LSMS calculation is still fairly good: the CEF-CPA underestimates the mean charges about 10 per cent and overestimates the widths of the charge distributions about 25 per cent. These errors somehow compensate giving a Madelung energy correct within 4 per cent. The comparison with LSMS for the charges at each sites, as displayed in Fig. (2), shows up small systematic errors with different signs on Cu and Zn sites. The histograms of the charge distributions present a small overlap around $q = 0$, as it is visible in Fig. (1). Also in this case, as above, preliminary tests [13] shows that the parameters obtained from the CPA+LF theory are *transferable*, in the sense that the size of the discrepancies between CEF-CPA and LSMS results appear independent on the amount of short range order in the alloy configurations considered. To maintain the same performances, even in the cases of ordered or segregated samples, is a very remarkable success for a theory, the CPA, originally proposed for random alloys. To my knowledge, this is the first time that a single site theory, free of adjustable parameters, is able to reproduce the charge distribution in metallic alloys. Better results have been obtained by the Polymorphous CPA (PCPA) [14] that, although based on the CPA theory, at similarity of 'exact' LSMS calculations, uses supercells and, hence, many different site potentials.

An important quantity, that is particularly relevant for its role in the energetics of metallic alloys [6], is the charge correlation function $g(\vec{r}_{ij})$. In Fig. (3), I plot $g(\vec{r}_{ij})$, as obtained from LSMS, CEF and CEF-CPA calculations, again for the 1024 atoms $Cu_{0.50}Zn_{0.50}$ random alloy sample of Ref. [8]. As it is apparent, the agreement between LSMS and CEF calculations is excellent, and very good also for the CEF-CPA. The test is particularly interesting since non

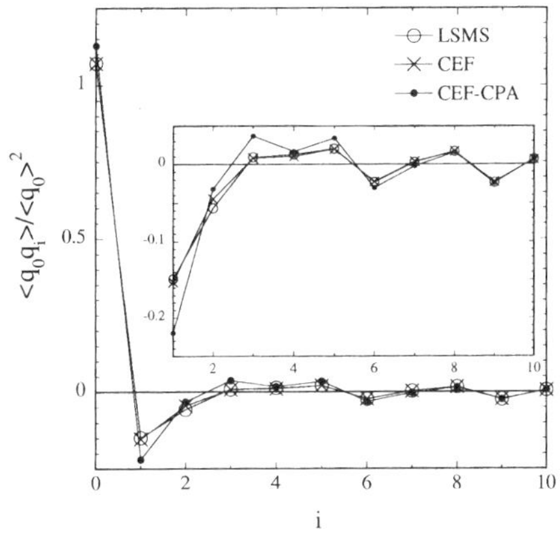

Figure 3. Normalised charge correlation function $\langle q_0 q_j \rangle / \langle q \rangle^2_{Cu}$ for the 1024 atoms bcc $Cu_{0.50}Zn_{0.50}$ random alloy sample (see Table I). Open circles: LSMS calculations (Ref.[8]), crosses: CEF calculations, filled circles: CEF-CPA calculations. In the inset, a detail of the same curve is plotted. The shell identifyer, i, is reported in abscissa. Lines joint the points.

correlated charge models would give $g(\vec{r}_{ij}) = 0$ for $r_{ij} > 0$. It could be therefore quite surprising to observe that, at least in the case at hand, the correlations are slightly *overestimated* by the CEF-CPA model.

3.2. Charge excesses versus local environments

The importance of local environments in determining the charge transfers in metallic alloys has been highlighted for the first time by Magri et al. [5]. Their model simply assumes the charge excesses to be proportional to the number of *unlike* nearest neighbours. When the development of order N calculation allowed a deeper investigation of the problem [3, 4], it was readily clear that such a simple model was not able to describe the details of the charge distributions. Later on, however, Wolverton et al. [6] generalised Magri's model by introducing additional terms proportional to the number of neighbour in outer shells and achieved appreciable improvements especially for fcc lattices.

The computational flexibility of the CEF method and the fact that it seems able to reproduce almost perfectly LSMS results allow to check the basic assumptions of the class of theories to which the models of Refs. [5, 6] belong. For this purpose I have evaluated the charge distributions in 40 $Cu_{0.50}Zn_{0.50}$ bcc random alloy samples, each containing 432 atoms. The CEF parameters have been extracted from the LSMS calculations of Ref. [8] (see Table I). The data obtained are analysed in Fig. (4). In the top frame the individual charge excesses are plotted vs. the number of nearest neighbours, n_1. The existence of correlation between the charge excesses, q, both with the site occupation and n_1, is evident. However, as it is apparent, the excesses of charge for atoms of the same chemical species and the same number of unlike nearest neighbours can take any value in intervals whose typical widths is about 0.05 electrons, moreover intervals corresponding to different values for n_1 present appreciable overlaps. The same observations have already been made in Ref. [3], the main difference is that I have considered a much larger number of configurations and used sample all corresponding to the same stoichiometry. The conclusion, however remains the same: n_1 is not sufficient to characterise the distribution of q. In order to check how much the consideration of the number of neighbours in the second or in the third shell, n_2 and n_3, can improve, I have selected all

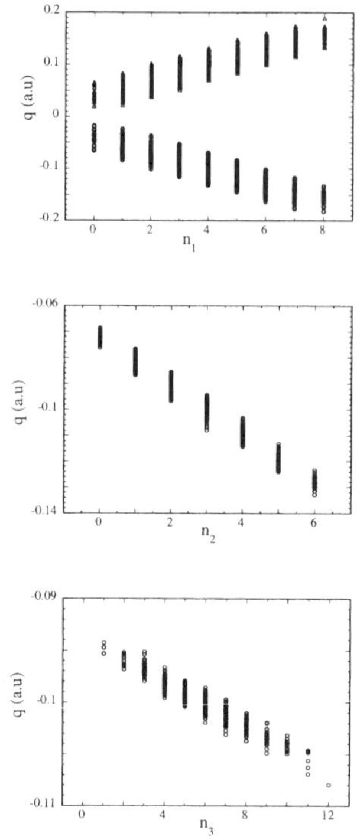

Figure 4. Top frame: charge excesses, q, vs. the number of unlike nearest neighbours, n_1, for bcc $Cu_{0.50}Zn_{0.50}$ random alloys. Middle frame: q vs. the number of unlike neighbours in the *second shell*, n_2, *only* for the Zn atoms that have 4 unlike neighbours in the first shell. Lower frame: q vs. the number of unlike neighbours in the *third shell*, n_3, *only* for the Zn atoms that have 4 and 3 unlike neighbours in the first two shells. Open circles and triangles identify, respectively, Zn and Cu sites. The data plotted have been obtained from CEF calculations on 40 random alloy samples each containing 432 atoms on a geometrical bcc lattice. The CEF parameters used are listed in Table I. Atomic units are used.

the Zn atoms with $n_1 = 4$ and plotted their charges vs. n_2 (Fig. (4), middle frame) and all the Zn site with $n_1 = 4$ and with $n_2 = 3$, the corresponding charges are plotted in the lower frame of the same Fig. (4). Although the qualitative picture is not changed, it is clear that, if the occupation of the neighbours in the first three shells is known, the uncertainty on the charge is reduced about one order of magnitude with respect to what can be obtained by considering n_1 only. In any case, trying to improve Magri's model by including more and more shells and more and more adjustable parameters, in my view, appears misleading in that it obscures the simple fact that a single number, the value of the Madelung field, is able to reduce the uncertainty to an amount comparable with numerical errors in order N calculations.

4. DISCUSSION

I like to conclude this paper by making, in a quite sparse order, some comments about several interesting aspects of the CEF model and discussing about possible future applications of the theory.

i) **A Coarse graining over the electronic degrees of freedom.** The CEF operates a *coarse graining* over the electronic degrees of freedom, that are reduced to *one for each atom,* the local excess of charge, *without any appreciable loss of accuracy* for the total energy. This is a consequence of the fact that, within theories like the CPA+LF [7] or the PCPA [14] any site diagonal property is a *unique function* of the Madelung potential, V_i, and the nuclear charge at the same site, Z_i. As noticed in [7] this uniqueness is due to the mathematical simplicity of the CPA projectors and, therefore, probably does not hold for more exact approaches, where some residual dependence on the site nearest neighbours environment is expected for. Nevertheless, the fact that the CPA theory accurately accounts for the spectral properties of metallic alloys [15] and the quantitative agreement with LSMS calculations found in Refs. [7, 14] suggest that the errors introduced by neglecting the nearest neighbours influence are probably not much larger than numerical errors in order N electronic structure calculations. The previous sentences require some clarification: when referring to the nearest neighbours environment, I mean the effects of the environment not already conveyed by V_i. The site Madelung potential, in fact, already contains much information about the occupations of near and far sites, each weighted as appropriate. Although the context was very different, I like to recall that a coarse graining over quantum degrees of freedom has precedents in the concepts of chemical valence and, more quantitatively, in that of electronegativity.

ii) **The CEF and the local environments.** Previous attempts to build theories dealing with charge transfers in metallic alloys, as, for instance the model of Magri et al. [5] or the charge-correlated CPA of Johnson and Pinski [16], have been focused on the number of unlike neighbours of each site. Subsequent extensions [6] included consideration for the occupations of outer shells. The qV linear relations suggest that the convergence of such schemes in the number of shells is slow, being basically related to the r^{-1} decay of the Coulombian interaction. The CEF model is more effective in that it accounts for these long ranged interactions. This notwithstanding, the models of Refs. [5, 6] suggest routes to possible future refinements of the CEF theory: improvements could be obtained, for instance, by including local fields in the charge correlated CPA model of Ref. [16].

iii) **Computational performances of the CEF theory.** Modern ab initio order N calculation require a number of operations directly proportional to the number of atoms in the supercell, N, unfortunately with huge praefactors. To fix the ideas, consider the case of LSMS calculations: the number of operations required is given by

$$n_{LSMS} \propto n_L^3 \; n_{LIZ}^3 \; n_E \; n_{it} \; N \approx 6 \cdot 10^9 N \tag{27}$$

where $n_L = (l_{max} + 1)^2$ is the size of the single site scattering matrices, n_{LIZ} the number of atoms in the local interaction zone, n_E and n_{it} the number of points in the energy mesh and the number of iterations that are necessary to solve the Kohn-Sham equation. The above estimate for the praefactor is quite optimistic and correspond to assuming $l_{max} = 3$, $n_{LIZ} = 24$, $n_E = n_{it} = 10$. The CEF requires $n_{CEF} = N^3$ operations when using conventional linear algebra algorithms. Accordingly, for a typical size of the supercell, $N = 1000$, the CEF is 3 to 4 orders of magnitude faster than LSMS. Of course, order N matrix inversion algorithms can be used for the CEF also, this would give $n_{CEF} = n_{LIZ}^3 N$, i.e. 5 orders of magnitude faster than LSMS, regardless of the size of the supercell.

iv) **A CEF-Monte Carlo mixed scheme.** The very remarkable speed up in electronic structure calculations that can be obtained using the CEF has a qualitative relevance because it opens unexplored possibilities. The minimum value of the CEF functional for a given alloy configuration, X, can be viewed as the total electronic energy corresponding to that configuration. Therefore, the same minimum value can be regarded as a *functional of the alloy configuration*, say:

$$Min_{\{q\}}\Omega([q]; X, c) = E_{el}(X; c) \tag{28}$$

If lattice vibrations and deformations are not considered, X is completely equivalent to the whole set of the atomic positions. If the validity of the Born-Oppenheimer approximation and of a classical approximation for the atomic degrees of freedom are assumed, then $E_{el}(X; c)$ can be regarded as a classical Hamiltonian for the alloy in study. Probably the functional dependence of $E_{el}(X; c)$ on the atomic degrees of freedom, X, is too much complicated for exact, even though approximate, statistical studies. My group is currently developing a mixed CEF-Monte Carlo scheme in which a Metropolis Monte Carlo algorithm is used to obtain ensemble averages for the classical Hamiltonian $E_{el}(X;c)$. The goal here is being able to determine *ab initio*, within *a non perturbative* method, the thermodynamics, the phase equilibria and the atomic correlation functions for metallic alloys. At the same time, taking advantage of the uniqueness of the site properties within CPA based approaches, such a scheme should allow for a careful determination of the electronic properties along the lines of the LSMS-CPA of Ref. [17].

v) **Improving the CEF-CPA.** In Section III, the distributions of the site charge excesses in a random alloy system have been studied. The validity of the qV laws implies that also the values of the Madelung field, V, at different sites can be described by two distributions $d_\alpha(V)$. With respect to these, a random alloy system can be viewed as a *charge glass*. In fact, the consequences of the q and V *polydispersivity* on the energetics of random alloys are similar to those of the polydispersivity of the bond lengths in ordinary glasses. It is easy to see that these distribution satisfy the following sum rules:

$$\int_{-\infty}^{\infty} dV\ d_\alpha(V) = 1$$

$$\sum_\alpha c_\alpha \int_{-\infty}^{\infty} dV\ V\ d_\alpha(V) = 0 \tag{29}$$

On the other hand, the standard CPA theory is based on the implicit assumption that

$$d_A(V) = d_B(V) = \delta(V) \tag{30}$$

i.e., the CPA considers random alloys are as *charge monodisperse* systems. Therefore, the CEF-CPA scheme presents the inconsistency that, while the parameters entering in the CEF are calculated by assuming the distribution in Eq. (30), the output distributions are typical of charge glasses. As it is well known, appreciable improvements over the standard CPA theory can be achieved by the SIM-CPA [18] or the screened CPA [16] models. Both theories are based on the prescription $d_A(V) = \delta(V - V_A)$, $d_B(V) = \delta(V - V_B)$, where the V_α are chosen in order to mimic the *mean* effect of the charge correlations, in such a way that the sum rules are obeyed. Although these are still *monodisperse* theories, displacing the centre of mass of the distributions allows for substantial improvements. The best CPA-based model to date available, the PCPA of Ujfalussy et al. [14] is a truly *polydisperse* theory in which the V distributions are defined self-consistently by the supercell used. As a theory based on specific supercells, however, the PCPA cannot (at least, without much labour) make predictions on the atomic correlations. We are currently developing an alternative approach that could maintain the advantages of the CEF-Monte Carlo scheme without paying the price of having non consistent charge distributions. The idea is simple: an approximation very similar to the PCPA theory can evaluate the polymorphous model defined by the V distributions obtained as an output of the CEF-Monte Carlo, rather those defined by a specified supercell. In this way a new set of improved coefficients for the CEF can be obtained. Thus, by iterating the above modified PCPA and the CEF-Monte Carlo, until convergence is obtained for the V distributions, one would obtain a completely ab initio non perturbative quantum theory of metallic alloys able to evaluate, at the same time, electronic and atomistic properties.

ACKNOWLEDGEMENTS

I wish to thank Sam Faulkner and Yang Wang that made available in digital form the data of Ref. [8]. I also aknowledge many interesting discussions with Sandro Giuliano, Antonio Milici and Leon Zingales.

REFERENCES

1. Y. Wang, G.M. Stocks, W.A. Shelton, D.M.C. Nicholson, Z. Szotek and W.M. Temmerman, Phys. Rev. Lett. **75**, 2867 (1995).
2. I.A. Abrikosov, A.M.N. Niklasson, S.I. Simak, B. Johansson, A.V. Ruban and H.L. Skriver, Phys. Rev. Lett. **76**, 4203 (1996).
3. J.S. Faulkner, Y. Wang and G.M. Stocks, Phys. Rev. B **52**, 17106 (1995).

4. J.S. Faulkner, Y. Wang and G.M. Stocks, Phys. Rev. B **55**, 7492 (1997).
5. R. Magri, S.H. Wei and A. Zunger, Phys. Rev. B **42**, 11388 (1990).
6. C.Wolverton, A. Zunger, S. Froyen and S.H. Wei, Phys. Rev. B **54**, 7843 (1996).
7. E. Bruno, L. Zingales and A. Milici, this conference; Phys. Rev. B to appear on December 15th, 2002; cond-mat/0206088.
8. J.S. Faulkner, B. Ujfalussy, N.Y. Moghadam, G.M. Stocks and Y. Wang, J. Phys. Condens. Matter **13**, 8573 (2001).
9. J.M. Cowley, J. Appl. Phys. **21**, 24 (1950).
10. J.F. Janak, Phys. Rev. B **9**, 3985 (1974).
11. J. M. Ziman, *Principles of the theory of solids*, Cambridge University Press (1969).
12. J.F. Janak, Phys. Rev. B **18**, 7165 (1978); R.M. Dreizler and E.K.U. Gross, *Density functional theory*, Springer-Verlag, Berlin (1990).
13. E. Bruno, L. Zingales and Y. Wang, unpublished.
14. B. Ujfalussy, J.S. Faulkner, N.Y. Moghadam, G.M. Stocks and Y. Wang, Phys. Rev. B **61**, 2005 (2000).
15. I.A. Abrikosov and B. Johansson, Phys. Rev. B **57**, 14164 (1998).
16. D.D. Johnson and F.J. Pinski, Phys. Rev. B **48**, 11553 (1993).
17. J.S. Faulkner, N.Y. Moghadam,, Y. Wang and G.M. Stocks, Phys. Rev. B **57**, 7653 (1998).
18. I.A. Abrikosov, Yu. Kh. Vekilov and A.V. Ruban, Phys. Lett. A **154**, 407 (1991); I.A. Abrikosov, Yu. Kh. Vekilov, P.A. Korzahavyi, A.V. Ruban and L.E. Shilkrot, Sol. St. Comm. **83**, 867 (1992).

LOCAL CHARGE DISTRIBUTIONS IN METALLIC ALLOYS: A LOCAL FIELD COHERENT POTENTIAL APPROXIMATION THEORY

Ezio Bruno, Leon Zingales and Antonio Milici

Dipartimento di Fisica and Unità INFM, Università di Messina, Salita Sperone 31, 98166 Messina, Italy

1. INTRODUCTION

In the last decade order N electronic structure calculations [1, 2] made possible the study of large supercells containing from 100 to 1000 atoms. Namely Faulkner, Wang and Stocks [2, 3] have shown that simple linear laws, the so called 'qV' relations, link the local charge excesses and the local Madelung potentials in metallic alloys. These qV linear laws have been obtained from the numerical analysis of data produced by Locally Self-consistent Multiple Scattering (LSMS) [1] calculations, while their formal derivation within the density functional theory has not yet been obtained. As a matter of fact, the above laws can be considered to hold at least within the approximations underlying LSMS calculations, i.e. the Local Density and the muffin-tin approximations.

In this paper we shall develop a new version of Coherent Potential Approximation theory (CPA). We apply a local external field and study the response of the mean field CPA alloy. Because of the fluctuation-dissipation theorem, the response to the external field must be equal to the internal field caused by electrostatic interactions. This new theoretical scheme, avoiding the consideration of specific supercells, will enable us to explore a broad range of fields and clarify certain aspects of the mentioned qV relations.

We shall find that, in a quite broad range of applied fields, Φ, the *integrated* charge excess at a given site, q, scales linearly with the field, in agreement with the findings of Refs [2, 3]. However, remarkably, in the same range of Φ values, the charge density at a given point does not obey a linear scaling. Our results for the CuPd and CuZn alloy systems compare favourably both with the LSMS and conventional superlattice multiple scattering theory calculations, as well as with the available experimental data. Our theory, when applied to random alloys, is computationally inexpensive in comparison with other approaches and can, in principle, be used, in conjunction with statistical methods, to describe ordering phenomena in metallic alloys.

In the next section 2, we shall discuss about charge transfers in multiple scattering theory calculations and CPA theory, while in section 3 we shall describe the above new version of the CPA theory that incorporates local fields (CPA+LF) and apply it to the study of fcc CuPd and fcc and bcc CuZn alloys. In the conclusion we shall summarize the most important features of our work.

2. CHARGE TRANSFERS IN METALLIC ALLOYS.

2.1. Charge transfers from LSMS calculations.

Faulkner, Wang and Stocks [2, 3] have analysed the distribution of charges in binary metallic alloys as obtained from LSMS calculations. They have studied large supercells with periodic boundary conditions containing hundreds of atoms and designed to simulate substitutional disorder. LSMS calculations are based on the local density approximation to the density functional theory [4, 5] and the muffin-tin approximation for the crystal potential; thus the results of their analysis hold *within* the same approximations. Below we shall summarize and comment the conclusions obtained in Refs. [2, 3] that are relevant for our present concerns:

i) *For a given alloy configuration*, the site charges q_i and the Madelung potentials V_i obtained from LSMS calculations for binary alloys *lie on two straight lines* of equations:

$$a_i q_i + V_i = k_i \tag{1}$$

where the quantities a_i and k_i take the values a_A and k_A if the i-th site is occupied by an A atom and a_B and k_B if it is occupied by B. The size of the deviations from linearity appears *comparable* with the numerical accuracy of LSMS calculations.

The Madelung potentials V_i entering in Eq. (1) are determined by the charges at all the crystal sites through the relationship:

$$V_i = 2 \sum_i M_{ij} q_j \tag{2}$$

where the factor 2 comes from using atomic units. The Madelung matrix elements, M_{ij}, are defined [6] as

$$M_{ij} = \sum_{\vec{R}} \frac{1}{|\vec{r}_{ij} + \vec{R}|} \tag{3}$$

in terms of the translation vectors from the i-th to the j-th site, $\vec{r}_{ij}$, and the supercell lattice vectors $\vec{R}$.

ii) For different alloy configurations corresponding *to the same molar fractions*, the four constants a_A, k_A, a_B and k_B in Eq. (1) take different values. This notwithstanding, the variations of the same constants when considering different samples at the same concentration appear much smaller than their variation with the concentration.

iii) The site charge excess corresponding to each chemical species in a random alloy configuration take *any* possible value in some interval $q_{min} \leq q_i \leq q_{max}$.

Faulkner, Wang and Stocks [2, 3] have stressed that the existence of a linear relation is not a trivial consequence of classical electrostatics. In fact, Eq. (1) is not verified at a generic Kohn-Sham iteration for the charge density in LSMS calculations, while it is found to hold *only* when convergence is achieved. Thus the linearity of the qV laws should be interpreted as a consequence of the screening phenomena that occur in metals. As shown by Pinski [7], linear qV laws can be obtained also by Thomas-Fermi density functional calculations. This circumstance strongly suggests that the linearity of the qV relations has little to do with the specific form of the density functional used in the calculation. The conclusions drawn in Refs. [2, 3] and summarized above are indirectly supported by photoemission experiments [8, 9]. Moreover, electronic structure calculations based on the Locally Self-consistent Green's Function method (LSGF) and the atomic sphere approximation for the crystal potential have also confirmed the linearity of the qV relations [10, 11, 12].

It should be clear that the definition of charge excess is based on the quite artificial partition of the crystal volume into "atomic" sites. This partition is accomplished using the muffin-tin approximation in Refs. [2, 3] or the atomic sphere approximation in Refs. [10, 11, 12]. Of course other procedures are possible, but even in the case in which no spherical approximation is made, as it could be for full potential calculations (that unfortunately are not yet available), the way in which the "atomic cells" are chosen would remain arbitrary. However, different partitions of the crystal volume *always* lead to linear laws. This has been shown, e.g., in Ref. [11] by changing the ratio r between the atomic radii associated with each

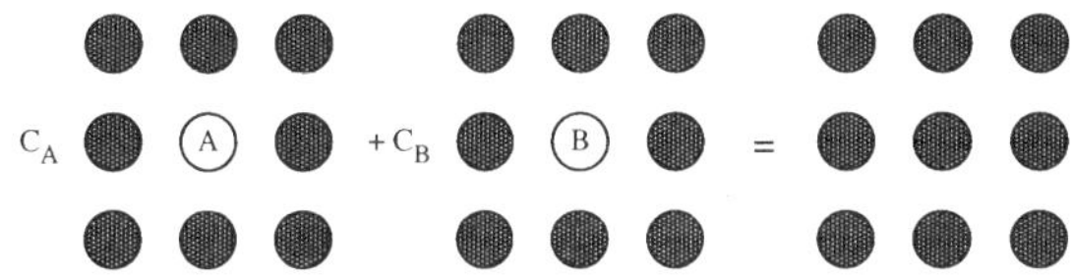

Figure 1. Schematic illustration of the CPA theory. Dark sites are occupied by the CPA coherent scatterer described by the single site scattering matrix t_C. The central impurity sites, labelled by A and B, are characterised by the single site matrices t_A and t_B.

chemical species [13]. To summarize: at least when a spherical approximation is used, the functional form of Eq. (1) is maintained while, of course, the actual values of the coefficients depend on the particular partition used.

As it is evident, the presence of the charge transfers leads to energy corrections that can be important in the physics of metallic alloys. The simple functional form in Eq. (1) allows an easy route for including such corrections [14]. An alternative way for accounting the electrostatic energy contribution due to charge transfers has been proposed by Gonis et al [13, 15]. It consists in choosing the dimensions of the atomic spheres for each alloying species in such a way to have zero charge transfers and, hence, zero contribution to the total energy. Of course, such a procedure could cause large overlap volumes (for simple lattices the overlap volume is minimum when equal atomic spheres are used) and, hence, large errors in density functional theory calculations.

Although, in principle, the quantities a_i, k_i in Eq. (1) can be influenced by the local environments, it is clear that *the consideration of the site chemical occupation only is sufficient to determine the same quantities* within an accuracy comparable with the numerical errors in LSMS calculations. This circumstance, as a matter of fact, suggests that a single site theory [16] as the CPA could be sufficient to determine the above a_i, k_i. In section 3 this suggestion shall be analysed.

2.2. Charge transfers in the CPA theory.

For many years the CPA theory [17] has been used for calculating the electronic properties of random metallic alloys. In fact, the CPA has allowed for very careful studies of spectral properties [10], Fermi surfaces [18], phase equilibria [19] and magnetic phenomena [20] in metallic alloys. Moreover, in spite of its simplicity, the theory has achieved remarkable successes in the calculation of properties related with Fermi liquid effects, such as spin [21] and concentration waves [22]. However, for the purpose of the present work two aspects of the theory are particularly relevant: its elegant formulation in terms of multiple scattering theory [23, 24] and the fact that it constitutes the natural first step for perturbative studies.

As it is well known [25], the CPA does not include the energetic contributions that derive from charge transfers in metallic alloys. In spite of this, the CPA is useful for understanding some physical properties related with these charge transfers. We will try to explain below the reasons for this apparent paradox.

The CPA theory (we shall use the multiple scattering theory formalism [23, 20] for a random binary alloy $A_{c_A}B_{c_B}$) consists of solving for t_C the so called CPA equation,

$$c_A G_A(t_A, t_C) + c_B G_B(t_B, t_C) = G_C(t_C) \tag{4}$$

The three Green's functions in Eq. (4), $G_A(t_A, t_C)$, $G_B(t_B, t_C)$ and $G_C(t_C)$, refer to the three different problems sketched in Fig. 1. In fact, $G_C(t_C)$ is the Green's function for an infinite crystal whose sites are all occupied by effective scatterers characterised by the single-site scattering matrix t_C. On the other hand, $G_{A(B)}(t_{A(B)}, t_C)$ is the Green's function for a single impurity 'atom' described by the single-site scattering matrix $t_{A(B)}$ and embedded in an infinite crystal with all the other sites characterised by the single-site scattering matrix t_C.

While the homogeneous effective crystal, the 'coherent' medium of the CPA theory, let us call it C, is electroneutral, the two impurity Green's functions lead to net charge excesses, q_A^0 and q_B^0, in the sites occupied by the A and B impurities. On behalf of Eq. (4), these charge excesses satisfy the condition,

$$c_A q_A^0 + c_B q_B^0 = 0 \tag{5}$$

In C there is no charge transfers from one site to the others and, thus, Eq.(5) cannot be interpreted as an ordinary electroneutrality condition. However $q_{A(B)}^0$ can be considered as the charge that the impurity A(B) attracts from the mean medium C, in the sense that there is an indirect charge transfer from A to B, through the mean medium C. The last can be reinterpreted as a reference system and plays a role similar to that of the Hydrogen atom for molecules, in the formulation of the electronegativity theory by Pauling [27].

In summary: we could say that the CPA 'charge transfers' $q_{A(B)}^0$ *reflect* the difference of electronegativity between A and B. Of course the CPA theory, being a single site and a mean field theory, cannot account for the complex charge relaxation phenomena that are expected to make non equivalent sites occupied by same species and surrounded by different local chemical environments. In order to have a picture in which sites occupied by the atoms of the same kind are no longer equivalent, it is necessary to renounce to a single-site picture. Non single-site formulations of the CPA theory have been proposed several times in the literature. Here we mention the charge-correlated CPA by Johnson and Pinski [28] and the Polymorphous Coherent Potential Approximation (PCPA) by Ujfalussy et al [29]. In this paper, we shall develop a different approach by introducing an external local field in a single-site CPA picture; this will allow to maintain all the mathematical simplicity of a single-site theory, nevertheless the presence of external fields will be sufficient to lead to polymorphous site potentials.

3. RESPONSE TO LOCAL FIELDS OF THE 'CPA ALLOY'.

3.1. The local field CPA (CPA+LF) model

In this section, we develop a new version of CPA theory by introducing an external local field Φ. It will formally enter in the theory as a parameter that can be varied at will. We shall focus on the response of the system due to the resulting rearrangement of the charge distribution.

We imagine an A impurity atom in a otherwise homogeneous crystal with all the other sites occupied by C scatterers. We suppose that the single site scattering matrix of the CPA medium, t_C, and its Fermi energy, E_F, have been determined by the CPA theory for the binary alloy $A_{c_A}B_{c_B}$. The local external field, Φ, takes a constant value within the impurity site volume and is zero elsewhere [30]. This situation is pictorially represented in Fig. 2. To simplify our discussion we shall solve the problem using the Atomic Sphere Approximation (ASA). However, the following considerations hold for any cellular method, and, with minor modifications, also for the muffin-tin approximation.

We shall refer to the impurity A in the presence of the external field Φ as to (A, Φ). When $\Phi = 0$, the site Green's function associated with it, $G_A^\Phi(t_A^\Phi, t_C)$, reduces to the usual CPA Green's function, $G_A(t_A, t_C)$. When $\Phi \neq 0$, $G_A^\Phi(t_A^\Phi, t_C)$ can be readily obtained using the multiple scattering theory impurity formula [26]:

$$G_A^\Phi(E, \vec{r}, \vec{r}') = \sum_{L,L'} [Z_L^\Phi(E, \vec{r}) \tau_{A,LL'}^\Phi Z_{L'}^\Phi(E, \vec{r}') - Z_L^\Phi(E, \vec{r}) J_{L'}^\Phi(E, \vec{r}') \delta_{LL'}] \tag{6}$$

where

$$\tau_A^\Phi = D_A^\Phi \tau_C = \left[1 + \tau_C \left((t_A^\Phi)^{-1} - t_C^{-1}\right)\right]^{-1} \tau_C \tag{7}$$

In Eqs. (6) and (7), E is the energy, t_C and τ_C are the CPA single site scattering matrix and scattering-path operator, *as determined by a standard CPA calculation*, i.e. $\Phi = 0$, for the alloy at hand. The single site scattering matrix corresponding to (A, Φ), t_A^Φ, is to be determined from the site potential $V_A^\Phi(\vec{r}) + \Phi$, D_A^Φ is the CPA projector relative to the

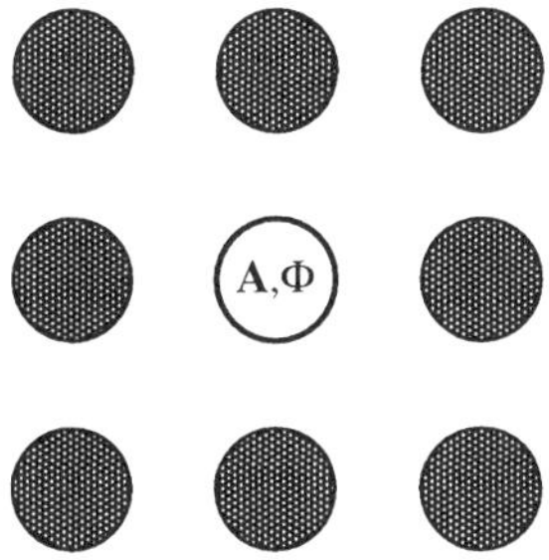

Figure 2. Schematic illustration of the CPA+LF method. As in Fig. 1, dark sites are occupied by the CPA coherent scatterer described by t_C. In the central site, occupied by A, acts also a constant field Φ.

same site potential, $Z_L^\Phi(E,\vec{r})$ and $J_L^\Phi(E,\vec{r})$ are two orthogonal solutions of the Schroedinger equation for the same potential, the first of which is regular at $r = 0$. In our notation $L = (l, m)$ labels the angular momentum quantum numbers and, for sake of simplicity, we omit the energy dependence of all the scattering matrices. A complete account of the notation can be found in Ref. [26].

The charge density corresponding to (A, Φ) is obtained integrating Eq. (6) over the energy up to the Fermi level,

$$\rho_A^\Phi(\vec{r}) = -\frac{1}{\pi} Im \left\{ \int_{-\infty}^{E_F} dE\, G_A^\Phi(E, \vec{r}, \vec{r}\,' = \vec{r}) \right\} \tag{8}$$

The corresponding site potential, $V_A^\Phi(\vec{r})$, can be reconstructed by solving the appropriate Poisson equation and adding the exchange-correlation contribution [31, 32]. Unless $\Phi = 0$, it will be different from the site potential obtained from the zero field CPA theory, $V_A(\vec{r}) = V_A^{\Phi=0}(\vec{r})$, due to the charge relaxations expected to screen in part the external field. In a numerical implementation of the theory, Eqs. (6-8) and the potential reconstruction need to be iterated starting from a convenient initial guess, until convergence is achieved for $V_A^\Phi(\vec{r})$ or, equivalently, for $\rho_A^\Phi(\vec{r})$. Hereafter we shall refer to the above model as to the Local Field CPA (CPA+LF).

Once convergence is obtained for the charge density, the net charge on the site A can be obtained by integrating over the atomic sphere volume and subtracting the nuclear charge, Z_A,

$$q_A(\Phi) = \int d\vec{r} \rho_A^\Phi(\vec{r}) - Z_A \tag{9}$$

It is important to realise that, while the above self-consistent determination of $V_A^\Phi(\vec{r})$ or $\rho_A^\Phi(\vec{r})$ allows for full charge relaxation at the impurity site, the CPA+LF does not modify the properties of the CPA medium C: these remain specified by the quantities t_C and E_F determined at zero external field. The resulting lack of self-consistency in the CPA+LF is not a serious drawback if one is interested, as in the present case, to the investigation of trends and general aspects of the screening phenomena.

3.2. CPA+LF results for CuZn and CuPd alloys: the site charges

We have implemented the CPA+LF theory within our well tested KKR-CPA code [33]. If t_C and τ_C from a previous standard KKR-CPA calculation are stored on a convenient energy mesh, the extra computational efforts required by the CPA+LF calculation are negligible.

In this paper we discuss results for fcc CuPd and for bcc and fcc CuZn random alloys at several concentrations. In all the cases we have used the Local Density approximation (LDA) for the exchange-correlation potential [4], the ASA approximation for the site potentials and the angular momentum expansions have been truncated at $l_{MAX} = 3$. We use a fully

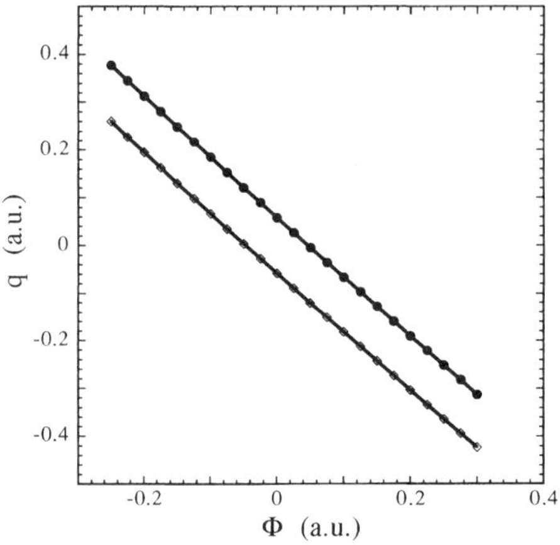

Figure 3. Site charge excesses q_α (α = Cu,Zn) vs. the external field, Φ, from CPA+LF calculations for $Cu_{0.50}Zn_{0.50}$ bcc random alloys. Circles and diamonds, respectively, indicate Cu and Zn impurities.

relativistic treatment for core electrons and a scalar relativistic approximation for valence electrons. For all the alloy systems considered in this paper, the lattice parameters have been kept fixed on varying the concentration. In particular, we set $a = 5.5$ a.u. and $a = 6.9$ a.u. for bcc and fcc CuZn, and $a = 7.1$ a.u. for fcc CuPd. With this choice, the atomic volumes in fcc and bcc CuZn alloys differ only about 1.3 per cent.

As we shall discuss in the next subsection, the charge relaxation occurring at the impurity site in presence of the external field phenomena are quite complex. Nevertheless, the CPA+LF model gives a simple linear relation between the potential Φ and the corresponding site charges. In Fig. 3 we report q_α (α =Cu, Zn) vs. Φ for a $Cu_{0.50}Zn_{0.50}$ bcc random alloy. As it is evident, the data can be fitted very well by two straight lines, one for each atomic species (with correlations that differ from one by less than one part over a million). Interestingly, the slopes of the two lines are different by a relatively small but statistically relevant amount, slightly less than 2 per cent.

We notice that in Fig. 3 we have considered also Φ values considerably larger that those observed in LSMS calculation or likely to occur in real systems; so according to our data the linear relations seem to be valid in a quite broad field range. We have fitted the q_α vs. Φ curves at each molar fraction for fcc Cu_cPd_{1-c}, fcc Cu_cZn_{1-c} and bcc Cu_cPd_{1-c} random alloys, at a number of alloy concentrations, using the linear relationships

$$q_\alpha(\Phi) = q^0_\alpha - R_\alpha \, \Phi \tag{10}$$

However, at $\Phi = 0$, our CPA+LF model satisfies the CPA 'electronegativity' condition, Eq. (5), and we have:

$$c_A q^0_A + c_B q^0_B = 0 \tag{11}$$

Henceforth, q^0_A and q^0_B are not independent quantities and we have chosen as the parameters of our fit only the three quantities R_A, R_B and

$$\Delta = q^0_A/c_B = -q^0_B/c_A = q^0_A - q^0_B \tag{12}$$

The results of these fits are reported in Table 1. The trends found for the fitting parameters vs. the alloy molar fractions are shown in Fig. 4. The dependence on the concentration is appreciable for all the fitting parameters, as expected on the basis of the arguments in section 2. Remarkably, the dependences on the alloy system and on the concentration appear at least as much important as that on the atomic species. Thus, for instance, for a given alloy system

TABLE 1. Fit parameters for the q vs. Φ relationships from CPA+LF calculations in fcc Cu_cPd_{1-c}, bcc Cu_cZn_{1-c} and fcc Cu_cZn_{1-c} random alloys [34] . The 'electronegativity difference', Δ, and the response coefficients, R_α, are defined in Eqs. (10) and (12), RMS is the root mean square deviation. The 'renormalized' response coefficients, $\tilde{R}_\alpha$ are defined in Eq. (14). On the right we report Δ and R_α from the LSMS calculations of Refs. [2, 3].

Alloys	c	Δ	R_{Cu}	$R_{Pd(Zn)}$	RMS $\times 10^4$	$\tilde{R}_{Cu}$	$\tilde{R}_{Pd(Zn)}$	Δ	$\tilde{R}_{Cu}$	$\tilde{R}_{Pd(Zn)}$
fcc Cu_cPd_{1-c}	0.10	0.183	1.093	1.156	1.8	0.762	0.792	0.238	0.833	0.843
	0.25	0.175	1.124	1.187	2.1	0.776	0.806	0.229	0.838	0.851
	0.50	0.160	1.184	1.244	1.9	0.805	0.832	0.219	0.843	0.851
	0.75	0.150	1.243	1.288	2.4	0.831	0.851	0.212	0.838	0.853
	0.90	0.148	1.267	1.307	4.4	0.842	0.860	0.211	0.836	0.853
bcc Cu_cZn_{1-c}	0.10	0.109	1.206	1.232	10	0.800	0.812	0.155	0.536	0.581
	0.25	0.114	1.237	1.255	10	0.814	0.822	0.159	0.526	0.554
	0.50	0.116	1.237	1.251	6.9	0.814	0.820	0.156	0.545	0.549
	0.75	0.116	1.247	1.255	5.0	0.819	0.822	0.155	0.567	0.564
	0.90	0.116	1.248	1.254	3.2	0.819	0.822	0.158	0.582	0.577
fcc Cu_cZn_{1-c}	0.10	0.106	1.202	1.223	8.2	0.805	0.815	0.145	0.575	0.628
	0.25	0.111	1.220	1.237	8.1	0.813	0.821	0.150	0.580	0.618
	0.50	0.116	1.222	1.241	5.5	0.814	0.822	0.151	0.600	0.622
	0.75	0.117	1.247	1.256	5.2	0.825	0.829	0.150	0.615	0.632
	0.90	0.118	1.249	1.256	3.3	0.826	0.829	0.152	0.616	0.630

and concentration, there are relatively small differences between the values of R corresponding to sites occupied by different atoms. On the other hand, we find much larger variations for R_{Cu} throughout the alloy systems considered. It is interesting to observe that the trends for the slopes, R_{Cu} and R_{Zn}, and for Δ are very similar in *both* fcc and bcc CuZn alloys. We notice also that Δ, a measure of the electronegativity difference between the alloying species, exhibits, at least for CuPd alloys, non negligible variations vs. the concentration. In the model of Ref. [25], the same quantity is assumed independent on the concentration. As we see from Table 1, the values for Δ from our theory are systematically smaller than those from LSMS. This fact has not to do with the presence of external fields and it is a feature of the standard CPA theory already discussed in the literature [35]. This notwithstanding, the CPA is able to catch the qualitative trends of Δ vs. the concentration for all the systems considered.
Although the CPA+LF model gives for q vs. Φ the same linear functional form as that obtained for q vs. V from LSMS calculations, the differences between the two different sets of calculations forbid, at this stage, a direct comparison of the fit coefficients. In fact, as we have already stressed, our CPA+LF model does not account for charge relaxations outside the impurity site volume. By its construction, the CPA medium C is able to screen the impurity charge at $\Phi = 0$, i.e. q^0_α. We can think that this amount of charge is screened by the infinite volume of C. The introduction of the local field at the impurity site causes a local excess of charge, $q_\alpha(\Phi) - q^0_\alpha$, with respect to the standard CPA. In order to have global electroneutrality in the CPA+LF theory, it is necessary to introduce, somewhere outside the impurity site, an opposite amount of charge, $q^0_\alpha - q_\alpha(\Phi)$. Here it will be accomplished using the arguments of the screened impurity model (SIM-CPA) model by Abrikosov et al. [36]. We suppose that the excess (with respect to the standard CPA) charge at the impurity site, $q_\alpha(\Phi) - q^0_\alpha$, is *completely screened* at some distance, ρ, of the order the nearest neighbours distance, r_1. Accordingly, in the mean, each of the n nearest neighbours of the impurity cell has a net charge excess $(q^0_\alpha - q_\alpha(\Phi))/n$. This, in turn, induces an extra field $\Phi_1 = n(2/\rho)(q^0_\alpha - q_\alpha(\Phi))/n = 2(q^0_\alpha - q_\alpha(\Phi))/\rho$ on the impurity site. The total field at the impurity site is then the sum of the external field Φ and of the above extra term, in formulae,

$$V_\alpha = \Phi + 2(q^0_\alpha - q_\alpha(\Phi))/\rho \tag{13}$$

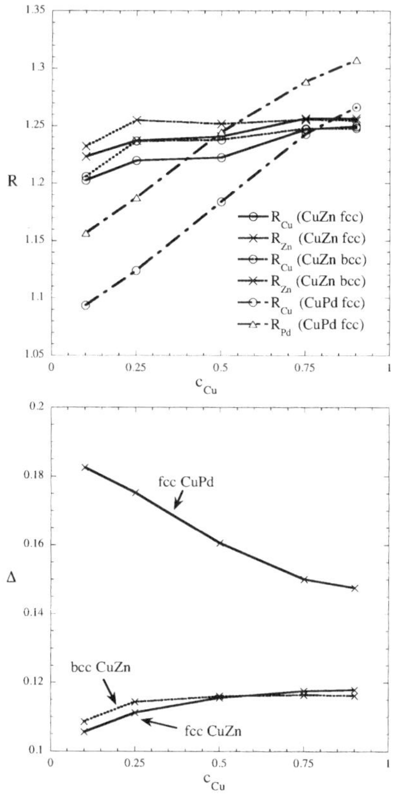

Figure 4. Fit coefficients of the linear law q vs. Φ from CPA+LF calculations for fcc CuPd and fcc and bcc CuZn alloys plotted vs. the Cu concent. Upper frame: response coefficients R_α, (α refers to the alloying species); lower frame: 'electronegativity difference', Δ. The various alloy systems are indicated by labels.

Then, by solving for Φ the last equation and substituting in Eq. (10), we find

$$q_\alpha(\Phi) = q_\alpha^0 - \frac{R_\alpha}{1 + 2R_\alpha/\rho} V_\alpha = q_\alpha^0 - \tilde{R}_\alpha V_\alpha \tag{14}$$

The coefficients $\tilde{R}_\alpha$ can be compared directly with the slopes of the qV relations from LSMS calculations. However, the comparison, reported in Table 1, requires a caveat: we have assumed $\rho = r_1$, i.e. a complete screening at the distance of the nearest neighbours.

Actually, the screening lengths in metals are of the order of this distance [37], but our estimate is too crude to expect for a very good quantitative agreement with LSMS calculations in which the charge relaxation is allowed at all the length scales. However, the agreement found is quite satisfactory, within 10 per cent, for CuPd alloys, while larger discrepancies are found for CuZn. Again, the trends for $\tilde{R}_\alpha$ vs. the concentration are qualitatively reproduced.

3.3. CPA+LF results for CuZn and CuPd alloys: the charge relaxation

We have already said, in spite of the qV linear laws, the relaxation phenomena occurring in presence of an external field are complicated. The CPA+LF model allows for the determination of the response to an external potential field by the electrons *inside* the atomic sphere A.

More specifically, the difference

$$\Delta V_A^{\Phi}(\vec{r}) = V_A^{\Phi}(\vec{r}) + \Phi - V_A^{\Phi=0}(\vec{r}) \tag{15}$$

can be interpreted as the sum of the external field, Φ and the internal screening field inside the atomic sphere. Some typical trends for this quantity are shown in Fig. 5. There we report $\Delta V_\alpha^{\Phi}(\vec{r})$, $(\alpha = Cu, Zn)$, for an bcc $Cu_{0.50}Zn_{0.50}$ random alloy, that we have selected as a typical case. At the Wigner-Seitz radius, $r_{WS} \approx 2.71$ a.u., the internal field is able to screen about one half of the external field, both for Cu and Zn impurities, while the screening is almost complete for about $r < 1$ a.u.. Apparently, the effect of the screening is far from being just a constant shift of the local chemical potential: if that was the case, in Fig. 5 we would have just equally spaced horizontal lines. What we observe is much more complicated. For instance we observe that the screening for small r is greater in Cu than in Zn. The complex nature of the screening phenomena is further confirmed by a look at the electronic densities. In Fig. 6 we plot the excess charge density induced by the external field

$$\Delta\rho_A^{\Phi}(\vec{r}) = \rho_A^{\Phi}(\vec{r}) - \rho_A^{\Phi=0}(\vec{r}) \tag{16}$$

both for Cu and Zn sites, again for random bcc $Cu_{0.50}Zn_{0.50}$.

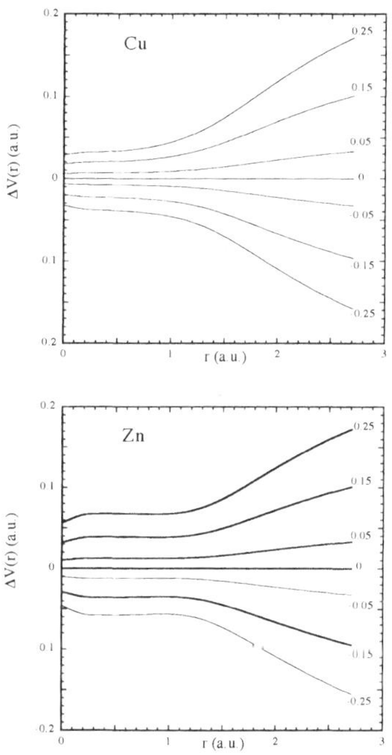

Figure 5. Calculated total field $\Delta V_\alpha^{\Phi}(r)$, α =Cu, Zn (see Eq. 15) in $Cu_{0.50}Zn_{0.50}$ bcc random alloys. The labels indicate the values of the external field, Φ. At the Wigner-Seitz radius, $r_{WS} \approx 2.71$ a.u., the total field results to be about one half of the external field, while the electronic screening is almost complete at $r = 0$.

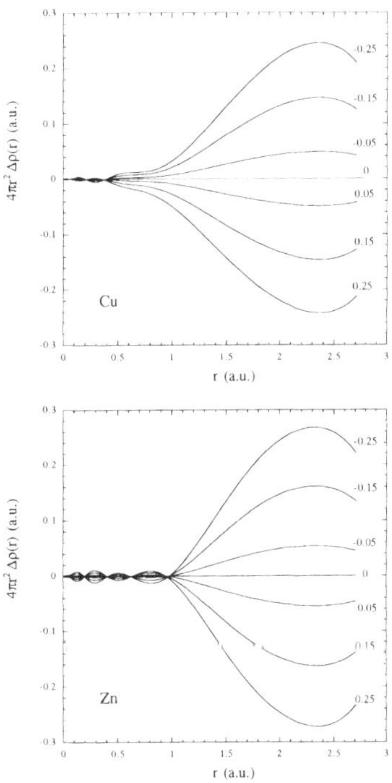

Figure 6. Calculated excess charge density, $4\pi r^2 \Delta\rho_\alpha^{\Phi}(r)$ (α =Cu, Zn) (see Eq. 16) in $Cu_{0.50}Zn_{0.50}$ bcc random alloys. The labels indicate the values of the external field, Φ.

The largest effects come from the large r region, where the electron density decreases on increasing Φ (everywhere in this paper the expressions "electronic density" or "charge density" are used indifferently with the meaning of "electron number density", i.e. the charge factor, $-e$, is *not* included). In the innermost part of the atomic spheres, the variations of the charge density sometimes may have opposite sign with respect to that observed close to the cell boundary. We have considered also the quantity,

$$b_\alpha^\Phi(r) = \frac{\rho_\alpha^\Phi(r) - \rho_\alpha^{\Phi=0}(r)}{\Phi\ \rho_\alpha^{\Phi=0}(r)} \approx \frac{\partial}{\partial\Phi} log \rho_\alpha^\Phi(r) \tag{17}$$

that, in the limit $\Phi \to 0$ reduces to the logarithmic derivative of $\rho_\alpha^\Phi(r)$ and that, on the basis of a formal scattering theory analysis [38] is expected to have a weak dependence on Φ. As we can see from Fig. 7, where we plot $b_\alpha^\Phi(r)$ for a bcc $Cu_{0.50}Zn_{0.50}$ random alloy the residual dependence on Φ is about a few per cent in a relatively small r interval not far from r_{WS} and less then 1 per cent in most of the atomic sphere. Moreover this feature appears to be more or less pronounced depending on the system considered, for this reason we plot in Fig. 8 $b_\alpha^\Phi(r)$ for a fcc $Cu_{0.50}Pd_{0.50}$ random alloy. Although the information contained in Figs. 7 and Fig. 8 can be valuable for the purpose of improving the initial guesses for the charge densities, however the dependence of $b_\alpha^\Phi(r)$ on r appears still quite complicated.

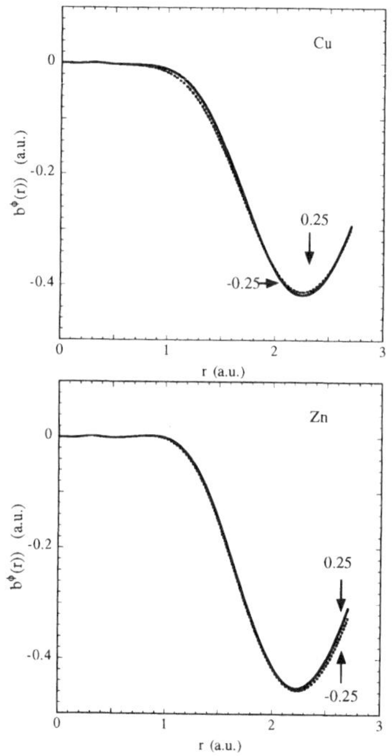

Figure 7. The 'logarithmic derivative' with respect to the external field (Eq. 17), $b_\alpha^\Phi(r)$ (α=Cu,Zn) in $Cu_{0.50}Zn_{0.50}$ bcc random alloys. The continuous and the dotted lines refer, respectively, to $\Phi = -0.25$ and $\Phi = -0.25$, the lowest and the highest Φ values considered.

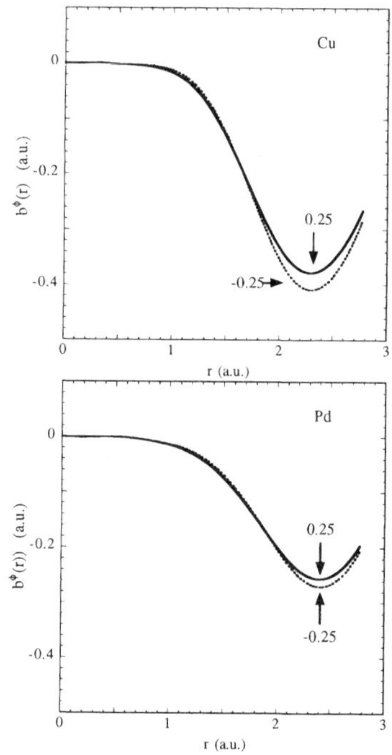

Figure 8. The 'logarithmic derivative' with respect to the external field (Eq. 17), $b_\alpha^\Phi(r)$ (α=Cu,Pd) in $Cu_{0.50}Pd_{0.50}$ fcc random alloys. The continuous and the dotted lines refer, respectively, to $\Phi = -0.25$ and $\Phi = 0.25$.

4. CONCLUSIONS

The most important result of this work is the reproduction of the linear laws between local charge excesses and local electrostatic fields, in good quantitative agreement with order N electronic structure calculations [2, 3]. This is very remarkable if one considers the *single site* nature of our CPA+LF model, that, hence, requires really modest computational efforts. The only non first-principles input of our theory has been the inclusion of a screening length that we have fixed to the nearest neighbours distance. Work is in progress to build a new, completely *ab initio*, version. The simple mathematical structure of our model has allowed the investigation of a range of fields much broader than that accessible by order N calculations. On this basis, we can conclude that the above linear relations have little to do with the *size* of the external field. On the other hand, our study shows that, in the same range of fields, non linear trends are clearly observable for other site properties, including the charge density $\rho(r)$ (see, e.g. Figs. 6, 7, 8).

As we have already noticed, the CPA+LF theory fixes the reference medium, the CPA alloy, or, in a more mathematical language, the system Green's function that depends only on the mean molar fractions. Thus, for a given concentration, any site physical observable depends only on the CPA projectors and the site wavefunctions (see Eqs. (6) and (7)), which, in turn, are completely determined by the nuclear charge on the impurity site and the coupling potential entering in the corresponding Schroedinger-Kohn-Sham equation. Thus, in the CPA+LF theory, *any* site property is a *unique* function of the chemical species and of Φ. A question arises: could this uniqueness be maintained in the more realistic multiple scattering theory treatment? We argue that, also in this case, there is a well defined system Green's function and, in principle 'site projectors' D_i, could be defined relating the site diagonal part at the site i to the system Green's function. The excellent performance of the CPA theory about the spectral properties of many alloy systems [10], the present results and those of Ref. [29] suggest that these generalized projectors should be very close to their CPA counterparts, D_α, but, in principle, they should also be affected by the chemical environment of the first few neighbours shells of i-th site. These effects, if they are important, could be studied, for instance by including local fields in the charge-correlated CPA scheme by Johnson and Pinski [28]. Of course, all the above does not solve the problem of a formal derivation of the qV laws within the density functional scheme, it simply offers a not too difficult mathematical ground in which, we hope, such a derivation could be obtained.

A further advantage of the CPA+LF model is that, in conjunction with the Charge Excess Functional theory [14], it is able to give a good description of the charge distribution in excellent agreement with order N calculations. This, and the flexibility of the scheme, that does not require the use of specific supercells and is then able to deal on the same footing with ordered or disordered configurations, suggest that it constitutes a first step towards an *ab initio* non perturbative theory of ordering phenomena in metallic alloys.

ACKNOWLEDGEMENTS

We thanks Professor J.S. Faulkner and Dr. Y. Wang for having made available in digital form the data of Refs [2, 3]. We acknowledge also discussions with Professor E.S. Giuliano.

REFERENCES

1. Y. Wang, G.M. Stocks, W.A. Shelton, D.M.C. Nicholson, Z. Szotek and W.M. Temmerman, Phys. Rev. Lett. **75**, 2867 (1995).
2. J.S. Faulkner, Y. Wang and G.M. Stocks, Phys. Rev. B **55**, 7492 (1997).
3. J.S. Faulkner, Y. Wang and G.M. Stocks, Phys. Rev. B **52**, 17106 (1995).
4. P. Hohenberg and W. Kohn, Phys. Rev. **136**, B864 (1964); W. Kohn and L.J. Sham, Phys. Rev. **140**, A1133 (1965).
5. Dreizler R.M., Gross E.K.U., *Density Functional Theory*, Spinger-Verlag (1990).
6. J.M. Ziman, *Principles of the theory of solids*, Cambridge University Press (1969).
7. F.J.Pinski, Phys. Rev. B **57**, 15140 (1998).
8. R.J. Cole, N.J. Brooks and P. Weightman, Phys. Rev. Lett. **78**, 3777 (1997).
9. J.S. Faulkner, Y. Wang and G.M. Stocks, Phys. Rev. Lett. **81**, 1905 (1998).
10. I.A. Abrikosov and B. Johansson, Phys. Rev. B **57**, 14164 (1998).

11. A.V. Ruban, S.I. Simak, P.A. Korzhavyi and H.L. Skriver, Phys. Rev. B **66**, 024201 (2002).
12. A.V. Ruban and H.L. Skriver, Phys. Rev. B **66**, 024202 (2002).
13. As noticed in Ref. [15], it exists a special value of r at which the charge transfers are zero for all the alloying species. This is not in contrast with the existence of linear laws: for the above value of r, the range $q_{min} \leq q_i \leq q_{max}$ collapses in a single point [11]. Also in this case qV linear relationships hold with finite values for a_A and a_B and with k_A=k_B=0 as it can be established by following the methods illustrated in section 3.
14. E. Bruno, this conference; E. Bruno, L. Zingales and Y. Wang, to be published.
15. Gonis A.,Turchi P.E.A., Kudrnovsky J., Drchal V. and Turek I., J. Phys. Cond. Mat. **8**,7883 (1996).
16. B. Velicky, S. Kirkpatrick and H. Ehrenreich, Phys. Rev. **175**, 747 (1968).
17. P. Soven, Phys. Rev. **156**, 809 (1967).
18. E. Bruno, B. Ginatempo, E.S. Giuliano, A.V. Ruban and Yu. Kh. Vekilof, Phys. Rep. **249**, 3353 (1994).
19. B.L. Gyorffy and G.M. Stocks, Phys. Rev. Lett **50**, 374 (1983); J.B. Staunton, D.D. Johnson and F.J. Pinski, Phys. Rev. B **50**, 1450 (1994).
20. J.B. Staunton, F.J. Pinski and D.D. Johnson, J. Appl. Phys. **61**, 3715 (1987); J.B. Staunton, J. Poulter, F.J. Pinski, B. Ginatempo and E. Bruno, Phys. Rev. Lett. **82**, 3340 (1999).
21. S.S.A. Razee, J.B. Staunton, B. Ginatempo, F.J. Pinski and E. Bruno, Phys. Rev. Lett. **82**, 5369 (1999).
22. I. Wilkinson, R.J. Hughes, Zs. Major, S.B. Dugdale, M.A. Alam, E. Bruno, B. Ginatempo and E.S. Giuliano, Phys. Rev. Lett. 87, 216401 (2001).
23. B.L. Gyorffy, Phys. Rev. B **5**, 2382 (1972).
24. J. Korringa, Physica (Amsterdam) **13**, 392 (1947); W. Kohn and N. Rostoker, Phys. Rev. **94**, 111 (1954).
25. R. Magri, S.H. Wei and A. Zunger, Phys. Rev. B **42**, 11388 (1990).
26. J.S. Faulkner and G.M. Stocks, Phys. Rev. B **21**, 3222 (1980); A. Gonis, *Green functions for ordered and disordered systems*, North-Holland Elsevier Science Publishers, Amsterdam, The Netherlands (1992).
27. L. Pauling, *The Nature of Chemical Bond*, Cornell University Press, Ithaca (1960).
28. D.D. Johnson and F.J. Pinski, Phys. Rev. B **48**, 11553 (1993).
29. B. Ujfalussy, J.S. Faulkner, N.Y. Moghadam, G.M. Stocks and Y. Wang, Phys. Rev. B **61**, 2005 (2000).
30. This steplike behaviour of the field Φ in our model corresponds to the assumptions made for the Madelung field in atomic sphere or muffin-tin approximation calculations for periodic systems.
31. J.F. Janak, Phys. Rev. B **9**, 3985 (1974).
32. H. Winter and G.M. Stocks, Phys. Rev. B **27**, 882 (1983).
33. E. Bruno and B. Ginatempo, Phys. Rev. B **55**, 12946 (1997).
34. E. Bruno, L. Zingales and A. Milici, Phys. Rev. B **66** to appear on December 15th, 2002; cond-mat/0206088.
35. E. Bruno and B. Ginatempo, Europhys. Lett. **42**, 649 (1998).
36. I.A. Abrikosov, Yu. Kh. Vekilov and A.V. Ruban, Phys. Lett. A **154**, 407 (1991); I.A. Abrikosov, Yu. Kh. Vekilov, P.A. Korzahavyi, A.V. Ruban and L.E. Shilkrot, Sol. St. Comm. **83**, 867 (1992).
37. D. Pines, Solid State Phys. **1** (1955).
38. E. Bruno, unpublished.

ON THE DEVELOPMENT OF ALLOY THEORY

A. Gonis and P. E. A. Turchi
Lawrence Livermore National Laboratory
P. O. Box 808, L-371
Chemistry and Materials Science
Livermore, CA 94551

A number of conditions are presented for assessing the integrity and viability of a theoretical construct in the physical science in general, and in the realm of alloy physics in particular. These conditions are obtained from mathematical, logical, and experimental requirements, and are discussed in connection with a number of formal schemes currently in use for understanding and interpreting alloy phenomena. Both older methodologies and more recent attempts at the construction of a satisfactory theory of alloys are considered.

INTRODUCTION

Alloy theory is aimed at understanding the physical, chemical, and mechanical properties of alloys, that is to say materials that contain atoms of more than one species. The most common understanding of the term materials in the present context confines it to solids, although liquid alloys must also be included in general (and often are.) The part of the theory that is based on the electronic structure of an alloy is concerned with identifying the electronic origins of these properties. The modern version of this theory can be traced at least to the early 1930's, in the book by Mott and Jones [1]. Based on the formal methodology of the at the time still newly discovered quantum mechanics, the book made a serious attempt at understanding the properties of alloys through the introduction of insightful, and initially seemingly successful, models describing some fundamental alloy properties at the electronic level. One of these models is worth recalling.

In the so-called rigid-band model (RBM), it is assumed that the electronic spectrum, describing the distribution of electronic states in energy of the alloy is the concentration-weighted average of the spectra of the pure constituents formed separately by the species present in the alloy. The model involves a not too unreasonable supposition, one that seemed to be consistent with experimental data, at least in its initial applications. However, its downfall came when it was shown experimentally that real systems in general do not subscribe to this model, except possibly in some limiting cases. We will make more references to the RBM mostly for the purposes of illustrating various concepts as they arise in the discussion. It will emerge from that discussion that the RBM lacked many of the requirements of a satisfactory theory but this should not be taken as approbation for the conceptual effort that led to its introduction. Faced with puzzling phenomena in the physics of alloys, one is compelled to begin by testing out any conjecture that may seem reasonable. And the RBM is eminently reasonable (if ultimately not correct). However, we are no longer in the 1930's and the nature of theory in natural philosophy has been clarified substantially since that time. So, we will attempt to illustrate the features of such a theory, as pertaining to alloys, and will use the RBM or other, more recently developed concepts, to carry through this illustration.

Other models were subsequently introduced, each ultimately judged against experimental results. Most were discarded when they too led to predictions inconsistent with data, or mathematical requirements. As theory became more and more sophisticated, fairly rigorous formalisms were developed for studying the electronic origins of alloy behavior. It is somewhat disconcerting that after all this great effort there exists today no single fully satisfactory alloy theory, at least in the sense of the term to be explained in the following paragraphs. But the effort has not been in vain. At least it has led to the development of a set of criteria that a satisfactory alloy theory should satisfy. Some of these are of a general sort, generic to the nature of theories in modern science [2], while others are more specific pertaining more closely to alloys and their properties [3]. Our purpose here is to list these criteria and make some comments about how well they are satisfied by some of the better known approaches to alloy theory, including both methodologies of long standing as well as more recent ones.

The physics of alloys is a vast field whose various aspects are far too numerous for anything like a complete consideration to be attempted here. A proper alloy theory must lend itself readily to the study of a number of general areas of alloy physics, such as the energetics of alloy formation, the spectra of alloys, the effects of short-range order on alloy properties, transport, both electronic and mass, magnetism, and many others. It must yield results that are easily and accurately interpretable in terms of experimental information thus allowing the possibility of a deep understanding of alloy phenomena. It must also be easily coupled and work in unison with other well developed conceptual frameworks relevant to the study of alloys, such as thermodynamics and statistical mechanics. A theory that is founded on experimental observation and the analysis of data, that possesses these features, and also satisfies the necessary mathematical conditions discussed below has a good chance of providing a reliable understanding of alloy physics.

CRITERIA FOR A SATISFACTORY ALLOY THEORY

In this section, we list a number of conditions to be met by a fully satisfactory scientific theory. The first five of these, stated explicitly, are of quite a general kind pertaining to physical theory in any domain. Other conditions, subsidiary to the main ones and pertaining to alloy theory in particular are also mentioned but quite often they are the byproduct of the general ones. In the sections that follow, we will use these conditions as criteria against which to judge the viability and integrity of some commonly used methodologies to study alloy phenomena.

A. General Conditions

General Condition 1

The first condition is related to the *mathematical character of modern physics.* This term implies the usage of mathematics as a way of offering *convincing* arguments about the relevant aspects of the physical world that are of concern to the theory. The theory does not require an epistemological identification of each of its elements with a part of nature. But it does require that mathematical formalism be employed in justifying a priori methodologies and procedures and their predictions about alloy phenomena. In this regard, (and only as an illustration), one does not have any mathematical basis for proposing the RBM. And had the model been found to be a sound representation of alloy physics, a mathematical justification a priori would have been in order. It follows that even a model found to be successful by means of its application must ultimately be given a solid mathematical basis in order to be convincing.

The ability to trace a theory to its *mathematical* foundations can be used to distinguish an *ab initio* or first-principles theory from other types. In the context of alloy theory, for example, a theory remains first-principles as long as any approximations made as one proceeds from a formally exact expression of the Schrödinger equation are well understood in mathematical terms. Alternatively, the introduction of parameters (obtained through fits to experimental data or the consideration of certain limits in the mathematical expressions of the theory) in designing a model system break the smooth running of the mathematics. Theories so constructed may indeed provide a viable phenomenological description of natural phenomena but they cannot claim a priori justification of their content.

General Condition 2

In addition to the mathematical requirement just mentioned, a theory must satisfy the *principle of consistency.* This exceeds the mathematical requirements above and implies the logical consistency of the arguments based on the theory and leading to the interpretation of the results obtained in it. The emphasis here is on the word logical. Logic must be not too much below the surface when one makes an argument about different sets of data corresponding to different properties of the same physical system. Logical inference must be at the forefront when invoking a given set of mechanisms to explain an ever-widening set of phenomena. This requirement demands that an alloy theory must be of sufficient generality to be brought to bear on a set (actually an exhaustive set as stated below) of physical properties.

One of the requirements placed on a theory is that it satisfies the condition that logicians and mathematicians call *modus ponens.* Somewhat crudely, this means that one should be able to trace the discussion back to its original and scientifically justified assumptions from any point in the discussion. This condition is designed to eliminate the use of arbitrary assumptions in setting up the "first-principles" or starting base of the theory. The need for a starting point that is itself logically justifiable is something that is possibly easier seen by counterexamples. The first involves the beleaguered RBM. The point is that there is no justification in the assumption that the alloy spectra are the concentration-weighted average of the materials constituting the alloy. And it can get much worse.

In certain more recent approaches [15] to alloy theory, one studies structures that are so constructed as to reproduce the first few moments of the spatial correlation functions of the real system. There is no logical justification, however, that such structures reflect any of the physical properties of the real material, in particular the nature of the electronic states. Further discussion on this point is given in a subsequent section.

General Condition 3

The third condition involves the *principle of unambiguous communication*. The theory, in its mathematical and non-mathematical aspects alike, must allow for the (sufficiently) unambiguous communication of both the experimental results and theoretical findings. To the extent possible, there must be an unambiguous definition of variables and concepts generated by or involved in the theory. This condition can be also considered as the condition of *definitions*, whereby terms are defined and understood in a particular and constant manner throughout any discourse based on the theory.

General Condition 4

The fourth, and possibly the most evident, of the conditions characterizing modern scientific theory, and hence a theory of alloys, is the *principle of experimental rigor*. In short, a viable theory is demonstrably physical. It must be of a nature that allows confrontation with experimental data and hence provides a means for its possible falsification. This is, of course, the reason that no physical theory can ever be proved correct, but can be refuted by a single disagreement with an experiment. However, once a falsifiable theory is proposed, this *condition of physicality*, as this requirement can also be called, can be used to sharpen its credibility.

According to this principle, a theory must correspond, agree with, and, within its limits, exhaust the experimental data for which it aims to account, even if the data itself is subject to interpretation. This condition hinges greatly on the concept of measurable quantities, and the extent to which one has a clear understanding of what is being measured. However, regardless of the efficacy of experimental procedures, a theory must make justifiable statements, justifiable a priori, about data in its purview. The RBM, for example, is severely limited in this regard, as there seems to be no justifiable way of applying it to phenomena other than those involving spectra.

The last point is a crucial one. It is not so much that the theory must provide an accurate prediction of experimental findings. As desirable as this may be, the physical aspects of the theory are to be judged on the basis of their correspondence to the physics of alloys a priori, and to the generality of phenomena covered by the theoretical construct. One must know why a theory is expected to work and why it is expected to yield (accurate) predictions before testing the accuracy of these predictions. It is to be noted that agreement with experiment alone does not provide proof of validity of a conceptual construct, nor does it necessarily lead to deep understanding of alloy phenomena. For this reason, theories that are directed at the matching of a set of data - by means of adjustable parameters, say - run the risk of diminishing the viability of the most crucial step in the verification of a theoretical model; namely comparison with experiment. Furthermore, theories that "predict" a restricted set of data, while possibly leading to inaccurate results or have nothing to say with regard to related data sets are to be viewed with skepticism.

Therefore, one must caution against placing undue trust on "theory" that has been manipulated to fit certain aspects of an admittedly more general problem. The rigid-band model provides a prime example of this situation. The model gave initial results that seemed to describe a restricted set of experimental data on the densities of states (DOS's) rather well until further investigation revealed the lack of agreement with even DOS data, as well as a broader spectrum of alloy physics. It would have been inappropriate and ultimately useless to attempt to fix the problem by means of adjustable parameters of one sort or another, and quite rightly the model, in so far as it provides a broad picture of alloy physics, has been put to rest.

A corollary of the last condition is that the theory should give results that are physically meaningful, such as non-negative spectra (DOSs), in both real and momentum space. It can be shown that negative spectra fail to satisfy causality and are sufficient to disqualify a certain formal construct as a viable theory of alloys. In addition, the methodology should yield results that are analytic in a mathematical sense and satisfy fundamental sum rules. For example, one set of such sum rules involves the integral over energy of the partial DOSs associated with different alloy species.

General Condition 5.

The theory must satisfy the principle of *conceptual minimalism*, in the sense that it must cover as large a domain of physical experience as possible with the smallest possible number of conceptual elements. Formal approaches that are consistent with this principle allow the determination of a large number of physical parameters and the explanation of various sets of data through computations based on a small number of formal quantities (like the electronic self-energy). The conceptual effort expanded in reducing the study of the phenomena to that of a small number of elements can yield immense rewards in terms of a deep understanding of the phenomena under consideration.

B. Subsidiary conditions

We now turn to a set of conditions that are more closely geared to the domain of alloy physics than the general considerations enumerated above. As already mentioned, these conditions are not necessarily independent of those just enumerated. Rather, they are specific applications of the general conditions whose form is decided through the specific issues pertaining to alloys.

Subsidiary Condition 1.

The theory should be unique in the sense that it can be derived within various formalisms and independent points of view. Possibly the most important aspect in this regard is the ability to derive the theory within both the real and reciprocal spaces defined by the lattice of the system under study.

The importance of this condition can hardly be overstated. Correlations are defined with reference to *real space*, and a real-space derivation would ensure that the spatial relationships among the fluctuating "sites" are taken properly into account. On the other hand, the symmetry of the system, in particular its translational symmetry, are most efficiently reflected in *reciprocal space*, and a derivation in such a space would tend to preserve the symmetry of the averaged system. This leads to the following, related, condition:

Subsidiary Condition 2.

The theory should preserve all relevant symmetries such as the translational periodicity of the underlying lattice. In the case of random alloys, the symmetry to be preserved is that of the configurationally averaged system. In liquids, and materials with topological disorder, symmetry must be defined in terms of alternative quantities, e.g., in terms of pair distribution functions.

The necessity of this condition should be strongly coupled with the condition of physicality mentioned above. A self-energy that vanishes everywhere preserves the translational invariance of a lattice, but its evident lack of physical meaning renders it unacceptable.

Subsidiary Condition 3.

The theory should become exact in all physical limits, such as the limit in which the concentration of a species in the alloy approaches zero (or one), or as the difference in the scattering strength characterizing the different alloy constituents vanishes.

Subsidiary Condition 4.

The theory should be applicable to different descriptions of the alloy Hamiltonian, such as that of phenomenological tight-binding (TB) theory, or *ab initio* methods relying on the direct solution of the single-particle Schrödinger equation for the alloy potential. This condition guarantees that the methodology can be applied to realistic descriptions of materials, and not just to particular model systems.

Subsidiary Condition 5.

In the case of disordered systems, it should describe the various kinds of disorder likely to be encountered within various descriptions of the Hamiltonian, such as diagonal, off-diagonal, and topological disorder.

Subsidiary Condition 6.

The theory should allow the treatment of short-range order, and lend itself to physically meaningful applications of thermodynamics and statistical mechanics in the study of alloys. This condition can be fulfilled only when, in the construction of the self-energy, specific alloy configurations in real space are taken into account.

Subsidiary Condition 7.

The theory should allow the calculation of one- and two-particle properties within an equal footing, thus allowing the reliable calculation of energetics, spectra, and transport properties in alloys.

Subsidiary Condition 8.

The theory should reproduce fundamental microscopic properties as accurately as possible. The moments of the density of states that are exactly reproduced by a theory, for example, as compared to those of model calculations provides one such measure.

Subsidiary Condition 9.

The theory should be able to accommodate various physical constraints imposed on a system, such as the so-called Goldstone sum rule in disordered magnetic alloys.

Subsidiary Condition 10.

The methodology should allow a treatment of the *embedding* problem, that is to say, the study of a finite cluster of impurities embedded in a host lattice. Of particular importance in this context are the values of the Hamiltonian parameters that describe the coupling of the impurity cluster to the host medium.

It is important to note that the embedding problem must be solved before a self-energy is determined. In general, a self-energy that possesses site non-diagonal terms precludes the treatment of impurities embedded in the effective medium defined by this self-energy because the coupling of the impurity to the medium is not well defined.

There is one more condition related to computational efficiency. Namely, the method should be computationally practical. Although one can hardly argue against computational efficiency, it must be realized that in and of itself computational ease does not validate a theory. However, one must realize that this is a double-edged sword. Although a conceptually *perfect* theory that is computationally impossible is of no use, a theory that is computationally feasible, even easy to implement, equally lacks usefulness if it is conceptually flawed.

We now use these conditions to judge the efficacy and reliability of various theories, some of which are older while others are more recently proposed to study the physical properties of alloys. We begin with a brief overview of some of the older methodologies.

BRIEF REVIEW OF PREVIOUS THEORIES

Our discussion will be focused primarily on so-called self-consistent theories in which the electronic self-energy, and hence the Green function, is determined by means of a condition on the scattering properties of an electron or a lattice wave propagating through the system. In spite of their historical significance, some of the non-self-consistent methods, such as the virtual-crystal approximation and the average t-matrix approximation, the embedded cluster method, or certain more recently developed methodologies will be reviewed only briefly. The interested reader can refer to previous publications [3] where lengthier discussions may be found. Also, we will not concern ourselves with numerical techniques designed to give "exact" results through application to large (but finite) collections of atoms. We are interested in self-consistent methods that provide a self-energy thus reducing the study of alloys to a small set of quantities (to only one, the self-energy, in this case.)

A. The coherent potential approximation

Possibly the best known and most widely used self-consistent alloy theory is based on the coherent-potential approximation (CPA) [4, 5] developed for the study of disordered (random) alloys. The properties of the CPA are by now well understood, and have been discussed in a number of publications [6] and books [7].

The CPA is a single-site theory that, in spite of its relative simplicity, provides a remarkably accurate tool for determining the physical properties of real materials. And it holds up exceedingly well when judged against the conditions outlined above.

Within the CPA, the real disordered material is replaced by a uniform effective medium that is determined in a self-consistent way through the requirement that the additional scattering resulting from embedding a real atom of the alloy into this medium vanishes when averaged over the components of the alloy. This eminently physical requirement is attended to by a great number of further desirable properties.

The CPA yields results such as self-energies and Green functions that possess the proper analytic properties and satisfy causality and the sum rules on the total and component densities of states. The CPA yields quantities that preserve the symmetry of the underlying lattice, it has the correct limiting properties when the concentration or the scattering strength approaches zero, and can be applied to Hamiltonians of the TB kind as well as to those derived within a first-principles framework. It has been extended to apply to alloys with both diagonal and off-diagonal disorder, and can be used to calculate transport properties, where it is shown to satisfy all fundamental sum rules, like the relevant Ward identities (particle conservation) (for review see [7].) It gives correctly the first seven moments of the density of states as easily verified through direct numerical calculation on model systems, and it can be derived within a great number of different formal frameworks attesting to its unique qualities. In particular, it can be derived within both a real-space and a reciprocal-space formalism. Finally, it allows the treatment of impurities embedded in the CPA host medium, and even the treatment of impurity clusters, although not necessarily in a self-consistent way. Also, the formalism of the CPA can be used to provide a fairly accurate treatment of the effects of short-range order, and there are a number of calculations [8] of the soft X-ray scattering amplitude that attest to this property.

However, in spite of its many desirable properties the CPA does have a number of shortcomings. Although it yields results with the proper analytic properties, it fails to satisfy the Lifshitz condition on the tails of the density of states. Whereas general and exact arguments [3] require the DOS to decay exponentially outside the band region, the CPA DOSs exhibit a logarithmic singularity at the band edge. The formalism of the CPA has not been satisfactorily extended to the case of topological disorder, and the CPA Green functions fail to satisfy the Goldstone sum rules for lattice vibrations with off-diagonal disorder, (and certain systems with magnetic disorder.) Furthermore, the formalism contains no hint on how it can be extended beyond the restrictions of statistical fluctuations confined to a single site.

This restriction is possibly the most vexing and disturbing shortcoming of the CPA. The method is based on a single-site mean-field theory and cannot account for the effects of long range inter-site statistical fluctuations in a disordered system. Thus, the density of states obtained within the CPA is oblivious to local environment effects on the scattering from a given site in the material. The site potential encountered by an electron or wave propagating through the material is bound to depend in a non-trivial way on the occupation of nearby sites by atoms of specific alloy species. This effect is completely ignored in the CPA in which all atoms of a given kind are treated as being identical throughout the system. The treatment of statistical fluctuations, whether short or long range, has provided an incredible intellectual challenge to alloy theorists over the last thirty years or so. This challenge is as present today as it was when it was first put forward at the time of the introduction of the CPA.

To treat the effects of statistical correlations on alloy properties, one invariably needs to go beyond the single-site approximation [3]. A number of non-self-consistent cluster theories were introduced for this purpose that allowed the treatment of a finite cluster of atoms embedded in a medium determined in some way – usually through the self-consistent condition of the CPA [3]. These methodologies, however, do not yield an improved self-energy and are not of further interest to us in this review.

Self-consistent cluster theories were also proposed in the attempts to improve on the CPA in the treatment of statistical fluctuations. In spite of intense and concentrated effort, it can be safely stated that no methodology has been found that improves significantly and uniformly over the CPA when judged against the entire set of criteria mentioned in the previous section. Improvements were indeed obtained in specific areas, but usually at the expense of various physical requirements. A discussion of many of these extensions, both the self-consistent and non-self-consistent variety can be found in published work [3], along with references to original papers.

However, one of these extensions, the molecular coherent potential approximation (MCPA)

[9] is of particular interest and we will provide a brief review.

B. The molecular coherent potential approximation

The molecular CPA, or MCPA, was introduced by Tsukada [9] along lines identical to those of the single-site CPA. In this case, however, the material is divided into non-overlapping clusters and a cluster-diagonal self-energy is obtained through the condition that the additional scattering introduced when a real cluster is embedded in the MCPA medium vanishes on averaging over all cluster configurations at the concentration of the alloy.

It can be shown that the MCPA is a physical theory that retains many of the desirable properties of the single-site CPA, such as analyticity and the satisfaction of fundamental sum rules. It gives correctly more moments of the DOS than the single site CPA, the number increasing as the size of the cluster treated increases. However, the MCPA leads to an effective medium that is a "superstructure" with cluster periodicity, and thus violates the requirement that the averaged medium must be translationally invariant.

This violation is not only conceptually unacceptable but has realistic ramifications. A superstructure with cluster rather than point periodicity can lead to inaccuracies in the determination of phonon spectra, and of transport coefficients. This is because the reciprocal space corresponding to the cluster periodic system contains zone boundaries that cause unphysical momentum reflections in the consideration of phonon and transport properties. Furthermore, in the MCPA the edges of the DOSs contain a logarithmic singularity, as in the CPA, because the MCPA, treating statistical fluctuations in a compact cluster, cannot account for the long-range fluctuations that cause the exponential decay of band edges.

After this admittedly brief review of the CPA and its molecular extension, we turn to recent developments.

A CLUSTER EXTENSION OF DMFT

Attempts to provide a treatment of electron correlations in the electronic structure calculation of solids have led to the rediscovery of the single-site CPA [10] and also given rise to cluster generalizations of it [11, 12]. The CPA and the MCPA have been reviewed above, but a particular cluster extension has recently been introduced by Jarrell [11, 13] that needs special attention.

A. The dynamical cluster approximation

Jarrell's methodology [11, 13], referred to as the dynamical cluster approximation (DCA), is quite novel as it leads to analytic self-energies that are **k** dependent while preserving lattice periodicity.

The seminal element of the methodology is to replace the exact disordered material by a fictitious cluster of finite size whose sites are assumed to possess the disorder characteristics of the real material. It is to be understood that the sites of the cluster, $\mathbf{R}_I$, can indeed possess coordinates that correspond to sites in the real lattice. Their fictitious nature arises from the fact that the inter-site Green functions for this cluster are considerably warped and deviate substantially from the corresponding Green functions of the averaged system. Now, this cluster is treated exactly and hence leads to a self-energy that has all the required analytic properties and preserves the real-space periodicity of the averaged lattice. The replacement of the original infinite system by one of finite size is carried out through an ingenious construct of coarse graining the reciprocal space of the real system.

Let the first Brillouin zone of the reciprocal lattice of a given real lattice be divided into a number of finite regions, $M_{\mathbf{K}}$, whose "centers", denoted by $\mathbf{K}$, form a cluster structure whose group point symmetry is the same as that of the real lattice. Corresponding to these points in reciprocal space, one also defines a cluster of points in direct configuration space by means of the relation,

$$\frac{1}{N}\sum_{\mathbf{K}} \exp \mathrm{i}\mathbf{K}\cdot(\mathbf{R}_i - \mathbf{R}_J) = \delta_{IJ}. \tag{1}$$

The last expression has profound consequences with respect to the topological interrelationships of the sites of the cluster. It is to be noted that all sites $\mathbf{R}_I$ are now topologically equivalent in contrast to the sites of a cluster embedded in an (infinite) medium. Even if the sites $\mathbf{R}_I$ correspond to sites in the original lattice, the condition defining their connection to a finite reciprocal space warps the cluster into a translationally invariant structure. For example, on a linear chain, a cluster of n sites is turned into a ring of n sites. The reader can be convinced of this effect in higher-dimensional systems; on a square lattice, the cluster is turned into a torrous in three dimensions while in three-dimensional system one obtains a torrous in four dimensions. This topological distortion is a consequence of the last condition on the sites of the cluster and their reciprocal space. (In an infinite medium, such a condition also holds but the summation is over all the points in the Brillouin zone.) Mathematical quantities with a dependence on reciprocal space are now coarse-grained over each region $M_{\mathbf{K}}$, and the result assigned to $\mathbf{K}$, the center of the region. Of particular interest are the Green function and the self-energy. A Green function, $G(\mathbf{K},\omega)$, is defined for each $\mathbf{K}$ by summing $G(\mathbf{k},\omega)$ over all $\mathbf{k}$ in the region $M_{\mathbf{K}}$,

$$\mathbf{G}(\mathbf{K},\omega) = \sum_{\mathbf{k}\in M_{\mathbf{K}}} G(\mathbf{k},\omega), \tag{2}$$

where ω is an energy parameter. This definition now leads naturally to a Green function $G_{IJ}(\omega)$ defined over the sites I and J of a finite cluster obtained as the reciprocal structure of the cluster formed by the vectors $\mathbf{K}$. The explicit expressions for the inters-site Green functions are the usual ones,

$$G_{IJ}(\omega) = \frac{1}{N_c}\sum_{\mathbf{K}} G(\mathbf{K},\omega) \exp \mathrm{i}\mathbf{K}\cdot(\mathbf{R}_i - \mathbf{R}_J). \tag{3}$$

This expression connects cluster sites that are topologically equivalent and hence incongruent with the topological distinctions between clusters in the real material. Therefore, the DCA cluster cannot be thought of a being "embedded" in an infinite medium. More importantly, the correlations between fluctuations on different sites of the cluster do not correspond to correlations between fluctuations in the sites embedded in such a medium. This is only accomplished in the strict limit of an infinite cluster, but at no intermediate step toward that limit.

For any given self-energy, $\Sigma(\mathbf{K},\omega)$, we also define the subsidiary quantity in reciprocal space,

$$\mathtt{G}^{-1}(\mathbf{K},\omega) = \mathbf{G}^{-1}(\mathbf{K},\omega) + \Sigma_{\mathbf{K}}, \tag{4}$$

that follows from the general expression

$$\mathbf{G}(\mathbf{K},\omega) = \mathtt{G}(\mathbf{K},\omega)[1 + \Sigma(\mathbf{K},\omega)\mathbf{G}(\mathbf{K},\omega)]. \tag{5}$$

An analogous definition holds in real space,

$$\mathtt{G}^{-1} = G^{-1}(\omega) + \Sigma. \tag{6}$$

Now, the sites of the cluster defined by the space reciprocal to the vectors $\mathbf{K}$ are assumed to have the disorder characteristics of the real material. For example, in the case of a disordered binary alloy, $A_{1-c}B_c$, the sites of the alloy are taken to be randomly occupied by atoms of type A or B, with corresponding probabilities $1 - c$ and c. Making the usual assumption that all atoms of a given species are identical throughout the system, each site carries a potential V_A or V_B in accordance with the type of atom occupying the site.
The Green function, $G_{IJ}(\omega)$ is defined as the Fourier transform of $G(\mathbf{K},\omega)$ as indicated above. Similarly, the real-space counterparts of $\mathtt{G}(\mathbf{K},\omega)$ and $\Sigma(\mathbf{K},\omega)$ can also be defined by Eq. (6), completing the loop of self-consistent equations to be solved for the self-energy.
These real-space quantities, defined over the sites of a finite cluster of sites, can be used to provide an "exact" treatment of the disorder in the cluster. Thus, the Green function G satisfies the expression, (using operator notation and suppressing the energy parameter from now on)

$$G = \mathtt{G}[1 + VG]. \tag{7}$$

Denoting the configurational average of G over the cluster configurations by $\langle G \rangle = \bar{G}$, a cluster self-energy is obtained through the expression,

$$\Sigma = \bar{G}^{-1} - \mathtt{G}^{-1}. \tag{8}$$

A Fourier transformation leads to $\Sigma(\mathbf{K})$, and a new value for the Green function. For every k in a domain $M_{\mathbf{K}}$, we have

$$G(\mathbf{k}) = [G_0^{-1}(\mathbf{k}) - \Sigma(\mathbf{K})]^{-1}. \tag{9}$$

This new Green function can be used to obtain a new coarse-grained value, $G(\mathbf{K})$, as shown in Eq. (2). An iterative process can be established that terminates when the change in the self-energy (or the Green function) falls within a predetermined tolerance.

B. The properties of the DCA

We now examine the DCA against the set of conditions to be met by a satisfactory alloy theory. By construction, the DCA yields an analytic self-energy and a Green function that take account of statistical fluctuations in the fictitious real-space cluster corresponding to the set of reciprocal lattice vectors $\mathbf{K}$. The self-energy is periodic with the point symmetry of the real lattice, and vanishes in the limit $c \to 0$ and as the scattering strength approaches zero. Its behavior for small but non-zero concentration is not known. This behavior would be of relevance in applications of the theory to ordered systems.

Perhaps the greatest practical drawback of the DCA is that it fails to treat correctly inter-site correlations in the real system (its logical restrictions are mentioned below.). The finite, real-space cluster representing the disordered material is marginally connected to the real system, failing to preserve the inter-site topological relationships of the true disordered material. Alternatively, the cluster inter-site Green functions do not correspond to the Green functions for sites of a real cluster embedded in a fluctuating medium. Thus, it cannot yield spectra associated with distinct configurations of sites in the material. Such information is often useful in analyzing physical properties, e.g., magnetism, which can be explicit functions of local environment. Consequently, the method leads to a physical single-site Green function (and hence single-site spectra) but not to multi-site spectral functions rendering difficult the study of short-range order effects in the real system.

It may be useful to expand on the discussion of the DCA Green functions. To see the difficulties encountered within the DCA consider the determination of Green function matrix elements that connect sites in a cluster to points in the surrounding medium. These elements are readily determined within the CPA or the MCPA but they are not defined in the DCA. This is because the topological relationships of the DCA cluster are not those characterizing a cluster embedded in a medium, but one standing isolated in real space.

We now examine more closely the formal justification of the DCA and contrast it with the CPA. The latter approximation yields a site-diagonal self-energy that can be considered both within real space and reciprocal space. Within real space, it can be viewed as being associated with *every* site in the real lattice, thus yielding an effective medium that preserves the translational invariance of that lattice. By construction, this self-energy accounts for statistical fluctuations confined only to a single site. Within reciprocal space it can be thought of as being independent of **k** vector in the Brillouin zone. To extend the CPA can also be thought of along real-space or reciprocal-space terms.

Within real space, one seeks an approximation that accounts for fluctuations associated with clusters larger than a single site. This necessity has led to an immense effort [3] to generalize the CPA to a self-consistent cluster theory. Judged against the criteria set up in Section II, this effort has met only with partial success. No extension of the CPA has been constructed that possesses the required analytic properties while retaining the translational symmetry of the underlying lattice while accounting exactly for inter-site correlations in real space clusters. The theory that comes the closest to meeting these criteria is the MCPA that, however, endows the effective medium with the super-periodicity of a cluster of sites. At the same time, the MCPA leads to a self-energy that properly accounts for inter-site fluctuations within compact clusters in the material.

The elements touched upon in the previous discussion culminate to a point of logic. A physical theory is usually characterized by both necessary and sufficient conditions. And as is well known one set cannot replace the other. It is necessary for the self-energy to have a **k** dependence to conforms to the periodicity of the underlying lattice and to possess certain analytic properties in the complex energy plane. But these conditions are not sufficient to guarantee that the self-energy satisfying them also accounts properly for fluctuations in the real, disordered system. Thus the fact that the DCA self-energy possesses certain necessary properties is no proof that it provides a correct description of a fluctuating system.

The DCA yields an analytic self-energy that preserves the translational invariance of the lattice. However, it results in a real-space system of strictly finite size whose Green functions represent fluctuation propagation from site to site only approximately. In the MCPA, the cluster Green functions are exact for each choice of cluster, being approximate in that they differ from those of the exact configuratioanlly averaged system. In the DCA the Green functions are not exact even for the finite size cluster used to obtain the self-energy. In the case of the MCPA, one is confident that the effects of short-range order can be accounted for through the treatment of cluster of appropriate size. The confidence is diminished in the case of the DCA because the cluster treated cannot be viewed as embedded in the real, infinite system, and the cluster Green functions are not those of the bare real-space cluster employed in determining the self-energy. In both cases, the treatment of transport and of phonons, both involving long-range effects, is problematic.

NON SELF-CONSISTENT THEORIES

A. The Quasi-Random Structure Method

In this approach [15] to the study of alloy electronic structure and energetics, one simulates a substitutionally disordered alloy by ensuring that the distribution of atoms of different species over the sites of a lattice preserve the first select few spatial correlation functions of the real disordered system. For example, the distribution may preserve the overall concentration and the nearest-neighbor pair correlation function.

This methodology, which is also called SQS (for special quasi-random structures) has been extended to incorporate volume effects and other physical features entering the description of an alloy, such as charge-transfer effects.

When judged against the criteria required for a satisfactory alloy theory, the SQS methodology can be seen to fail in a number of important areas. There is no mathematical justification for the method, as it has not been proven that the properties of a quasi-random structure have any connection to those of a manifestly disordered system. In particular, the electronic states in such a structure have a very different character than those in a random alloy. The latter are characterized by spreading due to the presence of randomness whereas the states in an ordered structure have sharp $E(\mathbf{k})$ dispersion relations. The primary aim of alloy theory in the context discussed here is the determination of the electronic states in the alloy. This would allow the calculation of expectation values of various operators (observables) to be compared with experiment. Because a formal justification establishing the equivalence of the states in a random system with those in a quasi-random structure is lacking, there is no way of producing logical arguments connecting the results of calculations to properties of real alloys.

Nor is the methodology exhaustive in terms of experimental data, as required of a proper alloy theory. Indeed, it produces results that are known to be contrary to the behavior of physical systems. For example, it yield a "self-energy" that is real whereas the dispersion in the experimental spectra of disordered alloys suggests a complex quantity. Because the method lacks scientific foundation, its generalization to incorporate a broader spectrum of physical reality is unjustified.

Again, the logical justification rears its demanding head. The observed spatial correlation functions are a result of the underlying fluctuating nature of the system. They are, by definition, an effect of these fluctuations. There is no logic to the statement that two systems that show identical effects are also identical at a fundamental level. Furthermore, there is no justification that a theory based on a particular effect as its starting point captures any of the physics of the system even that which is responsible for the effect. Reversing the roles of cause and effect, as is done in the SQS, is illogical in itself.

B. The Connolly-Williams Method

In this procedure [16] one extracts so-called cluster interactions that describe the contribution of a cluster of atoms to the energy of a system. This is done by solving a set of linear equations for these interactions. In these equations the interactions are the coefficients of an expansion in cluster basis functions of the energies of various configurations and for various concentrations.

Of all methodologies proposed thus far for the study of alloys, the CWM is the one that fails most dramatically in meeting even the rudiments of logical or mathematical rigor. (There are other failures as well, but of a subsidiary nature to the main ones.) In its proposed form, the method combines the energies of alloys at *different* concentrations thus excluding itself from making any statements whatsoever about any particular alloy, (for the fairly obvious reason that a particular alloy is defined by a specific concentration and its properties are often strong functions of concentration.) For example, the methodology allows the simultaneous treatment of alloys that are known to be magnetic in a given region of concentration and non-magnetic in others. The notion that the parameters that one obtains have the same value throughout the concentration range is not only unjustified but also non-physical.

Possibly most important, and most damaging, is the methodology's renunciation of some well-established mathematical laws. Expansions over a vector space are mediated by the use of a complete (often orthonormal) basis set. The totality of configurations at each concentration defines a complete vector space whose basis is formed by the cluster functions used in the CWM. For an infinite system each of these configurational vector spaces has the same dimensionality and allows a complete expansion of the energies in terms of the cluster functions that form the basis in the space. Mixing different concentrations corresponds to the mixing of different vector spaces, a procedure that is mathematically disallowed. (The orhonormality and completeness relations of the basis set cannot be shown to hold over such a mixture and the expansion, as a whole, is non-sensible.)

Furthermore, the resulting set of linear equation in the CWM is finite resulting in uncontrolled effects on the solutions (the cluster interactions) that are by definition of infinite number and generally of infinite extent. The argument that is often produced to counter this last defect, namely that the interactions converge fast enough so that higher terms can be neglected beyond a point carries no weight because the interactions are wrong in the first place as just pointed out. One could, of course, try an application of the methodology within a single concentration. Although identifying and computing a sufficient number of such configurations within a single concentration is by no means easy, the procedure would alleviate somewhat the mathematical and logical inconsistencies resulting from the mixing of different vector spaces (although the approximation of finite size expansions would remain). In addition, a serious physical inconsistency would also remain in that the calculations of the energies used would correspond to ordered systems. As such the electronic

structure of these systems is different from that of a truly random alloy at the same concentration and the failing of the conditions of physicality, mathematical and logical integrity, and experimental rigor follow in precipitous manner.
The logical consequence of the preceding analysis is that the extension of this sort of methodology to a broader domain of alloy physics is not justified.

THE SINGLE-CONFIGURATION AND FINITE-SIZE ARGUMENTS

There are two notions that have found some favor among scientists concerned with the development of alloy theory. The notions are, first, that the materials one deals with and subject to a theoretical understanding actually exist in a given, fixed configuration, and that these materials are of finite rather than infinite size. These two concepts are strongly related, with the former being inapplicable without the latter. Taken together, they deal a devastating blow to the very essence of the conceptual challenge offered by fluctuating systems as the first one does away with the need for treating fluctuations at all and the latter dispenses with thermodynamics. Although both notions can be discounted solely on the basis of their poor conceptual standing, there is something to be learned from looking at them both a bit more closely.

A. Single-configuration argument

The notion that, after all, a given piece of a real material actually exists in a given configuration is based on the possibility that one can establish the nature (chemical species) of the atoms occupying the Wigner-Seitz cells in the material. Of course, this can be done but only if the material is finite. For an infinite system, (of interest to thermodynamics that is necessary in the study of bulk alloys) this is not possible even in principle, so that the notion in question has no standing in the development of alloy theory. (The intricate connection to thermodynamics is addressed in the next section.) The properties of a random alloy (whether physical, chemical, or mechanical) are determined by and are a strong function of its fluctuating nature and the elimination of fluctuations removes subsequent considerations from the realm of alloy theory. The situation gets on even murkier logical ground when this train of thought proceeds to eliminate the relevance of concentration from the picture altogether.
Now, the argument runs along the following lines. In an alloy of infinite extent there are regions (finite or infinite) characterized by effective concentrations other than the alloy concentration. Therefore, in describing alloy properties (energies, strength, etc.) one is allowed to employ concentration-independent parameters (as is done, for example, in certain types of expansions such as the CWM.) Thus, the single-configuration argument can be used in either its weak form where a dependence on concentration is retained, or in its strong form in which the dependence of alloy properties on concentration is eliminated from the argument. In examining further the merits of the single-configuration concept, it is most convenient to begin with the strong form.
First, we consider the argument that an alloy contains regions of concentrations c' other than the alloy concentration, c. Actually, this argument is correct, postulating the existence in an infinite system of regions, also possibly infinite, in which the effective concentration fluctu-

ates. This is a simple consequence of the fluctuating nature of the atoms occupying different sites in the alloy, but does not imply that alloy properties are concentration-independent nor that they can be described in terms of concentration-independent parameters. The physical system behaves in a way that is consistent with the presence of a fixed concentration. For example, the chemical potentials for adding or removing atoms of a given species, or the Fermi level, the electronic chemical potential, and the physical properties of the system, e.g., magnetism, depend on and vary with alloy concentration. In other words, the system "knows" its concentration and its behavior is a function of it as observed experimentally and as demanded by thermodynamics. Therefore, representing an alloy by a collection of configurations at various concentrations is not justified on either physical or mathematical grounds.

The weak form of the argument considers an alloy as represented by a single configuration, or set of configurations, all at a given concentration. Compared to the strong form this method is greatly relieved of both the physical and mathematical difficulties connected with that approach. However, care should be taken to apply this methodology to quantities that do vary with configuration, and to include enough configurations for a proper statistical sampling. The energy, for example, is a good candidate for such a treatment, but microscopic quantities that only depend on concentration (the effects of configurations having been integrated out) such as chemical potentials, and the self-energy, say, are not.

There is, however, one case that demands particular attention, and that is the case of an ordered structure, say the so-called $L1_0$ ordered configuration exhibited by some alloys based on a face-centered cubic lattice. Such ordered structures do exist both as one of the many configurations allowed for a random system, as well as the only configuration allowed when the system undergoes a phase transition under particular external conditions on temperature and pressure. In the first case, they are counted in any proper theory of alloys (in which all configurations are treated in principle according to their statistical weight). In the latter, the passage to a specific configuration materializes under the loss or gain of energy in a singular fashion, and the properties of the particular phase so obtained usually have nothing to do with those of the disordered system. For example, a disordered alloy may be ductile or non-magnetic, while its ordered phases may become quite brittle, or exhibit magnetic behavior. Mathematically, this change in properties is described through the presence of singular behavior in various thermodynamic functions, such as the specific heat, a behavior that is also manifest experimentally.

It is, of course, evident that no fluctuations exist in an ordered structure in which the nature of any atom in the system is known throughout, even if the system is of infinite extent. With fluctuations gone, the system behaves quite differently from a random material. It is for this reason that the use of a single, ordered configuration is disallowed, both physically and logically, as a representative of a disordered alloy at a given concentration.

B. The finite-size argument

Within the context of materials science, and particularly the area of alloys, the realization that the materials we deal with are invariably of finite size has two different connotations. Finite-size materials that conform to and can be described by the laws of thermodynamics, and those that do not. It is indeed a remarkable fact that nature manages to integrate microscopic effects so that beyond a certain size, materials behave as if indeed they were of infinite extent. It took a great leap in theoretical development to realize that this is indeed the case, thus separating out from the plethora of materials those that can be studied and understood in the thermodynamic limit. Otherwise, materials whose size is not sufficiently large must be studied as function of dimension and their properties classified accordingly. Bulk alloys, as commonly perceived, lie in the first category. In cases in which one or more dimensions are finite so that size effects become manifest the usual laws of thermodynamics must be modified so as to accommodate these effects.

Once the discipline of thermodynamics has been introduced in the study of materials properties it becomes illogical to invoke the concept of size. The purpose of thermodynamics is to provide a model whereby human genius can probe certain classes of physical phenomena. The concept of the thermodynamic limit provides a means of examining the properties of materials that although in actuality finite behave in accordance with this limit. The formal study of such systems now must proceed on the basis of systems of infinite extent. However, instead of attempting to describe the nature of every atomic cell at infinity, one uses a rule that provides such a description in terms that are amenable to analysis. We can envision an ordered system of infinite extent by stating that "each cell is occupied by an atom of a given kind." We can envision a random system by stating that "every cell has a probability of being occupied by an atom of a given kind."

This has an immediate implication to the single-configuration argument discussed in the previous section. Namely, in a random system of infinite size it is not possible to specify the chemical nature of every atom in any given cell. This impossibility becomes obvious when it is recognized that the specification of the coordinates of a cell places that cell in a finite region of the otherwise infinite system. Because of this it is necessary to use a rule in describing the nature of infinite systems. And because it is not possible to approach infinity from the finite side, fluctuation in the system can never be eliminated. For this reason, the single-configuration argument has no intellectual standing in the study of alloys.

In other words, it is illogical to replace an alloy by a system of finite size-where the specification of a single configuration is, of course, possible-or by an infinite system in an ordered configuration. (Occasionally, use is made of a finite system averaged over all possible configuration. The disadvantages of that approach are those discussed previously in connection with the systems of finite size.) The finite-size and/or single-configuration constructs, whether used separately or in conjunction may provide some guidance in the study of thermodynamic systems. However, by themselves cannot be made into a formal basis for the study of thermodynamic quantities.

MODIFICATIONS OF THE MCPA AND A FUNDAMENTAL DIFFICULTY

A. Averaging the self-energy

One way of restoring the translational invariance that is lost in the breaking up of the lattice into the clusters of the MCPA is to average the self-energy and some attempts in that direction have been made [14]. However, such direct combinations or averages of the MCPA self-energy, although possibly preserving analyticity and restoring translational invariance, are beset with a number of formal problems.

First, averaging the self-energy wipes away the relation of the self-energy to distinct alloy configurations in the determination of spectral properties (density of states). In addition, it has the unphysical effect of producing a medium whose self-energy contains the fluctuations of incompatible configurations of the fluctuations in the system. For example, adding the self-energy at the center of a cluster to that at a border site creates an effective site that represents two different and incompatible local configurations in the system.

The difficulties are multiplied when extra conditions of translational invariance are placed upon the self energy obtained through the treatment of a finite cluster in an effective medium [12]. Now, non-analytic behavior can arise. All objections raised above regarding the treatment of embedding remain valid in the present case. Finally, and quite importantly, one would need to average the elements of the self-energy with elements of the Hamiltonian of the system in an ill-defined procedure given that the Hermitian character of the latter is absent in the former.

Some of the problems connected with averaging the self-energy, although not all, can be circumvented through an averaging of the Green function and a subsequent identification of a self-energy through inversion of the Green function over the whole space. Let G^{MCPA} denote the Green function obtained in an application of the molecular CPA to a disordered system and consider the averaged Green function, $\bar{G}$, defined through its matrix elements,

$$\bar{G}_{ij} = \frac{1}{N_{ij}} \langle G_{ij}^{\mathrm{MCPA}} \rangle. \tag{10}$$

Here, the angular brackets denote a summation over all elements of the MCPA Green function whose indices are connected to i, j by a single translation operation, and N_{ij} are the number of distinct such elements. For example, when $i = j$ one sums over all site-diagonal Green functions in the cluster defined in the MCPA and divides by the number of sites in the cluster. Similarly, one can obtain the elements of a Green function for all inter-site vectors, $\mathbf{R}_i - \mathbf{R}_j$.

It is clear that the procedure defined by the last equation yields a Green function that reflects the translational symmetry of the underlying lattice. It is also analytic, being the finite sum of analytic expressions. The self-energy corresponding to this translationally invariant medium is now obtained to any desired approximation through the inverse of $\bar{G}$,

$$\sigma_{ij} = [z - W - \bar{G}^{-1}]_{ij}, \tag{11}$$

where W denotes the translationally invariant part of the Hamiltonian describing the original disordered material. This expression can also be cast into reciprocal space,

$$\sigma(\mathbf{q}) = z - W(\mathbf{q}) - \bar{G}(\mathbf{q})^{-1}, \tag{12}$$

where

$$\bar{G}(\mathbf{q}) = \frac{1}{N} \Sigma_{ij} \bar{G}_{ij} \{\exp i\mathbf{q} \cdot (\mathbf{R}_i - \mathbf{R}_j)\}, \tag{13}$$

with N denoting the number of sites in the lattice.

The construction just described alleviates many of the undesirable features connected with the averaging of the self-energy. At the same time, even this approximation contains an unsatisfactory element. The final self-energy, in spite of its analyticity and translational invariance, is not the direct result of a self-consistent treatment of statistical fluctuations, but a derived one depending on the subsequent treatment of the Green function obtained in the MCPA. Therefore, it cannot be used to study the effects of vertex corrections on transport properties because it does not allow a study of specific clusters of sites embedded in the effective medium. On the other hand, it does treat correctly the effects of short-range order because these can be incorporated into the MCPA averaging over cluster configurations and are properly reflected in the final Green function and self-energy. It is easy to see that the procedure defined by Eq. (10) is an identity within the single-site CPA. It becomes exact, as does the MCPA, and also reduces to an identity as the cluster used in the MCPA approaches the size of the system.

B. A fundamental difficulty

The great deal of effort that has been expended in the as yet unsuccessful effort to produce a proper cluster extension of the single-site CPA - that is to say an approximation that treats inter-site fluctuation while preserving analyticity and lattice periodicity - can be traced to the violation of a basic formal requirements of scattering theory. Consider again the embedding problem in the case of a **k**-dependent self-energy. Although discussed above, the problem is worth further mention. In order to embed an impurity in the corresponding effective medium one must remove in a formal way the on-site element of the medium at the site of embedding, as well all inter-site elements that emanate from it. That done, these elements must now be replaced with the corresponding quantities connected with the real potential characterizing the impurity to be embedded and its connection to the surrounding medium.

The difficulties that arise with this replacement can be most prominently illustrated in the case of a first-principles, first-quantization application of the CPA, such as the KKR-CPA. In such a formalism one uses scattering theory to treat a free wave propagating through the lattice, being scattered by the on-site potentials connected with the chemical species occupying the sites of the lattice. In its most basic, scattering theory is founded on the construct that a wave in free space is scattered by a spatially bounded potential. The scattering matrix associated with that potential describes repeated scattering by a free wave by that potential. This construct is fully observed in implementations of the CPA or the MCPA. The propagation of a free wave is described by the structure constants that remain unaltered between sites (in the case of the CPA) or between clusters of sites (in the case of the MCPA).

Now, consider the case in which the self-energy is **k**-dependent, and hence can have, in principle, infinite extent in real space. Now, the propagation through the lattice is not that corresponding to a free wave, but a wave that has been modified by the potential of the effective medium as described by the self-energy. If an impurity is to be embedded, its scattering matrix must be determined with reference to that medium. Using the t-matrix derived with reference to free space introduces an inconsistency that cannot be removed through self-consistency conditions and that, indeed, prohibits their implementation. The proper procedure, of course, would be to calculate a t-matrix with respect to the medium but such an alternative is precluded as the self-energy is known only numerically. Equally important is the ignorance of how a real potential couples to the medium. That coupling is a source of scattering and it must be itself consistently determined with respect to the medium parameters. It is clear that propagation to and off the impurity can be described neither by the free-space structure constants nor by the inter-site elements of the self-energy. And there are, in principle, an infinite number of such elements. The reader may wish to contemplate how parameters should be chosen to describe the embedding of a single impurity at the central site of a cluster determined in the MCPA.

The insidious nature - or the strictness of this requirement of consistency with scattering theory - can be glimpsed when one realizes that removing the original cluster from a medium results in the alteration of the medium parameters. This is not the case in the CPA or the MCPA where free space surrounding an impurity or a vacancy retains its scattering properties.

There seems to be no way of alleviating this fundamental problem. It is the problem that must be overcome in order to allow for a unified treatment within real and reciprocal space. The CPA and the MCPA do allow such a treatment but neither of them yields a properly **k**-dependent self-energy. In all other cases, particularly in connection with attempts to preserve lattice periodicity, the discrepancies with respect to scattering theory provide a formidable stumbling block to their development.

CONCLUSIONS

If there is one message that the present paper is intended to send, is that physical theory must be consistent with the logic of mathematics and the logic that connects the mathematics to reality (the world of experiment and data gathering.) Each of these two different kinds of logic plays an indispensable role in establishing the viability and predictive power of a formal construct - a theory - in science. Both must be present and demonstrably acting *a priori* before a theory can be assessed as to the accuracy of its statements.

A number of models presented previously for the study of random alloys - CWM, SQS - were examined vis-a-vis these logical demands and found to lack, often severely, in compliance to them. As such, these formal constructs cannot be viewed as occupying a place in physical theory. Other formalisms, such as the DCA, were shown to have a weak formal basis, providing no justification that clusters of points in reciprocal space allow the treatment of fluctuations in the real material.

There can be, of course, no formal proof that a fully satisfactory alloy theory can never be constructed. (This would be the case if, for example, the conditions enumerated in the body of the paper were mutually contradictory.) However, all attempts towards the construction of such a theory reported thus far, including the modified form of the MCPA discussed in this paper, have failed to achieve this goal. The difficulties with the consistent application of scattering theory play a crucial role in this regard. It seems unlikely that a theory will emerge that allows one to consider the embedding problem in terms of a previously determined self-energy. So, for those demanding perfection, the outlook is not particularly encouraging. But, as always, there is plenty of room for improvement. And in the absence of a formal proof of non-existence, why not continue playing the game? Just as long as we keep an eye on the rules.

ACKNOWLEDGMENT

In-depth conversations by one of the authors (AG) with Alfredo Caro are gratefully acknowledged. This work was performed under the auspices of U. S. Department of Energy by the University of California, Lawrence Livermore National Laboratory, under Contract No. W-7405-ENG-48.

REFERENCES

[1] N. F. Mott and H. Jones, *The Theory of the Properties of Metals and Alloys*, Dover, New York (1958).

[2] Arkady Plotnitsky, *The Knowable and the Unknowable*, The University of Michigan Press, (Ann Arbor, MI, 2002).

[3] A. Gonis, *Green Functions for Ordered and Disordered Systems*, North Holland, Amsterdam New York, (1992).

[4] P. Soven, Phys. Rev. *156*, 809 (1967).

[5] D. W. Taylor, Phys. Rev. *156*, 1017 (1967).

[6] R. J. Elliot, J. A. Krumhansl, and P. Leath, Rev. Mod. Phys. *46*, 465 (1974).

[7] A. Gonis, *Theoretical Materials Science*, Materials Research Society, Warrendale, PA (2000).

[8] B. L. Gyorfy, D. D. Johnson, F. J. Pinski, D. M. Nicholson, and G. M. Stocks, in *Alloy Phase Stability*, G. M. Stocks, and A. Gonis (eds.) NATO-ASI Series, Vol. 163, Kluwer Academic Publishers, Dordrecht (1988), p. 421.

[9] M. Tsukada, J. Phys. Soc. (Japan) *32*, 1475 (1972).

[10] A. Georges, G. Kotliar, W. Krauth, and M. Rozenberg, Rev. Mod. Phys. *68*, 13 (1996).

[11] M. Hettler, A. Tahvildar-Zadeh, M. Jarrel, T. Pruschke, and H. Krishnamurthy, Phys. Rev. *58*, 7475 (1998).

[12] A. I. Lichtenstein, M. I. Katsnelson, G. Kotliar, in *Electron Correlations and Materials Properties 2*, A. Gonis, N. Kioussis, and M. Ciftan (eds.), Kluwer Academic/Plenum Publishers, New York (2003), p. 75

[13] D. A. Rowlands, J. B. Staunton, and B. L. Gyorffy, DCA to alloys

[14] A. Gonis, Thesis, University of Illinois, Chicago, 1978, (unpublished).

[15] Alex Zunger, S.-H. Wei, L. G. Ferreira, and James E. Bernard, Phys. Rev. Lett. *65*, 353 (1990).

[16] J. W. Connolly and A. R. Williams, Phys. Rev. B*27*, 5169 (1983).

MICROSCOPICAL DERIVATION OF GINZBURG-LANDAU-TYPE FUNCTIONALS FOR ALLOYS AND THEIR APPLICATION TO STUDIES OF ANTIPHASE AND INTERPHASE BOUNDARIES

I. R. Pankratov and V. G. Vaks

Russian Research Centre "Kurchatov Institute", Moscow 123182, Russia

INTRODUCTION

Studies of inhomogeneous alloys attract interest from both fundamental and applied points of view, in particular, in connection with the microstructural evolution during phase transformations [1-14]. Typical inhomogeneities in such problems are antiphase or interphase boundaries (APBs or IPBs) which separate the differently ordered domains or the different phases. Both, experimental and theoretical studies show that in situations of practical interest the APB or IPB width usually notably exceeds the interatomic distance [5-14]. Therefore, Ginzburg-Landau (GL) type gradient expansions can be used to describe the free energy of such states even though the order parameters and concentration variations here are typically not small, contrary to assumptions of the standard GL theory. Employing such generalized GL functionals (suggested first by Cahn and Hilliard [1]) is now referred to as the phase-field method, and it is widely used for most different systems, see e. g. [7-9]. However, a number of simplifying asumptions are usually employed in this phenomenological approach, and so its relation to more consistent theoretical treatments remains unclear. Recently, microscopical cluster methods have been developed for inhomogeneous alloys [10-15]. Below we use these methods to derive the GL functionals and then apply them for studies of APBs and IPBs.

DERIVATION OF GINZBURG-LANDAU FUNCTIONALS FROM CLUSTER EXPANSIONS FOR FREE ENERGY

To be definite, we consider a binary alloy A_cB_{1-c} at $c \leq 1/2$. Various distributions of atoms over lattice sites i are described by the mean occupations $c_i = \langle n_i \rangle$ where n_i

is unity when the site i is occupied by atom A and zero otherwise, while averaging is taken, generally, over the space- and time-dependent distribution function [10]. The free energy $F\{c_i\}$ in the cluster description can be written as a series [15]:

$$F = \sum_i f^i = \sum_i \Big(F_1^i + \sum_j F_2^{ij} + \cdots \sum_{j,\ldots k} F_m^{ij\ldots k} \Big). \tag{1}$$

Here f^i is the free energy per site i; $F_1^i = T[c_i \ln c_i + (1-c_i)\ln(1-c_i)]$ is the mixing entropy contribution; $F_l^{i\ldots k} = F_l(c_i, \ldots c_k)$ is the contribution of interactions within l-site cluster of sites $i, \ldots k$; and m is the maximum cluster size considered. The simplest mean-field approximation (MFA) and the pair-cluster one (PCA) correspond to neglecting many-site contributions $F_{m>2}$ in (1), while in a more refined, tetrahedron cluster approximation – TCA (that should be used, in particular, to adequately describe the $L1_2$ and $L1_0$-type orderings [10]) Eq. (1) includes also 4-site terms F_4^{ijkl} [15].

For the homogeneous ordered structure, the mean occupation $c_j = c(\mathbf{r}_j)$ at site j with the lattice vector $\mathbf{r}_j$ can be written as a superposition of concentration waves with some superstructure vectors $\mathbf{k}_s$ [9-13]:

$$c_j = c + \sum_s \eta_s \exp(i\mathbf{k}_s\mathbf{r}_j) \equiv \sum_p \eta_p \exp(i\mathbf{k}_p\mathbf{r}_j). \tag{2}$$

Here amplitudes η_s can be considered as order parameters; the last expression includes also term with $\eta_p = c$ and $\mathbf{k}_p = 0$; and for simplicity both parameters η_s and factors $\exp(i\mathbf{k}_s\mathbf{r}_i)$ are supposed to be real which is the case, in particular, for the B2, $L1_0$ and $L1_2$-type order. For weakly inhomogeneous states, amplitudes η_p in (2) are not constants but smooth functions of coordinates $\mathbf{r}_i$. Thus, functions $f^i\{c_j\}$ in (1) can be expanded in powers of differences $\delta c_j = \sum_p \delta\eta_p^j \exp(i\mathbf{k}_p\mathbf{r}_j)$ where

$$\delta\eta_p^j = \eta_p^j - \eta_p^i = \mathbf{r}_{ji}\nabla\eta_p + \frac{1}{2} r_{ji}^\alpha r_{ji}^\beta \nabla_{\alpha\beta}\eta_p. \tag{3}$$

Here $\nabla\eta_p = \partial\eta_p^i/\partial r_i^\alpha$, $\nabla_{\alpha\beta}\eta_p = \partial^2\eta_p^i/\partial r_i^\alpha \partial r_i^\beta$, $\mathbf{r}_{ji} = (\mathbf{r}_j - \mathbf{r}_i)$, and the summation over repeated Cartesian indices $\alpha, \beta = 1,2,3$ is implied. After substitution of these expressions into Eq. (1) one can proceed from the summation over i to the integration over continuous variable $\mathbf{r} = \mathbf{r}_i$. Making also standard manipulations with part-by-part integration of terms with $\nabla_{\alpha\beta}\eta_p$ [2], one obtains for the GL functional:

$$F = \frac{1}{v_a} \int d^3r \Big[\sum_{p,q} g_{pq}^{\alpha\beta} \nabla_\alpha\eta_p \nabla_\beta\eta_q + f\{\eta_p\} \Big]. \tag{4}$$

Here, v_a is volume per atom; $f\{\eta_p\}$ is function $f^i\{c_j\}$ in (1) averaged over all sublattices with c_j given by Eq. (2); and $g_{pq}^{\alpha\beta}$ is given by the expression:

$$g_{pq}^{\alpha\beta} = -\frac{1}{2} \sum_j r_{ij}^\alpha r_{ij}^\beta S_{ij}^i \exp[i(\mathbf{k}_p\mathbf{r}_j - \mathbf{k}_q\mathbf{r}_i)]$$
$$+ \frac{1}{2} \sum_{k,j;\, j\neq i} r_{ki}^\alpha (r_{ji}^\beta - r_{ki}^\beta) S_{kj}^i \exp[i(\mathbf{k}_p\mathbf{r}_j - \mathbf{k}_q\mathbf{r}_i)] \tag{5}$$

where $S_{kj}^i = \partial^2 f^i/\partial c_k \partial c_j$. Note that the last term of (5) is nonzero only when all three sites i, j and k are different, and so it is present only when the non-pairwise contributions $F_{m>2}^{ij\ldots k}$ in (1) are taken into account, such as the TCA terms F_4^{ijkl} [15].

For the $L1_2$ and $L1_0$ phases in FCC alloys, Eq. (2) includes waves with three vectors $\mathbf{k}_s$: $\mathbf{k}_1 = (100)2\pi/a$, $\mathbf{k}_2 = (010)2\pi/a$, and $\mathbf{k}_3 = (001)2\pi/a$ where a is the lattice constant [8-13]. The local order within APBs in these phases can be described by the distribution of amplitudes of these waves (η_1, η_2, η_3) of the type (ζ, η, η) and corresponds to the tetragonal symmetry [6,10,13]. To illustrate the form of terms $g_{pq}^{\alpha\beta}$ in (5), we present their MFA and PCA expressions for this type local order. Tensors $g_{pq}^{\alpha\beta}$ here can be described in terms of their "transverse" and "anisotropic" components, $g_{pq}^{\perp} = g_{pq}^{22} = g_{pq}^{33}$ and $g_{pq}^{a} = (g_{pq}^{11} - g_{pq}^{22})$. Using for simplicity the 2-neighbor-interaction model: $v_{n>2} = 0$, one obtains in the MFA:

$$\begin{aligned} g_{\zeta\zeta}^{\perp} &= -\frac{1}{2}a^2 v_2; \qquad g_{\zeta\zeta,\eta\eta}^{a} = \pm\frac{1}{2}a^2 v_1; \qquad g_{\eta\eta}^{\perp} = \frac{1}{2}a^2(v_1 - 2v_2); \\ g_{cc}^{\perp} &= -\frac{1}{2}a^2(v_1 + v_2); \qquad g_{cc}^{a} = g_{\eta c,\zeta c,\eta\zeta}^{\perp,a} = 0, \end{aligned} \tag{6}$$

while the PCA expressions for $g_{pq}^{\perp}$ and g_{pq}^{a} are:

$$\begin{aligned} g_{\zeta\zeta,cc}^{\perp} &= \frac{1}{4}a^2(-\varphi_1^+ \pm \psi_1^+ - \chi_2^+ - \varphi_2^{dd}); \qquad g_{\zeta c}^{a} = \frac{1}{4}a^2\varphi_1^-; \\ g_{\zeta\zeta,cc}^{a} &= \frac{1}{4}a^2(\varphi_1^+ \pm \psi_1^+); \qquad g_{\zeta c}^{\perp} = \frac{1}{4}a^2(-\varphi_1^- - \chi_2^+ + \varphi_2^{dd}); \\ g_{\eta\eta}^{\perp} &= \frac{1}{2}a^2(\varphi_1^{ab} - 2\chi_2^+); \qquad g_{\eta\eta}^{a} = -\frac{1}{2}a^2\varphi_1^{ab}; \\ g_{\eta\zeta,\eta c}^{\perp} &= \frac{1}{4}a^2(\pm\psi_1^- - 2\varphi_2^-); \qquad g_{\eta\zeta,\eta c}^{a} = \pm\frac{1}{4}a^2\psi_1^-. \end{aligned} \tag{7}$$

Here plus or minus in $(\pm)$ corresponds to the first or the second pair of lower indices in the left-hand-side of Eqs. (6)-(7);

$$\begin{aligned} \varphi_n^{ij} &= -Tf_n/R_n^{ij}; \qquad R_n^{ij} = [1 + 2f_n(c_i + c_j - 2c_ic_j) + f_n^2(c_i - c_j)^2]^{1/2}; \\ \varphi_n^{\pm} &= \frac{1}{2}(\varphi_n^{ab} \pm \varphi_n^{dd}); \qquad \psi_n^{\pm} = \frac{1}{2}(\varphi_n^{ad} \pm \varphi_n^{bd}); \qquad \chi_n^{\pm} = \frac{1}{2}(\varphi_n^{aa} \pm \varphi_n^{bb}); \end{aligned} \tag{8}$$

$f_n = \exp(-v_n/T) - 1$ is the Mayer function; and index i or j equal to a, b or d corresponds to the mean occupation c_i or c_j of one of three different sublattices:

$$c_a = c + \zeta + 2\eta; \quad c_b = c + \zeta - 2\eta; \quad c_d = c - \zeta. \tag{9}$$

If the TCA is used, the nearest-neighbor contributions φ_1^{ij} in Eqs. (7) are replaced by the relevant tetrahedron contributions S_{ij}^{i} presented in Ref. [16]. For the B2 order, there is only one order parameter $\eta_s = \eta$ [10]; terms $g_{pq}^{\alpha\beta}$ have a cubic symmetry: $g_{pq}^{\alpha\beta} = \delta_{\alpha\beta}g_{pq}$; and the MFA and PCA expressions for g_{pq} are similar to those for $g_{pq}^{\perp}$ in Eqs. (6) and (7).

EQUATIONS RELATING THE LOCAL COMPOSITION AND LOCAL ORDER WITHIN APB

Let us now consider the case of a plane APB (or IPB) when parameters η_p in (2) depend only on the distance $\xi = \mathbf{r}\mathbf{n}_0$ where $\mathbf{n}_0 = (\cos\alpha, \sin\alpha\cos\varphi, \sin\alpha\sin\varphi)$ is normal to the APB plane. To find the equilibrium structure, one should minimize the functional (4) with respect to functions $\eta_p(\xi)$ at the fixed total number of atoms [2]. Let us first consider the APB between two B2-ordered domains. Then the variational equations for the order parameter $\eta(\xi)$ and the concentration $c(\xi)$ have the form:

$$\begin{aligned} g_{\eta\eta}\eta'' + g_{\eta c}c'' + \frac{1}{2}g_{\eta}^{\eta\eta}(\eta')^2 + g_c^{\eta\eta}\eta'c' + (g_c^{\eta c} - \frac{1}{2}g_{\eta}^{cc})(c')^2 &= \frac{1}{2}f_{\eta}; \\ g_{\eta c}\eta'' + g_{cc}c'' + (g_{\eta}^{\eta c} - \frac{1}{2}g_c^{\eta\eta})(\eta')^2 + g_{\eta}^{cc}\eta'c' + \frac{1}{2}g_c^{cc} &= \frac{1}{2}(f_c - \mu). \end{aligned} \tag{10}$$

Here, prime means taking derivative with respect to ξ; the lower index η or c means taking derivative with respect to η or c; and μ is the chemical potential. At $\xi \to \infty$ functions c and η tend to their equilibrium values, c_0 and $\eta_0(c_0)$.

Multiplying the first and second Eq. (10), respectively, by η' and c'; summing them; and integrating the result, we obtain the first integral of this system of equations:

$$g_{\eta\eta}(\eta')^2 + 2g_{\eta c}\eta' c' + g_{cc}(c')^2 = \Omega(\eta, c) \tag{11}$$

where Ω is the non-gradien part of the local excess grand canonical potential per atom,

$$\Omega = f(\eta, c) - f_0 - \mu(c - c_0), \tag{12}$$

and index zero at the function means its value at $\eta = \eta_0$ and $c = c_0$.

In what follows, it is convenient to consider the order parameter η as an independent variable, while c and Ω as its functions determined by Eqs. (10)-(12). Then the dependence $\eta'(\eta)$ is determined by Eq. (11):

$$\eta' = d\eta/d\xi = (\Omega/G)^{1/2} \tag{13}$$

where G is $(g_{\eta\eta} + 2g_{\eta c}\dot{c} + g_{cc}\dot{c}^2)$, and $\dot{c}$ is $dc/d\eta$. Using Eq. (13) one can eliminate η' in the system of equations (10) and obtain the differential equation for $c(\eta)$, to be called the composition-order equation (COE):

$$[\ddot{c}\,(g_{\eta c}^2 - g_{cc}g_{\eta\eta}) + \Phi]\,2\,\Omega\,/G = (\mu - f_c)(g_{\eta\eta} + g_{\eta c}\dot{c}) + f_\eta(g_{\eta c} + g_{cc}\dot{c}) \tag{14}$$

where Φ is a linear function of derivatives g_r^{pq}:

$$\begin{aligned}\Phi = (g_{\eta c} \;\;+\;\; g_{cc}\dot{c})\Big[\frac{1}{2}g_\eta^{\eta\eta} + g_c^{\eta\eta}\dot{c} + \Big(g_c^{\eta c} - \frac{1}{2}g_\eta^{cc}\Big)\dot{c}^2\Big] - \\ -(g_{\eta\eta} \;\;+\;\; g_{\eta c}\dot{c})\Big(g_\eta^{\eta c} - \frac{1}{2}g_c^{\eta\eta} + g_\eta^{cc}\dot{c} + \frac{1}{2}g_c^{cc}\dot{c}^2\Big).\end{aligned} \tag{15}$$

Because of the equilibrium conditions: $f_\eta^0 = 0$, $f_c^0 = \mu$, function $\Omega(\eta)$ (12) at $\eta \to \eta_0$ is proportional to $(\eta - \eta_0)^2$. Therefore, the initial value $\dot{c}(\eta_0)$ can be found by taking the $\eta \to \eta_0$ limit of Eq. (14).

For the given solution $c(\eta)$ of COE, the coordinate dependence $\eta(\xi)$ is determined by integrating Eq. (13):

$$\xi = \xi_1 + \int_{\eta_1}^{\eta} d\eta\,(G/\Omega)^{1/2} \tag{16}$$

where the reference point ξ_1 is determined by the choice of value $\eta_1 = \eta(\xi_1)$. For the symmetrical APB for which $\eta \to \pm\eta_0$ at $\xi \to \pm\infty$, functions c, Ω and G are even in η, and it is natural to put $\xi_1 = 0$ at $\eta_1 = 0$. But for an IPB separating the ordered and the disordered phases, values $\eta_1 \to 0$ correspond to $\xi \to (-\infty)$, and so ξ_1 should be chosen at some intermediate value η_1.

The surface energy σ and the surface segregation Γ is the excess of the grand canonical potential and of B atoms, respectively, per unit area [2]. Taking into account Eqs. (11)-(13) one obtains for the surface energy:

$$\sigma = \frac{2}{v_a}\int_{\eta_{\min}}^{\eta_0} d\eta\,(G\,\Omega)^{1/2} \tag{17}$$

where $\eta_{\min}$ is $(-\eta_0)$ for an APB and zero for an IPB, while the surface segregation is given by the expression:

$$\Gamma = \frac{1}{v_a}\int_{-\eta_0}^{\eta_0} d\eta\,(c_0 - c)\,(G/\Omega)^{1/2}. \tag{18}$$

For a symmetrical APB, the integral in (17) or (18) can be written as twice the integral over positive η.

Relations similar to Eqs. (11)-(18) can also be derived for phases with several order parameters, such as the $L1_2$ or $L1_0$ phase. In particular, for an APB separating two $L1_2$-ordered domains, the order parameters (η_1, η_2, η_3) have the form (ζ, η, η) mentioned above with the limiting values (η_0, η_0, η_0) and $(\eta_0, -\eta_0, -\eta_0)$ at $\xi \to \pm\infty$. The variational equations and their first integral have the form analogous to Eqs. (10) and (11) but include three functions, $c(\xi)$, $\zeta(\xi)$ and $\eta(\xi)$. It is again convenient to consider c and ζ as functions of η and obtain a system of equations for $c(\eta)$ and $\zeta(\eta)$ analogous to COE (14). Equations for $\xi(\eta)$, σ and Γ preserve their form (16)-(18) but $G(\eta)$ now includes the derivative $\dot{\zeta} = d\zeta/d\eta$ and six functions g_{pq}, which are related to $g_{pq}^{\perp,a}$ in Eqs. (6), (7) as follows:

$$g_{pq}(\alpha) = g_{pq}^{\perp} + g_{pq}^{a}\cos^2\alpha, \tag{19}$$

where α is the angle between the APB orientation and the local tetragonality axis. When the nearest-neighbor interaction v_1 much exceeds the rest ones (as in CuAu-based alloys [10-13]), the functions $g_{pq}(\alpha)$ are highly anisotropic, which is illustrated by Eqs. (6): $g_{\eta\eta} \sim \sin^2\alpha$; $g_{\zeta\zeta} \sim \cos^2\alpha$. It results in a notable anisotropy of distributions of APBs, including the presence of many low-energy "conservative" APBs with $\alpha \simeq 0$ in the $L1_2$ phase and $\alpha \simeq \pi/2$ in the $L1_0$ phase, as well as a peculiar alignment of APBs in "twinned" $L1_0$ structures [10-13]. In more detail, COE and Eqs. (16)-(19) for the $L1_2$ and $L1_0$ phases will be discussed elsewhere.

Let us also note that for an APB separating two $L1_0$-ordered domains with the same tetragonality axis, the local order parameters (η_1, η_2, η_3) vary from values $(\zeta_0, 0, 0)$ at $\xi \to \infty$ to $(-\zeta_0, 0, 0)$ at $\xi \to -\infty$. Thus the GL equations always have the solution $(c, \eta_1, \eta_2, \eta_3)$ of the form $(c, \zeta, 0, 0)$ with just two nonzero functions, $c(\xi)$ and $\zeta(\xi)$. The "symmetry breaking " solution with non-zero η_2 and η_3, in particular, that of the type (c, ζ, η, η) mentioned above, can exist, too, and it seems usually to have the lower energy. However, the TCA-based simulations of kinetics of $L1_0$ ordering [11,12] show that these "not-necessary" components η_α are usually small: $|\eta_\alpha| \ll |\zeta|$ (see, for example, figure 12 in [11]), and so the "symmetrical" solution of COE of the form $(c, \zeta, 0, 0)$ can provide a sufficiently accurate description of such APBs. Thus, in this work for simplicity we describe APBs in the $L1_0$ phase with the simplest form of COE (14) which includes only one function $c(\zeta)$ to be determined.

COMPARISON WITH THE PHASE-FIELD DESCRIPTION

There are three main assumptions used in the conventional versions of the phase-field method (PFM) [7-9]:

(i) Coefficients g_{pq} at the gradient terms of GL functionals are supposed to not depend on the local values of parameters η_r.

(ii) Non-diagonal terms g_{pq}, such as $g_{\eta\zeta}$, $g_{\eta c}$, etc., as well as an "anisotropic" term g_{cc}^{a}, are put zero "on considerations of symmetry" (which is a formal consequence of (i)).

(iii) The free energy density $f(\eta_p)$ for a homogeneous alloy is approximated by a polynom with coefficients fitted to the equilibrium values η_p^0 and/or some other experimental parameters.

Eqs. (6)-(7) show that the assumptions (i) and (ii) may correspond only to the simplest MFA, while in more accurate approaches, such as the PCA and TCA, the

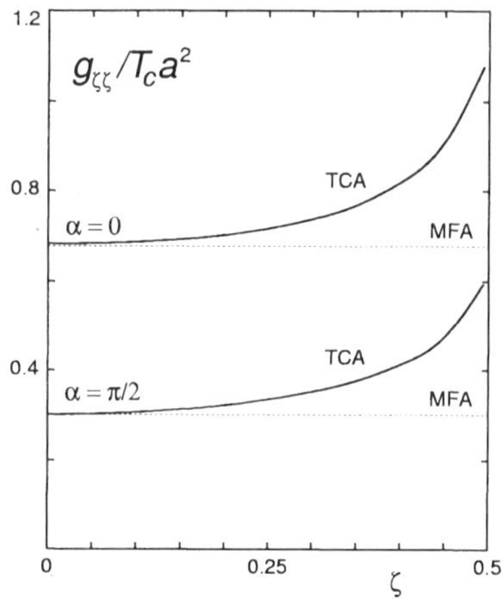

Figure 1: Reduced values $g_{\zeta\zeta}/T_c a^2$ versus order parameter ζ for an APB separating two $L1_0$-ordered domains with the same tetragonality axis for the Ni–Al-type alloy model at $c = 0.5$, the reduced temperature $T' = T/T_c = 0.7$, $\eta = 0$, and different angles α between the APB orientation and the tetragonality axis.

dependences $g_{pq}^{\alpha\beta}(\eta_r)$ can be significant. Validity of the polynomial interpolation (iii) at large order parameter values close to the saturation value should also be examined.

To get an idea about a possible scale of errors brought by the assumptions (i)–(iii), we calculated properties of APBs in the $L1_0$ phase for a realistic alloy model (to be referred to: "Ni–Al-type model") with the interaction parameters estimated by Chassagne et al. [17] from their experimental data for Ni–Al alloys: $v_2/v_1 = 0.125$; $v_3/v_1 = -0.021$; and $v_4/v_1 = -0.123$. Then we compared our results with those obtained under the PFM-type description of the same model. Some results of this comparison are presented in figures 1-3. The comparison shows:

(i) Variation of terms g_{pq} under variation of local parameters η_p within APBs can be significant. In particular, for the Ni-Al-type model at $T' = T/T_c = 0.7$, these variations reach 50-100 % for the $L1_0$ phase (see figure 1), and 20–30 % for the $L1_2$ phase.

(ii) The "non-diagonal" g_{pq} and g_{cc}^a terms can also be important. In particular, for the Ni-Al-type model at $T' = T/T_c = 0.7$, terms g_{cc}^a are typically not small in the $L1_0$ phase: $g_{cc}^a \sim (0.4 - 0.6)\, g_{cc}^{\perp}$, while $g_{\eta\zeta}$ are noticeable in the $L1_2$ phase at $c > 0.25$ where $g_{\eta\zeta} \sim (0.2 - 0.3)\, g_{\zeta\zeta}$.

(iii) Possible errors of polinomial inerpolations of $f(\eta_p)$ are illusrated in figures 2 and 3. Curve PFM-a in figure 2 corresponds to the interpolation of the TCA calculated $f = f_{\rm TCA}(\zeta)$ by the polynom $a(-\zeta^2 + \zeta^4/2\zeta_0^2)$ fitted to the "true" equilibrium value $\zeta_0^{\rm TCA}$ and the "true" curvature at $\zeta = 0$, while the curve PFM-b is fitted to the equilibrium value $\zeta_0^{\rm MFA}$ and the "true" MFA value of the coefficient b at ζ^4 in the Landau expansion at small ζ. Figure 3 shows that using such interpolations for the description of properties of APB can lead to notable errors.

The origin of the large distortion of the form of $f(\zeta)$ at large values $\zeta_0(T) \sim \zeta_0^{\max} = c$ by polynomial interpolations seems to be mainly related to the mixing entropy term $F_i(c_i)$ in the cluster expansion (1). At small $(c - \zeta)$ this term includes the "dilute solution" singularity $\sim (c - \zeta)\ln(c - \zeta)$ which describes a sharp variation of $f(\zeta)$ near

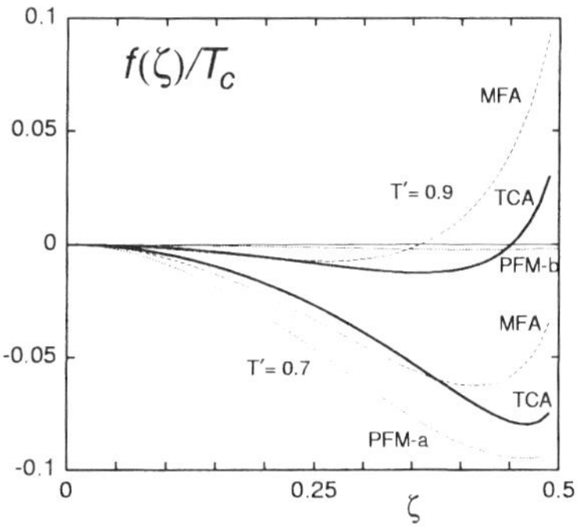

Figure 2: The reduced free energy density $f(\zeta)/T_c$ in the $L1_0$ phase at $c = 0.5$ and $\eta = 0$ for the Ni–Al-type model by Chassagne et al. (1989) at different reduced temperatures $T' = T/T_c$. The upper MFA and TCA curves correspond to $T' = 0.9$, and the lower ones, to $T' = 0.7$. Curve PFM-b corresponds to the polynomial interpolation $f = b(\zeta^4 - 2\zeta_0^2\zeta^2)$ with $b = b_{\mathrm{MFA}} = T_c/12$, and curve PFM-$a$, to the interpolation $f = a(-\zeta^2 + \zeta^4/2\zeta_0^2)$ with $a = -(\partial f_{\mathrm{TCA}}/\partial\zeta^2)_{\zeta=0}$ where both equilibrium value ζ_0 and f_{TCA} correspond to the TCA calculation at $T' = 0.7$.

ζ_0. At the same time, just this region is important for both the fit of the interpolation parameters and the calculations of characteristics of APB. Under the polynomial interpolation this sharp variation is lost, which can lead to significant errors in describing the APB properties. Therefore, for the PFM-type treatments, one may suggest to explicitly write the mixing entropy term in the free energy density f, using the polynomial interpolation only for the rest contributions being more smooth functions of local parameters η_p.

SEGREGATION AT APB NEAR THE SECOND-ORDER TRANSITION LINE

Let us discuss some applications of Eqs. (14)-(19). First, we consider the case when the equilibrium order parameter η_0 is small. Then the function f in (4) can be written as the Landau expansion:

$$f(\eta, c) = \varphi + a\eta^2 + b\eta^4 + d\eta^6 \tag{20}$$

where φ, a, b and d are some functions of concentration c and temperature T. The equilibrium value η_0 is determined by the equation $f_\eta^0 = 0$, while the ordering spinodal $T = T_s(c)$ (the disordered phase stability limit) is determined by the equation: $a(c,T) = 0$. Small values η_0 under consideration correspond to c_0, T points near the ordering spinodal where $a(c_0, T)$ is small.

It is clear from both physical considerations and the results below that the difference $(c - c_0)$ at small η_0 is also small. Therefore, functions $(f_c - \mu)$ and f_η in Eq. (14) can be expanded in powers of $(c - c_0)$, η and η_0:

$$f_c - \mu = a_c^0(\eta^2 - \eta_0^2) + \varphi_{cc}^0(c - c_0) + \dots \tag{21}$$
$$f_\eta = 4\eta\, b_0(\eta^2 - \eta_0^2) + 2\eta\, a_c^0(c - c_0) + \dots \tag{22}$$

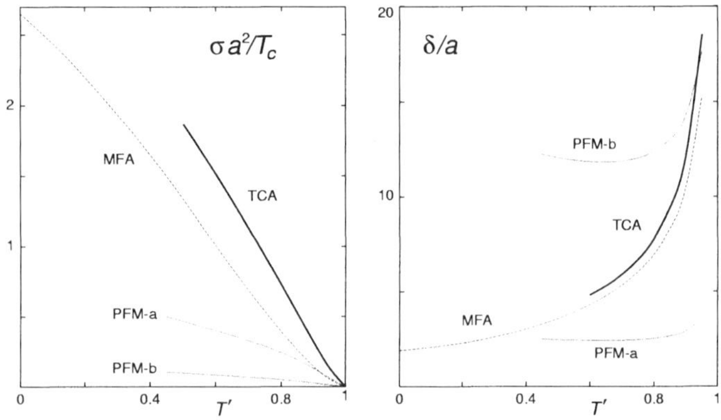

Figure 3: Left: Temperature dependence of the reduced surface energy $\sigma_r = \sigma a^2/T_c$ for the same APB as in figure 1 at $\alpha = \pi/4$. Curve PFM-b corresponds to the polynomial interpolation $f = b(\zeta^4 - 2\zeta_0^2\zeta^2)$ with $b = T_c/12$ and $\zeta_0 = \zeta_0^{\rm TCA}(T)$. Right: Same as in the left figure but for the APB width $\delta(T)$ defined as the distance between the points ξ_+ and ξ_- for which: $\eta(\xi_+) = 0.9\,\eta_0$, and $\eta(\xi_-) = -0.9\,\eta_0$.

Here φ_{cc}^0 is $(\partial^2\varphi/\partial c^2)_0$, and the dots mean terms of higher orders in η_0^2. The analogous expansion of $\Omega(c,\eta)$ starts with terms bilinear in $(c - c_0)$ and $(\eta^2 - \eta_0^2)$, while for functions $g_{\eta c}$ and $\dot{c}$ the expansions start with terms linear in η. Thus, the terms with Ω and f_η in (14) are proportional to η_0^4, being small compared to $(f_c - \mu) \sim \eta_0^2$, and COE is reduced to the equation $f_c = \mu$, which yields:

$$c_0 - c(\eta) = (\eta_0^2 - \eta^2)(-a_c^0)/\varphi_{cc}^0. \tag{23}$$

Taking the derivative of equation $a(c,T) = 0$, one obtains $(-a_c^0) = T_s'\alpha$, where α is $(\partial a/\partial T)_0$ and $T_s' = dT_s/dc$. Thus, the surface segregation at APB is proportional to the ordering spinodal slope $T_s'(c_0)$, and so it decreases with approaching the critical point where $T_s' = 0$.

Let us now suppose the alloy state c_0, T in the c, T plane to be close to the second-order transition line $T_s(c)$ far from the possible tricritical points. Then the higher-order terms in expansions (20)-(22) can be neglected, and for the function $\Omega(\eta)$ (12) such expansion yields:

$$\Omega(\eta) = \tilde{b}\,(\eta_0^2 - \eta^2)^2; \qquad \tilde{b} = b_0 - (\alpha\, T_s')^2/2\varphi_{cc}^0. \tag{24}$$

Using Eqs. (16)-(18) and (24) one obtains in this case for $\eta(\xi)$, $c(\xi)$, the APB energy σ and the segregation Γ:

$$\begin{aligned} \eta(\xi) &= \eta_0 \tanh(\xi/\delta); \qquad c_0 - c(\xi) = \eta_0^2\,\lambda \cosh^{-2}(\xi/\delta); \\ \sigma &= \frac{8}{3v_a}\eta_0^3 \left(g\,\tilde{b}\right)^{1/2}; \qquad \Gamma = \frac{2}{v_a}\,\eta_0\,\lambda(g/\tilde{b})^{1/2} \end{aligned} \tag{25}$$

where λ is $\alpha\, T_s'/\varphi_{cc}^0$; g is $g_{\eta\eta}^0$; and $\delta = (g/\eta_0^2\tilde{b})^{1/2}$ is the APB width. These expressions generalize the earlier MFA results [14] to the case of any GL functional. The temperature or the concentration dependence of the segregation Γ at small η_0 is more sharp than that of the APB energy: $\Gamma \propto \eta_0 \sim (T_s - T)^{1/2} \sim (c_0 - c_s)^{1/2}$

where $T_s = T_s(c_0)$ or $c_s = c_s(T)$ correspond to the second-order transition line, while $\sigma \propto \eta_0^3 \sim (T_s - T)^{3/2} \sim (c_0 - c_s)^{3/2}$. Eqs. (24)-(25) also show that the presence of segregation results in a renormalization of the Landau parameter b_0 entering characteristics of APB to the lesser value $\tilde{b}$ given by Eq. (24). It results in a decrease of the APB energy σ and an increase of its width δ and segregation Γ under decreasing temperature T along the ordering spinodal $T = T_s(c)$.

STRUCTURE OF APBS AND IPBS NEAR AND FAR FROM THE TRICRITICAL POINT

The point c_0, T at which both $a(c_0, T)$ in (20) and $\tilde{b}$ in (24) vanish corresponds to the tricritical point c_t, T_t. At $T < T_t$, the second-order transition line $T_s(c)$ in the c, T plane splits into two binodals, $c_{bo}(T)$ and $c_{bd}(T)$, delimiting the single-phase ordered and disordered field, respectively. Such tricritical point is observed, for example, in Fe–Al alloys [4]. At this point, the lowest order terms in Eqs. (22) and (24) vanish, so that one should consider the next-order terms, and function Ω (12) at small $x = (c_0 - c_t)$ and $t = (T - T_t)$ takes the form:

$$\Omega(\eta) = A\,(\eta_0^2 - \eta^2)^2(\eta^2 + h). \tag{26}$$

Here, $h = h(x, t)$ is a linear function of x and t, which can be written in terms of the binodal temperature derivative $c'_{bo} = dc_{bo}/dT$ as: $h = \nu\,(x - t\,c'_{bo})$, while A and ν are some positive constants. Using Eqs. (16)-(18) and (26), one obtains for the characteristics of APB near T_t:

$$\begin{aligned} \eta(\xi) &= \frac{\eta_0 \sinh y}{(\cosh^2 y + \alpha)^{1/2}}; \qquad c_0 - c(\xi) = \frac{\lambda\,\eta_0^2(1+\alpha)}{(\cosh^2 y + \alpha)}; \\ \sigma &= \frac{J(\alpha)}{v_a}(gA\eta_0^4)^{1/2}; \qquad \Gamma = \frac{2\lambda L(\alpha)}{v_a}(g/A)^{1/2}. \end{aligned} \tag{27}$$

Here y is $\xi/\tilde{\delta}$; α is η_0^2/h; while $\tilde{\delta}$, $L(\alpha)$ and $J(\alpha)$ are:

$$\begin{aligned} \tilde{\delta} &= [g/A\eta_0^2(h+\eta_0^2)]^{1/2}; \qquad L(\alpha) = \ln\left[(1+\alpha)^{1/2} + \alpha^{1/2}\right]; \\ J(\alpha) &= \frac{1}{2\alpha^2}\left[(1+4\alpha)L(\alpha) + (2\alpha - 1)(\alpha + \alpha^2)^{1/2}\right]. \end{aligned} \tag{28}$$

The function $h(x, t)$ in (26)-(28) is proportional to the distance in the c, T plane from point c_0, T to the binodal $c_{bo}(T)$, while η_0^2 is proportional to the distance to the ordering spinodal $T_s(c)$. Thus at small $\alpha \ll 1$, Eqs. (26)-(28) turn to (24)-(25) with $\tilde{b} = Ah$ and describe the critical behaviour of APB near $T_s(c)$ discussed above. The opposite case $\alpha \gg 1$ corresponds to the region of "wetting" APBs which has recently received much attention [3,6,18]. The value $h = 0$ corresponds to the ordered state with $c_0 = c_{bo}(T)$, and then COE (14) describes an IPB betwen this state and the disordered state with $\eta_d = 0$ and $c_d = c_{bd}(T)$. Substituting Eq. (26) with $h = 0$ into Eqs. (16) and (17), we obtain for this IPB:

$$\eta(\xi) = \frac{\eta_0}{(1 + e^{-z})^{1/2}}; \qquad c(\xi) - c_d = \frac{\lambda\,\eta_0^2}{(1 + e^{-z})}; \tag{29}$$

$$\sigma_{\mathrm{APB}} = 2\,\sigma_{\mathrm{IPB}}. \tag{30}$$

Here z is ξ/δ_1; $\delta_1 = (g/A)^{1/2}/2\eta_0^2$ is the IPB width; $\sigma_{\mathrm{IPB}} = (gA\eta_0^4)^{1/2}/2v_a$ is the IPB energy; c_d is $c_0 - \lambda\eta_0^2$; and the coordinate $\xi_1 = 0$ in (16) is chosen at $\eta_1 = \eta_0/\sqrt{2}$

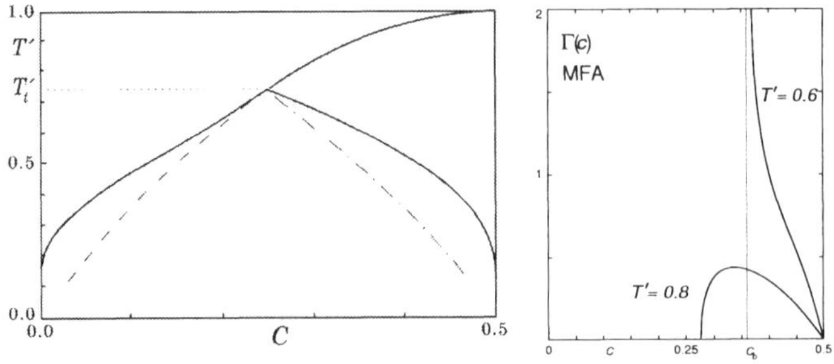

Figure 4: Left: Phase diagram for the alloy model with B2 ordering and tricritical point (B2-t model) for which: $v_2/v_1 = -2.6$, and $v_3/v_1 = 0.4$. Right: Surface segregation $\Gamma(c)$ for the B2-t model at $T > T_t$ (lower curve), and at $T < T_t$ (upper curve).

where $c(\eta_1)$ is $(c_0 + c_d)/2$. Equations (29) show that the order parameter η in the disordered phase decreases with moving from IPB much more slowly than the concentration deviation: $\eta \sim (c - c_d)^{1/2}$. Equations (27)-(30) also show that, in the "wetting" regime of large α, the profiles $\eta(\xi)$ and $c(\xi)$ in Eq. (27) correspond to the presence, at $\xi_\pm = \pm\delta_1 \ln \alpha$, of two almost independent IPBs described by Eqs. (29) (see figures 5a and 5b below). The total width $l \simeq (\xi_+ - \xi_-)$ and the segregation Γ for such APB are proportional to $\ln(1/h)$ while the energy difference $(\sigma_{\mathrm{APB}} - 2\sigma_{\mathrm{IPB}})$ is proportional to $h \ln(1/h)$, which are usual dependences for the wetting regime [19]. Eqs. (27)-(30) specify these relations for the vicinity of tricritical points and enable one to follow the transition from wetting to the critical behaviour of APBs under the variation of T or c_0.

Let us discuss the main qualitative features of segregation at APBs far from the phase transition lines. To this end we use the simple MFA model of an alloy with B2 ordering and tricritical point (B2-t model) for which $v_2/v_1 = -2.6$, $v_3/v_1 = 0.4$, while its MFA phase diagram is shown in the left figure 4. For this model, the coefficient g_{cc} in Eq. (14) vanishes, and so the COE is reduced to the simple nonlinear equation rather than to the differential one. The right figure 4 shows the concentration dependence of the segregation $\Gamma(c)$ at different temperatures. At the stoichiometric composition $c = 0.5$, the segregation is absent because of the symmetry of an alloy A_cB_{1-c} with pair interactions with respect to the interchange of all A and B atoms [2]. Figure 4 illustrates both the above-mentioned sharp increase of $\Gamma(c)$ near $T_s(c)$ at $T > T_t$, and the logarithmic divergence of $\Gamma(c)$ in the wetting regime at $c \to c_b(T)$. Figures 5, a–c illustrate the structure of IPBs between the ordered and disordered phase. We see, in particular, that in the disordered phase, the order parameter decreases with moving from IPB more slowly than the concentration deviation, both near and far from the tricritical point, and so it seems to be a general feature of order-disorder phase boundaries.

Let us now compare the results of the present approach with the more quantitative studies of APBs available. Schmid and Binder [5] applied Monte Carlo methods to investigate the structure of APB for the Fe-Al type model with the phase diagram shown in figure 6. In figures 6 and 7 we compare their results with our TCA and MFA calculations for the same model. Figure 7 shows that the GL approach describes the structure of APB for this model fairly well, though the APB width seems to be

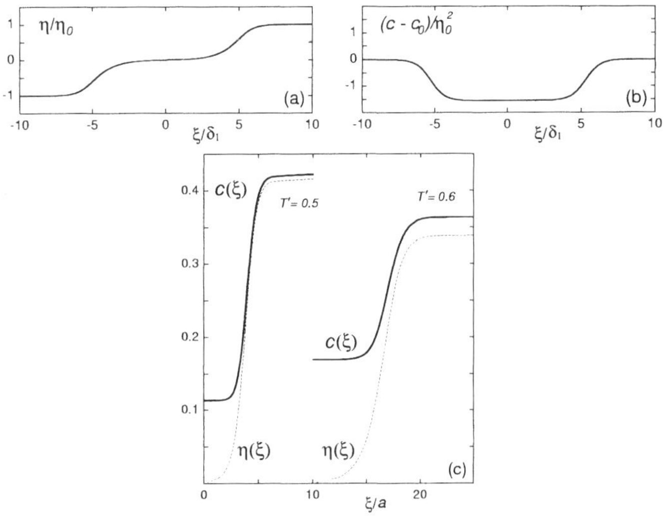

Figure 5: (a) Reduced order parameter η/η_0, and (b) reduced concentration deviation $(c - c_0)/\eta_0^2$, near tricritical point in the wetting regime described by Eqs. (17) at $\alpha = \eta_0^2/h = 10^4$. (c) Profiles $\eta(\xi)$ and $c(\xi)$ for IPBs in the B2-t model at different reduced temperatures $T' < T'_t \simeq 0.75$.

somewhat underestimated.

We also tried to compare our results with the first principle calculation by Asta and Quong [8] for Ti-Al alloys at 1300 K. As these authors did not describe details of their calculations, we used for comparison the above-mentioned Ni-Al-type model in the $L1_0$ phase with the same equilibrium order parameter value $\eta_0 \simeq 0.45$ as that found by Asta and Quong; it corresponds to the reduced temperature $T' = 0.62$ in our model. Figure 8 shows that the main features of structure and segregation at APB in the first-principle calculation for Ti-Al and in our calculation for the Ni-Al-type model seem to be similar, in spite of the difference of the alloys considered (as well as the approximations mentioned in the end of section 3). It may imply that the type of effective interactions in the Ti-Al and Ni-Al systems is not greartly different, while the APB structure, at the given concentration c and temperature T', is not very sensitive to the interaction details. Figure 8 also shows that the GL approach seems to again underestimate the APB width.

In figure 9 we show the concentration dependence of the APB energy $\sigma(c)$ at different temperatures calculated for the Ni-Al-type model and compare it with the similar dependence $\sigma_{\rm AQ}(c)$ found by Asta and Quong [8] for Ti-Al alloys at 1300 K. As the absolute value $\sigma_{\rm AQ}(c)$ depends on the interaction parameters not given in Ref. [8], in figure 9 we present the scaled quantity $\sigma'_{\rm AQ}(c) = \alpha\,\sigma_{\rm AQ}(c)$ with the scaling factor α fitted to our calculated value $\sigma_{\rm TCA}$ at $c = c_0 = 0.5$ and $T'_0 = 0.62$: $\alpha = \sigma_{\rm TCA}(c_0, T'_0)/\sigma_{\rm AQ}(c_0)$. Figure 9 shows the similarity of the first-principle and the TCA (but not MFA) calculated concentration dependences $\sigma(c)$. In addition to that, the comparison of dependences $\sigma_{\rm TCA}(c)$ at different temperatures T' shows that with lowering T' these dependences become more and more sharp, particularly near the stoichiometric concentration $c_0 = 0.5$. It il-

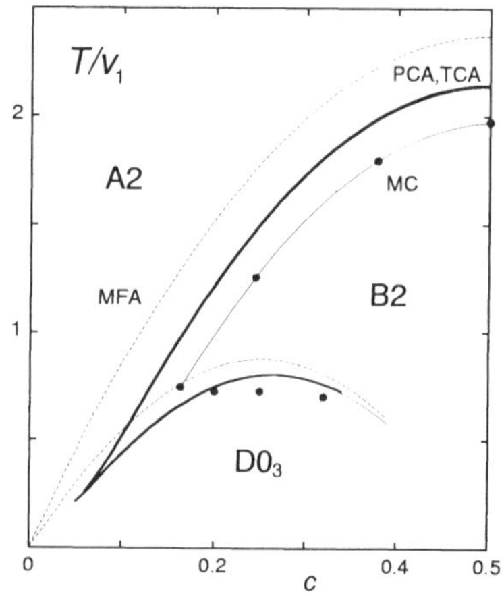

Figure 6: Phase diagram for the Fe–Al-type model by Schmid and Binder [5] for which: $v_2/v_1 = 0.167$, and $v_3/v_1 = -0.208$. Thick line, TCA and PCA; dashed line, MFA; circles and thin line, Monte Carlo results.

lustrates the low-temperature anomalies of APB energies in the short-range-interaction alloy systems, which are discussed below in section 8.

WETTING RELATIONS FOR SYSTEMS WITH SEVERAL ORDER PARAMETERS

Let us now apply Eqs. (14)-(17) to derive the wetting relation (30) for the phases with several order parameters, in particular, for the $L1_2$ phase in equilibrium with the $L1_0$ or the disordered FCC (A1) phase. This problem was discussed by a number of authors [3,6,18] but the general proof seems to be absent yet. Let us first note that in consideration of IPB, the initial condition to COE (14) can be put in either the ordered or the disordered phase, i.e. at (c, η) values equal to either (c_0, η_0) or $(c_d, 0)$, while the solution $c_{\rm IPB}(\eta)$ at either choice is the same and unique. Therefore, in consideration of APB with the same initial values c_0, η_0, the solution $c_{\rm APB}(\eta)$ coincides with $c_{\rm IPB}(\eta)$ at $\eta > 0$, it is $c_{\rm APB}(-\eta)$ at $\eta < 0$, and so Eq. (30) follows from Eq. (17). For the APB or IPB in the $L1_2$ phase, the local order can be described by the parameters (c, ζ, η) mentioned above, and their initial values in COE are (c_0, η_0, η_0), while the final ones are $(c_0, \eta_0, -\eta_0)$, $(c_d, 0, 0)$, and $(c_l, \eta_l, 0)$ for the case of an APB, IPB($L1_2$-A1), and IPB($L1_2$-$L1_0$), respectively, where c_l and η_l correspond to the second binodal $L1_2$-$L1_0$. Therefore, the wetting relation (30) for the $L1_2$-A1 or $L1_2$-$L1_0$ phase equilibrium follows from COE and Eq. (17), just as for the single-order-parameter case. Note, however, that at the given orientation $\mathbf{n}_0$ there are three types of an APB in the $L1_2$ phase with the local order (η_1, η_2, η_3) of the form (ζ, η, η), (η, ζ, η) or (η, η, ζ), and the structure and energy for each type is, generally, different. Therefore, there are at least three types of an IPB($L1_2$-A1) corresponding to "a half" of the relevant APB in the wetting limit. In the course of the kinetical wetting (for example, under A1→A1+$L1_2$ transformations studied in [7,10]) each APB first transforms into two "its own" IPBs, but later on these

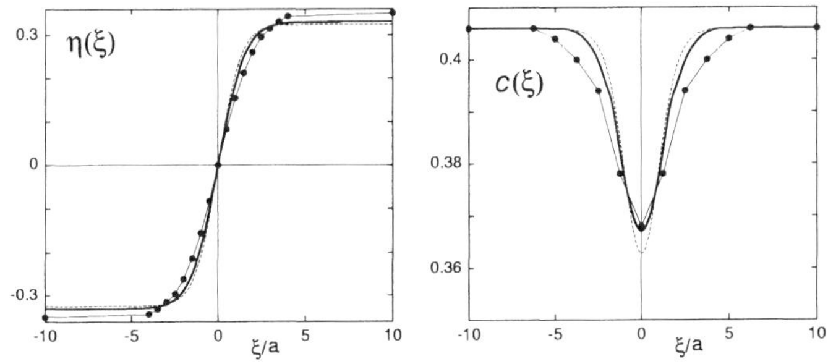

Figure 7: The order parameter $\eta(\xi)$ (left) and the concentration $c(\xi)$ (right) within APB at $T' = T/T_c = 0.76$ and $c_0 = 0.406$ for the same model and same notation as in figure 6.

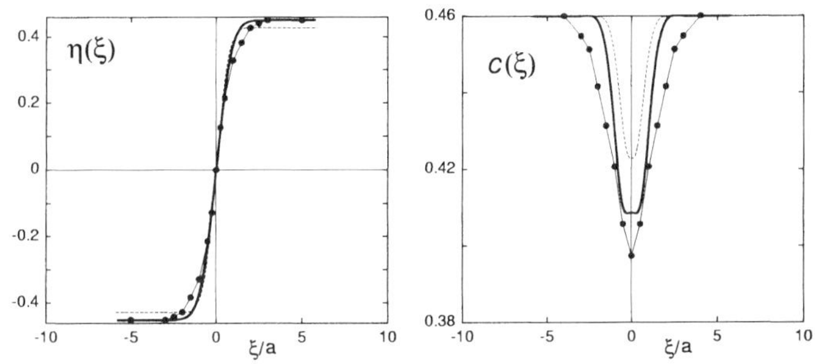

Figure 8: Local order parameter $\eta(\xi)$ (left) and local concentration $c(\xi)$ (right) within APB in the $L1_0$ phase at $c_0 = 0.46$. Thick and dashed line, TCA and MFA calculations for the Ni–Al-type model at $T' = 0.62$; circles, first-principle calculations by Asta and Quong [8] for Ti–Al alloys at $T = 1300$ K.

IPBs can evolve to other types.

The effects of anisotropy under wetting APBs in Eqs. (16)-(18) are described by a factor $G^{1/2}$, while the main contribution to $\Omega(\eta)$ here is determined by the thermodynamic relations. In particular, singular contributions to σ and Γ under wetting $L1_2$-APB by the A1 or $L1_0$ phase correspond to the region of small η where Ω has the same form as in Eq. (26), while $G(\eta)$ is reduced to its first term $g_{\eta\eta}$ as functions $g_{\eta\zeta}$, $g_{\eta c}$, $\dot{\zeta}$ and $\dot{c}$, being odd in η, vanish at small η. Thus the main contributions to the APB width and energy take the form:

$$l \sim g_{\eta\eta}^{1/2} \ln(1/h); \qquad (\sigma_{\text{APB}} - 2\sigma_{\text{IPB}}) \sim g_{\eta\eta}^{1/2}\, h \ln(1/h), \tag{31}$$

where the angular dependence $g_{\eta\eta}$ is given by (19). Therefore, the wetting effects can reveale a significant anisotropy, particularly in the short-range interaction systems. It agrees with some previous results [3,6,18].

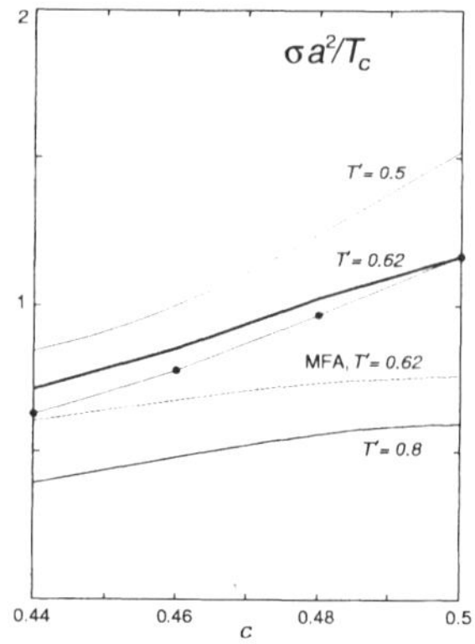

Figure 9: Concentration dependence of the reduced surface energy $\sigma_r(c)$ in the $L1_0$ phase. Thick and dashed line, TCA and MFA calculations for the Ni–Al-type model; circles, first-principle calculations by Asta and Quong [8] for Ti–Al alloys at $T = 1300$ K scaled by the value σ_{TCA} at $c = 0.5$, $T' = 0.62$ as explained in the text.

LOW-TEMPERATURE ANOMALIES OF APB ENERGY

Cahn and Kikuchi [2] noted that for the nearest-neighbor interaction model of a binary ordered alloy A_cB_{1-c} (they considered the B2 order in the BCC lattice), the APB energy $\sigma(c,T)$ at $T \to 0$ should vanish at any non-stoichiometric value $c \neq 0.5$, while at $c = 0.5$, σ monotonously increases with lowering T up to some finite value. This is because the addition of an extra B or A atom to the perfectly ordered stoichiometric structure leads to the breaking of the same number of "favourable bonds" A-B both within the ordered domain and near its APB. It can be easily understood considering, for example, the simplest case of the two-dimensional square lattice for which an addition of an extra B atom to the perfectly ordered "checkered" structure AB breakes 4 favourable bonds A-B irrespectively of the other extra atom positions. Then a vertical or horizontal APB can be formed from the initial single-domain ordered state with the randomly distributed "excess" B atoms in the following two steps: (i) transferring some of these excess B atoms from their random positions so as to form a continuous vertical or horizontal row of B atoms separating two "in-phase" ordered domains, and (ii) displacing these two domains with respect to each other along the row by one lattice constant. Both operations (i) and (ii) do not change the total alloy energy E, and so the final state with an APB and two antiphase-ordered domains has the same energy as the initial state. Therefore, the free energy loss, $\delta F = \delta(E - TS)$, under the creation of such APB is related just to the entropic contribution, which vanishes at $T \to 0$. It implies that the concentration dependence $\sigma(c)$ near the stoichiometric value $c = 0.5$ at low T should be very sharp becoming formally discontinuos at $T = 0$. It is illustrated by curves $\sigma(c)$ for the nearest-neighbor-interaction (NNI) model in figure 10.

To get an idea about the scale of similar low-temperature anomalies in more realistic systems in which the not-nearest-neighbor-interactions $v_{n>1}$ are also present, we used COE to calculate the concentration and temperature dependences $\sigma(c,T)$ for the two models:

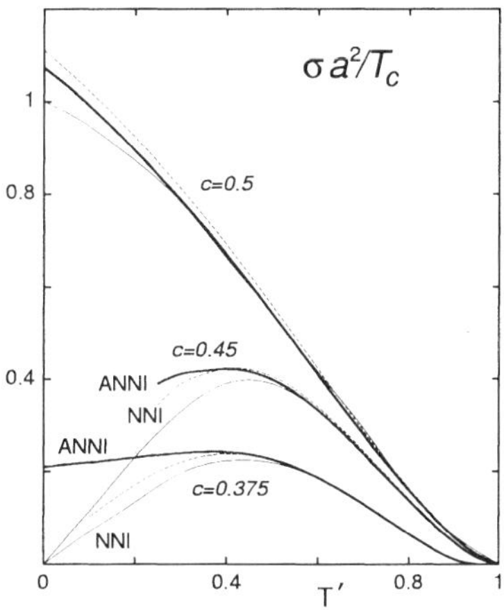

Figure 10: Temperature dependence of the reduced surface energy $\sigma_r(T)$ for some alloy models with B2 ordering calculated in the MFA. Thin lines, the discrete lattice calculations by Cahn and Kikuchi [2] for the nearest-neighbor-interaction (NNI) model; dashed lines, present calculations for the NNI-model; thick lines, present calculations for the almost-nearest-neighbor-interaction (ANNI) model for which: $v_2/v_1 = v_3/v_1 = 0.05$, and $v_4/v_1 = 0.01$.

(i) The "almost-nearest-neighbor-interaction" (ANNI) model of a B2-ordered alloy in the BCC lattice, with the following pair interaction parameters: $v_2/v_1 = v_3/v_1 = 0.05$, and $v_4/v_1 = 0.01$, being treated in the MFA for comparison with the analogous discrete lattice calculations by Cahn and Kikuchi [2] for the NNI model.

(ii) The above-mentioned Ni–Al type model of an $L1_0$-ordered alloy in the FCC lattice, being treated in both the MFA and TCA.

Some results of these calculations are presented in figures 10 and 11. Let us first discuss the results for the NNI model presented in figure 10. The comparison of our GL approach (dashed lines) with the discrete lattice calculations by Cahn and Kikuchi (thin lines) shows that the neglect of the dicrete lattice structure leads to some overestimation of the APB energy. However, at high and intermediate temperatures, $T' \gtrsim 0.5$, the relevant errors are virtually negligible, and they remain to be small up to rather low $T' \gtrsim 0.2$, particularly at small deviations of concentration c from the stoichiometric value $c_0 = 0.5$. The presence of not-nearest-neighbor interactions (characteristic of real alloys) corresponds to an increase of the interaction range with respect to to the NNI model, which should result in a higher accuracy of the continuous approximations. Therefore, for real alloys, the discrete lattice effects shoud be still less than those for the NNI model shown in figure 10.

The results for the ANNI model (thick curves in figure 10) show that the presence of even small not-nearest-neighbor interactions leads to the drastic change in the low-temperature behaviour of the APB energy: at $T \to 0$, the $\sigma(T, c)$ value becomes finite and concentration-dependent. However, with elevating temperature the influence of such small interactions rapidly weakens, and at $T' \gtrsim 0.3 - 0.4$, the APB energies for the NNI and ANNI models do virtually coincide.

Figure 11 (as well as figure 9) shows the temperature and concentration dependences of APB energies calculated for a more realistic, Ni-Al-type model with the $L1_0$ order. The nearest-neighbor interaction v_1 for this model much exceeds the other interactions

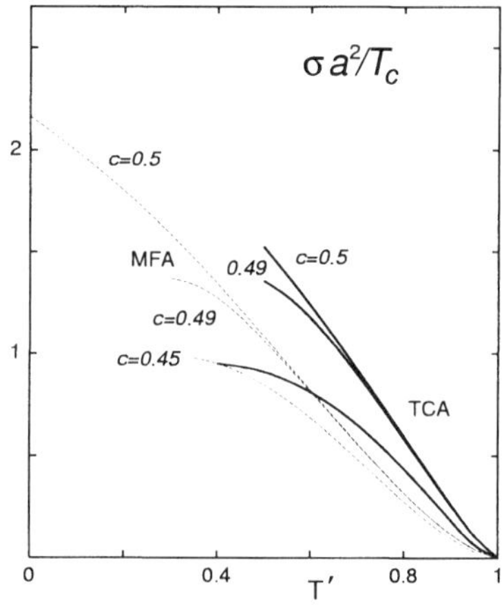

Figure 11: Temperature dependence of the reduced surface energy $\sigma_r(T)$ for (100)-oriented APBs under $L1_0$ ordering in the Ni–Al-type alloy model. Thick lines, TCA; dashed lines, MFA.

$v_{n>1}$ (see section 4), and this is also typical for most of real alloys. Therefore, the above-discussed anomalies in the dependences $\sigma(c, T)$ should be manifested for the Ni-Al-type model, and our calculations for this model can also illustrate the scale of similar anomalies for real alloys.

Figure 11 shows that the general form of these anomalies, both in the MFA and TCA calculations, appears to be the same as that for the ANNI model in figure 10. In particular, at all non-soichiometric concentrations $c \neq 0.5$, the characteristic maxima in the temperature dependences $\sigma(T)$ are also present for the Ni-Al-type model. Figure 11 also shows that in more accurate, TCA calculations, the anomalies have a larger scale and are manifested at higher temperatures compared to more crude, MFA calculations. In addtion to that, Figures 11 and 9 show that the "sharpening" of the concentration dependences $\sigma(c)$ with lowering T is quite pronounced even at not very low temperatures, $T' \gtrsim 0.5 - 0.6$, and so these effects probably can be observed in experiments with the properly relaxed and equilibrated APBs [8].

ACKNOWLEDGEMENTS

The authors are much indebted to Georges Martin for numerous stimulating discussions. The work was supported by the Russian Fund of Basic Research under Grants No. 00-02-17692 and 00-15-96709.

References

[1] J.W. Cahn and J.E. Hilliard, Free Energy of a Nonuniform System: I. Intefacial Free Energy, *J. Chem. Phys.* **28**(2), 258-267 (1958).

[2] J.W. Cahn and R. Kikuchi, Theory of Domain Walls in Ordered Structures: II and III, *J. Phys. Chem. Solids* **23**(1), 137-151 (1962), and **27**(5), 1305-1317 (1966).

[3] R. Kikuchi and J.W. Cahn, Theory of Interphase and Antiphase Boundaries in F.C.C. Alloys, *Acta Met.* **27**(5), 1337-1353 (1979).

[4] S.M. Allen and J.W. Cahn, Mechanisms of Phase Transformations within the Miscibility Gap of Fe-rich Fe-Al Alloys, *Acta Met.* **24**(2), 425-437 (1976).

[5] F. Schmid and K. Binder, Rough interphaces in a bcc-based binary allloy, *Phys. Rev.* B **46**(20), 13553-13564 (1992).

[6] R.J. Braun, J.W. Cahn, G.B. McFadden and A.A. Wheeler, Anisotropy of Interphaces in an Ordered Alloy: A Miltiple-Order-Parameter Model, *Phil. Trans. Roy. Soc. London A* **355**(1730), 1787-1833 (1997).

[7] Y. Wang, D. Banerjee, C.C. Su and A.G. Khachaturyan, Field kinetic model and computer simulation of precipitation of $L1_2$ ordered intermetallics from F.C.C. solid solution, *Acta Mater.* **46**(9), 2983-3001 (1998).

[8] M. Asta and A. Quong, The concentration and temperature dependences of antiphase-boundary energies in γ-TiAl: a first-principle study, *Phil. Mag. Lett.* **76**(5), 331-339 (1997).

[9] L. Proville and A. Finel, Kinetics of the coherent order-disorder transition in Al_3Zr, *Phys. Rev.* B **64**, 054104 – (1-7) (2001).

[10] K.D. Belashchenko, V.Yu. Dobretsov, I.R. Pankratov, G.D. Samolyuk and V.G. Vaks, The kinetic cluster-field method and its application to studies of $L1_2$-type orderings in alloys, and: Kinetic features of alloy ordering with many types of ordered domain: $D0_3$-type orderings, *J. Phys.: Condens. Matter* **11**(52), 10593-10620 and 10567-10592 (1999).

[11] I.R. Pankratov and V.G. Vaks, Kinetics of $L1_0$-type and $L1_2$-type orderings in alloys at early stages of phase transfotmations, *J. Phys.: Condens. Matter* **13**(32), 6031-6058 (2001).

[12] K.D. Belashchenko, I.R. Pankratov, G.D. Samolyuk and V.G. Vaks, Kinetics of formation of twinned structures under $L1_0$-type orderings in alloys, *J. Phys.: Condens. Matter* **14**(2), 565-589 (2002).

[13] V.G. Vaks, Ginzburg-Landau-type theory of antiphase boundaries in polytwinned structures, *Pis. Zh. Eksp. Teor. Fiz.* **73**(5), 274-278 (2001) [JETP Lett. **73**(5), 237-241 (2001)].

[14] V.Yu. Dobretsov, G. Martin, F. Soisson and V.G. Vaks, Effects of the Interaction between Order Parameter and Concentration on the Kinetics of Antiphase Boundary Motion, *Europhys. Lett.* **31**(7), 417-422 (1995).

[15] V.G. Vaks and G.D. Samolyuk, On accuracy of different cluster methods used in describing ordering phase transitions in fcc alloys, *Zh. Eksp. Teor. Fiz.* **115**(1) 158-179 (1999) [Sov. Phys.-JETP **88**(1) 89-100 (1999)].

[16] V.G. Vaks, N.E. Zein and V.V. Kamyshenko, On the cluster method in the theory of short-range order in alloys, *J. Phys. F: Metal Physics* **18**(8), 1641-1661 (1988).

[17] F. Chassagne, M. Bessiere, Y. Calvayrac, P . Cenedese and S. Lefebvre, X-Ray Diffuse-Scattering Investigation of Different States of Local Order in Ni-Al Solid-Solutions, *Acta Met.* **37**(9) 2329-2338 (1989).

[18] Y. Le Bouar, A. Loiseau, A. Finel and F. Ducastelle, Wetting behaviour in the Co-Pt system, *Phys. Rev.* B **61**(5) , 3317-3326 (2000).

[19] B. Widom, Structrure of the $\alpha - \gamma$ Interphace, *J. Chem. Phys.* **68**(8), 3878-3883 (1978).

INVESTIGATION OF STRUCTURES AND PROPERTIES OF C_3P_4 ALLOY USING FIRST-PRINCIPLES ELECTRONIC STRUCTURE CALCULATION

Adele Tzu-Lin Lim[1], Jin-Cheng Zheng[2] and Yuan Ping Feng[1]

[1] Department of Physics, National University of Singapore
2 Science Drive 3, Singapore 117542
[2] TCM, Cavendish Laboratory, University of Cambridge
Madingley Road, Cambridge, CB3 OHE, UK

INTRODUCTION

In the past, we have witnessed an increasing dependence of our technological and industrial base on advanced materials. There is every reason to believe that this trend will accelerate, and that progress in many areas, such as electronics, will be more dependent on the development of new materials and processing techniques. Meanwhile, in the process of searching and designing for new materials, numerical modelling has emerged as a powerful and widely used tool due to increasing computing resources and development of efficient algorithms. Compared to laboratory synthesis, computer modelling provides an inexpensive and efficient alternative for designing new materials, as the relative stability of various possible structures of a given compound can be studied and their physical properties explored by computer simulation before laboratory synthesis. This will at least provide some guidance to eventual laboratory synthesis of the compound. Such a strategy has been widely used in drug design.

Various computational techniques have been developed for studying materials, ranging from empirical potential or force field methods, semi-empirical calculation, and to the state-of-the-art first-principles method. First-principles simulations based on density functional theory (DFT) and particularly local density approximation (LDA) have proved to be the most reliable and computationally tractable tool in materials science. These simulations have now impacted virtually every area of this broad field, and have began to be used in the study of real materials. Compared to other computational techniques, the first-principles method requires no experimental parameters and is thus very powerful in predicting properties of new materials. Therefore, it is ideal for the study of new materials. Along with the advances in computing technology, there have been important algorithmic improvements, particularly for pseudopotential and plane-wave based methods. For certain classes of materials it is now feasible to simulate systems containing hundreds of atoms in a unit cell on a workstation. This opens the door for direct application of these techniques in studying a substantial set of real material problems.

One of the successes of first-principles methods in studying new materials is the prediction of superhard C_3N_4[1-4]. Using first-principles calculation based on pseudopotential and plane-wave method, Liu and Cohen predicted that the bulk modulus of C_3N_4 could be comparable to that of diamond which stimulated wide interests in this material. Films consisting of α-C_3N_4 and β-C_3N_4 with bulk moduli of up to 349 GPa have reportedly been synthesized [5].

In this work, we explore possible structures and properties of C_3P_4 using first-principles electronic structure calculation. C_3P_4 can be obtained by substituting N for P in C_3N_4. Since both C and P are group V elements in the periodic table, one can expect C_3P_4 to have properties similar to C_3N_4. Trend analysis according to mean atomic number suggests that C_3P_4 could have a wide band gap and could be a relatively hard material[6]. As a matter of fact, the bulk moduli of tetrahedrally bonded covalent solids can be estimated based on the following empirical model [7],

$$B = \frac{19.71 - 2.20\lambda}{d^{3.5}}$$

where d is the length and λ the ionicity of the bond. Since the sum of ideal tetrahedral covalent bond length between C and P atoms is 1.87 Å which incidentally is the same as that between Si and N atoms[8], it is expected that C_3P_4 will have a bulk modulus comparable to that of Si_3N_4.

Even though the existence of C_3P_4 was postulated together with C_3N_4 as early as in 1984[9], no research has been done on carbon phosphide besides a recent work on phosphorus-doped diamond-like carbon films[6]. The present study is therefore expected to provide basic information on the structure and properties of carbon phosphide. It is of interest to know whether C_3P_4 can form a stable alloy. And if it does, what is the possible structure? What physical properties does it have? The ultimate aim is to produce a stable form of carbon phosphide having potentially useful electronic properties.

COMPUTATIONAL DETAILS

In the absence of experimental data on crystalline carbon phosphide, we begin with possible configurations of C_3N_4 as suggested in Ref. [4] with N substituted by P. In this study, we only considered several carbon phosphide structures with stoichiometry C_3P_4. However, there is no reason to believe that only this stoichiometry is possible. Neither is this stoichiometry confined to the phases considered. The crystal structures considered are α-C_3P_4, β-C_3P_4, cubic-C_3P_4, pseudocubic-C_3P_4 and graphitic-C_3P_4. (see Figure 1)

We have performed ground state total-energy calculations using the pseudopotential plane-wave method based on density-functional theory in the local density approximation for exchange and correlation[10]. The Vanderbilt ultrasoft pseudopotentials were used [11-12].

The wave functions are expanded into plane-waves up to an energy cutoff of 310 eV. Special k points generated according to Monkhorst-Pack scheme[13] were used for integration over the irreducible wedge of the Brillouin zone for the various structures. We used 18 k points for α-C_3P_4, 36 for β-C_3P_4, 32 for cubic-C_3P_4, 32 for pseudocubic-C_3P_4 and 16 for graphitic-C_3P_4. Good convergence is achieved with this cutoff energy and number of k points for the various C_3P_4 structures considered. We also optimized each crystal geometry within the preselected space groups.

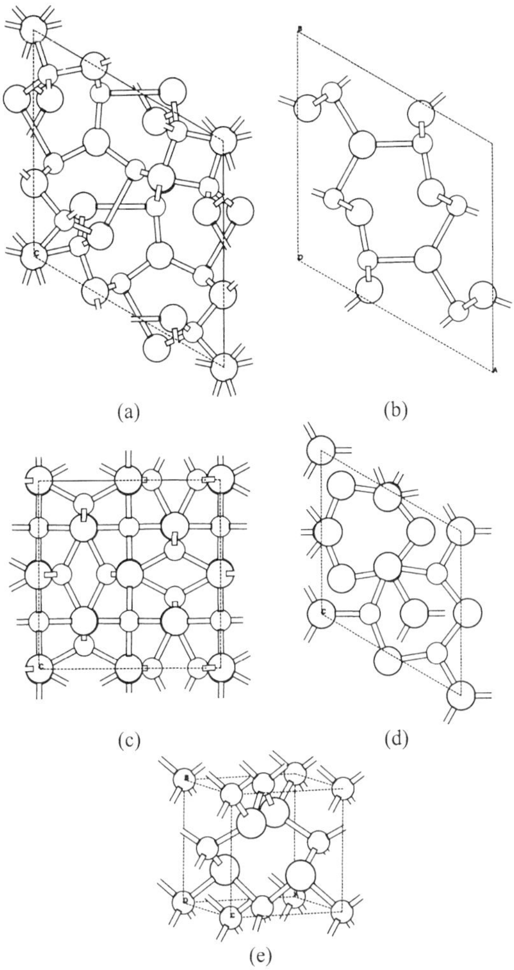

Figure 1. Unit cells of (a) α-C_3P_4, (b) β-C_3P_4, (c) cubic-C_3N_4, (d) graphitic-C_3N_4 and (e) pseudocubic-C_3P_4. The C atoms are shown using small spheres while the P atoms are shown using large spheres. All structures are fully optimized within the preselected space groups. The pseudocubic-C_3P_4 is energetically the most stable structure.

RESULTS AND DISCUSSION

Figure 2 shows the calculated total energy per C_3P_4 unit as a function of volume for the five structures considered. At each volume, the atomic positions were fully relaxed within the preselected space group for each structure. The optimized structures are shown in Figure 1.

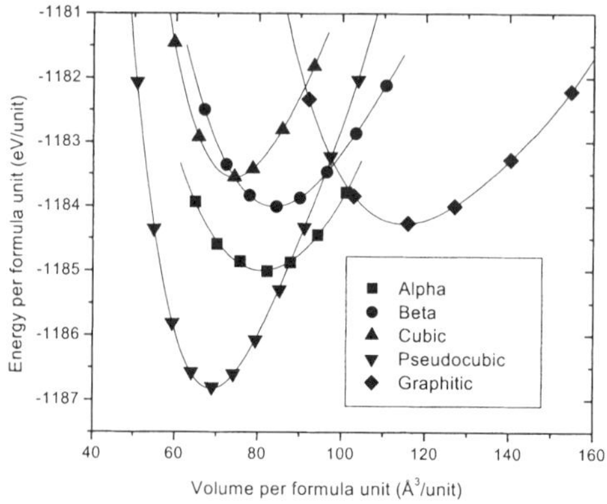

Figure 2. Total energy as a function of volume for the various structures investigated. The calculated data points (symbols) are fitted to fourth order polynomial expansion (lines) to obtain the bulk modulus for each structure.

Unlike C_3N_4, the calculations predict that pseudocubic-C_3P_4 is energetically favored relative to other phases. In particular, α-C_3N_4 and β-C_3N_4 are the stable phases for C_3N_4 but α-C_3P_4 and β-C_3P_4 are found to be less stable here. As a matter of fact, large deformations in internal coordinates were found during the structural relaxation in both α- and β-C_3P_4, which may indicate that they are unstable.

It is interesting to note that the stability of the pseudocubic phase of C_3P_4 is quite exceptional as the total energy per C_3P_4 unit of pseudocubic-C_3P_4 is almost 2 eV lower than that of the next most energetically favoured structure, *i.e.* α-C_3P_4.

In order to assess the mechanical stability of pseudocubic-C_3P_4, we further relaxed the structure under the P1 symmetry from the previously optimized geometry and from ideal zincblende coordinates. We found virtually no change in the structural parameters, internal coordinates and total energy from those of pseudocubic-C_3P_4. Hence, we predict that the pseudocubic phase is energetically and mechanically stable.

Table 1. Equilibrium structural parameters, bulk moduli and total energies calculated for α-C_3P_4, β-C_3P_4, cubic-C_3P_4, pseudocubic-C_3P_4 and graphitic-C_3P_4.

Structure	Space Group	*a* (Å)	*c* (Å)	B_0 (GPa)	E_0 (eV/C_3P_4 Unit)
α	$P3_1c$	8.120	5.755	85	-1185.006
β	$P6_3/m$	8.319	2.815	110	-1184.009
Cubic	$I\bar{4}3d$	6.670		142	-1183.552
Pseudocubic	$P\bar{4}2m$	4.102	4.101	203	-1186.801
Graphitic	$P\bar{6}m2$	5.723	8.168		-1184.264

For C_3N_4, the most stable structure among the phases that have been studied is graphitic-C_3N_4, followed by α-C_3N_4, β-C_3N_4 and cubic-C_3N_4, whereas pseudocubic-C_3N_4 is least stable[4]. It is known that in graphitic-C_3N_4, α-C_3N_4, β-C_3N_4 and cubic-C_3N_4, each C atom is approximately tetrahedrally-coordinated by N atoms and each N atom is nearly planarly-threefold coordinated to C atoms[1-4]. This suggests sp^3 hybrid on C atom and sp^2 hybrid on N atom. However, in pseudocubic-C_3N_4, the C–N–C bond angle is close to the sp^3 value of 109.47, indicating that the N atoms in the pseudocubic structure form sp^3 rather than sp^2 bonding orbitals.

Moving down the periodic table, sp^3 hybridization is energetically favourable and therefore, the lowest energy configuration for C_3P_4 corresponds to the structure in which P atoms form sp^3 bonding orbitals. The C–P–C bond angle of 103.4 in pseudoc\ubic-C_3P_4 is less than the ideal tetrahedral bond angle of 109.47 due to relaxation of the P atom towards an empty C site. The C–P bond length of 1.86 Å is close to the sum of ideal tetrahedral covalent bond length of 1.87 Å. In contrast, P atoms in cubic-C_3P_4 which is energetically least favourable among the phases considered, is constrained to form sp^2 bonding orbitals. Although the C–P bond length for this phase is also 1.86 Å, the C–P–C bond angle is 114.3. The stability of graphitic-, α- and β-C_3P_4, relies on the ability of P atoms to form sp^3 bonding orbitals. Therefore, the displacement of their atomic coordinates from ideal atomic coordinates of C_3N_4 for the respective phases is rather significant, with the α and β phases displaying the most significant changes, which is the cause of its structural instability.

Table 2. Optimized atomic positions of various phases of C_3P_4.

Phase	Atom	*u*	*v*	*w*
α	C1	0.4609	0.1065	0.1712
	C2	0.1625	0.2389	1.0158
	P1	0.0000	0.0000	0.1146
	P2	0.3333	0.6667	0.5906
	P3	0.3199	0.9958	0.9188
	P4	0.2634	0.3510	0.3035
β	C1	0.1754	-0.1828	0.2500
	P1	0.3653	0.0508	0.2500
	P2	0.3333	0.6667	0.2500
Cubic	C1	0.8750	0.0000	0.2500
	P1	0.2529	0.2529	0.2529
Pseudocubic	C1	0.0000	0.0000	0.0000
	C2	0.5000	0.0000	0.5000
	P1	0.2768	0.2768	0.2232
Graphitic	C1	0.3454	0.1727	0.0000
	C2	0.0121	0.5060	0.5000
	P1	0.0000	0.0000	0.0000
	P2	0.6667	0.3333	0.5000
	P3	0.1430	0.2860	0.5000
	P4	0.4759	0.5241	0.0000

Not only is the pseudocubic phase exceptionally stable, it also has the smallest volume per C_3P_4 unit and therefore the highest density among all the phases investigated. This implies that it is impossible for pseudocubic-C_3P_4 to undergo a phase transition under pressure. Of course, there may be other high-pressure phases of C_3P_4 besides the five structures investigated here. Pseudocubic-C_3P_4 can be classified as a defect-zincblende structure type. Based on the fact that the high-pressure phase of semiconductors such as GaAs and ZnS is the β-Sn structure, the corresponding defect β-Sn structure could be a candidate for the high-pressure phase of C_3P_4. We thus searched for possible metastable phases with a smaller volume per C_3P_4 unit by varying the *c/a* ratio in each case. However, the calculated total energy decreases monotonically with the *c/a* ratio and volume, and no energy minimum was found in the range of $V<V_0$ where V_0 is the equilibrium cell volume of pseudocubic-C_3P_4. We therefore conclude that there is no pathway for phase transition from the pseudocubic phase to other structural forms under pressure.

Table 3. Optimized atomic positions of various phases of C_3N_4.

Phase	Atom	*u*	*v*	*w*
α	C1	0.5169	0.0810	0.2033
	C2	0.1654	0.2547	0.9936
	N1	0.0000	0.0000	-0.0058
	N2	0.3333	0.6667	0.6303
	N3	0.3471	0.9511	0.9730
	N4	0.3148	0.3183	0.2447
β	C1	0.1786	0.7732	0.2500
	N1	0.3303	0.0332	0.2500
	N2	0.3333	0.6667	0.2500
Cubic	C1	0.8750	0.0000	0.2500
	N1	0.2839	0.2839	0.2839
Pseudocubic	C1	0.0000	0.0000	0.0000
	C2	0.5000	0.0000	0.5000
	N1	0.2451	0.2451	0.2549
Graphitic	C1	0.3499	0.1749	0.0000
	C2	0.0175	0.5087	0.5000
	N1	0.0000	0.0000	0.0000
	N2	0.6667	0.3333	0.5000
	N3	0.1699	0.3398	0.5000
	N4	0.5032	0.4968	0.0000

The calculated total energies are fitted to fourth order polynomials and the bulk modulus for each structure is then determined. Table 1 shows the equilibrium structural parameters, bulk moduli and total energies per C_3P_4 unit for the five structures of C_3P_4 being studied. For calculation of the bulk moduli, we have assumed uniform compression and expansion of the lattice. Incidentally, the pseudocubic phase also has the highest bulk modulus of 203 GPa.

In Table 2, the optimized atomic positions in each C_3P_4 structure are given in terms of the lattice translational vectors. For comparison, we have repeated similar calculations for the various phases of C_3N_4 and the optimized atomic positions are listed in Table 3. The calculated various physical quantities of C_3N_4 such as equilibrium lattice constants, bulk moduli, band structures, etc. are in good agreement with results of previous calculations[1-4]. As can be seen, both α-C_3P_4 and β-C_3P_4 are driven by significant atomic relaxation, resulting in the large differences between atomic positions in C_3P_4 and C_3N_4. Furthermore, there is no apparent symmetry in the atomic relaxation in α-C_3P_4 and β-C_3P_4. It is, therefore, reasonable to believe that these structures may not be stable at all. If the structures were allowed to relax without imposing any symmetry constraints, they may transform into other more stable structures such as the pseudocubic-C_3P_4. This will be explored in further study.

The P atom in the cubic phase relaxes along the cube diagonal in C_3P_4 relative to that in C_3N_4 while the C atom is constrained to its location by symmetry in both C_3N_4 and C_3P_4.

As mentioned earlier, the pseudocubic phase is a defect zincblende structure. Instead of tetrahedral bonding, each N or P atom is bonded to three C atoms. In the case of C_3N_4, the N atom relaxes away from the defect site by ~0.03 Å (2% of the C-N bond length). The P atom in C_3P_4, however, undergoes a larger relaxation of ~0.19 Å (11% of the C-P bond length) and it relaxes in the opposite direction, moving towards the defect site. This is due to the strong C-N bonding in C_3N_4 relative to the C-P bond in C_3P_4. Analysis on electronic properties shows that the nature of the C-N bond in C_3N_4 is more ionic with large charge transfer from C to N. In contrary, the C-P bond in C_3P_4 is more covalent and charge is transferred from P to C.

We have also investigated the electronic properties of the various phases of C_3P_4. It was found that within the LDA approximation, C_3P_4 is metallic. Even though it is known that LDA underestimates band gap, the overlap between the lowest conduction band and the highest valance band is almost 2 eV and correction of the band gap by other method may not be able to completely remove this large overlap. We therefore also conclude that C_3P_4 is metallic or at most a narrow gap semiconductor. The results of electronic properties of C_3P_4 will be published elsewhere.

CONCLUSION

In conclusion, we have investigated the possible structures and properties of C_3P_4 using first-principles electronic structure calculation. Our results show that C_3P_4 has unexpected properties. The most stable phase of C_3P_4 is the pseudocubic phase. This is in contrast to the structures of similar compounds, e.g. C_3N_4 and Si_3N_4. Furthermore, the stability of the pseudocubic-C_3P_4 is exceptional since the total energy of pseudocubic-C_3P_4 is almost 2 eV lower than the next most stable phase per C_3P_4 unit. The pseudocubic phase also has the highest density among the structures investigated. It may not be possible for pseudocubic-C_3P_4 to undergo a phase transition under pressure. The C_3P_4 alloy is also expected to be metallic.

REFERENCES

1. M. L Cohen, Phys. Rev. B **32**, 7988 (1985).
2. A. Y. Liu and M. L. Cohen, Science **245**, 841 (1989).
3. A. Y. Liu and M. L. Cohen, Phys. Rev. B **41**, 10727 (1990) and references therein.
4. D. M. Teter and R. J. Hemley, Science **271**, 53 (1996).
5. D. X. Shi, X. F. Zhang, L. Yuan, Y. S. Gu, Y. P. Zhang, Z. J. Duan, X. R. Chang, Z. Z. Tian and N. X. Chen, Appl. Surf. Sci. **148**, 50 (1999).
6. S. R. J. Pearce, P. W. May, R. K. Wild, K. R. Hallam and P. J. Heard, Diam. Relat. Mater. **11**, 1041 (2002).
7. M. L. Cohen, Phys. Rev. B **32**, 7988 (1985).
8. C. Kittel, *Introduction to Solid State Physics*, 7th ed. (Wiley, New York, 1996), p78.
C. M. Sung and M. Sung, Mater. Chem. Phys. **13**, 1 (1984)
10. M. C. Payne, M. P. Teter, D. C. Allan, T. A. Arias, and J. D. Joannopoulos, Rev. Mod. Phys. **64**, 1045 (1992).
11. D. Vanderbilt, Phys. Rev. B **41**, 7892 (1990).
12. K. Laasonen, R. Car, C. Lee and D. Vanderbilt, Phys. Rev. B **43**, 6796 (1991).
13. H. J. Monkhorst and J. D. Pack, Phys. Rev. B **13**, 5188 (1976).

INDEX